COMBINATORIAL PHYSICS

Combinatorial Physics

Combinatorics, quantum field theory, and quantum gravity models

Adrian Tanasa

University of Bordeaux, France

OXFORD
UNIVERSITY PRESS

OXFORD
UNIVERSITY PRESS

Great Clarendon Street, Oxford, OX2 6DP,
United Kingdom

Oxford University Press is a department of the University of Oxford.
It furthers the University's objective of excellence in research, scholarship,
and education by publishing worldwide. Oxford is a registered trade mark of
Oxford University Press in the UK and in certain other countries

First Edition published in 2021
Impression: 1

Published in the United States of America by Oxford University Press
198 Madison Avenue, New York, NY 10016, United States of America

British Library Cataloguing in Publication Data
Data available

Library of Congress Control Number: 2021932400

ISBN 978–0–19–289549–3

DOI: 10.1093/oso/9780192895493.001.0001

Printed and bound by
CPI Group (UK) Ltd, Croydon, CR0 4YY

To Luca, Brittany, and to our family

Contents

1 Introduction

The interplay between combinatorics and theoretical physics is a recent trend which appears to us as particularly natural, since the unfolding of new ideas in physics is often tied to the development of combinatorial methods, and, conversely, problems in combinatorics have been successfully tackled using methods inspired by theoretical physics. A lot of problems in physics are thus revealed to be enumerative. On the other hand, problems in combinatorics can be solved in an elegant way using theoretical physics-inspired techniques. We can thus speak nowadays of an emerging domain of **Combinatorial Physics**.

The interference between these two disciplines is moreover an interference of multiple facets. Thus, its most known manifestation (both to combinatorialists and theoretical physicists) has so far been the one between combinatorics and statistical physics, or combinatorics and integrable systems, as statistical physics relies on an accurate **counting** of the various states or configurations of a physical system.

However, combinatorics and theoretical physics interact in various other ways. One of these interactions is the one between combinatorics and quantum mechanics, because combinatorial tools can be used here for a better mathematical understanding of the algebras underlying quantum mechanics.

In this book, we mainly focus on yet another type of these multiple interactions between combinatorics and theoretical physics, the one between combinatorics and quantum field theory (QFT). We estimate that combinatorics is *built into* the mathematical formulation of QFT. This stems initially from the fact that the most popular tool of QFT is perturbation theory in the coupling constant of the model, which means that one considers Feynman *graphs*, with appropriate combinatorial weights, in order to encode the physical information of the respective system. Moreover, one elegant way of expressing the Feynman integrals associated with these graphs is to use the Kirchhoff–Symanzik polynomials of the parametric representation, polynomials which can be proven to be related to some multi-variate version of the celebrated Tutte polynomial of combinatorics. A particularly elegant way to prove this is to use the *Grassmann development* of the determinants and Pfaffians involved in these computations. Let us emphasize here that this Grassmann development uses Grassmann calculus, which were developed by physicists to express fermionic QFT. Grassmann calculus is further used in this book to give a simple proof of the celebrated Lingström–Gessel–Viennot (for graphs

Combinatorial Physics: Combinatorics, quantum field theory, and quantum gravity models. Adrian Tanasa, Oxford University Press (2021). © Adrian Tanasa. DOI: 10.1093/oso/9780192895493.003.0001

with cycles) and to further generalize some identities, initially proved by Stembridge, in the same context of graphs with cycles.

The so-called 0−dimensional QFT (called by some authors, combinatorial QFT), or more precisely the use of the intermediate field method in this setting, allows to establish a theorem concerning partial elimination of variables in the celebrated Jacobian conjecture (which concerns the global invertibility of polynomial systems).

Moreover, *analytic combinatorial* techniques are used on a regular basis in QFT computations. Thus, the propagator of any scalar model can be represented using the heat kernel. The Mellin transform technique can also be used in order to rapidly prove the meromorphy of Feynman integrands. The saddle point method is frequently used to tame the divergent behaviour of these integrals.

Last but not least, renormalization in QFT (which one can say lies at the very heart of QFT) has a highly non-trivial combinatorial core, and this has been recently presented in a *combinatorial Hopf algebra* form—the Connes–Kreimer Hopf algebra. Related to this, the Hochschild cohomology of this combinatorial Hopf algebra can be used to express the combinatorics of the Dyson–Schwinger equation (DSE) as a simple power series in some appropriate insertion operator of Feynman graphs.

All these combinatorial techniques (analytic or algebraic) generalize to more involved QFT models. Thus, non-commutative QFT (that is, QFT on a non-commutative space-time) also possesses most of these combinatorial properties. First, the graphs used in QFT are uplifted to ribbon graphs (or *combinatorial maps*). Furthermore, one can still use the heat kernel for propagators of the theories, but in order to have renormalizable models, one needs to use a more involved special function, the Mehler kernel or some non-trivial modification of the heat kernel. Moreover, the Mellin transform technique can again be used, as in the case of commutative QFT. The corresponding non-commutative Kirchhoff–Symanzik polynomials are proven to be a limit of a multi-variate version of the Bollobás–Riordan polynomial (which is a natural generalization for ribbon graphs of the universal Tutte polynomial). Finally, algebraic combinatorial techniques can also be used in non-commutative QFT. The corresponding combinatorial Connes–Kreimer Hopf algebra of ribbon Feynman graphs can be defined and related to non-commutative renormalization. Furthermore, the appropriate Hochschild cohomology then describes the combinatorics of the DSE of these models.

Non-commutative QFT can also be seen as a special case of the celebrated matrix models. Following this line of reasoning, one can naturally generalize random matrix models to random tensor models,

The combinatorics of tensor models per se is extremely involved. One cannot *just* use the genus to characterize the so-called large N expansion, N being the size of the matrix resp. of the tensor. It is worth emphasizing here that the large N expansion is, from a combinatorial point of view, a certain asymptotic expansion (corresponding to the limit $N \to \infty$).

However, in order to make the combinatorics simpler, several QFT-inspired simplifications of tensor models can be proposed. The first two such simplications were the coloured model and the multi-orientable models. For both of these models, one can implement the large N expansion and the double-scaling mechanism, which, in the case of matrix models, are very important mathematical physics tools. Several other models (based on $U(N)$ and then $O(N)$ models have also been studied.

Feynman graphs associated to tensor models, through the celebrated QFT perturbative expansions, are called tensor graphs and can be seen as a natural 3D generalization of maps or of ribbon graphs. The dominant term of the large N expansion of the previously mentioned tensor models are the so-called melonic graphs, which are, from a graph theoretical point of view, a particular case of series-parallel graphs.

The large N expansion is controlled in the 2D case, the matrix model case, by the genus of the corresponding combinatorial maps. In dimension higher than two, there is no direct analogue of the genus. Nevertheless, the tensor asymptotic expansion in N is controlled by an integer, called the degree, which is defined as the half-sum of the non-orientable genus of ribbon graphs canonically embedded in a tensor graph (called the jackets of the respective tensor graphs). The degree is thus a half integer, naturally generalizing the 2D notion of genus for tensor models.

In order to study the general term of the large N expansion of various such tensor models, we extensively use, in this book, various graph theoretical and *enumerative combinatorics* techniques to perform their enumeration and we establish which are the dominant configurations of a given degree.

It is worth emphasizing here that tensor models have recently been proven by Witten to be related to the celebrated holographic Sachdev–Ye–Kitaev (SYK) quantum mechanical model. This comes from the fact that, in the so-called large N expansion (N being in the case of the SYK model the total number of fermions of the model), both types of models are dominated by the melonic graphs. The large N expansion for various SYK-like tensor models is then studied using again graph theoretical and enumerative combinatorics techniques. These techniques allow us to asymptotically enumerate Feynman graphs of various SYK-like tensor models.

The book is organized as follows. In Chapter 2, we present some notions of graph theory that will be useful in the rest of the book. It is worth emphasizing that graph theorists and theoretical physicists adopt, unfortunately, different terminologies. We present here both terminologies, such that a sort of dictionary between these two communities can be established. We then extend the notion of graph to that of maps (or of ribbon graphs). Moreover, graph polynomials encoding these structures (the Tutte polynomial for graphs and the Bollobás–Riordan polynomial for ribbon graphs) are presented.

In Chapter 3, we briefly exhibit the mathematical formalism of QFT, which, as mentioned previously, has a non-trivial combinatorial backbone. The QFT setting can be understood as a quantum description of particles and their interactions, a description which is also compatible with Einstein's theory of special relativity. Within the framework of elementary particle physics (or high energy physics), QFT led to the Standard Model of Elementary Particle Physics, which is the physical theory tested with the best accuracy by collider experiments. Moreover, the QFT formalism successfully applies to statistical physics, condensed matter physics, and so on. We show in this chapter how Feynman graphs appear through the so-called QFT perturbative expansion, how Feynman integrals are associated to Feynman graphs, and how these integrals can be expressed via the help of graph polynomials, the Kirchhoff–Symanzik polynomials. Finally, we give a glimpse of renormalization, of the DSE, and of the use of the so-called intermediate field method. This chapter mainly focuses on the so-called Φ^4 QFT scalar model.

In Chapter 4, we define specific tree weights which appear natural when considering a certain approach to non-perturbative renormalization in QFT, namely constructive renormalization. Several examples of such tree weights are explicitly given in Appendix A.

Chapter 5 deals with a combinatorial QFT approach to the Jacobian conjecture. The Jacobian Conjecture states that any complex n-dimensional locally invertible polynomial system is globally invertible with a polynomial inverse. In 1982, Bass et al. proved an important reduction theorem stating that the conjecture is true for any degree of the polynomial system if it is true in degree three. We show, in this chapter, a result concerning partial elimination of variables, which implies a reduction of the generic case to the quadratic one. The price to pay is the introduction of a supplementary parameter, $0 \leq n' \leq n$ parameter, which represents the dimension of a linear subspace where some particular conditions on the system must hold. We exhibit a proof, in a QFT formulation, using the intermediate field method exposed in Chapter 3.

In Chapter 6, we use Grassmann calculus, used in fermionic QFT, to first give a reformulation of the Lingström–Gessel–Viennot lemma proof. We further show that this proof generalizes to graphs with cycles. We then use the same Grassmann calculus techniques to give new proofs of Stembridge's identities relating appropriate graph Pfaffians to a sum over non-intersecting paths. The results presented here go further than the ones of Stembridge, because Grassmann algebra techniques naturally extend (without any cost!) to graphs with cycles. We thus obtain, instead of sums over non-intersecting paths, sums over non-intersecting paths and non-intersecting cycles. In the fifth section of the chapter, we give a generalization of these results. In the sixth section of this chapter we use Grassmann calculus to exhibit the relationship between a multi-variate version of Tutte polynomial and the Kirchhoff–Symanzik polynomials of the parametric representation of Feynman integrals, polynomials already introduced in Chapters 1 and resp. 3.

In Chapter 7, we present how several analytic techniques, often used in combinatorics, appear naturally in various QFT issues. In the first section, we show how one can use the Mellin transform technique to re-express Feynman integrals in a useful way for the mathematical physicist. Finally, we briefly present how the saddle point approximation technique can be also used in QFT.

In Chapter 8, after a brief algebraic reminder, we introduce in the second section the Connes–Kreimer Hopf algebra of Feynman graphs and we show its relationship with the combinatorics of QFT perturbative renormalization. We then study the algebra's Hochschild cohomology in relation with the combinatorial DSE in QFT. In the fourth, section we present a Hopf algebraic description of the so-called multi-scale renormalization (the multi-scale approach to the perturbative renormalization being the starting point for the constructive renormalization programme).

In Chapter 9, we present the Phi^4 QFT model on the non-commutative Moyal space and the UV/IR mixing issue, which prevents it from being renormalizable. We then present the Grosse–Wulkenhaar Phi^4 QFT model on the non-commutative Moyal space, which changes the usual propagator of the Φ model (based on the heat kernel formula) to a Mehler kernel-based propagator. This Grosse–Wulkenhaar model is perturbatively renormalizable but it is not translation-invariant (translation-invariance being a usual property of high-energy physics models). We then show how the Mellin transform technique can be used to express the Feynman integrals of the Grosse–Wulkenhaar model. In the last part of the chapter, we present another Phi^4 QFT model on the non-commutative Moyal space, which is however both renormalizable and translation-invariant. We show the relation between the parametric representation of this model and the Bollobás–Riordan polynomial. Finally, we show how to define a Connes–Kreimer Hopf algebra for non-commutative renormalization and how to study its Hochschild cohomology in relation to the combinatorial DSE of these QFT models.

The last part of the book is dedicated to the study of combinatorial aspects of quantum gravity models. Thus, in Chapter 10, after a brief introductory section to quantum gravity, we mention the main candidates for a quantum theory of gravity: string theory, loop quantum gravity, and group field theory (GFT), causal dynamical triangulations, and matrix models. The next sections introduce some GFT models such as the Boulatov model, the colourable, and the multi-orientable model. The saddle point method for some specific GFT Feynman integrals is presented in the fifth section. Finally, some algebraic combinatorics results are presented: definition of an appropriate Connes–Kreimer Hopf algebra describing the combinatorics of the renormalization of a certain tensor GFT model (the so-called Ben Geloun–Rivasseau model) and the use of its Hochschild cohomology for the study of the combinatorial DSE of this specific model.

In Chapter 11, after a brief presentation of random matrices as a random surface QFT approach to 2D quantum gravity, we focus on two crucial mathematical physics results: the implementation of the large N limit (N being here the size of the matrix) and of the double scaling mechanism for matrix models. It is worth emphasizing that, in the large N limit, it is the planar surfaces which dominate. In the third section of the chapter, we introduce tensor models, seen as a natural generalization, in dimensions higher then two, of matrix models. The last section of the chapter presents a potential generalization of the Bollobás–Riordan polynomial for tensor graphs (which are the Feynman graphs of the perturbative expansion of QFT tensor models).

In Chapter 12, we first briefly present the $U(N)^D$-invariant tensor models (N being again the size of the tensor, and D being the dimension). The next section is then dedicated to the analysis of the Dyson–Schwinger equations (DSEs) in the large N limit. These results are essential to implement the double scaling limit mechanism of the DSEs which is done in the third section. The main result of this chapter is the doubly-scaled two-point function for a model with generic melonic interactions. However, several assumptions on the large N scaling of cumulants are made along the way. They are proved using various combinatorial methods.

Chapter 13 is dedicated to the presentation of the multi-orientable tensor model. After defining the model, the $1/N$ expansion and the large N limit are examined in the second section of the chapter. In the third section, a thorough enumerative combinatorial analysis of the general term of the $1/N$ expansion is presented. The implementation of the double scaling mechanism is then exhibited in the fourth section.

In Chapter 14, we define yet another class of tensor models, endowed with $O(N)^3$–invariance, N being again the size of the tensor. This allows to generate, via the usual QFT perturbative expansion, a class of Feynman tensor graphs which is strictly larger than the class of Feynman graphs of both the multi-orientable model and the $U(N)^3$-invariant models treated in the previous two chapters. We first exhibit the existence of a large N expansion for such a model with general interactions (not necessary quartic). We then focus on the quartic model and we identify the leading order (LO) and next-to-leading (NLO) Feynman graphs of the large N expansion. Finally, we prove the existence of a critical regime and we compute the so-called critical exponents. This is achieved through the use of various analytic combinatorics techniques.

In Chapter 15, we first review the SYK model, which is a quantum mechanical model of N fermions. The model is a quenched model, which means that the coupling constant is a random tensor with Gaussian distribution. The SYK model is dominated in the large N limit by melonic graphs, in the same way the tensor models presented in the previous three chapters are dominated by melonic graphs. We then present a purely graph theoretical proof of the melonic dominance of the SYK model. As already mentioned, it is this property which led E, witten to relate the SYK model to the coloured tensor model. In the rest of the chapter we deal with the so-called coloured SYK model, which is a particular case of the generalization of the SYK model introduced by D. Gross and V. Rosenhaus. We first analyse in detail the LO and NLO order vacuum, two- and four-point Feynman graphs of this model. We then exhibit a thorough asymptotic combinatorial analysis of the Feynman graphs at an arbitrary order in the large N expansion. We end the chapter by an analysis of the effect of non-Gaussian distribution for the coupling of the model.

In Chapter 16, we analyse in detail the diagrammatics of various SKY-like tensor models: the Gurau–Witten model (in the first section), and the multi-orientable and $O(N)^3$-invariant tensor models, in the rest of the chapter. Various explicit graph theoretical techniques are used.

2

Graphs, ribbon graphs, and polynomials

In this chapter, we present some notions of graph theory that will be useful in the rest of this book. Let us emphasize that graph theorists and quantum field theorists adopt, unfortunately, different terminologies. We present both here, such that a sort of **dictionary** between these two communities may be established.

We then extend the notion of graphs to that of maps (or of ribbon graphs). Moreover, graph polynomials encoding these structures (the Tutte polynomial for graphs and the Bollobás–Riordan polynomial for ribbon graphs) are presented.

In this chapter, we follow the original article (Thomas Krajewski et al. 2010) and the review article (Adrian Tanasa 2012).

2.1 Graph theory: The Tutte polynomial

For a general introduction to graph theory, the interested reader may refer to Claude Berge (1976). Let us now define a graph in the following way:

Definition 2.1.1 *A graph* Γ *is defined as a set of vertices* V *and of edges* E *together with an incidence relationship between them.*

Notice that we allow multi-edges and self-loops (see definition 2.1.2 4), but still use the term 'graph' (and not 'pseudograph').

The number of vertices and edges in a graph are also noted V and E for simplicity, since our context prevents confusion.

One needs to emphasize that in QFT a supplementary type of edge exists, **external edges**. These edges are only hooked to one of the vertices of the graph, the other end of the edge being 'free' (see Fig. 2.1 for an example of such a graph, with four external edges). In elementary particle physics, these external edges are related to the observables in some experiments.

Combinatorial Physics: Combinatorics, quantum field theory, and quantum gravity models. Adrian Tanasa, Oxford University Press (2021). © Adrian Tanasa. DOI: 10.1093/oso/9780192895493.003.0002

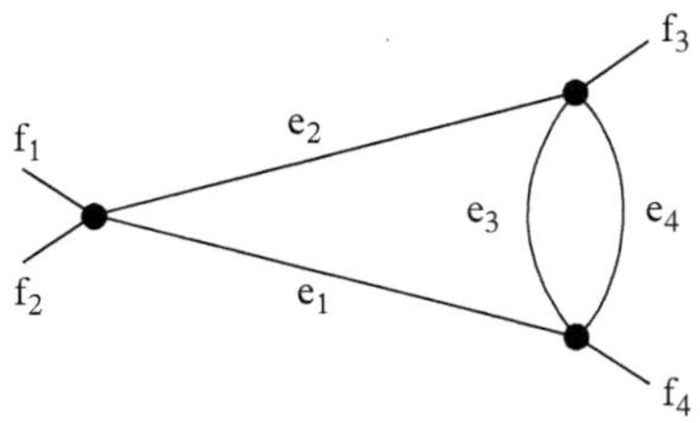

Figure 2.1 *A Φ^4 graph, with four internal edges and four external edges*

Let us now give the following definition:

Definition 2.1.2

1. *The number of edges at a vertex is called the* **degree** *of the respective vertex (field theorists refer to this as the* **coordination number** *of the respective vertex).*
2. *An edge whose removal increases the number of connected components of the respective graph is called a* **bridge** *(field theorists refer to this as a* **1-particle reducible** *edge).*
3. *A connected subset of equal number of edges and of vertices which cannot be disconnected by removing any of the edges is called a* **cycle** *(field theorists refer to this as a* **loop***).*
4. *An edge which connects a vertex to itself is called a* **self-loop** *(field theorists refer to this as a* **tadpole** *edge).*
5. *An edge which is neither a bridge nor a self-loop is called* **regular**.
6. *An edge which is not a self-loop is called* **semi-regular**.
7. *A graph with no cycles is called a* **forest**.
8. *A connected forest is called a* **tree**.
9. *A* **two-tree** *is a spanning tree without one of its edges.*
10. *The* **rank** *of a subgraph A is defined as*

$$r(A) := V - k(A), \tag{2.1}$$

where $k(A)$ is the number of connected components of the subgraph A.

11. *The* **nullity** *(or* **cyclomatic number***) of a subgraph A is defined as*

$$n(A) := |A| - r(A). \tag{2.2}$$

Remark 2.1.3 *For a connected graph, the nullity defined previously represents the number of independent circuits.*

In QFT, one often uses the term (number of) loops to denote (the number of) independent loops.

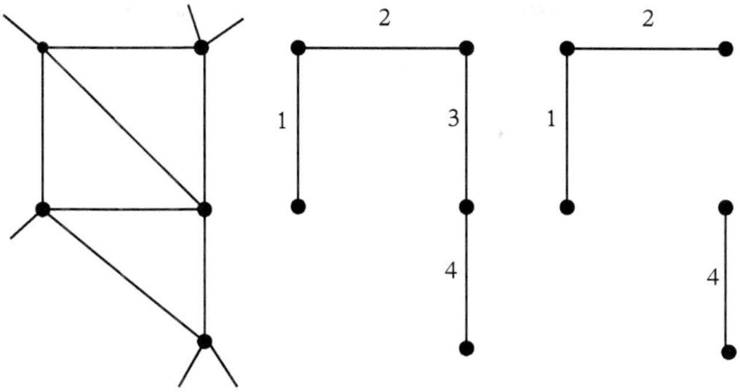

Figure 2.2 *An example of a graph (with seven edges and six external edges). We chose a spanning tree and labelled its edges by 1, …, 4. We then chose the edge 3 to be removed. The set {1, 2, 4} is a two-tree; one has two connected components (the first one formed by the edges 1 and 2 and the second one formed by edge 4). The external edges are attached to one of these two connected components*

Remark 2.1.4 *A two-tree generates two connected components on the respective graph. Let us also note that a two-tree can be defined as a spanning forest with two connected components (in this way, no relation with a tree is given).*

Let us illustrate this in Fig. 2.2. One can define two natural operations for an arbitrary edge e of some graph Γ:

1. the deletion, which leads to a graph noted $\Gamma - e$;
2. the contraction, which leads to a graph noted Γ/e. This operation identifies the two vertices v_1 and v_2 at the ends of e into a new vertex v_{12}, attributing all the edges attached to v_1 and v_2 to v_{12}; finally, the contraction operation removes e.

Remark 2.1.5 *If e is a self-loop, then Γ/e is the same graph as $\Gamma - e$.*

For an illustration of these two operations, one can refer to Fig. 2.3, where these operations are iterated until one reaches **terminal forms** namely graphs formed only of self-loops and bridges.

Let us now give a first definition of the Tutte polynomial:

Definition 2.1.6 *If Γ is a graph, then its Tutte polynomial $T_\Gamma(x,y)$ is defined as*

$$T_\Gamma(x,y) := \sum_{A\subset E} (x-1)^{r(E)-r(A)}(y-1)^{n(A)}. \tag{2.3}$$

A fundamental property of the Tutte polynomial is a deletion/contraction property:

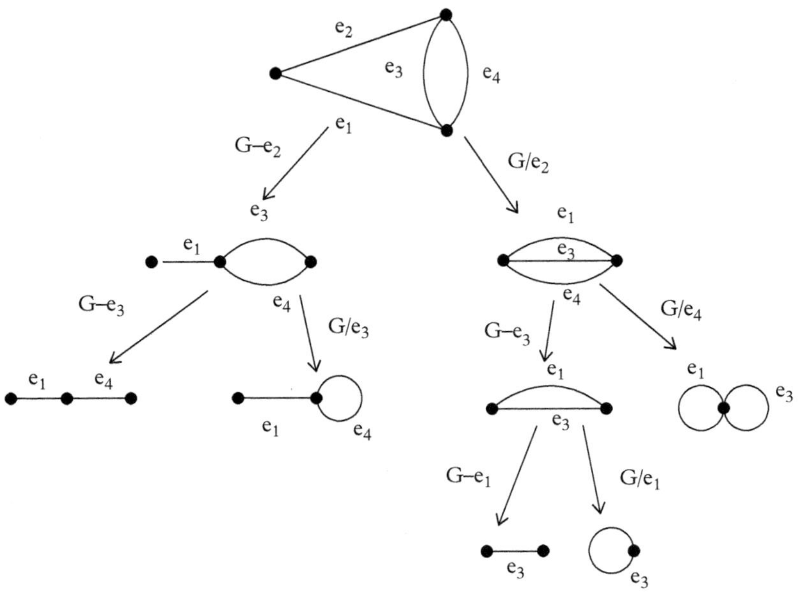

Figure 2.3 *The deletion/contraction of some graph. One is left with various possibilities (here five) of terminal forms (that is, graphs with only bridges or self-loops)*

Theorem 2.1.7 *If* Γ *is a graph, and e is a regular edge, then*

$$T_\Gamma(x,y) = T_{\Gamma/e}(x,y) + T_{\Gamma-e}(x,y). \tag{2.4}$$

This property of the Tutte polynomial is often used as its definition, if one completes it by giving the form of the Tutte polynomial on terminal forms:

$$T_\Gamma(x,y) := x^m y^n, \tag{2.5}$$

where m is the number of bridges and n is the number of self-loops.

Multi-variate (or weighted) versions of the Tutte polynomial exist in the literature. Thus, in the seminal paper, Alan D. Sokal (2005) analysed in detail such a multi-variate polynomial. The main idea is the following: one introduces a set of variables $\beta_1, \ldots, \beta_E$, one for each edge, and a variable q, instead of the couple of variables x and y of the Tutte polynomial.

Other multi-variate versions can also be found in the literature but they are essentially equivalent to the Sokal polynomial after appropriate changes of variables. Nevertheless, this is not the case for the polynomials defined in Zaslavsky (1992) and Bollobás and Riordan (1992) (see also Ellis-Monaghan and Traldi (2006) for generalizations).

Let us give the definition of the following multi-variate version of the Tutte polynomial:

Definition 2.1.8 *If Γ is a graph, then its multi-variate Tutte polynomial is defined as*

$$Z_\Gamma(q,\{\beta\}) := \sum_{A\subset E} q^{k(A)} \prod_{e\in A} \beta_e. \tag{2.6}$$

Similarly, one can prove that the multi-variate Tutte polynomial (2.6) satisfies the deletion/contraction relation, for any edge e. The definition of the polynomial on the terminal forms (graphs with v isolated vertices) is

$$Z_\Gamma(q,\{\beta\}) := q^v. \tag{2.7}$$

One can prove (through direct inspection) the relation between the Tutte polynomial (2.3) and its multi-variate counterpart (2.6):

$$\left[q^{-V} Z_\Gamma(q,\beta)\right]|_{\beta_e=y-1,q=(x-1)(y-1)} = (x-1)^{k(E)-V} T_\Gamma(x,y). \tag{2.8}$$

It is this version of the multi-variate Tutte polynomial (2.6) that we use in this book to prove the relation with the parametric representation of Feynman integrals in QFT (see Chapter 6.6).

Putting aside the Tutte polynomial, several graph polynomials have been defined and extensively studied in the literature. As we have seen previously, the Tutte polynomial is a two-variable polynomial. It has one-variable specializations, such as the chromatic polynomial or the flow polynomial.

The *chromatic polynomial* is a graph polynomial $P_\Gamma(k)$ ($k \in \mathbb{N}^\star$) which counts the number of distinct ways to colour the graph Γ with k or fewer colours, colourings being counted as distinct even if they differ only by permutation of colours. For a connected graph, this polynomial is related to the Tutte polynomial (2.3) by the relation

$$P_\Gamma(k) = (-1)^{V-1} k T_\Gamma(1-k,0). \tag{2.9}$$

In order to define the flow polynomial, we need a finite abelian group G. One can arbitrarily choose an orientation for each edge of the graph Γ, the result being independent of this choice (the same type of situation appears when computing Feynman integrals, see next section). A $G-$*flow* on Γ is a mapping

$$\psi : E \to G, \tag{2.10}$$

that satisfies current conservation at each vertex. A $G-$flow on Γ is said to be *nowhere-zero* if $\psi(e) \neq 0$ for all e. Let $F_\Gamma(G)$ be the number of nowhere-zero $G-$flows on Γ. One can prove that this number depends only on the order k of the group G; it can thus be written $F_\Gamma(k)$—it is the restriction to non-negative integers of a polynomial in k, the *flow polynomial.*

One has:

$$F_\Gamma(k) = (-1)^{E-V+1} T_\Gamma(0, 1-k). \tag{2.11}$$

A crucial property of the Tutte polynomial is its property of **universality**. This property states that any Tutte invariant (i.e. any graph polynomial satisfying deletion/contraction property and a multiplicative law on graph disjoint reunion and one-vertex joint) is an evaluation of the Tutte polynomial. This property is proved in the combinatorics literature by carefully using induction arguments on the edges of the graph.

Let us end this section by emphasizing that the Tutte polynomial (and its multi-variate version that we have presented here) extends in a natural manner to the more involved combinatorial notion of *matroids* (see again Sokal (2005)).

2.2 Ribbon graphs; the Bollobás–Riordan polynomial

In this section, we introduce a natural generalization of the notion of graphs—ribbon graphs or maps. For a general introduction to combinatorial maps, the interested reader may refer to Guillaume Chapuy's PhD thesis Chapuy (2009) or to Gilles Schaeffer's use (2009).

Let us define such a ribbon graph in the following way:

Definition 2.2.1 *A* **ribbon graph** Γ *is an orientable surface with its boundary represented as the union of closed disks, also called* **vertices**, *and ribbons also called* **edges**, *such that: the disks and ribbons intersect in disjoint line segments, each such line segment lying on the boundary of precisely one disk and one ribbon and finally, every ribbon containing two such line segments.*

Note that, as in the case of graphs, this definition can be extended by adding a new type of edge (not with two line segments, see the previous definition) such that **external edges** are allowed (see previous subsection). Examples of such graphs are given in Figs. 2.4 and 2.5.

Let us also mention that ribbon graphs can be defined as graphs equipped with a cyclic ordering of the incidence edges at each vertex or as graphs embedded in surfaces (the latter was actually the mathematical object on which B. Bollobás and O. Riordan defined their generalization of the Tutte polynomial in Bollobás and Riordan (2001) and Bollobás and Riordan (2002), see following section).

An interesting *connection between the Tutte polynomial of graphs and combinatorial maps* was proven in Bernardi (2008). He gave a characterization of the Tutte polynomial of graphs which is different to the initial one given by Tutte (which required some choice of linear order on the edge set, in order to write the Tutte polynomial as a function of

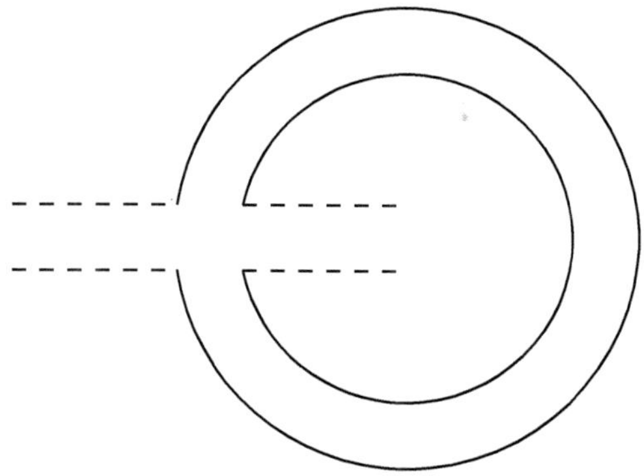

Figure 2.4 *An example of a ribbon graph with one vertex, one internal edge, and two external edges*

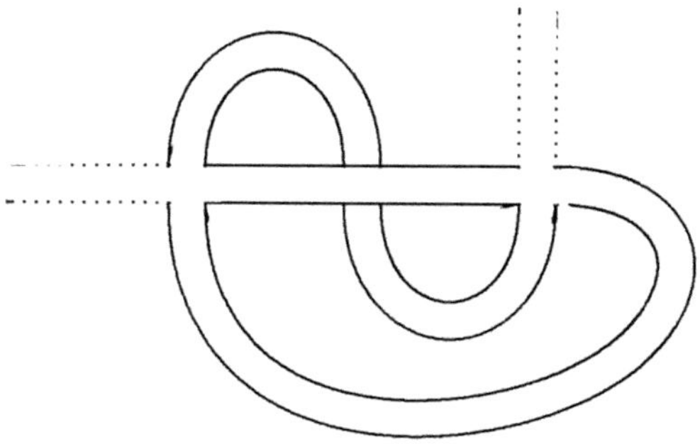

Figure 2.5 *An example of a ribbon graph with two vertices, three internal edges, and two external edges*

spanning trees). O. Bernardi proved that the Tutte polynomial can also be written as the generating function of spanning trees counted with the help of a cyclic order of the edges around each vertex.

Let us now give the following definition:

Definition 2.2.2 *A* **face** *of a ribbon graph is a connected component of its boundary as a surface.*

For example, the graph of Fig. 2.4 has two faces, while the one of Fig. 2.5 has a single face. If we glue disks along the faces we obtain a closed Riemann surface whose genus is also called the **genus** of the graph.

Definition 2.2.3 *The ribbon graph is called* **planar** *if it has vanishing genus.*

For example, the graph of Fig. 2.4 is planar while the one of Fig. 2.5 is non-planar (it has genus 1).

Definition 2.2.4 *A planar ribbon graph is called* **regular** *if the number of faces broken by external edges is equal to* 1.

For example, the graph of Fig. 2.4 is planar irregular, while the one of Fig. 2.5 is regular (it has only one face, which is also broken by both of the external edges).

Remark 2.2.5 *Planar regular ribbon graphs are also known in combinatorics literature as* **outer maps**.

Definition 2.2.6 *The Bollobás–Riordan polynomial of a ribbon graph G is defined as*

$$R_\Gamma(x,y,z) := \sum_{H\subset E} (x-1)^{r(G)-r(H)} y^{n(H)} z^{k(H)-F(H)+n(H)}. \tag{2.12}$$

Note that we have denoted by $F(H)$ the number of components of the boundary of the respective subgraph H (the number of faces). The supplementary variable z is required to keep track of the additional topological information.

Similar to the Tutte polynomial, the Bollobás–Riordan polynomial also obeys a deletion/contraction relation (see Fig. 2.6 for an example).

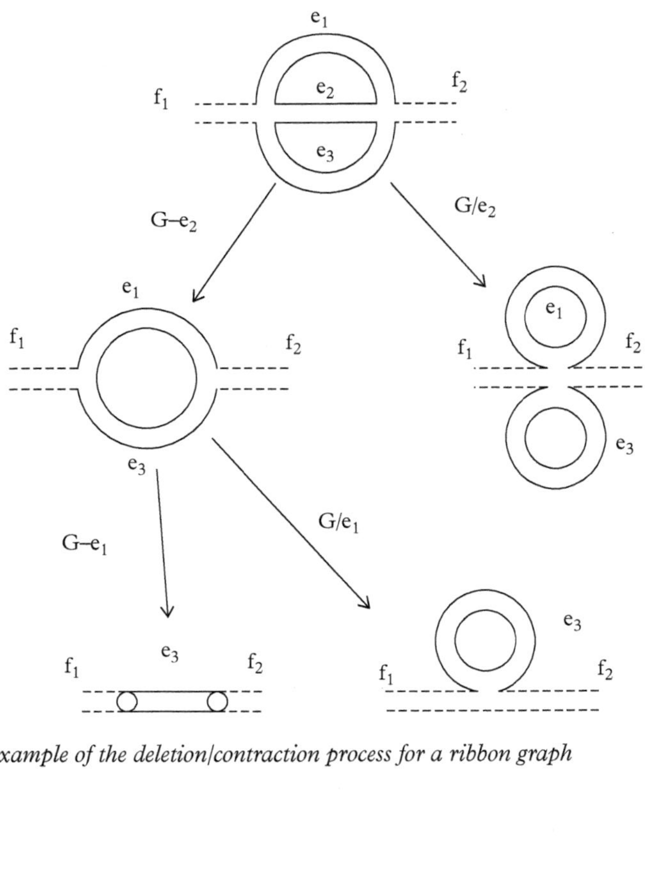

Figure 2.6 *An example of the deletion/contraction process for a ribbon graph*

Theorem 2.2.7 *Let* Γ *be a ribbon graph. One then has*

$$R_\Gamma = R_{\Gamma/e} + R_{\Gamma-e} \tag{2.13}$$

for any regular edge e of G. and

$$R_\Gamma = xR_{\Gamma/e} \tag{2.14}$$

for every bridge of Γ.

The situation is also analogous to that of the Tutte polynomial, in the sense that, defining the Bolobás–Riordan polynomial on terminal forms transforms the property (2.13) into a definition. On these terminal forms (that is, graphs with one vertex), the polynomial is defined as:

$$R_\Gamma(y,z) := \sum_{H\subset\Gamma} y^{E(H)} z^{2g(H)}, \tag{2.15}$$

since, in this case, $k(H) - F(H) + n(H) = 2g(H)$.

A multi-variate version of the Bollobás–Riordan polynomial exists in literature:

$$Z_\Gamma(x,\beta,z) = \sum_{H\subset E} x^{k(H)} \left(\prod_{e\in H} \beta_e\right) z^{F(H)}. \tag{2.16}$$

This version also satisfies a deletion/contraction relation.

Finally, let us mention that a signed version of the Bollobás–Riordan polynomial was also defined in Chmutov and Pak (2007); this is a three-variable polynomial defined on *signed ribbon graphs* (that is, ribbon graphs on which an element of the set $\{+,-\}$ is assigned to each edge). A partial duality with respect to a spanning subgraph was also defined Chmutov (2009); this allows to prove that the Kauffman bracket of a virtual link diagram is equal to the signed Bollobás–Riordan polynomial of some ribbon graph constructed from a state of the respective virtual link diagram (the interested reader may refer to Chmutov (2009) for details on this topic). Moreover, the properties of the multi-variate version of this signed Bollobás–Riordan polynomial were analysed in Vignes-Tourneret (2009) (namely, its invariance under the partial duality of Chmutov (2009) was proven). Finally, four-variable generalizations of the Bollobás–Riordan polynomial for ribbon graphs were defined in (Krushkal 2011; Krajewski, Rivasseau, and Vignes-Tourneret 2011).

2.3 Selected further reading

- Johanna Ellis-Monaghan and Criel Merino. (2010). Graph polynomial and their applications. In Johanna Ellis-Monaghan and Iain Moffatt (eds.) *The Tutte*

Polynomial. arXiv:0803.3079, invited chapter for Structural Analysis of Complex Networks, Birkhauser, 1–42.

A very good review on the Tutte polynomial.

- Johanna Ellis-Monaghan and Iain Moffatt, Graphs on surfaces. In *Springer Briefs in Mathematics.*

  A nice introduction to graphs on surfaces (and hence maps).
- *CRC Handbook on the Tutte Polynomial.* Johanna Ellis-Monaghan and Iain Moffatt (eds.). to appear.

  A vast collection of reviews on various aspects of the Tutte polynomial.

3

Quantum field theory (QFT)—*built-in combinatorics*

Let us now move to a (seemingly) completely different subject—QFT. We exhibit in this chapter that the mathematical formalism of QFT actually has a non-trivial *combinatorial backbone*.

Several (very) good books on QFT exist in (mathematical) physics literature. We only list a few of them here: that of Claude Itzykson and Jean-Bernard Zuber (2006), that of Michael Peskin and Daniel V. Schroeder (1995), that of Hagen Kleinert and Verena Schulte-Frohlinde (2001) and finally that of Jean Zinn-Justin (2002). Taking a more mathematical perspective, one can also name Alain Connes and Matilde Marcolli's encyclopaedic book (2008) (where QFT, although treated in detail, composes just the first chapter). Finally, let us also mention here A. Abdesselam's (2003) article, where a general introduction of Feynman graphs for combinatorists is provided.

QFT can generally be understood as a quantum description of particles and their interactions, a description which is also compatible with Einstein's theory of special relativity. Within the framework of elementary particle physics (or high-energy physics), QFT led to the Standard Model of Elementary Particle Physics, which is the physical theory tested with the best accuracy by experiments. Moreover, the QFT formalism successfully applies to statistical physics, condensed matter physics, and so on.

Let us conclude this short introduction by citing Claude Itzykson and Jean-Bernard Zuber (2006): 'QFT has remained throughout the years one of the most important tools in understanding the microscopic world.'

In this chapter, we focus mainly on the simplest QFT model, namely the scalar Φ^4 model. Nevertheless, some of the results exposed here extend to more involved models, such as gauge theories.

We show in this chapter how Feynman graphs appear through the so-called QFT perturbative expansion, how Feynman integrals are associated to Feynman graphs and how these integrals can be expressed via the help of graph polynomials, the Kirchhoff–Symanzik polynomials. Finally, we give a glimpse of renormalization, of the DSE and of the use of the so-called intermediate field method.

Combinatorial Physics: Combinatorics, quantum field theory, and quantum gravity models. Adrian Tanasa, Oxford University Press (2021). © Adrian Tanasa. DOI: 10.1093/oso/9780192895493.003.0003

3.1 Definition of the scalar Φ^4 model

A scalar field $\Phi(x)$ is a function

$$\Phi : \mathbb{R}^D \to \mathbb{K}, \tag{3.1}$$

where $D, \in, \mathbb{N}$, and $\mathbb{K}$ are taken to be $\mathbb{R}$ or $\mathbb{C}$. The parameter D represents the dimension of the space-time on which the field lives and is thus taken to be four (three spatial dimensions and one temporal dimension).

The field defined in such a way is said to live on an *Euclidean* D-dimensional space. Let us also note that one can perform the following analytical continuation:

$$t := \imath x_D. \tag{3.2}$$

This is known as a *Wick rotation* from Euclidean to Minkowskian *space-time*; the field $\Phi(x)$ then describes an (elementary) particle. In this book, we only work with the Euclidean signature.

A QFT model is defined by means of a *functional integral* representation of the *partition function Z*; from this partition function, *Green functions* (or Schwinger functions, see the following section) can then be obtained (by functional derivation).

Let us now explain what we mean by all these notions. One first needs to define the *action*, which, from a mathematical point of view, is a functional in the field $\phi(x)$. For the Φ^4 model previously given, the action is written:

$$S[\Phi(x)] = \int d^D x \left[\sum_{\mu=1}^{4} \left(\frac{\partial}{\partial x_\mu} \Phi(x) \right)^2 + \frac{1}{2} m^2 \Phi^2(x) + V[\Phi(x)], \right] \tag{3.3}$$

where the parameters m and λ are referred to as the mass and the coupling constant respectively. Moreover, the *interaction potential* is written:

$$V[\Phi(x)] = \frac{\lambda}{4!} \Phi^4(x). \tag{3.4}$$

Let us now introduce the notion of *functional integration* as the product of integrals at each space point x (up to some irrelevant normalization factor): $\mathcal{D}\phi(x) := \prod_x \int d\Phi(x)$. This infinite multiplication of Lebesgue measure is mathematically ill-defined; for the way in which to deal with this, the interested reader may refer for example to Manfred Salmhofer's book (1999). Another solution for this problem can be found, in Vincent Rivasseau's review article (2002) or Rivasseau (1992), where the quadratic part of the action is sent inside the measure.

Following Razvan Gurau, Vincent Rivasseau, and Alessandro Sfondrini., let us now present a yet different approach for this issue, which can be considered a different starting point for defining a QFT model. Probability measures are characterized by the expectations of polynomials in random variables—the moments of the measure. Thus, instead of 'bothering' with defining the measure itself, one can define only the moments of our measure, which, in QFT language correspond to the N-point functions of the Wick expansion (3.8).

The *partition function* is then defined as

$$Z := \int \mathcal{D}\Phi(x) e^{-S[\Phi(x)]}. \tag{3.5}$$

The physical information of a particle physics model is the cross-section—matrix elements of the diffusion matrix, matrix expressed (through the reduction formulas) from the *Green functions* announced previously. Note that Green functions are known for a Euclidean model under the name of *Schwinger functions*. Moreover, they can also be referred to in various textbooks as *N-point functions* or *correlation functions*.

These Schwinger functions are defined through the *Feynman–Kac formula*:

$$G^{(N)}(x_1,\ldots,x_N) := \frac{1}{Z}\int \mathcal{D}\phi(x)\Phi(x_1)\ldots\Phi(x_N)e^{-S[\Phi]}. \tag{3.6}$$

One can also write:

$$G^{(N)}(x_1,\ldots,x_N) = < \Phi(x_1)\ldots\Phi(x_N) > . \tag{3.7}$$

One can prove that *only correlation functions with an* **even** *number of fields N are non-vanishing* (see any of the mentioned textbooks for details). For $N = 2$ one refers to $G^{(2)}(x_1,x_2)$ as the *two-point function*. Furthermore, when the coupling constant λ is also taken to be vanishing, the two-point function $G^{(2)}(x_1,x_2)$ is denoted by $G_0^{(2)}(x_1,x_2)$ and is referred to as the *free two-point function* (or the *free propagator*). On a more general basis, the theory taken at vanishing value of the coupling constant λ is referred to as the *free theory*.

Another important result one can prove using appropriate combinatorics is that *any N-point function can be expressed as a sum of $(N-1)!!$ different products of $N/2$ propagators* $G_0^{(2)}$:

$$G_0^{(N)}(x_1,\ldots,x_N) = \sum_{i=1}^{(N-1)!!}\prod_{j=1}^{N/2} S_0^{(2)}(x_{\pi_i(2j-1)},x_{\pi_i(2j)}). \tag{3.8}$$

The indices $\pi_i(j)$ $(1 \le i \le (N-1)!!, 1 \le j \le N)$ enumerate the pair combinations in the so-called **Wick expansion** on the right-hand side of equation (3.8).

Translation-invariance

One can explicitly check (see again any of the mentioned textbooks for details) that the propagator is invariant under arbitrary translations $a \in \mathbb{R}^D$ of the arguments of the field $\Phi(x)$

$$x \to x + a. \tag{3.9}$$

It is thus easier to write the propagator with only a single argument

$$G_0^{(2)}(x_1, x_2) = G_0^{(2)}(x_1 - x_2). \tag{3.10}$$

This translation-invariance property further generalizes to the level of the free N-point function:

$$G_0^{(N)}(x_1, \dots, x_N) = G_0^{(N)}(x_1 - x_N, \dots, x_{N-1} - x_N, 0). \tag{3.11}$$

This can be proved using the Wick expansion (3.8) combined with the translation-invariance property (3.10) of the two-point function.

3.2 Perturbative expansion—Feynman graphs and their combinatorial weights

In general, one is unable to find an exact expression for a general Schwinger function (3.6). Therefore, theoretical physicists do a **perturbative expansion** in powers of the coupling constant λ:

$$G^{(N)}(x_1, \dots, x_N) = \frac{1}{Z} \sum_{p=0}^{\infty} G_p^{(N)}(x_1, \dots, x_N). \tag{3.12}$$

The expansion coefficients are sums of multiple integrals, referred to as *Feynman integrals*. Their organization is simplified by the use of *Feynman graphs*.

Similarly, the partition function Z is expanded:

$$Z = \sum_{p=0}^{\infty} Z_p, \tag{3.13}$$

where we have denoted by Z_0 the partition function Z taken at vanishing coupling constant λ. This is normalized to the unit: $Z_0 = 1$. The parameter p is referred to as the *order in perturbation theory*.

One can prove (see again any of the indicated textbooks) the following formula for the coefficient $G_p^{(N)}(x_1 \dots, x_N)$:

$$G_p^{(N)}(x_1, \dots, x_N) = \frac{1}{p!}\left(\frac{-\lambda}{4!}\right)^p \int d^D z_1 \dots d^D z_p \tag{3.14}$$

$$G_0^{(N+4p)}(z_1, z_1, z_1, z_1, \dots, z_p, z_p, z_p, z_p, x_1, \dots, x_N). \tag{3.15}$$

This formula expresses the N-point function as a function of the free-field N-point function. Furthermore, one can now use (3.8) in order to Wick-expand these free-field N-point functions into sums over products of propagators S_0.

Working all these formulae leads to the following result:

$$\begin{aligned} &G_p^{(N)}(x_1, \dots, x_N) \\ &= \frac{1}{p!}\left(\frac{-\lambda}{4!}\right)^p \int d^D z_1 \dots d^D z_p \int d^D y_1 \dots d^D y_{4p+N} \prod_{\ell=1}^{N} \delta^{(D)}(y_{4p+\ell} - x_\ell) \\ &\prod_{k=1}^{p} \left(\delta^{(D)}(y_{4k-3} - z_k)\delta^{(D)}(y_{4k-2} - z_k)\delta^{(D)}(y_{4k-1} - z_k)\delta^{(D)}(y_{4k} - z_k)\right) \\ &\sum_{i=1}^{(4p+N-1)!!} \prod_{j=1}^{\frac{4p+N}{2}} G_0(y_{\pi_i}^{(4p+N)}(2j-1), y_{\pi_i}^{(4p+N)}(2j)). \end{aligned} \tag{3.16}$$

Each of the products in the sum (3.16) can be depicted by a Feynman graph. Let us give a few more explanations on this issue. The N external points (to which the external edges hook) are

$$y_{4p+1} = x_1, \dots, y_{4p+N} = x_N$$

(through the δ-functions in the first line of equation (3.16)). One has a free propagator $G_0(y, y')$ connecting any of the points of the graph. The order p in perturbation theory is nothing but the number of vertices of the graph. For each such vertex one has four δ-functions (the second line in equation (3.16)); this represents the **locality** of the interaction. These $4p$ variables (equal by groups of four, each group for one of the vertices of the Feynman graph) are integrated against; they are the *internal points*. For each such vertex one then has a final integration over the point z. Let us recall that each such vertex comes with a coupling constant λ.

We have thus seen that integrals can be associated with these graphs; this is done through **Feynman rules**.

An **N-point (Feynman) graph** is a Feynman graph associated with the respective N-point function.

Thus, a four-point Feynman graph is given in Fig. 2.1 (it has four external edges). Following Hagen Kleinert and Verena Schulte-Frohlinde (2001), let us now give a few more definitions that are useful in the combinatorics of QFT:

Definition 3.2.1 *The* **multiplicity** M_Γ *of a Feynman graph* Γ *is the number of Wick contractions leading to the same Feynman integral.*

This multiplicity can be computed using the following considerations. The *symmetry factor of the* Φ^4 *interaction is* 4!. This can be seen in the fact that there are 4! ways of labelling the four incoming/outgoing edges of a vertex. (One can check this analytically by performing the integrations over $y_1,\dots,y_4$). Using the Feynman rules introduced earlier, the associated integral writes

$$\begin{aligned}&\int d^D y_1 \dots d^D y_4 \delta^{(D)}(y_1 - z)\dots\delta^{(D)}(y_4 - z) G_0(y_1,\bar{y}_1)\dots G_0(y_4,\bar{y}_4)\\&\quad = G_0(z,\bar{y}_1)\dots G_0(z,\bar{y}_4),\end{aligned} \tag{3.17}$$

where we have denoted by $\bar{y}_1,\dots,\bar{y}_4$ the four external points to which the four propagators hook to (previously mentioned). There are indeed 4! ways of labelling the $y_1,\dots,y_4$ variables inside the Feynman integrand of (3.17) which leave the result unchanged.

Nevertheless, this vertex factor can be reduced in the following cases:

1. The self-contraction of a vertex—two of the y points contract to each other. The integral is written:

$$\begin{aligned}&\int d^D y_1 \dots d^D y_4 \delta^{(D)}(y_1 - z)\dots\delta^{(D)}(y_4 - z) G_0(y_1,y_2) G_0(y_3,\bar{y}_3) G_0(y_4,\bar{y}_4)\\&\quad = G_0(z,z) G_0(z,\bar{y}_3) G_0(z,\bar{y}_4).\end{aligned} \tag{3.18}$$

   This is the QFT tadpole (or (self-)loop in graph theory, see Definition 2.1.2 4). Let us denote by S the number of tadpoles in the respective graph.

2. The double connection of a vertex—two of the y variables (say y_1 and y_2) contract to the same point $\bar{y}_1$. The integral thus is written

$$\begin{aligned}&\int d^D y_1 \dots d^D y_4 \delta^{(D)}(y_1 - z)\dots\delta^{(D)}(y_4 - z) S_0(y_1,\bar{y}_1) S_0(y_2,\bar{y}_1) S_0(y_3,\bar{y}_3) S_0(y_4,\bar{y}_4)\\&\quad = S_0(z,\bar{y}_1) S_0(z,\bar{y}_1) S_0(z,\bar{y}_3) S_0(z,\bar{y}_4).\end{aligned} \tag{3.19}$$

   The permutation y_1, y_2 is irrelevant and thus the appropriate factor is $4!/2$. Let us denote by D the number of double connections in a graph.

3. The triple connection—three of the y variables contract to the same point; the appropriate factor is $4!/3!$. Let us denote by T the number of triple connections in a graph.
4. The fourfold connection—all four of the y variables contract to the same point; the appropriate factor is 1. Let us denote by F the number of fourfold connections in a graph.

Another important combinatorial notion is the one of *identical vertex permutations (IVP)*. It corresponds to vertices that can be interchanged in the Feynman integrand such that the final result is not affected. We denote the number of such IVP by N_{IVP}.

One can now prove the following proposition:

Proposition 3.2.2 *The multiplicity of a general Feynman graph* Γ *is given by:*

$$M_\Gamma = \frac{(4!)^p p!}{(2!)^{S+D}(3!)^T(4!)^F N_{\text{IVP}}}. \tag{3.20}$$

Note that the entities appearing on the right-hand side (number of vertices, number of tadpoles and so on) refer to the respective graph Γ*. Nevertheless, in order to simplify the notations, we have not indexed them with* Γ*.*

Finally, one has the following definition:

Definition 3.2.3 *The weight factor* W_Γ *of a Feynman graph* Γ *with* p *vertices (or at order* p *in perturbation theory) is defined by:*

$$W_\Gamma := \frac{M_\Gamma}{(4!)^p p!}. \tag{3.21}$$

3.3 Fourier transform—the momentum space

The results of the previous two sections were written in *position space* (or *direct space*), generally used in statistical field theory or in condensed matter physics. One can have a Fourier transform to *momentum space*, which is most frequently used in elementary particle physics.

The Fourier transform of the action (3.3) thus is written:

$$\tilde{S}[\tilde{\Phi}] = \int_{\mathbb{R}^4} d^4p \left[\frac{1}{2}\sum_{\mu=1}^{4} p_\mu p_\mu \tilde{\Phi}^2 + \frac{1}{2}m\tilde{\Phi}^2 + \tilde{V}_{\text{int}}[\tilde{\Phi}]\right], \tag{3.22}$$

where we have denoted by $\tilde{V}_{\text{int}}[\tilde{\Phi}]$ the Fourier transform of the interaction potential (3.4).

As in position space, one has perturbative expansion in powers of the coupling constant λ; this leads to the Feynman graphs described in the previous section. One associates some orientation and some momentum to any edge (external or internal).

Each such momentum is a four-vector (it lives on $\mathbb{R}^4$). The **Feynman rules** that allow one to associate a Feynman integral with such a graph consist of:

- conservation of momenta—the sum of incoming momenta at a vertex must be equal to the sum of outgoing ones. This is realized with a $\delta-$function.
- Integration of the internal momenta against their propagators. For each such edge of momentum p, one has a propagator

$$C(p) = \frac{1}{p^2 + m^2}. \tag{3.23}$$

Moreover, one has to multiply the final result with λ^p.

Let us emphasize that the Feynman integral does not depend on the chosen orientation for the edges of the graph.

Let us now illustrate the previous with the example of the Feynman graph of Fig. 2.1, a graph at order three in perturbation theory with four internal edges and four external edges. Applying these rules, its associated Feynman integral can be written as:

$$\lambda^3 \int \left(\prod_{i=1}^{4} d^4 p_{e_i}\right) \left(\prod_{i=1}^{4} \frac{1}{p_{e_i}^2 + m^2}\right)$$
$$\delta(p_{f_1} + p_{f_2} - p_{e_1} - p_{e_2})\delta(p_{e_1} + p_{e_3} - p_{e_4} - p_{f_4})\delta(p_{e_2} - p_{e_3} + p_{e_4} - p_{f_3}). \tag{3.24}$$

We invite the interested reader to find out the orientation of the edges which were chosen for such a Feynman integral to occur. Because of the presence of the three $\delta-$functions, the number of remaining integrals is equal to the number of independent cycles of the Feynman graph (the number of (independent) loops, in QFT terminology, see again Definition 2.1.2 3). This is a property which is always valid in QFT.

In the end, the Feynman integral (3.24) leads to a logarithmic divergence in the high-energy regime ($|p| \to \infty$) of the internal momenta (the so-called *ultraviolet regime*). The appearance of such divergences is actually a very frequent phenomena in QFT; it is the renormalization process which, when possible, takes care of these infinities in a highly non-trivial way (see Section 3.6).

Note that, in the case presented here, the mass m prevents the integral from being divergent in the *infrared regime* (that is, $|p| \to 0$) as well. We will come back to these divergence issues in Section 3.6.

3.4 Parametric representation of Feynman integrands

Let us now introduce the *parametric representation* of a Feynman integral. The main idea is to write each of the internal propagators (3.23) of the integral under the form of an integral on some parameter α_e (called a *Schwinger parameter*):

$$C(p_e) = \int_0^\infty d\alpha_e e^{-\alpha_e(p_e^2+m^2)}, \ \forall e = 1,\ldots,E. \tag{3.25}$$

Inserting these formulae into the general expression of a Feynman integral allows to integrate out (through Gaussian integrations) the internal momenta p_e.

The Feynman integral then writes

$$\phi(\Gamma) = \int_0^\infty \frac{e^{-V_\Gamma(p_{\text{ext}},\alpha)/U_\Gamma(\alpha)}}{U_\Gamma(\alpha)^{D/2}} \prod_{\ell=1}^{E} (e^{-m^2\alpha_\ell} d\alpha_\ell), \tag{3.26}$$

where $U(\alpha)$, and respectively $V(p_{\text{ext}},\alpha)$, are polynomials in the set of parameters α, and in the set of external momenta p_{ext} and the set of parameters α respectively. They are called the Kirchoff–Symanzik polynomials. Their exact expression is proven to depend only on the structure of the respective graphs

$$U_\Gamma(\alpha) = \sum_{\mathcal{T}} \prod_{\ell \notin \mathcal{T}} \alpha_\ell, \tag{3.27}$$

and

$$V_\Gamma(p_{\text{ext}},\alpha) = \sum_{\mathcal{T}_2} \prod_{\ell \notin \mathcal{T}_2} \alpha_\ell \Big(\sum_{i \in E(\mathcal{T}_2)} p_i \Big)^2,$$

where we have denoted by $\mathcal{T}$ a tree of the graph and by $\mathcal{T}_2$—a two tree which, as already stated in the previous section (see remark 2.1.4), separates the graph in two connected components. We have denoted by $E(\mathcal{T}_2)$ one of the connected components thus obtained.

Remark 3.4.1 *By momentum conservation, the total momenta of one of these connected components (for example $E(\mathcal{T}_2)$) is equal to that of the other connected component.*

For the example of Fig. 2.1, one has:

$$\begin{aligned} U_\Gamma(\alpha) &= \alpha_3\alpha_4 + \alpha_2\alpha_4 + \alpha_2\alpha_3 + \alpha_1\alpha_3 + \alpha_1\alpha_4, \\ V_\Gamma(p_{\text{ext}},\alpha) &= (p_{f_1} + p_{f_2})^2 \alpha_1\alpha_2(\alpha_3+\alpha_4) + p_{f_4}^2 \alpha_1\alpha_3\alpha_4 + p_{f_3}^2 \alpha_2\alpha_3\alpha_4. \end{aligned} \tag{3.28}$$

Remark 3.4.2 *The parametric representation* (3.26) *of a Feynman integral has a trivial dependence of the space-time dimension D, which is now only a parameter. (This is of particular importance for the implementation of the dimensional renormalization scheme, see Subsection 3.6.5).*

Let us now rewrite the Kirchhoff–Symanzik polynomials as

$$U(\alpha)=\sum_j \prod_{\ell=1}^{L} \alpha_\ell^{u_{\ell j}} \equiv \sum_j U_j, \qquad V(\alpha)=\sum_k s_k \left(\prod_{\ell=1}^{L} \alpha_\ell^{v_{\ell k}}\right) \equiv \sum_k V_k, \tag{3.29}$$

where j runs over the set of spanning trees and k over the set of the two-trees,

$$u_{\ell j}=\begin{cases} 0 & \text{if the line } \ell \text{ belongs to the tree } j \\ 1 & \text{otherwise} \end{cases} \tag{3.30}$$

and

$$v_{\ell k}=\begin{cases} 0 & \text{if the line } \ell \text{ belongs to the two-tree } k \\ 1 & \text{otherwise.} \end{cases} \tag{3.31}$$

The form (3.29) of the Kirchhoff–Symanzik polynomials simply represents some splitting of these polynomials into a sum of monomials. As we will see later in this book, this form is particularly useful in applying the Mellin transform technique (see Section 7.1).

For the sake of completeness, we end this section by mentioning that the Kirchhoff-Symanzik polynomials allow a reformulation of Feynman integrals into an algebraic framework; studying QFT in the language of hypersurfaces leads to a motivic version of the Feynman rules that we have seen in this chapter (see, for example, (Aluffi and Marcolli 2009, 2010, 2011a,b; Bloch, Esnault, and Kreimer 2006), or the Matilde Marcolli (2010)).

3.5 The propagator and the heat kernel

Let us now briefly get back to position space. Using the inverse Fourier transform for formula (3.25), one has (up to irrelevant constants) the following expression for the propagator of the theory:

$$C(x,y)=\int_0^\infty \frac{da}{a^{D/2}} e^{-am^2-\frac{(x-y)^2}{4a}}. \tag{3.32}$$

One thus sees the integral representation of the heat kernel. This allows to establish a connection between the scalar propagator and the Gaussian probability distribution of a Brownian path going from x to y in time a. Moreover, this heat-kernel form of the Euclidean propagator permits the Schwinger functions G^N to satisfy the so-called Osterwalder–Schrader axioms which allow the analytic continuation in Minkowski space—see, for example, Vincent Rivasseau's course (2012).

For the sake of completeness, we also mention that new relationships between hypergeometric functions have recently been found by evaluating Feynman integrals in Kniehl and Tarasov (2011).

3.6 A glimpse of perturbative renormalization

Feynman integrals in QFT do not generally converge. As we have already seen, this can come from two integration regions:

1. the small-momentum region ($|p| \to 0$) (or long distances)—**infrared divergences**; and
2. the high-momentum region ($|p| \to \infty$) (or short distances)—**ultraviolet divergences**.

For the Φ^4 model discussed in this chapter, infrared divergences occur if and only if $m = 0$. On the other hand, ultraviolet divergences occur independently of the value of the mass parameter. (One can see, for example, the Feynman integral of equation (3.24), which is quadratically divergent for large values of the internal momenta). In this chapter the mass is taken as non-vanishing; we thus only have to deal with ultraviolet divergences.

This phenomenon has almost led theoretical physicists to abandon the mathematical formalism of QFT. The situation was saved later on by the discovery of renormalization, which deals with these divergences in an efficient way.

Here, let us elaborate on two issues. **Not all QFT models are renormalizable**, one has renormalizable and non-renormalizable models. Moreover, the finite quantities obtained after renormalization are related to quantities measured in physical experiments, such as the celebrated CERN's accelerator experiments. These quantities have been measured with extremely high accuracy and they correspond to the ones computed in these QFT calculations.

In order to have a renormalizable model, the following ingredients need to be present:

- a power counting theorem; and
- a principle of locality.

A supplementary crucial notion is the one of **scale**. We will get back to these key ingredients in Subsections 3.6.1, 3.6.2, and 3.6.3.

In theoretical physics, one has a *perturbative renormalization*, that is an order by order (in perturbation theory) renormalization, and a *non-perturbative renormalization*.

In this chapter, we focus on perturbative renormalization. Several approaches exist in the literature for non-perturbative renormalization. Here, we only mention **constructive renormalization**, which takes into consideration the summation of all the finite quantities remaining after the order by order perturbative renormalization. For more details, we invite the interested reader to consult Vincent Rivasseau (1992). For more recent developments on constructive renormalization, one may also refer to V. Rivasseau and

Z. Wang (2010a). We also refer the reader to the last section of this chapter for some considerations on DSE, whose solutions also represent a non-perturbative QFT result.

The first step in a renormalization process is the **regularization procedure**, which renders the Feynman integrals finite. The main idea of such a regularization procedure is to introduce a *regularization parameter*, such that the divergences of the Feynman integrals appear as singularities in this new parameter.

Several regularization schemes exist in the literature. Following again Hagen Kleinert and Verena Schulte-Frohlinde (2001), we give here a short list of the main regularization schemes:

- momentum space cut-off. The idea behind this procedure is to introduce a cut-off Λ such that the integrations on $|p|$ are carried out only until Λ. The Feynman integrals are now convergent, but they are of course divergent in the limit $\Lambda \to \infty$.
- Pauli–Villars regularization (1949). One changes the propagator in the following way:

$$\frac{1}{p^2+m^2} \to \frac{1}{p^2+m^2} - \frac{1}{p^2+M^2}. \tag{3.33}$$

  The modified propagator decreases faster within the limit $|p| \to \infty$; the role of the momentum space cut-off Λ is now played by the parameter M.
- analytic regularization Speer (1969). One has the following change of propagator:

$$\frac{1}{p^2+m^2} \to \frac{1}{(p^2+m^2)^{-z}}, \tag{3.34}$$

  where $z \in \mathbb{C}$ with $Re(z)$ being large enough to make the Feynman integrals converge. The result is then analytically continued to a region around the physical value $z=1$—the divergences now appear as poles for $z=1$.
- dimensional renormalization Hooft (1972, 1973). The idea behind this approach is to allow the space-time dimension D in Feynman integrals to be a *complex number*. This scheme is extensively used nowadays in elementary particle physics computations. (Its main interest for physicists comes from the fact it naturally preserves gauge invariance.) Moreover, it is interesting to mention here that a concrete geometric meaning of this renormalization scheme can be exhibited (see section 1.19 of Alain Connes and Matilde Marcolli (2008)).

The **multi-scale analysis** is more complex than all these, because it also takes into account the energy scale on which the internal propagators reside. Thus, the propagators are 'sliced'. This is an appropriate tool for implementing the constructive renormalization (see previous mention). For more details on the multi-scale analysis refer to Subsection 3.6.3.

The **Polchinski flow equation method** (see, for example, Manfred Salmhofer (1999)) is yet another way of performing renormalization.

It is, in our opinion, particularly interesting to mention that the Polchinski flow equation idea is also used in combinatorics to prove a result of E. M. Wright's which expresses the generating function of connected graphs under certain conditions (fixed excess). In this proof, one needs to choose an arbitrary edge in such a connected graph and then remove it. Two possibilities for this chosen edge can then appear (see, for example, Proposition II.6 of Philippe Flajolet and Robert Sedgewick (2008)), just as in the QFT Polchinski renormalization.

Finally, let us also mention that such a QFT-inspired Polchinski flow equation technique was recently used in a proof given by T. Krajewski for Postnikov's hook length formula (2012) (the *Postnikov hook length formula* gives the number obtained when one sums over all plane binary trees of a given order n on the product over v of $(1+1/h_v)$, v being a vertex, and h_v the hook length, i.e. the number of vertices below v).

3.6.1 The power counting theorem

In a generic QFT model, a special role in the process of renormalization is played by the primitively divergent graphs which are defined in the following way:

Definition 3.6.1 *A* **primitively divergent graph** *of a QFT model is a graph whose Feynman integral is divergent, but which does not contain any subgraph for which the Feynman integral is also divergent.*

All the divergences of a renormalizable model come from insertion of these Feynman graphs into 'larger' Feynman graphs. Furthermore, only a finite class of graphs should be primitively divergent, if one wants to have a renormalizable model.

Let us now give the following definition:

Definition 3.6.2 *The* **superficial degree of divergence** ω, *is the difference between the total number of powers in internal momenta between the denominator and the nominator.*

Thus, for the Feynman integral (3.24), after integration using the δ–function, one has two independent momenta in the denominator (each of them being a four vector) and four momenta, each of them squared) in the nominator. One thus has

$$\omega = 4\times 2 - 2\times 4 = 0. \tag{3.35}$$

The power counting theorem gives the superficial degree of divergence for a general Feynman graph (of a given QFT model). For the Φ^4 model, the power counting theorem is the following:

Theorem 3.6.3 (Power counting theorem) *The superficial degree of divergence is given by*

$$\omega = N - 4. \tag{3.36}$$

A direct way of proving this theorem is through a rescaling of the parameters α in the parametric representation (3.26) of the Feynman integrand.

Example 3.6.4 *The Φ^4 graph of Fig. 2.1 has four external edges; thus, its superficial degree of divergence is equal to* 0. *This corresponds to a logarithmic divergence in an ultraviolet momentum space cut-off.*

In order for a QFT model to be renormalizable, its superficial degree should not depend on the internal structure (number of edges, loops etc.) of the graph.

Let us also give the following definition:

Definition 3.6.5 *A* **superficially divergent graph** *of a QFT model is a graph with a negative or vanishing superficial degree of divergence.*

Remark 3.6.6 *A primitively divergent graph is a superficially divergent graph with no sub-divergences.*

3.6.2 Locality

The principle of locality states that the counterterms one requires to subtract the divergence of some Feynman integral are of the same type as the terms already presented in the bare (non-renormalized) action. These terms are *local*, hence the name of locality.

This principle can be most easily understood when computing Feynman integrals in position space (since space locality appears the most clearly in ... position space). Thus, for the example of a one-loop four-point graph of Fig. 3.1, (graph with two vertices, localized in two points x and y in position space), the Feynman integral is divergent when integrating on the sector:

$$x \sim y. \tag{3.37}$$

The rest of the integration domain does not lead to a divergence. One thus needs to subtract a **counterterm** which corresponds exactly to the *local* part of the Feynman integral—a *local counterterm*. This can be illustrated as is done in Fig. 3.2.

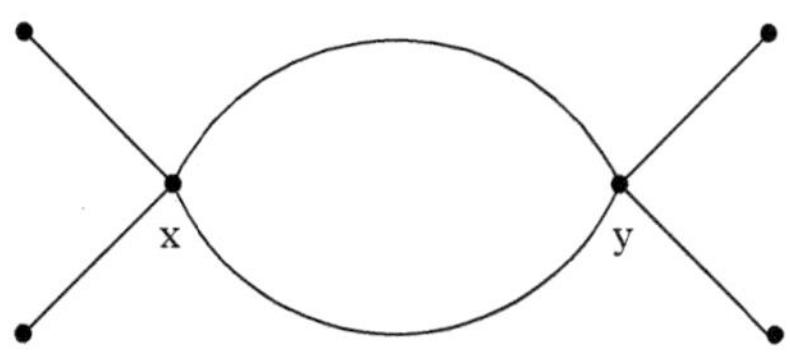

Figure 3.1 *A four-point one-loop Feynman graph. It has two vertices, localized, in position space, in x and y*

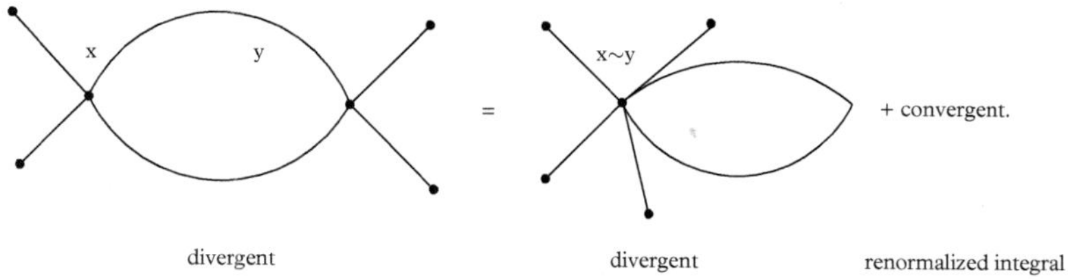

Figure 3.2 *Illustration of the local subtraction of the counterterm for the one-loop graph of Fig. 3.1. The divergent part of the integral corresponds to the region of the integration domain where $x \sim y$ (the locality principle)*

More concretely, writing the Feynman integral to renormalize gives (up to irrelevant normalization factors)

$$\phi(\Gamma) = \int dx dy C(x,y)^2 C(x,\bar{y}_1) C(x,\bar{y}_2) C(y,\bar{y}_3) C(y,\bar{y}_4) \Phi(x)^2 \Phi(y)^2, \tag{3.38}$$

where we have denoted the four external points that the external edges hook to by $\bar{y}_1, \ldots, \bar{y}_4$. Let us now leave aside for the moment the fields (for simplicity) and focus on the propagators of the formula (3.38). As previously discussed, we will subtract a term corresponding to the region $x \sim y$. This can be achieved using the expansion formula

$$f(t) = f(0) + \int_0^1 dt f'(t). \tag{3.39}$$

The previous integral is divergent and can be rewritten as

$$\int dx dy C(x,y)^2 C(x,\bar{y}_1) C(x,\bar{y}_2) \Phi(x)^2 \Phi(y)^2 \tag{3.40}$$
$$\left(C(x,\bar{y}_3) C(x,\bar{y}_4) + \int_0^1 dt (y-x) \frac{d}{dt} \left(C(x+t(y-x),\bar{y}_3) C(x+t(y-x),\bar{y}_4) \right) \right),$$

where we have completed the formula with the products of the corresponding external fields. One can prove that the second term in the right-hand side of (3.40) is finite. The first term in the sum correspond to the divergent part of the integral—it will be subtracted (the counterterm),

$$\tau_\Gamma \phi(\Gamma) = \int dx \, (\Phi(x))^4 \int dy C(x,y)^2. \tag{3.41}$$

This term is of the desired local-like form of the term present in the action. This term can thus be reabsorbed in a *redefinition of the coupling constant* λ).

The same type of redefinition of the other 'constants' appearing in the action can be done in the case of two-point graphs.

For a more general Feynman graph (not necessary a one-loop one), the locality of the counterterms can be seen as *sending all the external edges to the same point* (again, in position space).

We denote by Γ/γ the **cograph** obtained by shrinking the (divergent) subgraph γ inside the Feynman graph Γ:

Definition 3.6.7 *To* **shrink** *a subgraph means to erase its internal structure.*

Remark 3.6.8 *When shrinking a two-point subgraph, one should take into consideration whether the respective divergence contributes to the renormalization of the mass or of the wave function (the two parameters corresponding to the two terms in the quadratic part of the action). Nevertheless, this does not play an important part at a combinatorial level.*

Definition 3.6.7 allows us to mention the following general property:

$$\tau_\gamma \phi(\Gamma) = \phi(\Gamma/\gamma)\tau\phi(\gamma), \tag{3.42}$$

for γ a divergent subgraph of Γ.

Let us emphazise that locality principle is not only present in position space. Thus, in momentum space, using a similar analysis, one can prove that the required counterterms have the same form as the terms present in the action ($p^2\Phi^2$, Φ^2 or Φ^4) (see, for example, sections 1.5 or 1.6 of Alain Connes and Matilde Marcolli (2008)).

3.6.3 Multi-scale analysis

The notion of physical energy scale lies at the heart of the multi-scale approach. The main idea is to 'slice' the propagator in a discrete sum of contributions, each corresponding to an energy sector (scale). The starting point is the integral form (3.32) of the propagator in position space. One has

$$C(x-y) = \sum_{i=0}^{p} C_i(x-y), \tag{3.43}$$

where

$$C_i(x-y) = \int_{M^{-2i}}^{M^{-2(i-1)}} e^{-m^2 a - \frac{(x-y)^2}{4a}} \frac{da}{a^{D/2}}, \tag{3.44}$$

and

$$C_0(x-y) = \int_1^\infty e^{-m^2 a - \frac{(x-y)^2}{4a}} \frac{da}{a^{D/2}}. \tag{3.45}$$

The parameter M is taken greater than 1 (in appropriate units) and represents the 'thickness' of the slice.

Using this decomposition, one can, for example, give a proof of the power counting theorem (see Chapter 11.1 of Vincent Rivasseau (1992)).

The main advantage of this decomposition is that the propagator is now sensitive to energy scales. This allows for a precise subtraction of the counterterms—the subtraction is done if and only if the respective internal edges are in the ultraviolet sector (the so-called 'useful counterterms' because they actually cancel a divergence). Doing perturbative renormalization in this way allows to obtain better results for summing up the entire perturbative series—the constructive renormalization (previously given). Let us also mention that this multi-scale analysis idea is in some ways related to the one of the Hepp sectors (of the 'usual' BPHZ renormalization). Nevertheless, in the case of Hepp sectors, one has an order of relation between *all* Schwinger parameters of the graph, while in the multi-scale analysis, edges can be degenerated within some scale. This will then lead to crucial differences in the attempt to sum up the perturbative series (see again Vincent Rivasseau (1992)).

3.6.4 The subtraction operator for a general Feynman graph

In Subsection 3.6.2, we have seen an explicit example, for the bubble graph, of the extraction of the local counterterm (renormalization). Thus, equation (3.41) represents the *subtraction operator* for the respective bubble graph.

In order to define this operator for a general Feynman graph, one needs the following definition:

Definition 3.6.9 *A* **Zimmermann forest** $\mathcal{F}$ *of a Feynman graph* Γ *is a set of subgraphs of* Γ *such that*

$$\gamma\cap\gamma'=\emptyset \text{ or } (\gamma\subset\gamma' \text{ or } \gamma'\subset\gamma),\ \forall\gamma,\gamma'\in\mathcal{F}. \tag{3.46}$$

Remark 3.6.10 *The Zimmermann forest previously defined is a distinct notion from that of forest in graph theory (see Definition 2.1.2 8).*

Let us mention that the Zimmermann forests previously given are also called *inclusion forests* in Rivasseau (2012).

For a general Feynman graph, the (Bogoliubov) subtraction operator is written as

$$\bar{R}:=\sum_{\mathcal{F}}\prod_{\gamma\in\mathcal{F}}(-\tau_\gamma), \tag{3.47}$$

where the sum runs on all Zimmermann forests $\mathcal{F}$ of superficially divergent subgraphs γ (including the empty forest).

Applying this subtraction operator on the Feynman integral $\phi(\Gamma)$ extracts, in a local way, the divergences. The remaining part, the **renormalized Feynman integral**

$\phi\text{ren}(\Gamma)$ is thus finite—the Bogoliubov–Parasiuk–Hepp–Zimmermann theorem (Bogoliubov and Parasiuk 1957; Hepp 1966; Zimmermann 1969),

$$\phi\text{ren}(\Gamma) = \bar{R}(\phi(\Gamma)). \tag{3.48}$$

As already mentioned in Subsection 3.6.1, all the divergences in the Φ^4 model come from insertions of these divergent two- and four-point graphs into 'larger' graphs. Thus, the Feynman graph of Fig. 3.3 (which we denote by Γ), is divergent because of the sub-divergence represented by the four-point subgraph made of the internal edges 1 and 2 (which we denote by γ).

Once this (unique) sub-divergence is renormalized, the whole graph is renormalized. This can be explicitly seen from the following. The Zimmermann forests of this graph are

$$\emptyset, \{\gamma\}. \tag{3.49}$$

Note that, if the graph Γ would have had four (or two) external edges, then the previous list would have been: $\emptyset$, $\{\gamma\}$. $\{\Gamma\}$, $\{\gamma, \Gamma\}$. The renormalized Feynman integral is written as:

$$\phi\text{ren}(\Gamma) = \bar{R}(\phi(\Gamma)) = \phi(\Gamma) - R(\phi(\gamma))\phi(\Gamma/\gamma). \tag{3.50}$$

The operator R acting on a Feynman integral contains the divergent part of the respective integral; it is constructed via Taylor developments. Thus, in formula (3.50), the first term

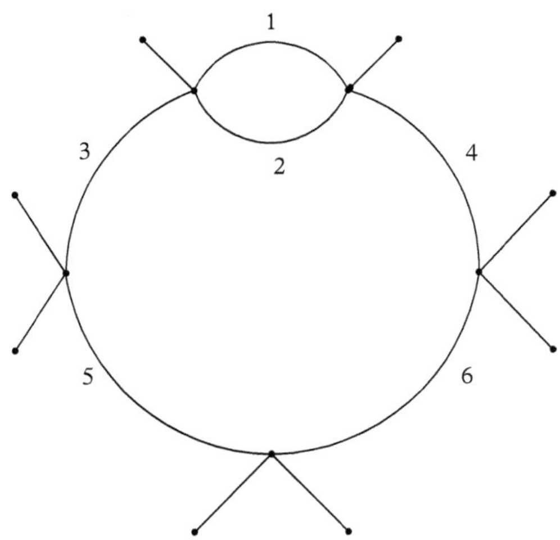

Figure 3.3 *A Feynman graph, with six internal edges and with a four-point sub-divergence (the subgraph given by the internal edges 1 and 2)*

represents the un-renormalized (or bare) divergent integral, while the second term is the product between:

1. the divergent part corresponding to the subgraph γ; and
2. a convergent part corresponding to the cograph Γ/γ.

This represents the general factorization appearing in (3.42). Since, for this example, γ happens to be a *primitive* divergence, the procedure stops here (otherwise it would have continued recursively). The difference between the two terms in the right-hand side of (3.50) can be proven to be finite—the renormalized Feynman integral (in the left-hand side of equation (3.50)).

Let us now mention one final complication. As one can see from a direct inspection of formula (3.47), the sum on Zimmermann forests in the subtraction operator does not depend on scales. Nevertheless, when the graph to be treated contains *overlapping divergences*, the sum in (3.47) should not be performed on all Zimmermann forests. One should use a scale decomposition—in each scale assignment (or Hepp sector) the forests can be classified in a different way (see again Vincent Rivasseau (1992) for details).

3.6.5 Dimensional renormalization

A particularly convenient starting point for the implementation of this scheme is the parametric representation (3.26). This comes from the fact that in this expression of the Feynman integral, the space-time dimension D is a simple parameter (see Remark 3.4.2), such that the analytic continuation in the complex plane can be trivially performed.

A first important result one needs to take into account is that a divergent Feynman integral of some graph Γ factories in two parts, as explained in the previous subsection. For a (rigorous) proof of this crucial result, the interested reader may report to M. Bergere and F. David's original articles (Bergere and David 1979, 1981). This property is the translation in this scheme of the locality principle, and it is needed in order to have a renormalizable model (as explained in the previous sections).

Moreover, it is this type of property which makes the definition of the Connes–Kreimer coproduct (see Section 8.2) relevant for an algebraic description of the combinatorics of renormalization in QFT.

Several steps are required in order to implement the renormalization:

- One needs to prove that the analytically continued $\phi(\Gamma)$ is a *meromorphic function of the space-time dimension D* (for the proof of this result, one can refer to M. Bergere and F. David's articles Bergere and David (1979) or Bergere and David (1981) or to the proof of Theorem 1.9 of Alain Connes and Matilde Marcolli (2008)).
- The next step is to use the Bogoliubov subtraction operator in order to get rid of the undesirable poles, thus obtaining the renormalized integral $\phi_r(\Gamma)$. The final result is that, after all these operations, *the renormalized integral $\phi_r(\Gamma)$ is an analytic function of D* (the interested reader may refer to the proof of Theorem 3 of Bergere and David (1981)).

3.7 Dyson–Schwinger equation

The DSE is a quantum equation of motion, i.e. the classical equation of motion (obtained from minimizing the action) corrected with quantum corrections. From a mathematical point of view, it can be seen as a manifestation of the general property stating that the integral of total derivative vanishes. For the Φ^4 model considered here, one can obtain the DSE:

$$G^{(2)}(x,y) = C(x,y) - 4\lambda \int dz C(x,z) G^{(4)}(z,z,z,y). \tag{3.51}$$

Note that $G^{(4)}$ previously mentioned is not necessarily associated with connected Feynman graph contributions. Similar equations can be written relating $G^{(N)}$ to $G^{(N+2)}$, with N higher than two.

Let us also mention that, when expanding the previous DSE one recovers perturbation theory equations.

For systems with additional symmetries, the Dyson–Schwinger system of equations closes—one has equations relating only the two-point function to itself and so on.

In the mathematical physics literature, one refers to the *combinatorial DSE*, which is, as we will see in the sequel, an equation written at a diagrammatic level (with the appropriate combinatorial weights)—see for example Karen Yeats (2007). In this book we will first present combinatorial DSE for QFT in section 8.3 and we will then show how this technique can be applied for both non-commutative QFT (see Subsection 9.8.2) and for tensor models (see Subsection 10.6.3). In order to obtain the *analytical DSE* one needs to apply the Feynman rules to the combinatorial DSE.

For an explicit example of the use of the DSE, we point the reader to Chapter 12, where DSE are thoroughly analysed for so-called the $U(N)^D$-invariant tensor model.

Let us end this section by emphazising that, solving the DSE means finding non-perturbative solutions, which is known to be an extremely difficult but important task in QFT. Another approach for non-perturbative QFT is given by *constructive renormalization*. The interested reader may refer to Vincent Rivasseau (2007) on this subject.

3.8 Combinatorial (or 0-dimensional) QFT and the intermediate field method

In this section, we introduce combinatorial (or 0-dimensional) QFT and we then give the general idea of the so-called intermediate field method. This section follows the review paper Tanasa (2019).

3.8.1 Combinatorial (or 0-dimensional) QFT

Usually, in QFT, the scalar field φ is function of space ($\mathbb{R}^D$), as previously mentioned. However, one can consider the case $D = 0$. In this case the scalar field φ_i is not a function of space (since there is no space anymore), and φ is simply a (real or complex) variable. Some authors refer to this simpler formulation of QFT as **combinatorial QFT**, and we will do the same here.

Thus, in the real case, the *partition function* (which, from a combinatorial point of view, is a generating function) of the so-called φ^4 model expresses the integral as

$$Z = \int_{\mathbb{R}} d\varphi \, e^{-\frac{1}{2}\varphi^2 + \frac{\lambda}{4!}\varphi^4}, \tag{3.52}$$

where the constant λ is the QFT coupling constant. The term φ^4 is called an interaction term of degree four.

It is worth emphasizing here that in combinatorial QFT, functional integrals (which are particularly involved to rigorously define for $D \geq 1$) become usual (real or complex) integrals. Combinatorial QFT is thus much easier to manipulate from a mathematical point of view.

However, combinatorial QFT still presents a certain interest for the mathematical physicist because it can be seen as some kind of 'laboratory' to test the usual QFT mathematical tools. Thus, one needs to evaluate integrals of the same type as in usual QFT, namely type integrals

$$\frac{\lambda^n}{n} \int d\varphi \, e^{-\varphi^2/2} \left(\frac{\varphi^4}{4!} \right)^n \tag{3.53}$$

coming from the so-called QFT perturbative expansion (which comes to Taylor expand the exponential in (3.52) and then dealing with each term one by one, instead of dealing with the integral as a whole). In order to evaluate this type of integral, one can use standard QFT techniques. Namely, one can define

$$Z_0(J) = \int_{\mathbb{R}} d\varphi \, e^{-\varphi^2/2 + J\phi}, \tag{3.54}$$

where J is the QFT source.

Using this QFT source technique, the $(2k)$-point correlation functions can be computed in the following way:

$$\int_{\mathbb{R}} d\phi \, e^{-\phi^2/2} \varphi^{2k} = \frac{\partial^{2k}}{\partial J^{2k}} \int_{\mathbb{R}} d\varphi \, e^{-\varphi^2/2 + J\varphi} |_{J=0} = \frac{\partial^{2k}}{\partial J^{2k}} e^{J^2/2} |_{J=0}. \tag{3.55}$$

3.8.2 The intermediate field method

We give now the general idea of the so-called intermediate field method. Note that we present this method in the case of combinatorial QFT, but this idea can generalize to arbitrary D.

Thus, the intermediate field method consists of introducing a new field, σ, used to rewrite the interaction in a way that allows for the degree of the interaction to be reduced.

In order to illustrate this, let us take the example of the φ^6 model, where φ is, as previously illustrated, a real 0-dimensional field (thus, a real variable).

The partition function of the φ^6 combinatorial QFT are written:

$$Z(\lambda) = \int_{\mathbb{R}} \frac{d\varphi}{\sqrt{2\pi}} e^{-\frac{1}{2}\varphi^2} e^{-\lambda\varphi^6}. \tag{3.56}$$

The intermediate field model consists of rewriting this partition function with the help of a supplementary integral on the intermediate field σ in the following way:

$$Z(\lambda) = \int_{\mathbb{R}} \frac{d\varphi}{\sqrt{2\pi}} e^{-\frac{1}{2}\varphi^2} \int_{\mathbb{R}} \frac{d\sigma}{\sqrt{2\pi}} e^{-\frac{1}{2}\sigma^2} e^{\imath\sqrt{2\lambda}\varphi^3\sigma}. \tag{3.57}$$

Note that this allowed to replace the interacting term φ^6 of (3.56) by a lower degree interacting term $\varphi^3\sigma$. This lowering of the degree of the interaction can be particularly useful in various contexts, such as the Jacobian conjectures (see Chapter 5).

3.9 Selected further reading

- Claude Itzykson and Jean-Bernard Zuber. (2006). *Quantum Field Theory*. Dover Publications.

  A standard textbook QFT for physicists. One can find a section on the parametric representation, which is related to the Tutte polynomial (see Section 6.6 of this book). The book doesn't only study scalar QFT models.
- Hagen Kleinert and Verena Schulte-Frohlinde. (2001). Critical properties of theories. *World Scientific*.

  A less known book on QFT, which is nevertheless a very good book, and which focuses, just like the chapter here, on the scalar Φ^4 model.
- Alain Connes and Matilde Marcolli. (2008). Noncommutative geometry, quantum fields and motives. *World Scientific*.

  A more recent book, whose Chapter 1 deals in detail with QFT, focusing on the Φ^3 model. Connes and Marcolli's book is most likely easier to read by a mathematician than by a physicist. Moreover, Connes and Marcolli's book analyses in detail the Connes–Kreimer algebraic approach to the combinatorics of renormalization (see Section 8.2 of this book).
- Vincent Rivasseau. (1992). *From Perturbative to Constructive Renormalization*. Princeton: Princeton University Press.

  A very good book which focuses on the renormalization, perturbative, and then constructive, of scalar QFT.

4

Tree weights and renormalization in QFT

A fundamental step in QFT is to compute the logarithm of functional integrals used to define the partition function of a given model (see Chapter 3). This comes from a fundamental theorem of enumerative combinatorics, stating the logarithm counts of the connected objects.

The main advantage of the perturbative expansion of a QFT into a sum of Feynman amplitudes is to perform this computation explicitly: the logarithm of the functional integral is the sum of Feynman amplitudes restricted to *connected* graphs. The main disadvantage is that the perturbative series indexed by Feynman graphs typically *diverges*. The constructive theory approach (see for example Rivasseau (1992)) is a certain compromise in this sense, because it allows to compute logarithms, hence connected quantities, but through *convergent series*.

Thus, perturbative QFT writes quantities of interest for the physicist as sums of amplitudes of connected graphs

$$\sum_G A_G. \tag{4.1}$$

However, such a formula (obtained by expanding in a power series the exponential of the interaction in Feynman functional integral and then 'illegally' commuting the power series and the functional integration) is *not* a valid definition since usually, even with cut-offs, even in zero dimension, we have $\sum_G |A_G| = \infty$. This divergence, known since Dyson (1952), is due to the more-than-exponential growth of the number of graphs with many vertices. We can say that Feynman graphs *proliferate too fast*. More precisely, the power series in the coupling constant λ corresponding to (4.1) has zero radius of convergence.[1] Nevertheless, for the stable Bosonic models which have been rigorously built by constructive field theory, the constructive answer is always the *Borel sum* of the

[1] This can be proved easily for ϕ^4, the Euclidean Bosonic QFT with quartic interaction treated in Chapter 3, see again Rivasseau (1992).

Combinatorial Physics: Combinatorics, quantum field theory, and quantum gravity models. Adrian Tanasa, Oxford University Press (2021). © Adrian Tanasa. DOI: 10.1093/oso/9780192895493.003.0004

perturbative series (see Vincent Rivasseau (1992) and references therein). Hence the perturbative expansion, although divergent, contains all the information of the theory; but it should be *reshuffled* into a convergent sum—see again Rivasseau (1992) for details.

The key point for the success of constructive theory is that *trees* do *not* proliferate as fast as graphs, and they are sufficient to compute QFT logarithms.

The combinatoric core of modern constructive QFT has been reformulated in a more transparent way in Rivasseau and Wang (2014). The basic idea is to define a set of positive weights $w(G,T)$ which are associated to any pair made of a connected graph G and a spanning tree $T \subset G$. They are normalized so as to form a probability measure on the spanning trees of G:

$$\sum_{T \subset G} w(G,T) = 1. \tag{4.2}$$

To compute constructively instead of perturbatively a QFT quantity one needs to use equation (4.2) to introduce a sum over trees for each graph, and then simply *exchange* the order of summation between graphs and their spanning trees

$$\sum_G A_G = \sum_G \sum_{T \subset G} w(G,T) A_G = \sum_T A_T, \quad A_T = \sum_{G \supset T} w(G,T) A_G. \tag{4.3}$$

However, in constructive QFT, if one uses the 'right' graphs and 'right' weights, then for certain Bosonic interactions we get

$$\sum_T |A_T| < +\infty, \tag{4.4}$$

which means that a QFT quantity is now well-defined; furthermore, the result is the one the mathematical physicist looks for, namely the Borel sum of the ordinary perturbation expansion, see again Rivasseau (1992) for details.

Consider from now on an interaction such as ϕ^4 and its intermediate field perturbation expansion. Not every probability measure $w(G,T)$ on the spanning trees of the corresponding graphs leads to a constructive reshuffling, namely one for which (4.4) holds. For instance, the trivial, equally distributed weights $1/\chi(G)$, where $\chi(G)$, the complexity of G, is the number of its spanning trees, form such a probability measure, but there is no reason to think they lead to a constructive reshuffling.

In Rivasseau and Wang (2014) symmetric tree weights were analysed. However, in the mathematical physics literature, non-symmetric forest formulas for constructive expansions were used before the symmetric Taylor forest formula of perturbative renormalization (see again Chapter 3). It thus appears that the symmetric weights are not the only ones leading to constructive QFT properties. In this chapter we explore this issue in detail. First, we define precisely this property of constructive positivity. We then identify a very general class of weights, associated to any non-trivial partition of the vertices of a graph, which have this constructive property. This is the main aim of this chapter.

Examples are then given in Appendix A. In particular, weights for *rooted* and *multi-rooted graphs* (which just correspond to particular cases of vertex partitions, where one considers one or more singletons and the rest of vertices are grouped into a single remaining block) are defined and compared to the symmetric weights.

This chapter follows the original research article Rivasseau and Tanasa (2014).

4.1 Preliminary results

Consider a fixed set V of vertices. We associate weights $w(G,T)$ (also called *amplitudes* in the QFT context) to any pair (G,T), where G is a labelled connected graph G with vertex set V and edge set E, and T is a spanning tree of G (where by spanning tree we mean an acyclic maximal subset of E, hence of cardinality $|V|-1$). Self-loops (otherwise known as tadpoles in the QFT language, see Chapter 3) and multiple edges in G are allowed, since they are a fundamental feature of QFT. From now on, we omit the word 'spanning', since throughout this chapter the trees considered are always spanning for a related graph $G \supset T$.

We write $\sum_{T\subset G} w(G,T)$ to indicate summation over the finite family of trees of a fixed G of such weights $w(G,T)$. We can also consider trees T as particular labelled connected graphs themselves with vertex set V. There is an infinite family of graphs obtained by adding an arbitrary number L of edges between the vertices of T, since self-loops and multiple edges are allowed. Such graphs have $|E(G)| = |V|-1+|L|$ edges and nullity L (that is L independent loops, since we deal with connected graphs, as already stated). In that case we write $W(T) = \sum_{G\supset T} w(G,T)$ to indicate summation over the infinite family of such G's.[2] The corresponding series may of course be divergent or convergent depending on the exact weights considered.[3]

A complete ordering of the $|E(G)|$ edges of G is called a **Hepp sector** in QFT terminology. The set of such orderings, $S(G)$, has $|E(G)|!$ elements. For any such Hepp sector $\sigma \in S(G)$, *Kruskal greedy algorithm* defines a particular tree $T(\sigma)$, which minimizes $\sum_{\ell\in T} \sigma(\ell)$ over all trees of G. We call it the *leading tree* for σ for short. Let us briefly explain how the algorithm works. The algorithm simply picks the first edge ℓ_1 in σ which is not a self-loop. The algorithm then picks the next edge ℓ_2 in σ that does not add a cycle to the (disconnected) graph with vertex set V and edge set ℓ_1 and so on. Another way to look at it is through a deletion-contraction recursion: following the ordering of the

[2] This family could possibly be enlarged or restricted by additional 'Feynman rules' in the context of a QFT with a particular set of edges (propagators) and interactions (vertices), but this issue is not important at the level of generality of this paper, which does not deal with a particular QFT model.

[3] The category of Feynman graphs to consider for the constructive applications illustrated has, in fact, slightly more structure, since Feynman graphs have also labelled half-edges, according to Wick's theorem. It provides them, in particular, with a canonical ciliated ribbon structure, by labelling the half-edges starting from the cilium. These additional (important) subtleties are not considered here, as the half-edge labelling will play no role in *this* paper.

sector σ, every edge is either deleted if it is a self-loop or contracted if it is not. The set of contracted edges is exactly the leading tree for σ.

Note that this leading tree $T(\sigma)$ has been considered intensively in the context of perturbative and constructive renormalization in QFT (Rivasseau 1992), as it plays an essential role to get sharp bounds on renormalized quantities.

Consider also that given any Hepp sector σ the (unordered) tree $T(\sigma)$ comes naturally equipped with an *induced ordering* τ (the order in which the edges of $T(\sigma)$ are picked by Kruskal's algorithm). The corresponding *ordered* tree will be denoted as T_τ.

Definition 4.1.1 *A* **probability measure** *on trees is a set of positive weights $w(G,T)$ for any labelled connected graph G and tree $T \subset G$ such that*

$$\sum_{T \subset G} w(G,T) = 1. \tag{4.5}$$

The measure and the weights w are called **rational** *if all $w(G,T) \in \mathbb{Q}$ and they are called* **symmetric** *if $w(G,T) = w(G^\nu, T^\nu)$, where ν is any permutation of V, hence any relabelling of the vertices of G and T. The measure and the weights are called* **constructive** *if there exists a T-dependent probability space $(\Omega_T, \Sigma_T, \mu_T)$[4] and a (T,u) dependent real positive-definite symmetric matrix $X^T_{v,v'}(u)$ for any $u \in \Omega_T$, with diagonal $X^T_{v,v}(u) = 1$ for any u, such that:*

$$w(G,T) = \int_{\Omega_T} d\mu_T(u) \prod_{\ell \notin T} X^T_{v(\ell)v'(\ell)}(u), \tag{4.6}$$

where $v(\ell)$ and $v'(\ell)$ denote, by a slight abuse of notation, the two vertices incident to the edge ℓ.

Note that the order of the two previously mentioned vertices $v(\ell)$ and $v'(\ell)$ plays no role, since the matrix X is symmetric. From a QFT perspective, this comes from the fact that one can endow the internal edges of a Feynman graph with any orientation.

This constructive property is exactly what allows, in the case of stable Bosonic interactions such as ϕ,[4] to rewrite any tree amplitude A_T of the loop vertex expansion in (4.3) as an integral over Ω_T for the measure $d\mu_T$ of a functional integral over a positive Gaussian measure of covariance $X^T_{ij}(u)$ of a well-bounded integrand. Hence, it is exactly the property necessary for inequality (4.4) to hold. For non-constructive weights, there is no such functional integral representation.

Let us recall here the definition of the symmetric weights $w_s(G,T)$ (see again Rivasseau and Wang (2014)):

[4] In all concrete examples Ω_T is a topological space, and the sigma-algebra Σ_T is its Borel sigma-algebra and will play no further role.

Definition 4.1.2 *The symmetric weights $w_s(G,T)$ are the percentage of Hepp sectors for which the tree T is a leading tree*

$$w_s(G,T) = \frac{1}{|E(G)|!} \sum_{\sigma \in S(G)} \chi(T(\sigma) = T) \tag{4.7}$$

where $\chi(T(\sigma) = T)$ is 1 if the leading tree for the Hepp sector σ is the tree T, and 0 otherwise.

Normalization and rationality of these weights are obvious. We then recall the main result of Rivasseau and Wang (2014):

Theorem 4.1.3 *The symmetric weights $w_s(G,T)$ are constructive.*

4.2 Partition tree weights

Consider again a fixed vertex set V.

A partition of V into k nonempty disjoint subsets $V = V_1 \cup \cdots \cup V_k$ is called **trivial** if $k = 1$ and **non-trivial** if $k \geq 2$. The subsets of the partition are also called **blocks** in what follows. From now on, we suppose we made a choice of a fixed such partition Π. Our goal is to define, for any graph G with vertex set V, an associated rational constructive measure for the trees of G.

An edge $\ell \in G$ with ends i and j is called **trans-block** for the partition Π if the vertices i and j belong to two distinct blocks $V_{k(i)}$ and $V_{k(j)}$ of Π. Note that a self-loop is never trans-block, for any partition.

Given the graph G and a trans-block edge $\ell \in G$ with ends i and j, we can consider the contracted graph G/ℓ in which the vertices i and j are replaced by a single contracted vertex $\widehat{ij}$ and the edge ℓ is removed. This contracted graph is naturally equipped with a **contracted partition** Π/ℓ defined by the blocks $V_1, \cdots, V_{k(i)} - \{i\}, \cdots, V_{k(j)} - \{j\}, \cdots, V_k, V_{k+1} = \{\widehat{ij}\}$, *any empty block being omitted.* Hence, it is a partition into k' blocks, with $k - 1 \leq k' \leq k + 1$. Notice that this reduced partition has always at least a singleton block, namely V_{k+1}. See also that it can be trivial only if the graph G has exactly two vertices; indeed, the equation $k' = 1$ implies that the initial partition was solely made of $V_{k(i)} = \{i\}$ and $V_{k(j)} = \{j\}$.

Iterating this construction, we arrive at the definition of a trans-block ordered forest:

Definition 4.2.1 *An ordered forest $\mathcal{F} = \{\ell_1, \cdots \ell_p\}$ $p \leq |V| - 1$ is called* **trans-block** *for the partition Π if the edge ℓ_1 is trans-block for the partition Π, the edge ℓ_2 is trans-block for the contracted graph G/ℓ_1 and its contracted partition Π/ℓ_1 and so on until the last edge ℓ_p of $\mathcal{F}$ which is trans-block for the contracted graph $G/\ell_1/\ell_2/\cdots/\ell_{p-1}$ and the contracted partition $\Pi/\ell_1/\ell_2/\cdots/\ell_{p-1}$.*

The sequence of graphs $\{G_0 = G, G_1 = G/\ell_1, \cdots, G_p = G/\ell_1/\cdots/\ell_p\}$ and the sequence of partitions $\{\Pi_0 = \Pi, \Pi_1 = \Pi/\ell_1, \cdots, \Pi_p = \Pi_{p-1}/\ell_p\}$ is noted $\mathcal{S}(G, \Pi, \mathcal{F})$.

Notice that the sequence $\mathcal{S}(G,\Pi,\mathcal{F})$ indeed depends on the ordering of the forest in a critical way.

The maximal trans-block ordered forests are the trans-block ordered trees:

Lemma 4.2.2 *For any ordered trans-block forest $\mathcal{F}$, the last partition in $\mathcal{S}(G,\Pi,\mathcal{F})$ is trivial (i.e. made of a single block) if and only if $\mathcal{F}$ is a tree, i.e. has exactly $|V|-1$ edges.*

Proof Consider a trans-block ordered forest $\mathcal{F}=\{\ell_1,\cdots\ell_p\}$ and the sequence of $p+1$ reduced graphs $G_0=G,G_1,\cdots G_p$. The number of vertices decreases by exactly one in each step of this sequence, so the last graph has a single vertex, hence we reach a trivial partition if and only if $p=|V|-1$, hence if and only if $\mathcal{F}$ is a (trans-block) *tree*.

Definition 4.2.3 *An ordering τ of a given tree T is called* **admissible** *for the partition Π if the* ordered *tree T_τ is trans-block for the respective partition. The set of such admissible orderings for a given tree T is denoted by $\mathcal{A}^\Pi(T)$.*

The set of admissible orderings is never empty if the respective partition Π is non-trivial. Any admissible ordering τ of T defines a sequence of contracted graphs and partitions $\mathcal{S}(G,\Pi,T_\tau)$. Do not confuse orderings τ and the Hepp sectors for the *full* graph G considered in the previous section. Observe, however, that the orderings τ can be considered as Hepp sectors for *the tree T*.

The partition weight $w^\Pi(G,T)$ will be defined in formula (4.10) as a sum over all admissible orderings $\tau\in\mathcal{A}^\Pi(T)$ of certain finite dimensional simple integrals. Their definition requires first that we define the so-called **contact indices**. These indices are defined for any pair of vertices (v,v') of G (including the case $v=v'$) and any ordered trans-block tree T_τ:

Definition 4.2.4 (Contact Indices) *Consider an ordered trans-block tree $T_\tau=\{\ell_1,\ldots,\ell_{|V|-1}\}$ and its associated sequence of reduced graphs and partitions $\mathcal{S}(G,\Pi,T_\tau)=\{G_p,\Pi_p\}$ with $0\le p\le|V|-1$. We define the* ***first contact index*** *$i^\Pi_{T_\tau}(v,v')$ as the smallest value of p such that the two vertices v and v' belong to different blocks for Π_p, and the* ***second contact index*** *$j^\Pi_{T_\tau}(v,v')$ as the smallest value of p for which v and v' are collapsed into a single reduced vertex in G_p. If $v=v'$ we set by convention: $i^\Pi_{T_\tau}(v,v')=-1$ and $j^\Pi_{T_\tau}(v,v')=0$.*

Let us make the following remark. If the two vertices v and v' belong to distinct blocks of the partition Π, we have therefore $i^\Pi_{T_\tau}(v,v')=0$.

Lemma 4.2.5 *The two contact indices obey $i^\Pi_{T_\tau}(v,v')<j^\Pi_{T_\tau}(v,v')$.*

Proof This follows directly from the previous definition.

Definition 4.2.6 (Contact Matrices) *For any graph with vertex set V, any given vertex set partition Π, any given ordered tree T_τ and $u=\{u_1,\ldots,u_{|V|-1}\}$ in $[0,1]^{|V|-1}$, we define* *the* $|V|$ *by* $|V|$ *u-dependent real symmetric matrix, called the* **contact matrix,** $X^{\Pi,T_\tau}(u)$ *by the following formula:*

$$X^{\Pi,T_\tau}_{v,v'}(u) := \prod_{i^{\Pi}_{T_\tau}(v,v')<k\leq j^{\Pi}_{T_\tau}(v,v')} u_k, \quad \forall v,v' \in V. \tag{4.8}$$

Moreover, we define the **tree edge factor,** *by the following formula:*

$$Y^{\Pi,T_\tau}_{\ell}(u) := \prod_{i^{\Pi}_{T_\tau}(v(\ell),v'(\ell))<k<j^{\Pi}_{T_\tau}(v(\ell),v'(\ell))} u_k. \quad \forall \ell \in T. \tag{4.9}$$

Note that in (4.9) *we have again denoted by* $v(\ell)$ *and* $v'(\ell)$ *(by the same slight abuse of notation) the two vertices incident to the (tree) edge* ℓ.

Notice that $X^{\Pi,T_\tau}_{v,v}(\sigma)(u) = 1$, since by convention an empty product is one, hence the matrix $X^{\Pi,T_\tau}(u)$ has diagonal entries all equal to one.

We are now finally in position to define the partition weights associated to an admissible sector σ.

Definition 4.2.7 (Partition Tree Weights) *For a set* V *and a partition* Π *fixed, the partition weights* $w^{\Pi}(G,T)$ *associated to any tree* T *of the graph* G *(whose vertex set is* V*) are defined as*

$$w^{\Pi}(G,T) := \sum_{\tau\in\mathcal{A}^{\Pi}(T)} \int_0^1 du_1 \cdots \int_0^1 du_{|V|-1} \Big[\prod_{\ell\in T} Y^{\Pi,T_\tau}_{\ell}(u)\Big] \Big[\prod_{\ell\notin T} X^{\Pi,T_\tau}_{v(\ell),v'(\ell)}(u)\Big], \tag{4.10}$$

where $v(\ell)$ *and* $v'(\ell)$ *are the two vertices incident to the edge* ℓ.

Each partition tree weight $w^{\Pi}(G,T)$ being obviously a sum over $\mathcal{A}^{\Pi}(T)$ of positive rational numbers is a positive rational number. We first prove a lemma stating the normalization of these weights, for a given graph and a given partition of its vertex set.

Lemma 4.2.8 *We have*

$$\sum_{T\subset G} w^{\Pi}(G,T) = 1. \tag{4.11}$$

Proof The key idea is to Taylor expand the function $1 = [\prod_{\ell\in G} x_\ell]|_{x_\ell=1\,\forall\ell}$ according to a Π-dependent recursion with $|V|-1$ steps which, at each step i, uses an elementary first-order Taylor expansion with integral remainder

$$f(1) = f(0) + \int_0^1 f'(u_i)du_i, \quad i = 1,\ldots,|V|-1. \tag{4.12}$$

This recursion 'builds' step by step a sum over all trans-block ordered trees.

Let $F_0(x_1,\ldots,x_{|E|}) := x_1\ldots x_{|E|}$. The first step of the induction requires first to multiply each variable x corresponding to a trans-block edge for the partition Π_0 with a dummy variable u_1. This amounts to consider the function

$$F_0(u_1x_1,\ldots,u_1x_{k_0},x_{k_0+1},\ldots,x_{|E|})|_{x_i=1,\forall i}, \tag{4.13}$$

where, without any loss of generality we have placed on the first k_0 positions the k_0 variables associated to the trans-block edges for the partition Π_0. Note that, since Π_0 is non-trivial and G is connected, one has $k_0>0$. Using the multi-variable version of the Taylor expansion (4.12) leads to

$$\begin{aligned} 1 &= F(0,\ldots,0,1,\ldots,1) \\ &+\sum_{\ell_1}\int_0^1 du_1\left(\frac{\partial}{\partial x_{\ell_1}}F_0\right)(u_1x_1,\ldots,u_1x_{k_0},x_{k_0+1},\ldots,x_{|E|})|_{x_i=1,\forall i}, \end{aligned} \tag{4.14}$$

where the sum over ℓ_1 runs over the k_0 trans-block edges for the partition Π_0. Note that the first term on the right-hand side of equation (3.17) vanishes (because of the definition of the function F_0 and because $k_0\neq 0$, as already noticed). For the sake of simplicity, let us fix the trans-block edge in the sum (4.14). It will be the first edge in the recursive construction of our ordered tree. In the right-hand side of equation (4.14) the derivative leads to

$$1=\sum_{\ell_1}\int_0^1 du_1u_1^{k_0-1}F_1. \tag{4.15}$$

In function F_1 we fix definitely $x_{\ell_1}=1$, so that F_1 no longer depends on it. One thus has, as previously mentioned, $F_1(x_2,\ldots,x_{|E|})=x_2\ldots x_{|E|}$. We can now proceed with the next induction, which is identical but for the graph $G_1=G/\ell_1$ and the partition $\Pi_1=\Pi/\ell_1$.

At a generic step n, this expansion leads to

$$1=\sum_{\ell_1,\ldots,\ell_n}\int_0^1 du_1\ldots du_n u_1^{k_0-1}\ldots u_n^{k_{n-1}-1}F_n, \tag{4.16}$$

where the exponent k_j represents the number of trans-block edges for the partition Π_j in our sequence of partitions. As previously illustrated, the factor F_n does not depend on the variables associated to the contracted edges $\ell_1,\ldots,\ell_n$. This process continues until there are no more trans-block edges for the respective partition; this happens at the $(|V|-1)$th step. The corresponding factor $F_{|V|-1}$ is then no longer interpolated, hence equal to 1. The expansion therefore results in a sum over all trees and all their admissible sectors, renaming dummy variables u:

$$1 = \sum_{T,\tau \in A^{\Pi}(T);\, T_\tau = \ell_1,\dots,\ell_{|V|-1}} \int_0^1 du_1 \dots du_{|V|-1} u_1^{k_0-1} \dots u_{|V|-1}^{k_{|V|-2}-1}. \tag{4.17}$$

Let us now identify, for a fixed ordered tree T_τ in this sum, the integrand $u_1^{k_0-1} \dots u_{|V|-1}^{k_{|V|-2}-1}$ with the two products appearing in (4.10):

$$\prod_{\ell \in T} Y_\ell^{\Pi,T_\tau} \prod_{\ell \notin T} X_{v(\ell),v'(\ell)}^{\Pi,T_\tau} = u_1^{k_0-1} \dots u_{|V|-1}^{k_{|V|-2}-1}. \tag{4.18}$$

Indeed, within the left-hand side of (4.18)—which is, by construction, a product of various powers of the (tree) edge variables u_k $(k = 1,\dots,|V|-1)$—the variable u_1 appears as many times as there are edges ℓ with contact indexes $i = 0$ and $j \geq 1$. This is unimportant, but the number of edges which are trans-block for the partition Π_0 minus 1, the minus 1 correction coming from the fact that, for ℓ_1, the contact indices are 0 and 1, and the factor u_1 is not taken into account, by definition, within the tree factor $Y_{\ell_1}^{\Pi,T_\tau}$. The analogous reasoning holds for u_k $(k = 2,\dots,|V|-1)$ and this concludes the proof.

Let us now rewrite formula (4.10) as

$$w^{\Pi}(G,T) = \sum_{\tau \in A^{\Pi}(T)} w^{\Pi}(G,T^{\tau}), \tag{4.19}$$

where we define:

$$w^{\Pi}(G,T^{\tau}) := \int_0^1 du_1 \cdots \int_0^1 du_{|V|-1} \Big[\prod_{\ell \in T} Y_\ell^{\Pi,T_\tau}(u)\Big] \Big[\prod_{\ell \notin T} X_{v(\ell),v'(\ell)}^{\Pi,T_\tau}(u)\Big]. \tag{4.20}$$

One has:

Corollary 4.2.9 *For a given graph G, vertex partition Π and ordered tree T^{τ}, the weights defined in* (4.20) *can be written as*

$$w^{\Pi}(G,T^{\tau}) = \prod_{i=0}^{|V|-2} \frac{1}{k_i}, \tag{4.21}$$

where k_i give the number of trans-block edges for the partition Π_i in the corresponding partition sequence.

Proof This is a direct consequence of equation (4.16); the factors $1/k_i$ result from integrating over the u_i's.

Let us mention here that a somehow similar use of the Taylor formula with integral reminder was made in Abdesselam (2004), within the cluster expansion framework. The use of the integral reminder is one possible way of not dealing with the whole perturbative series, since this series most often diverges in QFT.

Let us now prove the following lemma:

Lemma 4.2.10 *The symmetric matrix* $X^{\Pi,T_\tau}_{v,v'}(u)$ *is positive semi-definite.*

Proof We prove the positivity for any fixed ordered tree T_τ again by a recursion based upon the sequence of graphs $G_0,\dots,G_{|V|-1}$ and the associated partitions $\Pi_0,\dots,\Pi_{|V|-1}$.

We define first for any n by n matrix X and any partition Π of $[1,\dots,n]$, the projected matrix X^Π which has elements $X^\Pi_{ij} = X_{ij}$, if i and j belong to the same block of Π and 0. We then remark that for any symmetric positive X, X^Π is positive and for any real $u \in [0,1]$ the interpolated matrix $X(u) = uX + (1-u)X^\Pi$ is also positive, as a barycentric combination of two positive matrices.

Let us now define the positive $|V|$ by $|V|$ matrix X_0, which has matrix elements all equal to one.

We then define:

$$
\begin{aligned}
X_1(u_1) &= u_1 X_0 + (1-u_1) X_0^{\Pi_0} \\
X_2(u_1,u_2) &= u_2 X_1(u_1) + (1-u_2) X_1^{\Pi_1}(u_1) \\
&\dots \\
X_{|V|-1}(u_1,\dots,u_{|V|-1}) &= u_{|V|-1} X_{|V|-2}(u_1,\dots,u_{|V|-2}) \\
&\quad + (1-u_{|V|-1}) X u_{|V|-2}^{\Pi_{|V|-2}}(u_1,\dots,u_{|V|-2}).
\end{aligned}
\tag{4.22}
$$

Note that all these matrices are positive, again as barycentric combinations of two positive matrices. This mechanism thus leads that to a 'final' $|V|$ by $|V|$ matrix $X_{|V|-1}$, for which the matrix element corresponding to the entry (v,v') $(v,v' = 1,\dots,|V|)$ is a product of variables u_j $(j = 1,\dots,|V|-1)$. Each such variable u_j is present in this product if and only if the edge (or the edges, since multi-edges are allowed) connecting the vertices v and v' in G_{j-1} is a trans-block edge for the corresponding partition Π_{j-1}.

We have thus 'constructed' the matrix given by the formula (4.8), since the combinatorial definition of the contact indices lead to the same product of the tree variables u_j. This concludes the proof.

The main result of this chapter is the following theorem:

Theorem 4.2.11 *For any set* V *and partition* Π *the partition weights* $w^\Pi(G,T)$ *on the trees* T *of any graph* G *with vertex set* V *form a rational constructive probability measure.*

Proof The claim follows from Lemmas 4.2.8 and 4.2.10. Indeed, the set Ω_T is the disjoint union for all admissible τ's of a distinct copy of $[0,1]^{|V|-1}$ with measure

$$
d\mu_{T_\tau} = \prod_{\ell \in T} Y_\ell^{\Pi,T_\tau}(u) \prod_{k=1}^{|V|-1} du_k. \tag{4.23}
$$

It is important to note that the normalization of the full measure $d\mu_T$ is nothing but equation (4.11) for the particular case when $G = T$ concludes the proof.

Before ending this section, let us mention that an important particular example of our results can be obtained when the vertex partition is made of a singleton plus a single block containing all the remaining vertices. As already mentioned in the introduction, the mechanism exposed in this paper leads to constructive tree weights for a *rooted graph* (the singleton being the root of the graph). An example of such tree weights for a rooted graph is given in Subsection A.2. This is then generalized for *multi-rooted graphs*, which correspond to vertex partitions made of several singletons and a remaining block containing the rest of the vertices. An example of a double-rooted graph is analysed in detail in Subsection A.3.

4.3 Selected further reading

- Rivasseau, Vincent and Wang, Zhituo. (2014). How to resume Feynman graphs. *Annales Henri Poincare*, 15: 2069–2083.

  This article presents a similar analysis to the one in this chapter, however, the case of symmetric tree weights is analysed, unlike this chapter, where the tree weights are taken to be non-symmetrical. This article can thus be seen as a companion article for the results exposed in this chapter.

- Vincent Rivasseau. (1992). From perturbative to constructive renormalization. Princeton: Princeton University Press.

  A very good book which focuses on the renormalization, perturbative, and then constructive properties of scalar QFT.

5

Combinatorial QFT and the Jacobian Conjecture

In 1939, Keller formulated in Keller (1939) the **Jacobian Conjecture**, as a remarkably simple and natural conjecture on the invertibility of polynomial systems. The conjecture states that a polynomial function $F:\mathbb{C}^n \to \mathbb{C}^n$ is invertible, and its inverse is polynomial, if, and only if, the determinant of the Jacobian matrix $J_F(z) = \left(\mathrm{d}F_j(z)/\mathrm{d}z_i\right)_{1\le i,j\le n}$ is a non-zero constant.[1]

In order to illustrate this celebrated conjecture, let us give a simple example. Let $n=2$, $F(z_1,z_2) = (z_1 + z_2^3, z_2)$. The Jacobian is written as

$$\det\begin{pmatrix} \frac{dF_1}{dz_1} & \frac{dF_1}{dz_2} \\ \frac{dF_2}{dz_1} & \frac{dF_2}{dz_2} \end{pmatrix} = det\begin{pmatrix} 1 & 3z_2^2 \\ 0 & 1 \end{pmatrix} = 1. \tag{5.1}$$

One easily finds: $F^{-1}(z_1,z_2) = (z_1 - z_2^3, z_2)$.

Despite several efforts (of mathematicians such as Gröbner, Zariski, Oda, to namely only a few), and various promising partial results, the Jacobian Conjecture remains unsolved. An introduction to the problem, the context, and the state of the art up to 1982, can be found in the seminal paper Bass, Connell, and Wright (1982).

The function $F:\mathbb{C}^n \to \mathbb{C}^n$ is said to be a *polynomial system* if all the coordinate functions F_j's are polynomials. Let us call $\mathcal{P}_n$ the set of such functions. For a function F, define

$$J_F(z) = \left(\frac{\mathrm{d}}{\mathrm{d}z_i}F_j(z)\right)_{1\le i,j\le n,} \tag{5.2}$$

the corresponding Jacobian matrix, and the two subspaces of $\mathcal{P}_n$:

[1] In Blaschke et al. (1939), Keller formulated this conjecture for $n=2$ and polynomials with integral coefficients, but this conjecture was shortly after generalized to the form previously given.

Combinatorial Physics: Combinatorics, quantum field theory, and quantum gravity models. Adrian Tanasa, Oxford University Press (2021). © Adrian Tanasa. DOI: 10.1093/oso/9780192895493.003.0005

Definition 5.0.1

$$\mathcal{J}_n^{\text{lin}} := \{F \in \mathcal{P}_n \,|\, \det J_F(z) = c \in \mathbb{C}^\times\}, \tag{5.3}$$
$$\mathcal{J}_n := \{F \in \mathcal{P}_n \,|\, F \text{ is invertible}\}. \tag{5.4}$$

It will often be sufficient to analyse the subset of $\mathcal{J}^{\text{lin}}$ such that $\det J_F(z) = 1$. The Jacobian Conjecture rewrites as:

Conjecture 5.0.2 (Jacobian Conjecture, Bass, Connell, and Wright (1982))

$$\mathcal{J}_n^{\text{lin}} = \mathcal{J}_n \qquad \forall n. \tag{5.5}$$

Define the *total degree* of F, $\deg(F)$, as $\max_j \deg(F_j(z))$, and introduce the subspaces

$$\mathcal{P}_{n,d} = \{F \in \mathcal{P}_n \mid \deg(F) \leq d\}; \tag{5.6}$$

and similarly, for $\mathcal{J}$ and $\mathcal{J}^{\text{lin}}$.

Let us now mention two positive results on the conjecture:

1. A theorem for the quadratic case ($d = 2$), established first in Wang (1980), and then in Oda (1980) (see also (Wright 1989; Lemma 3.5; Bass, Connell, and Wright 1982) Thm. 2.4, pag. 298).

   Theorem 5.0.3 (Wang (1980))

   $$\mathcal{J}_{n,2}^{\text{lin}} = \mathcal{J}_{n,2} \qquad \forall n. \tag{5.7}$$

2. A reduction theorem, from the general case to the cubic case, established by Bass et al. Bass, Connell, and Wright (1982): sec. II]).

   Theorem 5.0.4 (Bass, Connell, and Wright (1982))

   $$\mathcal{J}_{n,3}^{\text{lin}} = \mathcal{J}_{n,3} \quad \forall n \qquad \Longrightarrow \qquad \mathcal{J}_n^{\text{lin}} = \mathcal{J}_n \quad \forall n. \tag{5.8}$$

The result of de Goursac, Sportiello, and Tanasa (2016) that we will review here from a QFT perspective is a reduction theorem, somewhat analogous to the previous one, that tries to 'fill the gap' between the cases of degrees two and three. However, we need to introduce an adaptation of the statement of the Jacobian conjecture with one more parameter.

Definition 5.0.5 *For $n' \leq n$ and $F \in \mathcal{P}_{n,d}$, we write $z = (z_1, z_2)$ and $F = (F_1, F_2)$ to distinguish components in the two subspaces $\mathbb{C}^{n'} \times \mathbb{C}^{n-n'} \equiv \mathbb{C}^n$. We set $R(z_2; z_1)$ as synonymous for $F_2(z_1, z_2)$, emphasizing that, in R, we consider z_2 as the variables in a polynomial system, and z_1 as parameters. Then, we define the subspaces of $\mathcal{P}_{n,d}$:*

$$\mathcal{J}_{n,d;n'} := \{F \in \mathcal{P}_{n,d} \,|\, F^{-1} \text{ is defined on } \mathbb{C}^{n'} \times \{0\}\}$$
$$\mathcal{J}^{\text{lin}}_{n,d;n'} := \{F \in \mathcal{P}_{n,d} \,|\, R \in \mathcal{J}_{n-n',d} \text{ and } (\det J_F)(z_1, R^{-1}(0,z_1)) = c \in \mathbb{C}^{\times},\ \forall z_1 \in \mathbb{C}^{n'}\}$$

Let us now state the reduction theorem we will review in this paper:

Theorem 5.0.6 (de Goursac, Sportiello, and Tanasa (2016)) *For $n \in \mathbb{N}$ and $d \geq 3$, there exists an injective map $\Phi : \mathcal{P}_{n,d} \mapsto \mathcal{P}_{n(n+1),d-1}$ satisfying*

$$\Phi(\mathcal{J}^{\text{lin}}_{n,d}) \equiv \mathcal{J}^{\text{lin}}_{n(n+1),d-1;n} \cap \Phi(\mathcal{P}_{n,d})\,;\, \Phi(\mathcal{J}_{n,d}) \equiv \mathcal{J}_{n(n+1),d-1;n} \cap \Phi(\mathcal{P}_{n,d})\,. \tag{5.9}$$

Combining Theorem 5.0.4 and the previous theorem, the full Jacobian Conjecture reduces to the question whether

$$\mathcal{J}^{\text{lin}}_{n(n+1),2;n} = \mathcal{J}_{n(n+1),2;n}. \tag{5.10}$$

This chapter follows the review paper Tanasa (2019).

5.1 The Jacobian Conjecture as combinatorial QFT model (the Abdesselam–Rivasseau model)

In Abdesselam (2003a), the Jacobian Conjecture was re-expressed as combinatorial QFT model which we call here the Abdesselam–Rivasseau model.

Let $n, d \geq 1$, and let $F \in \mathcal{P}_{n,d}$. The coordinate functions of F can be written as

$$F_i(z) = z_i - \sum_{k=2}^{d} \sum_{j_1,\dots,j_k=1}^{n} w^{(k)}_{i,j_1\dots j_k} z_{j_1} \dots z_{j_k} =: z_i - \sum_{k=2}^{d} W^{(k)}_i(z), \tag{5.11}$$

for $i \leq n$ and $w^{(k)}_{i,j_1\dots j_k}$ some coefficients which are nothing but the combinatorial QFT coupling constants (see Section 3.8).

The partition function of the Abdesselam–Rivasseau combinatorial QFT model is

$$Z(J,K) = \int_{\mathbb{C}^n} \mathrm{d}\varphi \mathrm{d}\varphi^{\dagger} e^{-\varphi^{\dagger}\varphi + \varphi^{\dagger}\sum_{k=2}^{d} W^{(k)}(\varphi) + J^{\dagger}\varphi + \varphi^{\dagger}K},$$

where J, K are vectors in $\mathbb{C}^n$ (and they are nothing but the combinatorial QFT sources, see again Section 3.8). The measure of the previous integral is

$$\mathrm{d}\varphi\mathrm{d}\varphi^{\dagger} := \prod_{i=1}^{n} \frac{\mathrm{dRe}\varphi_i \,\mathrm{dIm}\varphi_i}{\pi}, \tag{5.12}$$

and we have used the standard QFT notations:

$$\varphi^\dagger K := \sum_{i=1}^{n} \varphi_i^\dagger K, \quad J^\dagger \varphi := \sum_{i=1}^{n} J_i^\dagger \varphi_i. \tag{5.13}$$

Note that the Abdesselam–Rivasseau model has very particular Feynman graphs obtained through perturbative expansion (see the original article Abdesselam (2003) for details on this).

Independently of this, setting the coupling constants to zero (the free theory), the partition function is calculated by Gaussian integration:

$$\int_{\mathbb{C}^n} \mathrm{d}\varphi \mathrm{d}\varphi^\dagger e^{-\varphi^\dagger \varphi + J^\dagger \varphi + \varphi^\dagger K} = e^{J^\dagger K}.$$

Let us now stress the following two crucial points (see again Abdesselam (2003) for details):

1. The partition function Z coincides with the inverse of the Jacobian:

$$Z(0,u) = \det(\partial G(u)) = JG(u) = \frac{1}{JF(G(u))}. \tag{5.14}$$

2. The inverse G of F corresponds to the one-point correlation function:

$$G_i(u) = \frac{\int_{\mathbb{C}^n} \mathrm{d}\varphi \mathrm{d}\varphi^\dagger \varphi_i e^{-\varphi^\dagger \varphi + \varphi^\dagger \sum_{k=2}^{d} W^{(k)}(\varphi) + \varphi^\dagger u}}{\int_{\mathbb{C}^n} \mathrm{d}\varphi \mathrm{d}\varphi^\dagger e^{-\varphi^\dagger \varphi + \varphi^\dagger \sum_{k=2}^{d} W^{(k)}(\varphi) + \varphi^\dagger u}}. \tag{5.15}$$

The sets of polynomial functions involved in the Jacobian Conjecture can be written in this framework:

$$\mathcal{J}_{n,d}^{\text{lin}} = \{F \in \mathcal{P}_{n,d} \,|\, Z(0,u) = 1, \forall u \in \mathbb{C}^n\},$$
$$\mathcal{J}_{n,d} = \{F \in \mathcal{P}_{n,d} \,|\, G_i(u) \text{ given by (5.15) is in } \mathcal{P}_n\}.$$

5.2 The intermediate field method for the Abdesselam–Rivasseau model

As we have seen in subsection 3.8.2, applying the combinatorial QFT intermediate field method will reduce the degree d of F.

We thus add n^2 intermediate fields σ_{ij} $(i,j = 1,\dots,n)$ to the model. Indeed, we have, from the general Formula (5.14) of Gaussian integration,

$$e^{(\varphi_i^\dagger\varphi_j)\left(\sum_{j_2,\dots,j_d=1}^n w^{(d)}_{i,j,j_2\cdots j_d}\varphi_{j_2}\cdots\varphi_{j_d}\right)}$$
$$= \int_{\mathbb{C}^{n^2}} \mathrm{d}\sigma_{i,j}\mathrm{d}\sigma^\dagger_{i,j} e^{-\sigma^\dagger_{i,j}\sigma_{i,j}+\sigma^\dagger_{i,j}\left(\sum_{j_2,\dots,j_d=1}^n w^{(d)}_{i,j,j_2\cdots j_d}\varphi_{j_2}\cdots\varphi_{j_d}\right)+(\varphi_i^\dagger\varphi_j)\sigma_{i,j}}. \quad (5.16)$$

We now use Identity (5.16), for each pair (i,j), in the partition function of the model with n dimensions and degree d, in order to re-express the monomials of degree d in the fields φ. This leads to:

$$Z(J,K) = \int_{\mathbb{C}^n} \mathrm{d}\varphi\mathrm{d}\varphi^\dagger \int_{\mathbb{C}^{n^2}} \mathrm{d}\sigma\mathrm{d}\sigma^\dagger e^{-\varphi^\dagger\varphi+\varphi^\dagger\sum_{k=2}^{d-1}W^{(k)}(\varphi)+J^\dagger\varphi+\varphi^\dagger K}$$
$$e^{\sum_{i,j=1}^n\left(-\sigma^\dagger_{i,j}\sigma_{i,j}+\sigma^\dagger_{i,j}\sum_{j_2,\dots,j_d=1}^n w^{(d)}_{i,j,j_2\cdots j_d}\varphi_{j_2}\cdots\varphi_{j_d}+\varphi_i^\dagger\varphi_j\sigma_{i,j}\right)}.$$

For a graphical representation of the intermediate field method leading to the model in dimension $n(n+1)$, see Fig. 5.1.

We define the new vector ϕ of $\mathbb{C}^{n+n^2}$ by $\phi = (\varphi_1,\dots,\varphi_n,\sigma_{1,1},\dots,\sigma_{1,n},\cdots,\sigma_{n,1},\dots,\sigma_{n,n})$. We further define the interaction coupling constants $\tilde{w}$ as:

- for $k = d-1$, we set $\tilde{w}^{(d-1)}_{i,j,j_2\cdots j_d} := w^{(d-1)}_{i,j,j_2\cdots j_d}$ and $\tilde{w}^{(d-1)}_{i\cdot n+j,j_2\cdots j_d} = w^{(d)}_{i,j,j_2\cdots j_d}$ with $i,j,j_2,\dots j_n \leq n$; and
- for $k\in\{3,\dots,d-2\}$, we set $\tilde{w}^{(k)}_{i,j,j_2\cdots j_k} := w^{(k)}_{i,j,j_2\cdots j_k}$ with $i,j,j_2,\dots j_n\leq n$; and
- for $k=2$, we set $\tilde{w}^{(2)}_{i,j,j_2} := w^{(2)}_{i,j,j_2}$ and $\tilde{w}^{(2)}_{i,j,i\cdot n+j} = 1$ with $i,j,j_2\leq n$.

The remaining coefficients of $\tilde{w}$ are set to 0.

In the same way, the external sources are defined to be $\tilde{J} = (J,0)$ and $\tilde{K} = (K,0)$, where, of course, the number of extra vanishing coordinates is n^2. It is important to note that these external sources have fewer degrees of freedom than coordinates (n vs. $n(n+1)$). We also remark that, for generic d, in order to have a relation adapted to

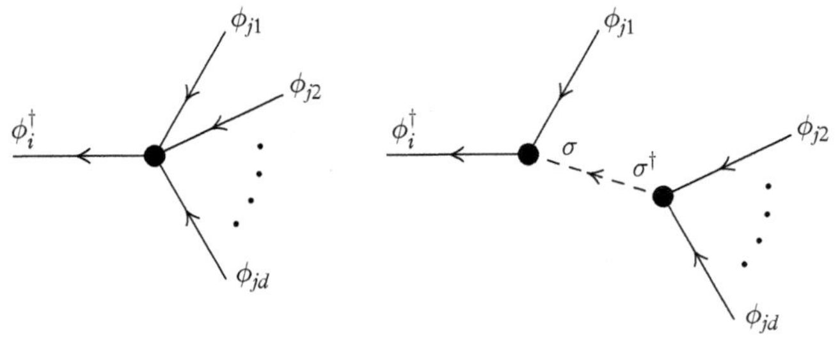

Figure 5.1 *Illustration of the intermediate field method. On the left-hand side, the interaction term before applying the intermediate field method, and on the right-hand side, the interaction term after applying the method*

induction, it is crucial to consider an inhomogeneous model, since the intermediate field method originates terms of degrees $d-1$ and 3.

The resulting QFT model, after applying the intermediate field method, has partition function

$$Z(J,K) = \int_{\mathbb{C}^{n+n^2}} \mathrm{d}\phi \mathrm{d}\phi^\dagger e^{-\phi^\dagger\phi+\phi^\dagger\sum_{k=2}^{d-1}\tilde{W}^{(k)}(\phi)+\tilde{J}^\dagger\phi+\phi^\dagger\tilde{K}}$$

and one-point correlation function

$$G_i(u) = \frac{\int_{\mathbb{C}^{n+n^2}} \mathrm{d}\phi \mathrm{d}\phi^\dagger \phi_i e^{-\phi^\dagger\phi+\phi^\dagger\sum_{k=2}^{d-1}\tilde{W}^{(k)}(\phi)+\phi^\dagger\tilde{u}}}{\int_{\mathbb{C}^{n+n^2}} \mathrm{d}\phi \mathrm{d}\phi^\dagger e^{-\phi^\dagger\phi+\phi^\dagger\sum_{k=2}^{d-1}\tilde{W}^{(k)}(\phi)+\phi^\dagger\tilde{u}}},$$

for $i \in \{1,\ldots,n\}$.

We have thus shown, in a QFT way, that the partition function (resp. the one-point correlation function) of the model with dimension $n \in \mathbb{N}$ and degree $d \in \mathbb{N}\setminus\{1,2\}$ is equal to the partition function (resp. the n first coordinates of the one-point correlation function) of the model with dimension $n(n+1)$ and degree $d-1$, up to a redefinition of the coupling constant $w \mapsto \tilde{w}$ and a trivial redefinition of the external sources. Since, as previously announced, the partition function corresponds to the inverse of the Jacobian (resp. the one-point correlation function corresponds to the formal inverse), this gives a QFT proof of Theorem 5.0.6. This represents a reduction theorem to the quadratic case for Jacobian conjecture, up to the addition of a new parameter, related to the introduction of additional intermediate fields σ. An algebraic proof (not using any QFT arguments) of Theorem 5.0.6 was also derived in the original article de Goursac, Sportiello, and Tanasa (2016).

5.3 Selected further reading

- Abdesselam, A., Annales Henri Poincaré. (2003). The Jacobian conjecture as a problem of perturbative quantum field theory, 4: 199–215. 199, 215.

  A very nice article introducing the Abdesselam–Rivasseau model is used in the chapter above.

6

Fermionic QFT, Grassmann calculus, and combinatorics

Grassmann algebra (see, for example Feldman, Konörer and E. Trubowitz for a general reference) is extensively used in mathematical physics to describe fermions, which are elementary particles obeying the so-called Fermi-Dirac statistics (let us note that, somehow related to the algebraic construction we work with here, the Grassmann-Cayley algebra was introduced to combinatorics and used to study the invariant theory, see (Doubilet, Rota, and Stein 1976; Rota and Stein 1976)). Let us emphasize here that we have not delt with fermions in Chapter 3 of this book, where only bosons were analysed.

The generators of the Grassmann algebra are called by physicists Grassmann variables and they obey anti-commuting relations (unlike real or complex variables, which obey commuting relations, and which are used in mathematical physics to describe bosons, see again Chapter 3). One can then define the so-called Grassmann integral, which is a linear form on the Grassmann algebra.

The rules of Grassmann calculus lead to very interesting developments outside the original field of mathematical physics. Thus, in combinatorics, Grassmann algebra is used to derive elegant expressions for determinants and Pfaffians. These expressions have already been used to prove the contraction-deletion formula of the Tutte and Bollobás–Riordan polynomials Krajewski, Rivasseau, Tanasa, and Wang (2010a), matrix-tree-like theorems Abdesselam (2004), generalizations of the classical Cauchy–Binet formula Caracciolo, Sokal, and Sportiello (2014), and so on.

In Carrozza, Krajewski, and Tanasa (2016), these techniques were used to give a new proof of the Lindström–Gessel–Viennot (LGV) formula for graphs with cycles (formula first found by means of purely combinatorial techniques in Talaska (2012)—the interested reader should also consult Lalonde (1996) for a combinatorial proof of Jacobi's equality relating a cofactor of a matrix with the complementary factor of its inverse).

In the first section of this chapter, we use Grassmann algebra to give, first, a reformulation of the LGV lemma proof initially proposed in the original research article Carrozza, Krajewski, and Tanasa (2016). Our proof is identical to that of Zinn-Justin (2008), but we show that it generalizes to graphs with cycles. We then use the same

Combinatorial Physics: Combinatorics, quantum field theory, and quantum gravity models. Adrian Tanasa, Oxford University Press (2021). © Adrian Tanasa. DOI: 10.1093/oso/9780192895493.003.0006

Grassmann techniques to give new proofs of Stembridge's identities relating appropriate graph Pfaffians to sum over nonintersecting paths—see Stembridge (1990). Our results go further than the ones of Stembridge (1990), because Grassmann algebra techniques naturally extend (without any cost!) to graphs with cycles. We thus obtain, instead of sums over nonintersecting paths, sums over nonintersecting paths and nonintersecting cycles. In the fifth section, we give a generalization of these results.

It is worth mentioning here that our proofs do not require any clever involution argument (as it is the case for the 'classical' proofs of (Gessel and Viennot 1985; Talaska 2012; Stembridge 1990)). The vanishing contribution of intersecting paths and cycles is an immediate consequence of the basic properties of Grassmann algebras.

The results of the first five section of this chapter follow the article Carrozza and Tanasa (2018).

In the sixth section of this chapter we use Grassmann calculus to exhibit the relation between the multi-variate version of the Tutte polynomial and the Kirchhoff–Symanzik polynomials of the parametric representation of Feynman integrals, notions already introduced in this book. This section follows the article Krajewski, Rivasseau, Tanasa, and Wang (2010b).

6.1 Grassmann algebras and Grassmann calculus

6.1.1 The Grassmann algebra

Let R be a unital commutative ring such that $\mathbb{Q} \subseteq R$, and let $\chi_1, \dots, \chi_m$ $(m \in \mathbb{N})$ be a collection of letters.

Definition 6.1.1 *The* **Grassmann algebra** $R[\chi_1, \dots, \chi_m]$, *or simply* $R[\chi]$, *is the quotient of the free noncommutative* R*-algebra with generators* $\chi_1, \dots, \chi_m$, *by the two-sided ideal generated by the expressions* $\chi_i\chi_j + \chi_j\chi_i$, *for all* $i, j = 1, \dots, m$.

Thus, the generators of the Grassmann algebra $\chi_1, \dots, \chi_m$, also called Grassmann (or anti-commuting) variables satisfy anticommutation relations:

$$\chi_i\chi_j + \chi_j\chi_i = 0, \quad \forall i, j = 1, \dots, m. \tag{6.1}$$

As an immediate consequence (since 2 is invertible in R), one has the following crucial identity (which encodes Pauli's exclusion principle in the physics literature):

$$\chi_i^2 = 0, \quad \forall i = 1, \dots, m. \tag{6.2}$$

One can then easily prove that a general element of the Grassmann algebra can be uniquely written as

$$f(\chi) = \sum_{n=0}^{m} \sum_{1 \le i_1 < \ldots < i_n \le m} a_{i_1 \ldots i_n} \chi_{i_1} \cdots \chi_{i_n} \,, \tag{6.3}$$

with $a_{i_1 \ldots i_n} \in R$.

The multiplication rule for Grassmann monomials is

$$(\chi_{i_1} \cdots \chi_{i_n})(\chi_{j_1} \cdots \chi_{j_p}) = \begin{cases} 0 & \text{if} \quad \{i_1, \ldots, i_n\} \cap \{j_1, \ldots, j_p\} \neq \emptyset \\ \operatorname{sgn}(k)\, \chi_{k_1} \cdots \chi_{k_{n+p}} & \text{otherwise} \end{cases}, \tag{6.4}$$

where $1 \le i_1 < \ldots < i_n \le m$, $1 \le j_1 < \ldots < j_p \le m$, and k is the permutation which maps $(i_1, \ldots, i_n, j_1, \ldots, j_p)$ to $(k_1, \ldots, k_{n+p})$ where $k_1 < \ldots < k_{n+p}$ is the non-decreasing rearrangement of $i_1, \ldots, i_n, j_1, \ldots, j_p$. The multiplication rule for general Grassmann algebra elements follows by distributivity. Note that Grassmann variables and elements of the ring R commute.

The exponential of a Grassmann function f is defined by the usual formula:

$$\exp(f(\chi)) := \sum_{p=0}^{+\infty} \frac{1}{p!} (f(\chi))^p \,. \tag{6.5}$$

Note that this is a polynomial in χ (as a consequence of (6.1) and of the fact that the number of generators is finite). For example, one has

$$\exp(\chi_{i_1} \cdots \chi_{i_n}) = 1 + \chi_{i_1} \cdots \chi_{i_n} \,. \tag{6.6}$$

6.1.2 Grassmann calculus; Pfaffians as Grassmann integrals

Following Berezin (1966), we define the **Grassmann integral** $\int d\chi \equiv \int d\chi_m \ldots d\chi_1$ as the unique linear map from $R[\chi_1, \ldots, \chi_m]$ to R such that

$$\int d\chi \, \chi_1 \ldots \chi_m = 1 \,, \qquad \text{and} \qquad \int d\chi \, \chi_{i_1} \cdots \chi_{i_n} = 0 \quad \text{if} \quad n < m \,. \tag{6.7}$$

Thus, for $f \in R[\chi_1 \ldots \chi_m]$ written as in (6.3), one has:

$$\int d\chi \, f(\chi) = a_{12 \ldots m} \,. \tag{6.8}$$

Example 6.1.2 *Let χ_1 and χ_2 be two independent Grassmann variables. One has:*

$$\int d\chi_1 d\chi_2 \exp(-\frac{1}{2}\chi_1 a \chi_2 + \frac{1}{2}\chi_2 a \chi_1) = \int d\chi_1 d\chi_2 (1 - \frac{1}{2}\chi_1 a \chi_2 + \frac{1}{2}\chi_2 a \chi_1) = a \,. \tag{6.9}$$

If m is an even integer and A is an $m \times m$ skew-symmetric matrix with coefficients in R, then its Pfaffian is written

$$\mathrm{Pf}(A) = \int d\chi_1 \dots d\chi_m \exp(-\frac{1}{2}\sum_{i,j=1}^{m} \chi_i A_{ij} \chi_j). \tag{6.10}$$

Example 6.1.2 is thus an illustration of this formula for the matrix $A = \begin{pmatrix} 0 & -a \\ a & 0 \end{pmatrix}$.

Finally, let now M be an n-dimensional square matrix, again with entries in R, and consider $2n$ Grassmann variables $\bar{\chi}_i, \chi_i, i = 1, \dots, n$. Note that the conjugate notation has nothing to do with complex conjugation. One has

$$\det M = \int d\bar{\chi}_N d\chi_N \dots d\bar{\chi}_1 d\chi_1 \exp\left(-\sum_{i,j=1}^{N} \bar{\chi}_i M_{ij} \chi_j \right). \tag{6.11}$$

Formulas (6.10) and (6.11) are of course compatible with the standard relation between determinants and Pfaffians: $\mathrm{Pf}\begin{pmatrix} 0 & M \\ -{}^tM & 0 \end{pmatrix} = (-1)^{\frac{N(N-1)}{2}} \det M$.

6.2 On Grassmann Gaussian measures

In this section we present a preliminary identity, which we will use extensively in the rest of the paper.

Let S be a $2n \times 2n$ skew-symmetric, invertible matrix. Since the Pfaffian is a square-root of the determinant, this implies in particular that $\int d\chi \exp\left(-\frac{1}{2}\sum_{i,j=1}^{2n} \chi_i S_{ij}^{-1} \chi_j\right) \neq 0$. Following Feldman, Knörer and Trubowitz (2002), one defines a notion of (normalized) Gaussian measure associated to S:

Definition 6.2.1 *The* **Gaussian measure** $d\mu_S$ *associated to* S *is given by the following formula:*

$$d\mu_S(\chi) := d\chi \frac{\exp\left(-\frac{1}{2}\sum_{i,j=1}^{2n} \chi_i S_{ij}^{-1} \chi_j\right)}{\int d\eta \exp\left(-\frac{1}{2}\sum_{i,j=1}^{2n} \eta_i S_{ij}^{-1} \eta_j\right)}. \tag{6.12}$$

Let us introduce the following notation: for any square matrix M of size p, and any subsets $\mathcal{A}, \mathcal{B} \subset \{1, \dots, p\}$, we will denote by $M_{\mathcal{AB}}$ the matrix obtained from M by deleting all rows (resp. columns) which are not in $\mathcal{A}$ (resp. $\mathcal{B}$).

The moments of Grassmann Gaussian measures are related to Pfaffians in the following way:

Lemma 6.2.2 *(Proposition I. 19 of Feldman, Knörer and Trubowitz (2002)). Let S be a $2n \times 2n$ skew-symmetric and invertible matrix. For any even $k \geq 2$, and for any subset $\mathcal{I} = \{i_1, \ldots, i_k\}$ with $1 \leq i_1 < \ldots < i_k \leq 2n$, one has:*

$$\int d\mu_S(\chi)\chi_{i_1} \cdots \chi_{i_k} = \mathrm{Pf}\,(S_{\mathcal{I}\mathcal{I}})\,. \tag{6.13}$$

As already noticed in Feldman, Knörer and Trubowitz (2002), one may extend the notion of Gaussian measure to non-invertible skew-symmetric matrices S, in which case $d\mu_S$ is *defined* by (6.2.2).

6.3 Lingström–Gessel–Viennot (LGV) formula for graphs with cycles

In this section, we present a simplified version of the proof given in Carrozza, Krajewski, and Tanasa (2016) of the LGV lemma for graphs with cycles.

Let G be a digraph (or a directed graph) which allows loops or multiple edges. Let $V = \{v_1, \ldots, v_n\}$ be the set of vertices of the digraph, which we equip with an order relation $<$ such that $v_1 < \ldots < v_n$.[1] To each edge e of the digraph, we assign a **weight** denoted by w_e. One further assumes that the variables w_e thus defined are commuting variables.

Let us consider a collection of edges $P = (e_1, e_2, \ldots, e_k)$ such that one can reach a vertex v' from a vertex v by successively traversing the edges of the collection P in the specified order. We then say that P is a **path** in the digraph, from the vertex v to the vertex v'.[2]

Let the product $\mathrm{wt}(P) := \prod_{k=1}^{m} w_{e_k}$ denote the **weight** of a given path $P = (e_1, \ldots, e_k)$. We then construct the **weight path matrix** $M_{ij} := \sum\limits_{P: v_i \rightsquigarrow v_j} \mathrm{wt}(P)$, where $1 \leq i, j \leq n$ and the sum runs over paths P connecting the vertex v_i to the vertex v_j. Note that the entries of the matrix M are considered as formal power series in the edge weights w_e. Matrix equations such as $M = (1 - A)^{-1}$, where A is the weighted adjacency matrix,[3] are also meant in a formal power series sense.

A **cycle** is defined as a path from a vertex v to itself (up to a change of source vertex v along the path). The set of all possible collections of self-avoiding and pairwise vertex disjoint cycles, including the empty collection, is denoted by $\mathcal{C}$. Let the weight and sign of $\mathbf{C} = (C_1, \ldots, C_k) \in \mathcal{C}$ be given by the following formulas:

[1] To simplify notations, we will often identify the ordered set V with $\{1, \ldots n\}$ itself.

[2] In graph theory terminology, such defined $P = (e_1, e_2, \ldots, e_k)$ is a walk; if there are no repeated internal vertices, then it is a path. However, in this paper, we do not use this graph theory terminology, but the terminology defined in the body text, which is the same terminology as the one used in previous papers on this topic, such as (Gessel and Viennot 1985; Stembridge 1985; Talaska 2012). The correspondences are: (path, walk), (self-avoiding path (see following text), path).

[3] For any $1 \leq i, j \leq n$, $A_{ij} := \sum\limits_{e: v_i \rightsquigarrow v_j} w_e$, where the sum runs over edges connecting v_i to v_j.

$$\mathrm{wt}(\mathbf{C}) := \prod_{i=1}^{k} \mathrm{wt}(C_i), \qquad \mathrm{sgn}(\mathbf{C}) := (-1)^k. \tag{6.14}$$

We further set $\mathrm{wt}(\mathbf{C}) = 1$ and $\mathrm{sgn}(\mathbf{C}) = 1$ for the empty collection $\mathbf{C} = \emptyset$.

One considers now subsets of vertices $\mathcal{A} = \{a_1, \ldots, a_p\}$ and $\mathcal{B} = \{b_1, \ldots, b_p\}$, with $a_1 < \ldots < a_p$ and $b_1 < \ldots < b_p$. A collection of paths $\mathbf{P} = (P_1, \ldots, P_p)$, where the path P_i connects the vertex a_i to the vertex $b_{\sigma_{\mathbf{P}}(i)}$ (where $\sigma_{\mathbf{P}}$ is some permutation), is called a **p-path** from the subset $\mathcal{A}$ to the subset $\mathcal{B}$. Let

$$\mathrm{wt}(\mathbf{P}) := \prod_{i=1}^{k} \mathrm{wt}(P_i), \qquad \mathrm{sgn}(\mathbf{P}) := \mathrm{sgn}(\sigma_{\mathbf{P}}), \tag{6.15}$$

denote the weight and resp. the sign of the p-path $\mathbf{P}$. One says that $\mathbf{P}$ is **self-avoiding** if: each path P_i is self-avoiding and if, for each pair of paths P_i and P_j with $i \neq j$, P_i and P_j are vertex disjoint.

Let us now give the following definition:

Definition 6.3.1 *A* **self-avoiding flow** *from $\mathcal{A}$ to $\mathcal{B}$ is a pair $(\mathbf{P}, \mathbf{C})$ such that $\mathbf{C} \in \mathcal{C}$, $\mathbf{P}$ is a self-avoiding p-path from $\mathcal{A}$ to $\mathcal{B}$, and $\mathbf{P}$, and $\mathbf{C}$ are vertex disjoint. The set of self-avoiding flows from $\mathcal{A}$ to $\mathcal{B}$ is denoted by $\mathcal{F}_{\mathcal{A},\mathcal{B}}$. We furthermore identify $\mathcal{F}_{\emptyset,\emptyset}$ with $\mathcal{C}$.*

Given $2n$ independent Grassmann variables θ_i and $\bar{\theta}_i$ with i running from 1 to n, we introduce the notations

$$[d\theta \cdot d\bar{\theta}] := d\theta_n d\bar{\theta}_n \ldots d\theta_1 d\bar{\theta}_1, \qquad [\bar{\theta}_{\mathcal{B}} \cdot \theta_{\mathcal{A}}] := \bar{\theta}_{b_1}\theta_{a_1} \ldots \bar{\theta}_{b_p}\theta_{a_p}, \tag{6.16}$$

where $\mathcal{A}$ and $\mathcal{B}$ are sets as previously illustrated.

One then has:

Lemma 6.3.2 *Let $\mathcal{A}$ and $\mathcal{B}$ be two subsets of $\{1, \ldots, n\}$ with cardinal p. Denote by $a_1 < \ldots < a_p$ the elements of $\mathcal{A}$ and $b_1 < \ldots < b_p$ the elements of $\mathcal{B}$. Then:*

$$\int [d\theta \cdot d\bar{\theta}]\,[\bar{\theta}_{\mathcal{B}} \cdot \theta_{\mathcal{A}}] \exp\left(\sum_{i,j=1}^{N} \bar{\theta}_i M^{-1}_{ij} \theta_j\right) = \sum_{(\mathbf{P},\mathbf{C}) \in \mathcal{F}_{\mathcal{A},\mathcal{B}}} \mathrm{sgn}(\mathbf{P})\mathrm{wt}(\mathbf{P})\,\mathrm{sgn}(\mathbf{C})\mathrm{wt}(\mathbf{C}). \tag{6.17}$$

Proof Let us give the main idea of the proof. As noticed previously, $M^{-1} = 1 - A$, and one can rewrite the exponential appearing in the integrand as a product of terms of the form $(1 + \bar{\theta}_i\theta_i)$ and $(1 - \bar{\theta}_i A_{ij} \theta_j)$. After expansion of this product, the monomials which do not vanish after Grassmann integration must contain each Grassmann variable once, and only once. This implies that they are indexed by elements of $\mathcal{F}_{\mathcal{A},\mathcal{B}}$. One furthermore checks that each flow is weighted as in the right-hand side of formula (6.17).

The LGV formula for a digraph with cycles is:

Theorem 6.3.3 *(Theorem 2.5 of Talaska (2012)). One has:*

$$\det(M_{\mathcal{AB}}) = \frac{\sum\limits_{(\mathbf{P},\mathbf{C})\in\mathcal{F}_{\mathcal{A},\mathcal{B}}} \mathrm{sgn}(\mathbf{P})\mathrm{wt}(\mathbf{P})\,\mathrm{sgn}(\mathbf{C})\mathrm{wt}(\mathbf{C})}{\sum\limits_{\mathbf{C}\in\mathcal{C}} \mathrm{sgn}(\mathbf{C})\mathrm{wt}(\mathbf{C})}. \tag{6.18}$$

In particular, if G is acyclic, one recovers the LGV Lemma (Lindström 1973; Gessel and Viennot 1985).

Proof Let us define the skew-symmetric matrix S (and its inverse) as

$$S = \begin{pmatrix} 0 & -M \\ {}^tM & 0 \end{pmatrix}, \qquad S^{-1} = \begin{pmatrix} 0 & {}^tM^{-1} \\ -M^{-1} & 0 \end{pmatrix}. \tag{6.19}$$

Define $\overline{\mathcal{B}} := \{n+b|b\in\mathcal{B}\}$. We now use Lemma 3.8 with the Matrix (6.19) and $\mathcal{I} = \mathcal{A}\cup\overline{\mathcal{B}}$. Using the explicit expression of S, one finds

$$S_{\mathcal{II}} = \begin{pmatrix} 0 & -M_{\mathcal{AB}} \\ {}^tM_{\mathcal{AB}} & 0 \end{pmatrix}, \tag{6.20}$$

which implies that $\mathrm{Pf}\,(S_{\mathcal{II}}) = (-1)^{\frac{p(p+1)}{2}}\det(M_{\mathcal{AB}})$. One then has

$$\det(M_{\mathcal{AB}}) = (-1)^{\frac{p(p+1)}{2}} \int d\mu_S(\chi)\,(\chi_{a_1}\cdots\chi_{a_p})\,(\chi_{n+b_1}\cdots\chi_{n+b_p}). \tag{6.21}$$

Using the explicit form (6.12) of the Gaussian measure, we can rewrite this identity as

$$\det(M_{\mathcal{AB}}) = \frac{1}{Z}\int [d\theta\cdot d\bar{\theta}]\,[\bar{\theta}_{\mathcal{B}}\cdot\theta_{\mathcal{A}}]\exp\left(\sum_{i,j=1}^{n}\bar{\theta}_i M^{-1}_{ij}\theta_j\right), \tag{6.22}$$

where we have introduced the Grassmann variables $\theta_i = \chi_i$ and $\bar{\theta}_i = \chi_{n+i}$ $(i\in\{1,\ldots,n\})$, and the normalisation

$$Z = \int [d\theta\cdot d\bar{\theta}]\exp\left(\sum_{i,j=1}^{n}\bar{\theta}_i M^{-1}_{ij}\theta_j\right). \tag{6.23}$$

To conclude the proof, we apply Lemma 6.3.2 to both the numerator and denominator of $\det(M_{\mathcal{AB}})$. The evaluation of Z reduces to a sum over non-intersecting cycles, while the numerator contains all non-intersecting flows in $\mathcal{F}_{\mathcal{A},\mathcal{B}}$.

6.4 Stembridge's formulas for graphs with cycles

In this section, we consider the situation in which the paths start from a specified set $\mathcal{A}$ of p vertices (as previously stated) but their endpoints are not fixed. They are only restricted to lie within the graph region $\mathcal{I} \subset \{1,\ldots,n\}$, where the cardinal of $\mathcal{I}$ can be larger than p. Moreover, we assume the cardinal of $\mathcal{A}$ to be even: $p = 2m$.

Following Stembridge (1990), let us define the $n \times n$ matrix $Q^{\mathcal{I}}$ whose entries are

$$Q^{\mathcal{I}}_{ij} := \sum_{k,l \in \mathcal{I} | k<l} (M_{ik}M_{jl} - M_{il}M_{jk}) . \tag{6.24}$$

Let us furthermore denote by

$$\mathcal{F}^{\mathcal{I}}_{\mathcal{A}} := \bigcup_{\mathcal{B} \subset \mathcal{I}, |\mathcal{B}| = |\mathcal{A}|} \mathcal{F}_{\mathcal{A},\mathcal{B}} \tag{6.25}$$

the set of self-avoiding flows from $\mathcal{A}$ to $\mathcal{I}$ (defined as in Definition 6.3.1). In this definition, the union over subsets $\mathcal{B} \subset \mathcal{I}$ is disjoint; we can thus naturally extend the weight and sign functions previously introduced to $\mathcal{F}^{\mathcal{I}}_{\mathcal{A}}$. Note that the sign is always computed relative to the natural order on the set of integers $\mathcal{A}$ and $\mathcal{B}$.

We can now state the main result of this section:

Theorem 6.4.1 *Let $\mathcal{A}$ and $\mathcal{I}$ be subsets of $\{1,\ldots,n\}$ such that $|\mathcal{A}|$ is even. One has:*

$$\mathrm{Pf}\left(Q^{\mathcal{I}}_{\mathcal{A}\mathcal{A}}\right) = \frac{\sum_{(\mathbf{P},\mathbf{C}) \in \mathcal{F}^{\mathcal{I}}_{\mathcal{A}}} \mathrm{sgn}(\mathbf{P})\mathrm{wt}(\mathbf{P})\,\mathrm{sgn}(\mathbf{C})\mathrm{wt}(\mathbf{C})}{\sum_{\mathbf{C} \in \mathcal{C}} \mathrm{sgn}(\mathbf{C})\mathrm{wt}(\mathbf{C})} . \tag{6.26}$$

Proof As before, we denote by $a_1 < \ldots < a_{2m}$ the elements of $\mathcal{A}$. Let us also introduce the $2n \times 2n$ skew-symmetric and invertible matrix

$$S^{-1} = \begin{pmatrix} 0 & {}^tM^{-1} \\ -M^{-1} & B \end{pmatrix}, \tag{6.27}$$

where the n-dimensional and skew-symmetric square matrix B is defined by: $B_{ij} = 1$ if $i,j \in \mathcal{I}$ and $i < j$, $B_{ij} = -1$ if $i,j \in \mathcal{I}$ and $i > j$, and $B_{ij} = 0$ otherwise. We have then:

$$S = \begin{pmatrix} MB^tM & -M \\ {}^tM & 0 \end{pmatrix} = \begin{pmatrix} Q^{\mathcal{I}} & -M \\ {}^tM & 0 \end{pmatrix} . \tag{6.28}$$

As in the previous section, we make use of Lemma 3.8, which now implies:

$$\mathrm{Pf}\left(Q^{\mathcal{I}}_{\mathcal{A}\mathcal{A}}\right) = \mathrm{Pf}\left(S_{\mathcal{A}\mathcal{A}}\right) = \int d\mu_S(\chi)\chi_{a_1}\cdots\chi_{a_{2m}} . \tag{6.29}$$

Using the Grassmann variable conjugate notation $\{\theta_i, \bar{\theta}_i\}$, we can rewrite this equality in the form,

$$\mathrm{Pf}\left(Q^{\mathcal{I}}_{\mathcal{A}\mathcal{A}}\right) = \frac{1}{Z}\int [d\theta \cdot d\bar{\theta}]\, \theta_{a_1} \ldots \theta_{a_{2m}} \exp\left(\sum_i^n \bar{\theta}_i \theta_i - \sum_{i,j=1}^n \bar{\theta}_i A_{ij} \theta_j - \sum_{i,j\in\mathcal{I}|i<j} \bar{\theta}_i \bar{\theta}_j\right), \tag{6.30}$$

where:

$$Z = \int [d\theta \cdot d\bar{\theta}] \exp\left(\sum_i^n \bar{\theta}_i \theta_i - \sum_{i,j=1}^n \bar{\theta}_i A_{ij} \theta_j - \sum_{i,j\in\mathcal{I}|i<j} \bar{\theta}_i \bar{\theta}_j\right). \tag{6.31}$$

It is easy to see that, when expanding the integrand of Z into monomials, all terms containing a factor $\bar{\theta}_i \bar{\theta}_j$ coming from the third sum in the exponential will vanish, since such terms will contain strictly more $\bar{\theta}$ variables than θ variables. Hence the third sum in the exponential does not contribute, and Z is again a sum over non-intersecting cycles. As for the numerator of (6.30), it can be expanded as a sum over subsets $\mathcal{B} \subset \mathcal{I}$ of size $2m$

$$\sum_{\mathcal{B}\subset\mathcal{I},|\mathcal{B}|=2m} (-1)^m \int [d\theta \cdot d\bar{\theta}]\, \theta_{a_1} \ldots \theta_{a_{2m}} F_{\mathcal{B}}(\bar{\theta}) \exp\left(\sum_{i,j=1}^n \bar{\theta}_i M^{-1}_{ij} \theta_j\right), \tag{6.32}$$

where $F_{\mathcal{B}}(\bar{\theta})$ is a weighted sum of monomials. More precisely, denoting by $b_1 < \ldots < b_{2m}$ the elements of $\mathcal{B}$, we have

$$F_{\mathcal{B}}(\bar{\theta}) = \sum_{(i_1 i_2)\ldots(i_{2m-1} i_m)} \prod_{k=1}^m \bar{\theta}_{b_{i_{2k-1}}} \bar{\theta}_{b_{i_{2k}}}, \tag{6.33}$$

where the sum runs over all possible fixed-point free involutions (also known as *matchings* or *pairings* in the literature) on $\{1, \ldots, 2m\}$, and we have furthermore adopted the convention that $i_{2k-1} < i_{2k}$. We can then reord the Grassmann variables appearing in each monomial to obtain

$$F_{\mathcal{B}}(\bar{\theta}) = \sum_{\text{pairing}\,\sigma} (-1)^{cr(\sigma)} \bar{\theta}_{b_1} \ldots \bar{\theta}_{b_m}, \tag{6.34}$$

where $cr(\sigma)$ is the number of crossings in the fixed-point free involution σ. One may, furthermore show, by induction on m, that

$$\sum_{\text{pairing}\,\sigma} (-1)^{cr(\sigma)} = 1. \tag{6.35}$$

Note that this formula can also be obtained as a particular case of the Touchard–Riordan formula (see (Touchard 1950; Riordan 1975)). The numerator of (6.30) thus reduces to

$$\sum_{\mathcal{B}\subset\mathcal{I},|\mathcal{B}|=2m} \int [d\theta\cdot d\bar{\theta}]\,[\bar{\theta}_{\mathcal{B}}\cdot\theta_{\mathcal{A}}]\exp\left(\sum_{i,j=1}^{n}\bar{\theta}_i M^{-1}_{ij}\theta_j\right). \tag{6.36}$$

Now using Lemma 6.3.2 complete the proof.

Let us now consider the case in which some of the paths' endpoints are fixed. More precisely, for any $\mathcal{B}\subset V$ such that $\mathcal{B}\cap\mathcal{I}=\emptyset$, we define $\mathcal{F}^{\mathcal{I}}_{\mathcal{A},\mathcal{B}}$ as the set of flows from $\mathcal{A}$ to $\mathcal{B}\cup\mathcal{D}$, where $\mathcal{D}$ is an arbitrary subset of $\mathcal{I}$:

$$\mathcal{F}^{\mathcal{I}}_{\mathcal{A},\mathcal{B}} := \bigcup_{\mathcal{D}\subset\mathcal{I},|\mathcal{B}\cup\mathcal{D}|=|\mathcal{A}|} \mathcal{F}_{\mathcal{A},\mathcal{B}\cup\mathcal{D}}\,. \tag{6.37}$$

This is again a disjoint union of subsets; the functions wt and sgn are extended to $\mathcal{F}^{\mathcal{I}}_{\mathcal{A},\mathcal{B}}$ in a natural way. For simplicity, we will only consider subsets $\mathcal{B}$ and $\mathcal{I}$ such that any element of $\mathcal{B}$ is strictly smaller than any element of $\mathcal{I}$, which we denote $\mathcal{B}<\mathcal{I}$. Note that this restriction does not affect the generality of the following Theorem 6.4.2, since one is free to relabel the vertices of the graph in order to insure that $\mathcal{B}<\mathcal{I}$ holds.

Theorem 6.4.2 *Let $\mathcal{A}$, $\mathcal{B}$, and $\mathcal{I}$ be subsets of $\{1,\ldots,n\}$ such that: $|\mathcal{A}|=r$ and $|\mathcal{B}|=s$ with $r+s$ even, $s\le r$, and $\mathcal{B}<\mathcal{I}$. One has:*

$$\operatorname{Pf}\begin{pmatrix} Q^{\mathcal{I}}_{\mathcal{A}\mathcal{A}} & -M_{\mathcal{A}\mathcal{B}} \\ {}^{t}M_{\mathcal{A}\mathcal{B}} & 0 \end{pmatrix} = \frac{\sum_{(\mathbf{P},\mathbf{C})\in\mathcal{F}^{\mathcal{I}}_{\mathcal{A},\mathcal{B}}} \operatorname{sgn}(\mathbf{P})\operatorname{wt}(\mathbf{P})\operatorname{sgn}(\mathbf{C})\operatorname{wt}(\mathbf{C})}{\sum_{\mathbf{C}\in\mathcal{C}}\operatorname{sgn}(\mathbf{C})\operatorname{wt}(\mathbf{C})}. \tag{6.38}$$

Proof We introduce the same matrix S as in the proof of Theorem 6.4.1, and compute

$$\operatorname{Pf}\begin{pmatrix} Q^{\mathcal{I}}_{\mathcal{A}\mathcal{A}} & -M_{\mathcal{A}\mathcal{B}} \\ {}^{t}M_{\mathcal{A}\mathcal{B}} & 0 \end{pmatrix} = \operatorname{Pf}(S_{\mathcal{A}\cup\bar{\mathcal{B}}\mathcal{A}\cup\bar{\mathcal{B}}})\,. \tag{6.39}$$

Using again Lemma 3.8 yields the same denominator as before, denominator which expands as a sum over nonintersecting cycles. The numerator can be written in the form

$$\sum_{\mathcal{D}\subset\mathcal{I},|\mathcal{D}|=r-s} (-1)^{\frac{r-s}{2}} \int [d\theta\cdot d\bar{\theta}]\,\theta_{a_1}\ldots\theta_{a_r}\bar{\theta}_{b_1}\ldots\bar{\theta}_{b_s}F_{\mathcal{D}}(\bar{\theta})\exp\left(\sum_{i,j=1}^{n}\bar{\theta}_i M^{-1}_{ij}\theta_j\right),$$

where $F_{\mathcal{D}}(\bar{\theta}) = \bar{\theta}_{d_1} \ldots \bar{\theta}_{d_{r-s}}$ and $d_1 < \ldots < d_{r-s}$ are the elements of $\mathcal{D}$. Since $\mathcal{B} < \mathcal{D}$ for any $\mathcal{D}$ appearing in the previous formula, no extra sign needs to be included. The proof then goes along the lines of the proof of Theorem 6.4.1.

6.5 A generalization

In this section we provide a generalization of the results of the previous sections. This generalization is, up to our knowledge, not existing in the combinatorics literature, even for acyclic digraphs.

Let us now consider two subsets $\mathcal{I}, \mathcal{J} \subset V$ such that $\mathcal{A} < \mathcal{I}$ and $\mathcal{B} < \mathcal{J}$. We denote by $\mathcal{F}_{\mathcal{A},\mathcal{B}}^{\mathcal{I},\mathcal{J}}$ the set of self-avoiding flows which respect the following conditions: for each vertex $a \in \mathcal{A}$, there exists a path starting at a; for each vertex $b \in \mathcal{B}$, there exists a path ending at b; and, finally, all other paths' starting points (resp. endpoints) are in $\mathcal{I}$ (resp. in $\mathcal{J}$). This means that we can define $\mathcal{F}_{\mathcal{A},\mathcal{B}}^{\mathcal{I},\mathcal{J}}$ by the disjoint union:

$$\mathcal{F}_{\mathcal{A},\mathcal{B}}^{\mathcal{I},\mathcal{J}} := \bigcup_{\mathcal{A}' \subset \mathcal{I}, \mathcal{B}' \subset \mathcal{J}, |\mathcal{A} \cup \mathcal{A}'| = |\mathcal{B} \cup \mathcal{B}'|} \mathcal{F}_{\mathcal{A} \cup \mathcal{A}', \mathcal{B} \cup \mathcal{B}'} . \tag{6.40}$$

Once again, the sign and weight functions are naturally extended to $\mathcal{F}_{\mathcal{A},\mathcal{B}}^{\mathcal{I},\mathcal{J}}$. Following the notation of the previous section, for any $\mathcal{K} \subset V$ we denote by $B^{\mathcal{K}}$ the $n \times n$ matrix such that: $B_{ij}^{\mathcal{K}} = 1$ if $i, j \in \mathcal{K}$ and $i < j$, $B_{ij}^{\mathcal{K}} = -1$ if $i, j \in \mathcal{I}$ and $i > j$, and $B_{ij}^{\mathcal{K}} = 0$ otherwise.

We can now state the main result of this section:

Theorem 6.5.1 *Let $\mathcal{A}, \mathcal{B}, \mathcal{I}$, and $\mathcal{J}$ be subsets of $\{1, \ldots, n\}$ such that: $|\mathcal{A}| = r$, $|\mathcal{B}| = s$ with $r + s$ even, $\mathcal{A} < \mathcal{I}$, and $\mathcal{B} < \mathcal{J}$. One has*

$$\operatorname{Pf}\begin{pmatrix} P_{\mathcal{A}\mathcal{A}} & -R_{\mathcal{A}\mathcal{B}} \\ {}^t R_{\mathcal{A}\mathcal{B}} & Q_{\mathcal{B}\mathcal{B}} \end{pmatrix} = \frac{\sum_{(\mathbf{P},\mathbf{C}) \in \mathcal{F}_{\mathcal{A},\mathcal{B}}^{\mathcal{I},\mathcal{J}}} \operatorname{sgn}(\mathbf{P}) \operatorname{wt}(\mathbf{P}) \operatorname{sgn}(\mathbf{C}) \operatorname{wt}(\mathbf{C})}{\sum_{(\mathbf{P},\mathbf{C}) \in \mathcal{F}_{\emptyset,\emptyset}^{\mathcal{I},\mathcal{J}}} \operatorname{sgn}(\mathbf{P}) \operatorname{wt}(\mathbf{P}) \operatorname{sgn}(\mathbf{C}) \operatorname{wt}(\mathbf{C})}, \tag{6.41}$$

where the $n \times n$ matrices P, Q, R are defined by

$$R := M\left(1 + B^{\mathcal{J}\,t} M B^{\mathcal{I}} M\right)^{-1}, \qquad P := M B^{\mathcal{J}\,t} R, \qquad Q := {}^t M B^{\mathcal{I}} R. \tag{6.42}$$

Proof Let us give here the main idea of the proof. We use again Lemma 3.8, but this time we take:

$$S^{-1} = \begin{pmatrix} B_{\mathcal{I}} & {}^t M^{-1} \\ -M^{-1} & B_{\mathcal{J}} \end{pmatrix}, \qquad S = \begin{pmatrix} P & -R \\ {}^t R & Q \end{pmatrix}. \tag{6.43}$$

The proof then goes along the lines of the proof of Theorem 6.4.1.

Corollary 6.5.2 *Making the same assumptions as in Theorem 6.5.1 and further assuming that there exists no path connecting $\mathcal{I}$ to $\mathcal{J}$, one has*

$$\mathrm{Pf}\begin{pmatrix} Q^{\mathcal{J}}_{\mathcal{AA}} & -M_{\mathcal{AB}} \\ {}^tM_{\mathcal{AB}} & {}^tQ^{\mathcal{I}}_{\mathcal{BB}} \end{pmatrix} = \frac{\sum\limits_{(\mathbf{P},\mathbf{C})\in\mathcal{F}^{\mathcal{I},\mathcal{J}}_{\mathcal{A},\mathcal{B}}} \mathrm{sgn}(\mathbf{P})\mathrm{wt}(\mathbf{P})\,\mathrm{sgn}(\mathbf{C})\mathrm{wt}(\mathbf{C})}{\sum\limits_{\mathbf{C}\in\mathcal{C}} \mathrm{sgn}(\mathbf{C})\mathrm{wt}(\mathbf{C})}. \tag{6.44}$$

Proof We remark that $B^{\mathcal{I}}MB^{\mathcal{J}} = 0$. Hence the matrix S given in (6.43) reduces to

$$S = \begin{pmatrix} Q^{\mathcal{J}} & -M \\ {}^tM & {}^tQ^{\mathcal{I}} \end{pmatrix}. \tag{6.45}$$

The result follows since $\mathcal{F}^{\mathcal{I},\mathcal{J}}_{\emptyset,\emptyset} = \mathcal{F}_{\emptyset,\emptyset}$, and is therefore identified with $\mathcal{C}$.

6.6 Tutte polynomial and the parametric representation in QFT

In this section, we exhibit the relation between two previously presented notions: a multivariate version of the Tutte polynomial and the Kirchhoff–Symanzik polynomials of the parametric representation of Feynman integrals. Note that this result can appear to be surprising, because, on the one hand, one has a graph theory notion and, on the other hand, one has a QFT notion, which *a priori* have no reason whatsoever to be related.

The contraction/deletion property of the Kirchhoff–Symanzik polynomials has appeared in various papers, like (Bloch, Esnault, and Kreimer 2006; Brown 2019). Nevertheless, the proof we give in this chapter—which follows the one in Krajewski et al. (2010b)—uses a different technique, namely the Grassmann developments of the determinants (see Subsection 6.1.2). Moreover, following again Krajewski et al. (2010b), we explicitly prove the way in which the Kirchhoff–Symanzik polynomial is obtained as an appropriate limit of a multi-variate version of the Tutte polynomial.

In this section we use a slightly different convention for Grassmann calculus with respect to the one used in the previous sections of this chapter. More precisely, for a set of Grassmann variables $\chi_1, ..., \chi_n$ the rules of Grassmann integrations are

$$\int d\chi = 0, \quad \int \chi d\chi = 1. \tag{6.46}$$

The determinant of any n-dimensional square matrix can then be expressed as

$$\det M = \int \prod d\bar{\psi}_i d\psi_i e^{-\sum_{ij}\bar{\psi}_i M_{ij}\psi_j}. \tag{6.47}$$

Let us first recall the following results:

Proposition 6.6.1 *(Proposition 3 of Abdesselam (2004)) The Pfaffian of an n-dimensional antisymmetric matrix A is written as*

$$\mathrm{Pf}(A) = \int \prod d\chi_i e^{-\sum_{i<j} \chi_i A_{ij} \chi_j} = \int d\chi_1 ... d\chi_n e^{-\frac{1}{2}\sum_{i,j} \chi_i A_{ij} \chi_j}. \tag{6.48}$$

Lemma 6.6.2 *(Lemma 2.1 of Krajewski et al. (2010b)). The determinant of a matrix $D + A$, where D is diagonal and A antisymmetric, is written as*

$$\det(D+A) = \int \prod d\chi_i d\omega_i e^{-\sum_i \chi_i D_{ii} \omega_i - \sum_{i<j} \chi_i A_{ij} \chi_j + \sum_{i<j} \omega_i A_{ij} \omega_j}. \tag{6.49}$$

The main result of this section is the following theorem:

Theorem 6.6.3 *For any semi-regular edge e,*

$$U_\Gamma(\alpha) = \alpha_e \, U_{\Gamma - e}(\alpha) + U_{\Gamma/e}(\alpha). \tag{6.50}$$

Moreover, one has the terminal form evaluation,

$$U_\Gamma(\alpha_e) = \prod_e \alpha_e \tag{6.51}$$

for Γ solely made of tadpole edges attached to isolated vertices.

Proof In order to prove this result, we use the Grassmann representation (see Appendix 6.1.2) of the polynomial U_Γ.

A key ingredient of this representation is the *incidence matrix* ϵ_{ev} of the graph (a matrix which has a number of rows equal to the number of edges of the graph, $e = 1, \ldots E$, and a number of columns equal to the number of vertices of the graph, $v = 1, \ldots, V$). The elements of this matrix are equal to 1 if the edge e is outgoing from the vertex v, -1 if the edge e is incoming at the vertex v and 0 if the edge e does not hook to the vertex v. Let us mention that this matrix encodes the complete information of a Feynman graph without self-loops. In order to also encode the information from self-loops (if they are present), one needs to define an additional matrix $\eta_{ev} = |\epsilon_{ev}|$ and with non-vanishing entries for the tadpoles (see Krajewski et al. (2010b) for details).

One has the following lemma:

Lemma 6.6.4 *The polynomial $U_\Gamma(\alpha)$ is given by*

$$\det Q_\Gamma, \tag{6.52}$$

where the graph matrix Q_Γ is a square matrix which is written

$$Q_\Gamma = \begin{pmatrix} \alpha_e & -\epsilon_{ev} \\ -^t\epsilon_{ev} & 0 \end{pmatrix}. \tag{6.53}$$

($^t\epsilon_{ev}$ being the transpose of the incidence matrix).

Proof One can prove this by performing the Gaussian integrations obtained by inserting the propagator form (3.25) in the general form of a Feynman integral (see the original paper Krajewski et al. (2010b) for details) (QED).

As a direct consequence of Lemma 6.6.4 and of Lemma 6.6.2, one has

$$U_\Gamma(\alpha_e) = \int \prod_{v,e} d\chi_v d\omega_v d\chi_e d\omega_e e^{-\alpha_e \chi_e \omega_e} e^{-\chi_e \epsilon_{ev} \chi_v + \omega_e \epsilon_{ev} \omega_v}. \tag{6.54}$$

Note that we have split each of the two sets of Grassmann variables (χ and ω) required for the representation of such a determinant (see Lemma 6.6.2) into two subsets, one of cardinal E (indexed by e) and one of cardinal V (indexed by v); this splitting corresponds to the block form of the total matrix Q_Γ.

Now, let e be a semi-regular edge connecting vertices v_1 and v_2. One has (see the definition (6.1) for a single Grassmann variable):

$$e^{-\alpha_e \chi_e \omega_e} = 1 + \alpha_e \omega_e \chi_e. \tag{6.55}$$

We denote the two terms of this sum by $\det Q_{\Gamma,e,1}$ and by $\det Q_{\Gamma,e,2}$, respectively. Let us first closely investigate the term $\det Q_{\Gamma,e,1}$. By a direct inspection of the various terms coming from similar expansions of the other Grassmann exponentials in (6.54), one can prove (using the Grassmann integration rules (6.46)) the following formula (see Krajewski et al. (2010b) for details):

$$\det Q_{\Gamma,e,1} = \int \prod_{e' \neq e, v} d\chi_{e'} d\omega_{e'} d\chi_v d\omega_v (\chi_{v_1} - \chi_{v_2})(\omega_{v_1} - \omega_{v_2})$$
$$e^{-\sum_{e' \neq e} \alpha'_e \chi_{e'} \omega_{e'}} e^{-\sum_{e' \neq e, v} \chi_{e'} \epsilon_{e'v} \chi_v + \sum_{e' \neq e, v} \omega_{e'} \epsilon_{e'v} \omega_v}. \tag{6.56}$$

Let us now perform the following (triangular) change of variables (with unit Jacobian):

$$\hat{\chi}_{v_1} = \chi_{v_1} - \chi_{v_2}, \qquad \hat{\chi}_v = \chi_v \ \ \forall\, v \neq v_1, \tag{6.57}$$

and the same for the set of ω Grassmann variables. This leads to the identification (see again Krajewski et al. (2010b) for details):

$$\det Q_{\Gamma,e,1} = \det Q_{\Gamma/e}. \tag{6.58}$$

The second term of the sum (6.55) similarly leads to

$$\det Q_{\Gamma,e,2} = \alpha_e \det Q_{\Gamma-e} \tag{6.59}$$

We again refer the interested reader to Krajewski et al. (2010b) for obtaining the terminal form evaluation (QED).

Let us now show how the polynomial $U_\Gamma(\alpha)$ can be obtained as a limit of the multi-variate Tutte polynomial $Z_\Gamma(q,\beta)$. One needs to consider

$$q^{-k(\Gamma)} Z_\Gamma(q,\beta). \tag{6.60}$$

Taking the limit $q \to 0$ one obtains a sum over *maximally spanning subgraphs* A, that is subgraphs with $k(A) = k(G)$:

$$S_\Gamma(\beta) = \sum_{A \text{ maximally spanning } E} \prod_{e\in A} \beta_e. \tag{6.61}$$

If one now retains only the lowest degree of homogeneity in β one obtains a sum over maximally spanning graphs with lowest number of edges, i.e. maximally spanning acyclic graphs (or spanning forests, see Definition 2.1.2 7) of Γ.

$$F_\Gamma(\beta) = \sum_{\mathcal{F} \text{ spanning forest of } \Gamma} \prod_{e\in\mathcal{F}} \beta_e. \tag{6.62}$$

We denote this lowest number of edges by p'.

Finally, if one divides $F_\Gamma(\beta)$ by $\prod_{e\in E}\beta_e$ and makes the change of variables to $\alpha_e = \beta_e^{-1}$, the polynomial $U(\alpha)$ is obtained. Since the Kirchhoff–Symanzik polynomial is defined on connected graphs (see the previous chapter), the sum in (6.62) restricts to spanning trees of the respective graphs.

All of this is summarized through the formula

$$U_\Gamma(\alpha) = \left[\left(\prod_{e\in E}\beta_e\right) \lim_{q'\to 0} \frac{1}{(q')^{p'(\Gamma)}} \lim_{q\to 0} q^{-k(G)} Z_\Gamma(q, q'\beta)\right]_{\beta_e^{-1}=\alpha_e}. \tag{6.63}$$

Note that the variable q' was introduced to take the limit retaining only the lowest degree of homogeneity in β from $S_\Gamma(q'\beta)$ (see previous points).

Let us also point out the universality of the Tutte polynomial. This property, also known as the 'recipe theorem', means that any Tutte–Gröthendick invariant must be an evaluation of the Tutte polynomial, with the necessary substitutions given by the recipe (see, for example, the survey papers Ellis-Monaghan and Merino (2010a); Ellis-Monaghan and Merino 2010b). The relation between the Symanzik polynomial U_Γ and the multi-variate Tutte polynomial Z_Γ can thus be deduced from this result and Theorem 6.6.3.

The situation is similar for the second polynomial $V(p_{\text{ext}},\alpha)$. One thus has:

Theorem 6.6.5 *For any semi-regular edge e,*

$$V_\Gamma(\alpha_e, p_f) = \alpha_e V_{\Gamma - e}(\alpha_e, p_f) + V_{\Gamma/e}(\alpha_e, q_v, p_f). \tag{6.64}$$

Moreover, one has the terminal form evaluation

$$V_\Gamma(\alpha_e, p_f) = \prod_e \alpha_e \sum_v p_v^2 \tag{6.65}$$

for Γ *solely made of self-loops attached to isolated vertices.*

Let us end this chapter by mentioning that, for various other results using Grassmann techniques, the interested reader may consult Abdelmalek Abdesselam's paper Abdesselam (2004); or Sergio Caracciolo's, Alan D. Sokal and A. Sportiello's paper (2014), to name only a few.

6.7 Selected further reading

- Joel S. Feldman, Horst Knorrer, Eugene Trubowitz. (2002). *Fermionic Functional Integrals and the Renormalization Group*. CRM Monograph Series.

  A very well-written monography introducing Grassmann variables and the use of Grassmann integrals for fermionic QFT.

7

Analytic combinatorics and QFT

The first phrase of Philippe Flajolet and Robert Sedgewick's encyclopedic (2008) book on analytic combinatorics gives the reader a first glimpse of what analytic combinatorics deals with: 'Analytic combinatorics aims at predicting precisely the properties of large structural combinatorial configurations, through an approach based extensively on analytic methods'.

In the following chapter, we present how several analytic techniques, often used in combinatorics, appear naturally in various QFT issues. In the first section, we show how one can use the Mellin transform technique to re-express Feynman integrals in a useful way for the mathematical physicist. Finally, we briefly present how the saddle point approximation technique can be also used in QFT.

7.1 The Mellin transform technique

For any function $f(u)$, piecewise smooth for $u > 0$, if the integral

$$g(x) := \int_0^\infty du\, u^{-x-1} f(u) \tag{7.1}$$

is absolutely convergent for $\alpha < \text{Re}\, x < \beta$, then for $\alpha < \sigma < \beta$, one can write:

$$f(u) = \frac{1}{2\pi i} \int_{\sigma - i\infty}^{\sigma + i\infty} dx\, g(x)\, u^x. \tag{7.2}$$

Let us recall that the Mellin transform technique plays an important role in analytic combinatorics. Thus, in Philippe Flajolet and Robert Sedgewick (2008) on analytic combinatorics, several problems are treated using this versatile technique.

In order to apply this technique in QFT, one needs to use the form (3.29) of the polynomials of the parametric representation of Feynman integrals. Thus, applying the Mellin transform for e^{-u}, where $u = V_k/U$ (see (3.29), one gets:

Combinatorial Physics: Combinatorics, quantum field theory, and quantum gravity models. Adrian Tanasa, Oxford University Press (2021). © Adrian Tanasa. DOI: 10.1093/oso/9780192895493.003.0007

$$e^{-V_k/U} = \int_{\tau_k} \Gamma(-y_k) \left(\frac{V_k}{U} \right)^{y_k}, \tag{7.3}$$

where $\int_{\tau_k}$ is a short notation for $\int_{-\infty}^{+\infty} \frac{d(\text{Im } y_k)}{2\pi}$, with Re y_k fixed at $\tau_k < 0$. This equation introduces a set of parameters y_k which are referred to as *Mellin parameters*.

Let us now recall the identity:

$$\Gamma(u)\,(A+B)^{-u} = \int_{-\infty}^{\infty} \frac{d(\text{Im } x)}{2\pi} \Gamma(-x) A^x \Gamma(x+u) B^{-x-u}. \tag{7.4}$$

Taking $A \equiv U_1(x)$ and $B \equiv U_2 + U_3 + \cdots$ and using iteratively the previous identity, leads, for $u = \sum_k y_k + D/2$, to

$$\Gamma\left(\sum_k y_k + \frac{D}{2}\right) U^{-\sum_k y_k - \frac{D}{2}} = \int_\sigma \prod_j \Gamma(-x_j) U_j^{x_j}, \tag{7.5}$$

with Re $x_j = \sigma_j < 0$, Re $\left(\sum_k y_k + \frac{D}{2}\right) = \sum_k \tau_k + \frac{D}{2} > 0$, and $\int_\sigma$ stands for $\int_{-\infty}^{+\infty} \prod_j \frac{d(\text{Im } x_j)}{2\pi}$ with $\sum_j x_j + \sum_k y_k = -\frac{D}{2}$. Then, using (7.3) and (7.5), the Feynman integral of a graph denoted now by G (to avoid confusion with the Γ function) is written as

$$\mathcal{A}_G(s_k, m_l^2) = \int_\Delta \frac{\prod_j \Gamma(-x_j)}{\Gamma(-\sum_j x_j)} \prod_k s_k^{y_k} \Gamma(-y_k) \int_0^\infty \prod_l d\alpha_l\, \alpha_l^{\phi_l - 1}\, e^{-\sum_l \alpha_l m_l^2}, \tag{7.6}$$

where

$$\phi_i \equiv \sum_j u_{ij} x_j + \sum_k v_{ik} y_k + 1. \tag{7.7}$$

The symbol $\int_\Delta$ stands for integration over the independent variables $\frac{\text{Im } x_j}{2\pi}$, $\frac{\text{Im } y_k}{2\pi}$ in the convex domain Δ defined by (σ and τ standing respectively for Re x_j and Re y_k):

$$\Delta = \left\{ \sigma, \tau \left| \begin{array}{l} \sigma_j < 0;\ \tau_k < 0;\ \sum_j x_j + \sum_k y_k = -\frac{D}{2}; \\ \forall i,\ \text{Re } \phi_i \equiv \sum_j u_{ij}\sigma_j + \sum_k v_{ik}\tau_k + 1 > 0 \end{array} \right. \right\}. \tag{7.8}$$

This domain Δ is non-empty as long as D is positive and small enough so that every subgraph of G has convergent power counting de Calan and Malbouisson (1980). Hence, it is always non-empty for any graph for $0 < D < 2$, for the Φ^4 theory.

After all these manipulations, the α integrations can be *performed explicitly*. One thus has:

$$\int_0^\infty d\alpha_l\, e^{-\alpha_l m_l^2} \alpha_l^{\phi_l - 1} = \Gamma(\phi_l)\left(m_l^2\right)^{-\phi_l}. \tag{7.9}$$

The complete Mellin representation of the Feynman integral is finally written:

$$\mathcal{A}_G(s_k, m_l^2) = \int_\Delta \frac{\prod_j \Gamma(-x_j)}{\Gamma(-\sum_j x_j)} \prod_k s_k^{y_k} \Gamma(-y_k) \prod_l \left(m_l^2\right)^{-\phi_l} \Gamma\left(\phi_l\right). \tag{7.10}$$

This representation can now be extended to complex values of D, the dimension of space-time. The Mellin transform technique thus allows for a quick *proof of the meromorphy* of the Feynman integrand in D (because of the meromorphy of the Γ functions). Let us recall from Section 3.6 that this property of meromorphy is a key ingredient in the proof of the renormalization of the respective QFT model, when one uses the dimensional renormalization scheme.

Moreover, it is important to emphasize that the Mellin transform technique in QFT is the only one that provides asymptotic expansions of Feynman integrals in powers and powers of logarithms in rescaling regimes of arbitrary subsets of external variables.

7.2 The saddle point method

The saddle point method is used in combinatorics to obtain the asymptotic behaviour for the number of *involutions* (that is permutations which, when squared, lead to the identical permutation) or the number of partitions of a set of a given number of elements. This asymptotic counting information is obtained in an analytic combinatorial way, that is, using extensively generating functions (see again Philippe Flajolet and Robert Sedgewick (2008), or more precisely, its section 4). Obviously, this list of combinatorial examples is far from being exhaustive.

On the other hand, in QFT also (as in quantum mechanics), one can build up such a list. Thus, a first such example is the appearance of non-constant solutions of the classical field equation (classical here meaning that the quantum corrections that we have discussed in Chapter 3 are not taken into consideration). Under certain conditions (for more details, the interested reader may consult, for example, Jean Zinn-Justin (2002)), these solutions are called, in an Euclidian framework, *instanton solutions*. These solutions are given by the *saddle points* which are relevant for a computation by the steepest descent method.

Let us illustrate this with the zero-dimensional Φ^4 QFT example, following section 39.2 of Jean Zinn-Justin (2002). The number D of space-time dimensions is vanishing; the consequences of this are multiple: the field becomes a number, and there is no more integration in the expression (3.3) of the action (for an implementation of constructive renormalization techniques on zero-dimensional scalar QFT, the interested reader may refer to V. Rivasseau and Z. Wang (2010b)). The integral that one has to deal with in QFT (the partition function, see Section 3.1) becomes:

$$\int_{-\infty}^{\infty} d\Phi e^{-\left(\frac{m^2\Phi^2}{2} + \frac{\lambda}{4!}\Phi^4\right)}. \tag{7.11}$$

Let us first consider the case when the coupling constant λ is taken to be positive and small (the case of positive coupling, which is the case we have studied so far in this book).

One can check that the function

$$f_\lambda(\Phi) = \frac{m^2\Phi^2}{2} + \frac{\lambda}{4!}\Phi^4 \tag{7.12}$$

admits a vanishing minimum for any λ, minimum which becomes more and more narrow (for these conditions, see, for example, section 5.5 of Walter Appel (2002)).

Note that this vanishing minimum is not a soliton solution; one can then prove that the integral (7.11) is dominated by this saddle point at the origin (for details on the general proof of the saddle point approximation, the interested reader may refer, for example, to Chapter 1 of Philippe Dennery and André Krzywicki (1967)).

Non-trivial saddle points ($\pm\sqrt{\frac{m}{-\lambda}}$) are obtained for negative coupling constant. After some analysis (see again section 39.2 of Jean Zinn-Justin (2002) for details), one can prove that the integral (7.11) is dominated by these saddle points.

This instanton method generalizes to the analysis of the behaviour of QFT at large order in perturbation theory (see, for example, Chapter 42 of Jean Zinn-Justin (2002)). For the sake of completeness, let us end this chapter by mentioning that further examples for the use of the saddle point approximation method in random matrices models can be found in Margherita Disertori and Vincent Rivasseau (2008) or in Margherita Disertori's work (2011).

7.3 Selected further reading

- P. Flajolet and R. Sedgewick. (2009). Analytic Combinatorics (2015). Cambridge: Cambridge University Press.

  A book very well-known by combinatorists (it won the Leroy P. Steele Prize for Mathematical Exposition of the American Mathematical Society in 2019), but unfortunately not so well knwon by mathematical physicists. One can find, amongst many other topics, various aspects of the saddle point method treated in detail for combinatorial purposes.

8

Algebraic combinatorics and QFT

We have seen in the previous chapter some of the non-trivial interplay between analytic combinatorics and QFT. In this chapter, we illustrate how yet another branch of combinatorics, algebraic combinatorics, interferes with QFT.

Algebraic combinatorics can be defined as the domain of combinatorics which studies algebraic structures based on combinatorial objects (words, graphs, matroids, and so on). Within algebraic combinatorics, an important role is played by the so-called combinatorial Hopf algebras (CHAs), since Hopf algebras are a refined algebraic notion.

When one takes again the QFT point of view, algebraic combinatorics (and CHAs in general) play a particularly important role in describing the combinatorics of renormalization—the Connes–Kreimer algebra of Feynman graphs. Moreover, the combinatorial DSE (see Section 3.7) can be rewritten as a power series of some appropriate insertion operator of Feynman graphs, an operator which is a Hochschild one-cocycle of the Connes–Kreimer algebra, as we will show in the last section.

The Connes–Kreimer Hopf algebra, and its generalization to non-commutative Moyal QFT, see the following chapter, or to quantum gravity models, see Chapter) use a particular type of rule for defining the coproduct—a so-called **selection/contraction rule**. This means that, on the left-hand side of the coproduct, one *selects* a sub-structure (here a subgraph) and on the right-hand side one *contracts* this selected sub-structure within the initial combinatorial structure (here, the initial graph). We call this type of CHAs, **CHAs of type I.**

The CHAs used in algebraic combinatorics use a different type of rule for defining the coproduct—the **selection/deletion rule**. This means that, as previously mentioned, one *selects* a sub-structure and on the right-hand side; one *deletes* this selected sub-structure from the initial combinatorial structure. Several example of CHAs are given by the (non-commutative) (quasi-)symmetric function algebras Hopf algebras which have been extensively studied by combinatorialists (see (Duchamp et al. 2011; Novelli and Thibon 2010; Duchamp, Hivert, and Thibon 2012) and references within). We call this type of CHAs, **CHAs of type II.**

Let us also mention that the name of CHA is usually reserved by the Algebraic Combinatorics community for Hopf algebras generalizing the one of symmetric functions—this naturally includes what we call in this book CHAs of type II, but it does not include CHAs

Combinatorial Physics: Combinatorics, quantum field theory, and quantum gravity models. Adrian Tanasa, Oxford University Press (2021). © Adrian Tanasa. DOI: 10.1093/oso/9780192895493.003.0008

of type I. We chose here to refer to call CHAs any type of Hopf algebra (of type I or II, or even of a different type, not currently defined) which is implemented at the level of combinatorial objects, as already mentioned.

In this chapter, after a brief algebraic reminder, we introduce, in the second, section the Connes–Kreimer Hopf algebra of Feynman graphs and we show its relationship with the combinatoris of QFT perturbative renormalization. We then study the algebra's Hochschild cohomology in relation with the combinatorial DSE in QFT. In the fourth subsection we present a Hopf algebraic description of the so-called multi-scale renormalization (the multi-scale approach to the perturbative renormalization being the starting point for the constructive renormalization programme).

8.1 Algebraic reminder; Combinatorial Hopf Algebras (CHAs)

In this section, we briefly recall the definitions of algebraic notions that will be used later.

Definition 8.1.1 *A unital associative* **algebra** $\mathcal{A}$ *over a field* $\mathbb{K}$ *is a* $\mathbb{K}$*-linear space endowed with two vector space homomorphisms:*

- *a product* $m : \mathcal{A} \otimes \mathcal{A} \to \mathcal{A}$ *satisfying the* associativity *condition:*

$$m \circ (m \otimes \mathrm{id})(\Gamma) = m \circ (\mathrm{id} \otimes m)(\Gamma), \ \forall \Gamma \in \mathcal{A}^{\otimes 3}; \ \textit{and} \tag{8.1}$$

- *a unit* $u : \mathbb{K} \to \mathcal{A}$ *satisfying:*

$$m \circ (u \otimes \mathrm{id})(1 \otimes \Gamma) = \Gamma = m \circ (\mathrm{id} \otimes u)(\Gamma \otimes 1), \ \forall \Gamma \in \mathcal{A}. \tag{8.2}$$

Definition 8.1.2 *A* **pre-Lie algebra** $\mathcal{A}$ *over a field* $\mathbb{K}$ *is a* $\mathbb{K}$*-linear space endowed with a bilinear map* $\star$ *satisfying:*

$$(\Gamma_1 \star \Gamma_2) \star \Gamma_3 - \Gamma_1 \star (\Gamma_2 \star \Gamma_3) = (\Gamma_1 \star \Gamma_3) \star \Gamma_2 - \Gamma_1 \star (\Gamma_3 \star \Gamma_2), \ \forall \Gamma_i, \ i = 1,2,3. \tag{8.3}$$

Definition 8.1.3 *A (coassociative, counital)* **coalgebra** $\mathcal{C}$ *over a field* $\mathbb{K}$ *is a* $\mathbb{K}$*-linear space endowed with two linear homomorphisms:*

- *a coproduct* $\Delta : \mathcal{C} \to \mathcal{C} \otimes \mathcal{C}$ *satisfying the* coassociativity *condition:*

$$\forall \Gamma \in \mathcal{C}, \ (\Delta \otimes \mathrm{id}) \circ \Delta(\Gamma) = (\mathrm{id} \otimes \Delta) \circ \Delta(\Gamma); \ \textit{and} \tag{8.4}$$

- *a counit* $\varepsilon : \mathcal{C} \to \mathbb{K}$ *satisfying:*

$$\forall \Gamma \in \mathcal{C}, \ (\varepsilon \otimes \mathrm{id}) \circ \Delta(\Gamma) = \Gamma = (\mathrm{id} \otimes \varepsilon) \circ \Delta(\Gamma). \tag{8.5}$$

Definition 8.1.4 *A* **bialgebra** $\mathcal{B}$ *over a field* $\mathbb{K}$ *is a* $\mathbb{K}$*-linear space endowed with both an algebra and a coalgebra structure (see Definitions 8.1.1 and 8.1.3), such that the coproduct and the counit are unital algebra homomorphisms (or equivalently the product and unit are coalgebra homomorphisms):*

$$\Delta \circ m_{\mathcal{B}} = m_{\mathcal{B}\otimes\mathcal{B}} \circ (\Delta \otimes \Delta),\ \Delta(1_{\mathcal{B}}) = 1_{\mathcal{B}} \otimes 1_{\mathcal{B}}, \tag{8.6a}$$
$$\varepsilon \circ m_{\mathcal{B}} = m_{\mathbb{K}} \circ (\varepsilon \otimes \varepsilon),\ \varepsilon(1_{\mathcal{B}}) = 1. \tag{8.6b}$$

Let us also note that the problem of Sweedler's duals for bialgebras as essentially characterizing the domain of the transpose of the multiplication was described in Duchamp and Tollu (2007).

Definition 8.1.5 *A* **graded bialgebra** *is a bialgebra graded as a linear space:*

$$\mathcal{B} = \bigoplus_{n=0}^{\infty} \mathcal{B}^{(n)} \tag{8.7}$$

such that the grading is compatible with the algebra and coalgebra structures:

$$\mathcal{B}^{(n)}\mathcal{B}^{(m)} \subseteq \mathcal{B}^{(n+m)} \text{ and } \Delta\mathcal{B}^{(n)} \subseteq \bigoplus_{k=0}^{n} \mathcal{B}^{(k)} \otimes \mathcal{B}^{(n-k)}. \tag{8.8}$$

Definition 8.1.6 *A* **connected bialgebra** *is a graded bialgebra* $\mathcal{B}$ *for which* $\mathcal{B}^{(0)} = u(\mathbb{K})$.

Definition 8.1.7 *A* **Hopf algebra** $\mathcal{H}$ *over a field* $\mathbb{K}$ *is a bialgebra over* $\mathbb{K}$ *equipped with an antipode map* $S : \mathcal{H} \to \mathcal{H}$ *obeying:*

$$m \circ (S \otimes \mathrm{id}) \circ \Delta = u \circ \varepsilon = m \circ (\mathrm{id} \otimes S) \circ \Delta. \tag{8.9}$$

The antipode of a Hopf algebra is the inverse of the unit with respect to the coproduct.

A practical introduction to Hopf algebras for a combinatorial physicist can be found in Duchamp et al. (2008). For further details on this topic, the interested reader can refer to Christian Kassel (1995).

Let us end this section with the following definition regarding Hochschild cohomology.

Let A be a bialgebra and Δ its coproduct. We regard linear maps

$$L : A \to A^{\otimes n}$$

as n-cochains. We define a coboundary map b,

$$b^2 = 0,$$

by

$$bL := (id \otimes L) \circ \Delta + \sum_{i=1}^{n} (-1)^i \Delta_i \circ L + (-1)^{n+1} L \otimes 1_A, \tag{8.10}$$

where Δ_i denotes the coproduct applied to the i-th factor in $A^{\otimes n}$. This defines the cohomology of A.

8.2 The Connes–Kreimer Hopf algebra of Feynman graphs

We introduce in this section the Connes–Kreimer Hopf algebra of Feynman graphs and show its relation to the combinatorics of perturbative renormalization. Several other reviews on this topic exist in the literature (we direct, for example, the interested reader to section 1.6 of Alain Connes and Matilde Marcolli (2008); or to the review articles (Tanasa 2010b, 2012a).

We consider $\mathcal{H}$ the unital associative algebra freely generated by 1PI Feynman graphs of the Φ^4 model (including the empty graph which we denote by 1). The product m, given by operation of disjoint union, is bilinear and commutative.

Let us now give a formal, algebraic, definition of the Feynman rules introduced in Chapter 3:

Definition 8.2.1 ***Feynman rules*** *are a homomorphism ϕ from $\mathcal{H}$ to some target space $\mathcal{A}$.*

For the sake of completeness, let us also remark that this target algebra is naturally equipped with some Rota–Baxter algebraic structure. Recently, the algebraic implications one has when this condition is relaxed were studied in detail in Ebrahimi-Fard and Patras (2010, 2011).

We now recall, also from Chapter 3, that, in the particular case of the commutative Φ^4 model, the class of primitively divergent graphs is formed by the graphs with two or four external edges that do not contain as a subgraph any graph with two or four external edges.

Definition 8.2.2 *The* ***projection*** *R is a map from $\mathcal{A}$ to $\mathcal{A}$ which has to fulfil: $\forall \Gamma \in \mathcal{H}$, Γ divergent,*

$$(\mathrm{id}_{\mathcal{A}} - R) \circ \phi(\Gamma) < \infty. \tag{8.11}$$

Remark 8.2.3 *This means that if $\phi(\Gamma)$ is superficially divergent then its overall divergence is entirely included in $R \circ \phi(\Gamma)$.*

This is a formal definition of the subtraction operator R of QFT (See Subsection 3.6.4). Thus, for renormalization in position space, it represents the Taylor subtraction operator τ.

We denote from now on by $\underline{\Gamma}$ the set of two and four external edges subgraphs of a generic graph Γ, i.e. the superficially divergent subgraphs of Γ.

Lemma 8.2.4 *(Lemma 3.2 of Tanasa and Vignes-Tourneret (2008)). Let $\Gamma \in \mathcal{H}$. Provided:*

1. $\forall \gamma \in \underline{\Gamma}, \forall \gamma' \in \underline{\gamma}, \gamma, \gamma'$ *and* γ/γ' *are superficially divergent; and*
2. $\forall \gamma_1 \in \mathcal{H}, \forall \gamma_2 \in \mathcal{H}$ *such that* γ_1 *and* γ_2 *superficially divergent, there exists gluing data* G *(see the following Definition 8.2.5) such that* $(\gamma_1 \circ_G \gamma_2)$ *superficially divergent,*

the following coproduct is coassociative:

$$\Delta\Gamma = \Gamma \otimes 1 + 1 \otimes \Gamma + \Delta'\Gamma, \tag{8.12a}$$

$$\Delta'\Gamma = \sum_{\gamma \in \underline{\Gamma}} \gamma \otimes \Gamma/\gamma. \tag{8.12b}$$

Note that we have denoted by Γ/γ the cograph obtained by shrinking the subgraph γ in the graph Γ (see Definition 3.6.7). Furthermore, we denote by

$$\Delta'(\Gamma) = \Delta(\Gamma) - \Gamma \otimes 1 - 1 \otimes \Gamma, \ \forall \Gamma,$$

the non-trivial part of the coproduct Δ.

One has the following definition:

Definition 8.2.5 Gluing data *represents a bijection between the external edges of the graph to be inserted and the edges of the propagator (or vertex) where the insertion is done.*

Let us now illustrate how this works for the commutative Φ^4 model presented in this book. The first condition of the lemma is trivial, since when one shrinks a subgraph, the number of external edges is conserved. Let us also mention that since the subgraph to be shrinked has two or four external edges, the resulting cograph is still a Φ^4 graph. Let us now check condition 2 of Lemma 8.2.4. We consider two graphs γ_1 and γ_2 with 2 or 4 external edges; furthermore, we consider $\gamma_0 = \gamma_1 \circ_G \gamma_2$ for any gluing data G. Let V_i, I_i and E_i the respective numbers of vertices, internal and external edges of γ_i, $i \in \{0,1,2\}$. For all $i \in \{0,1,2\}$, we have

$$4V_i = 2I_i + E_i \tag{8.13a}$$

$$V_0 = \begin{cases} V_1 + V_2 & \text{if } E_2 = 2 \\ V_1 + V_2 - 1 & \text{if } E_2 = 4 \end{cases} \tag{8.13b}$$

$$I_0 = \begin{cases} I_1 + I_2 + 1 & \text{if } E_2 = 2 \\ I_1 + I_2 & \text{if } E_2 = 4, \end{cases} \tag{8.13c}$$

which proves that $E = E_1$. Then as soon as γ_1 is superficially divergent, so is γ_0. Concerning condition 1, note that $\gamma'' = \gamma/\gamma' \iff$ it exists G s. t. $\gamma = \gamma'' \circ_G \gamma'$ which allows to prove that condition 1 also holds and that the coproduct (8.12) is coassociative.

Let now the coproduct $\Delta : \mathcal{H} \to \mathcal{H} \otimes \mathcal{H}$ defined as

$$\Delta\Gamma = \Gamma \otimes 1 + 1 \otimes \Gamma + \sum_{\gamma \in \underline{\Gamma}} \gamma \otimes \Gamma/\gamma, \; \forall \Gamma \in \mathcal{H}. \tag{8.14}$$

Furthermore, let us define the counit $\varepsilon : \mathcal{H} \to \mathbb{K}$:

$$\varepsilon(1) = 1, \; \varepsilon(\Gamma) = 0, \; \forall \Gamma \neq 1. \tag{8.15}$$

This means that, for any non-trivial element of $\mathcal{H}$, the counit returns a trivial answer and, equivalently, the result returned by the counit is only non-trivial for the trivial element of $\mathcal{H}$ (the empty graph 1). Finally, the antipode is given recursively by

$$\begin{aligned} S : \mathcal{H} &\to \mathcal{H} \\ S(1) &= 1, \\ \Gamma &\mapsto -\Gamma - \sum_{\gamma \in \underline{\Gamma}} S(\gamma)\Gamma/\gamma. \end{aligned} \tag{8.16}$$

One then has the following:

Theorem 8.2.6 *(Theorem* 1 *of Connes and Kreimer (2000)). The quadruple* $(\mathcal{H}, \Delta, \varepsilon, S)$ *is a Hopf algebra.*

Let us also notice that $\mathcal{H}$ is graded by

1. the number of loops; and
2. number of edges; or
3. number of vertices minus one.

For a proof of this, we refer the interested reader to Proposition 1.31 of Alain Connes and Matilde Marcolli (2008).

Now, let $f, g \in \mathrm{Hom}(\mathcal{H}, \mathcal{A})$ where $\mathcal{A}$ is the range algebra of the projection T (see previous example). The **convolution product** $*$ in $\mathrm{Hom}(\mathcal{H}, \mathcal{A})$ is defined by

$$f * g := m_{\mathcal{A}} \circ (f \otimes g) \circ \Delta_{\mathcal{H}}. \tag{8.17}$$

Let $\phi_- \in \mathrm{Hom}(\mathcal{H}, \mathcal{A})$ the **twisted antipode**, defined recursively as: $\forall \Gamma \in \mathcal{H}$,

$$\phi_-(\Gamma) := -R\big(\phi(\Gamma)\big) + R\big(\sum_{\gamma \in \underline{\Gamma}} \phi_-(\gamma)\, \phi(\Gamma/\gamma)\big). \tag{8.18}$$

The renormalized Feynman integral ϕ_{ren} of a graph $\Gamma \in \mathcal{H}$ is given by:

$$\phi_{\text{ren}}(\Gamma) = \phi_- * \phi(\Gamma). \tag{8.19}$$

Moreover, Bogoliubov's subtraction operator is given by (see, for example Kreimer (2007) for details):

$$\bar{R}(\Gamma) = \phi(\Gamma) + \sum_{\gamma \in \underline{\Gamma}} \phi_-(\gamma)\phi(\Gamma/\gamma). \tag{8.20}$$

Let us now illustrate this with the example of the divergent Feynman graph of Fig. 3.3. The expression (8.20) becomes:

$$\bar{R}(\Gamma) = \phi(\Gamma) + \phi_-(\gamma)\phi(\Gamma/\gamma), \tag{8.21}$$

where the only sub-divergent graph γ is given by the internal edges 1 and 2 (see again Fig. 3.3.). The twisted antipode (defined in (8.18)) gives

$$\phi_-(\gamma) = -R(\phi(\gamma)), \tag{8.22}$$

(since the subgraph γ is primitive). Inserting the expression (8.22) of the twisted antipode in (8.21) finally gives

$$\bar{R}(\Gamma) = \phi(\Gamma) - R(\phi(\gamma))\phi(\Gamma/\gamma), \tag{8.23}$$

which is, as announced, the same expression as (3.50), the expression computed using non-algebraic formulae.

For the sake of completeness, let us mention that a combinatorial Hopf algebraic structure like the one described in this section, can also be defined for more involved QFT models—gauge theories (see (Kreimer 2006; van Suijlekom 2007)).

Let us also emphasize that the operation of insertion of graphs leads to a *pre-Lie algebra structure* on the linear space of 1PI Feynman graphs. Antisymmetrizing the pre-Lie product gives a Lie bracket, thus defining a Lie structure on the same space. Let us now consider the graded dual of the universal enveloping algebra of this Lie structure. This gives, through the *Milner–Moore Theorem*, the renormalization Connes–Kreimer Hopf algebra described in this section.

Finally, let us also state that a slightly different combinatorial Hopf algebra—the *core Hopf algebra*—was defined in Bloch and Kreimer (2008) (and independently in Krajewski and Martinetti (2011) for vacuum graphs). Its coproduct does not sum over the class of superficially divergent subgraphs but on *all* subgraphs of the respective graph:

$$\Delta\Gamma = \Gamma \otimes 1 + 1 \otimes \Gamma + \sum_{\gamma \subset \Gamma} \gamma \otimes \Gamma/\gamma, \ \forall \Gamma \in \mathcal{H}. \tag{8.24}$$

The definitions of the coproduct and of the counit follow in a straightforward manner. One can then verify that this structure is also a Hopf algebra. The cohomology of this core Hopf algebra was also investigated in detail (see (Kreimer 2010; Kreimer and van Suijlekom 2009)).

This core algebra contains the renormalization Hopf algebra as a quotient algebra. Moreover, it can be seen as the Connes–Kreimer Hopf algebra of a QFT model formulated in infinite dimension (see (Kreimer 2010; Kreimer and van Suijlekom 2009) for more details).

The Hopf algebra of graphs (where the coproduct is defined as a sum over all the subgraphs of the respective graph) generalizes in an elegant manner to a Hopf algebra of **matroids** (see (Schmitt 1994; Crapo and Schmitt 2005)). Let us note that matroids are a combinatorial notion which generalizes the notion of graphs or of matrices (see James Oxley (1992) for a pedagogical introduction to the subject). A matroid is defined in the following way:

Definition 8.2.7 *A* **matroid** M *is a pair* $(E,\mathcal{I})$ *consisting of a finite set* E *and a collection of subsets of* E *satisfying:*

(I1) $\mathcal{I}$ *is non-empty.*
(I2) Every subset of every member of $\mathcal{I}$ *is also in* $\mathcal{I}$.
(I3) If X *and* Y *are in* $\mathcal{I}$ *and* $|X| = |Y| + 1$, *then there is an element* x *in* $X - Y$ *such that* $Y \cup \{x\}$ *is in* $\mathcal{I}$.

8.3 The B_+ operator, Hochschild cohomology of the Connes–Kreimer algebra

In this section, following Kreimer (2006), (for a detailed presentation of all these aspects, the interested reader may also refer to the PhD thesis of Karen Yeats (2006)) we define the B_+ operator and its relation to the combinatorial DSE. In order to do this, we first need to define some appropriate *combinatorial factors for Feynman graphs*:

Definition 8.3.1 *The* **symmetry factor** $\mathrm{sym}(\gamma)$ *of a graph* Γ *is defined as the rank of the automorphism group of* Γ.

Definition 8.3.2 *The* **number of maximal forests** maxf *of a graph* Γ *is the number of ways to shrink subdivergencies such that the resulting cograph is primitive.*

Furthermore, we denote by

$$|\Gamma|_V$$

the number of distinct graphs Γ which are equal upon removal of external edges. Note that these graphs can be obtained from each other through permutations of the external edges.

Let now γ, X, and Γ be some Feynman graphs of our QFT model. We denote by

$$\mathbf{bij}(\gamma, X, \Gamma)$$

the number of bijections between the external edges of the graph X and adjacent edges of places in γ such that the graph Γ is obtained.

Finally, we denote by

$$[\gamma|X]$$

the number of such insertion places of the graph X in the graph γ.

Let us define the map

$$B_+^{k;r} := \sum_{\substack{|\gamma|=k \\ \mathrm{res}(\gamma)=r}} \frac{1}{\mathbf{sym}(\gamma)} B_+^{\gamma}, \tag{8.25}$$

where

$$B_+^{\gamma}(X) := \sum_{\Gamma} \frac{\mathbf{bij}(\gamma, X, \Gamma)}{|X|_\vee} \frac{1}{\mathrm{maxf}(\Gamma)} \frac{1}{[\gamma|X]} \Gamma, \tag{8.26}$$

and we have denoted by $|\gamma|$ the number of loops of the graph γ. In order for this definition to be complete, we set: $B_+^{\gamma}(1_{\mathcal{H}}) = \gamma$.

This B_+ operator thus generalizes the insertion operator used to define a pre-Lie algebra structure on the vector space of 1PI Feynman graphs (see previous section).

One can then prove that such an operator is a Hochschild $1-$cocycle:

Theorem 8.3.3 *(Theorem* 5 *ii) of Kreimer (2006))*

$$\Delta B_+^{k;r} = B_+^{k;r} \otimes 1_{\mathcal{H}} + (\mathrm{id} \otimes B_+^{k;r})\Delta. \tag{8.27}$$

Furthermore, one can prove the following:

Theorem 8.3.4 *(Theorem* 5 *i) of Kreimer (2006)). Any relevant graph in the perturbative expansion (with the right combinatorial factor) is in the range of a Hochschild one-cocycle* B_+.

The DSE can thus be seen as a formal construction based on the Hochschild cohomology of the renormalization Connes–Kreimer Hopf algebra.

Let us end this section by defining the sum of graphs of a given residue r and a given loop number k

$$c_k^r = \sum_{|\Gamma|=k} \Gamma. \tag{8.28}$$

One then has:

Theorem 8.3.5 *(Theorem 5 iii) of Kreimer (2006))*

$$\Delta(c_k^r) = \sum_{j=0}^{k} \mathrm{Pol}_j^r(c) \otimes c_{k-j}^r, \tag{8.29}$$

where $\mathrm{Pol}_j^r(c)$ *is a polynomial in the variables* c_m^r *of degree* j.

This last result ensures that the elements c_k^r form a Hopf subalgebra; this was used in Broadhurst and Kreimer (2001) to obtain some exact solutions of the DSE. More precisely, for the Yukawa theory in a four-dimensional space-time, an integro-differential DSE was obtained and solved in terms of the error function. A similar numerical result was obtained for a scalar theory in a six-dimensional space-time. Nevertheless, let us also mention here that these methods did not take into consideration the non-trivial choice of insertion places of subgraphs (see Broadhurst and Kreimer (2001) for more details).

For various other results on these combinatorial DSE in the Hopf algebra of decorated rooted trees, one may also refer to L. Foissy's papers (Foissy 2012, 2010). We also refer the interested reader to K. Yeats (2006).

8.4 Multi-scale renormalization, CHA description

In this section we present the implementation of a Hopf algebraic description of the combinatorics of multi-scale renormalization (see Subsection 3.6.3), description recently realized in Krajewski, Rivasseau, and Tanasa (2010).

For this purpose, we need to replace graphs with *assigned graphs*, that is graphs which have an integer associated to each edge, namely its energy scale (see again Subsection 3.6.3).

We this give the following definition:

Definition 8.4.1 *A* **scale assignment** μ *for a Feynman graph with labelled internal edges is a list of positive integers* i_ℓ, $\ell = 1, \ldots, E$ *associated to the internal edges of the respective Feynman graph.*

Let us emphasize here that the integers in the Definition 8.4.1 are bounded by some discrete cut-off ρ. One further has the following definition:

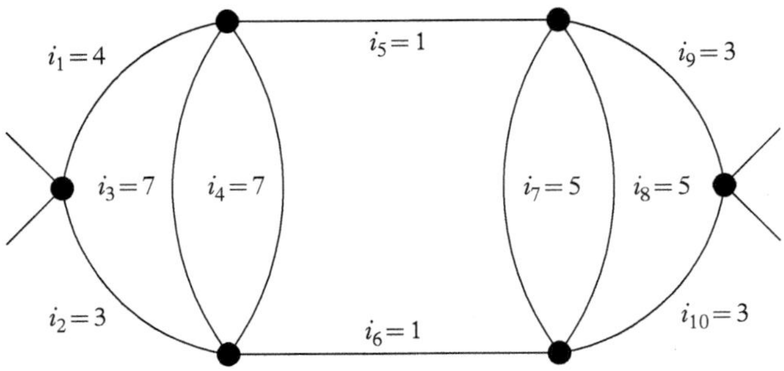

Figure 8.1 *A Feynman graph with a scale assignment*

Definition 8.4.2 *An* **assigned graph** (G,μ) *is an isomorphism class of couples formed by the one particle irreducible (1PI) edge labelled Feynman graph G, together with a scale assignment* μ.

An example of such an assigned graph is given in Fig. 8.1.

Obviously, assigned graphs correspond to graphs whose edges are labelled by scales; they can be seen as a particular class of decorated graphs.

Definition 8.4.3 *An* **assigned subgraph** (g,ν) *of a given assigned graph* (G,μ) *is constructed in the following way (see previous section). One considers a subgraph* g *of the Feynman graph* G*, in the usual QFT way. The scale assignment* ν *of* G *is given by the restriction of the scale assignment* μ *to the internal edges of* g *(which are also internal edges of* G*). Moreover, the external edges of* g *which are internal edges of* G *have the scale assignment which is attributed to them by* μ*. The same holds for the external edges of* g *which are external edges of* G.

Furthermore, one can define the usual graph theoretical notions (number of edges, vertices, (independent) loops etc.) of an assigned graph (G,μ) as the respective notions of the Feynman graph G. Moreover, we call (G,μ) an N-point assigned graph if G is an N-point 1PI Feynman graph.

We then define the internal and external index for a subgraph (g,ν) of an assigned graph (G,μ) as:

$$i_g(\mu) = \inf_{l\in g} \mu(l) \tag{8.30}$$

$$e_g(\mu) = \sup_{l \text{ external edge of } g} \mu(l) \tag{8.31}$$

(with the μ dependence sometimes omitted for shortness).

Let us now give the following definition:

Definition 8.4.4 *Let* (G,μ) *an assigned graph. We say that a subgraph* (g,ν) *is a* **high subgraph** *if*

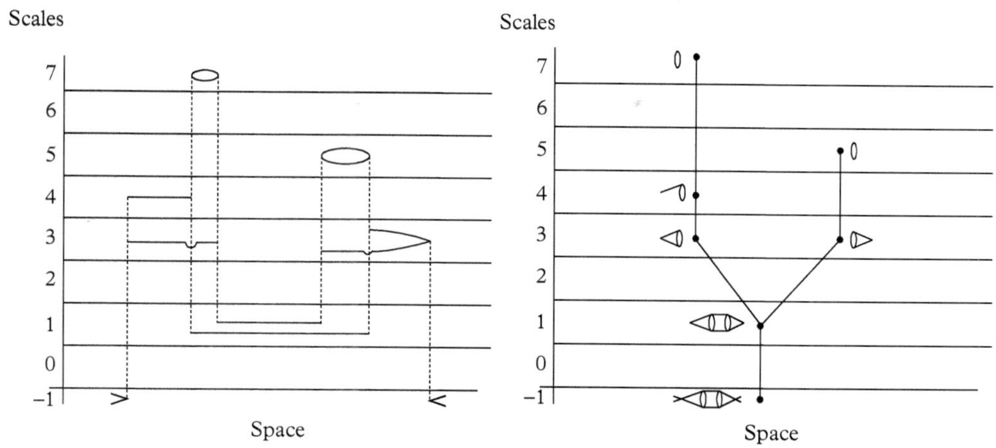

(a) Phase-space representation.
(b) Gallavotti–Nicolò tree.

Figure 8.2 *Phase-space representation of the graph of Fig. 8.1 and the associated Gallavotti–Nicolò tree*

- *g is connected; and*
- *the internal index of g is higher than its external index:*

$$e_g(\mu) < i_g(\mu) \qquad \text{(high condition).} \tag{8.32}$$

One can associate to a connected assigned graph (G,μ), the Gallavotti–Nicolò tree $T_{(G,\mu)}$, which is defined in the following way (see V. Rivasseau (1992) for more details):

Definition 8.4.5 *The Gallavotti–Nicolò tree $T_{(G,\mu)}$ is a rooted tree whose nodes at a distance i from the root are decorated with the connected high subgraphs G^i_c with scales $\geq i$ and whose arrows join the nodes decorated with G^i_c and $G^{i-1}_{c'}$ if, and only if, G^i_c is a high subgraph of $G^{i-1}_{c'}$.*

In order to represent the Gallavotti–Nicolò tree, it is convenient to adopt a phase-space representation with positions on the horizontal axis and scales on the vertical one. The phase space representation of the graph of Fig. 8.1 and its Gallavotti–Nicolò tree can be found on Fig. 8.2.

High subgraphs are partially ordered by inclusion. An essential result is that they form a Zimmermann forest in the following sense.

Lemma 8.4.6 *Let (G,μ) be a fixed graph and scale assignment. The set of high subgraphs is a forest, in the sense that if g_1 and g_2 are both high, we have either $g_1 \subset g_2$, or $g_2 \subset g_1$, or $g_1 \cap g_2 = \emptyset$.*

Proof Suppose we could find S_1 and S_2 with a non-trivial intersection; in this case S_1 would have an external edge in S_2 and conversely; but the scale of any of these two edges should be both strictly larger and strictly smaller than the other, which leads to a contradiction (QED).

In general, the final graph G is connected, and this inclusion forest of high subgraphs forms a tree; the so-called Gallavotti–Nicolò tree (the interested reader may refer to the article of Gallavotti and Nicolò (1985)).

Hopf algebra structures on the Gallavotti–Nicolò trees

One has:

Proposition 8.4.7 *Let $T_{(G,\mu)}$ be the Gallavotti–Nicolò tree associated with the assigned graph (G,μ).*

1. *The root of $T_{(G,\mu)}$ is decorated with G itself.*
2. *The leaves of $T_{(G,\mu)}$ all are at distance ρ from the root, with ρ the ultraviolet cut-off.*
3. *If the scales j and k are such that the scales i obeying $j \leq i \leq k$ do not appear in (G,μ), then any subtree of $T_{(G,\mu)}$ whose root is a at distance j and leaves at a distance k from the root of $T_{(G,\mu)}$ does not branch and has all its nodes decorated by the same graph.*

Proof The previously mentioned items follow as a direct consequence of the Definition 8.4.5 of the Gallavotti–Nicolò trees.

In order to define the Hopf algebra underlying multiscale renormalization on the Gallavotti–Nicolò trees, it is useful to introduce the following terminology. If $T' \subset T_{(G,\mu)}$ is a subtree, we define its completion $\overline{T'} = T_{(G',\mu')}$ to be the Gallavotti–Nicolò tree associated with its root (G',μ'). Furthermore, we define an admissible cut C to be a non-empty subset of $|C|$ arrows of $T_{(G,\mu)}$ that join nodes decorated by two different graphs, the graph farther form the root having two of four external edges and such that any path form the leaves to the root contains at most one arrow in C. Removing the arrows in C we get a subtree $T_<$ that contain the root and trees $T^n_>$ that do not contain the root.

Proposition 8.4.8 *The free commutative algebra $\mathcal{H}_{\text{GN}}$ generated by all Gallavotti–Nicolò trees is a graded Hopf algebra whose counit and coproduct are defined on the generators by $\epsilon(T) = 0$ and*

$$\Delta(T) = T \otimes 1 + 1 \otimes T + \sum_{\substack{C \\ \text{admissible cut}}} \Big(\prod_{1 \leq n \leq |C|} \overline{T^n_>} \Big) \otimes \overline{T_<}. \tag{8.33}$$

Its grading is $n(T) = \#\{\text{arrows joining nodes decorated with different graphs}\} + 1$.

Remark 8.4.9 *Let us observe that the number $n(T)$ given previously is simply the number of vertices, counting those decorated by the same graph only once per connected component.*

Proof The only non-trivial assertions to check are the coassociativity of the coproduct and the existence of the grading and the antipode. The first proof is analogous to the proof of the coassociativity of the coproduct in the algebra of rooted trees, see Connes and Kreimer (1998). The assertion pertaining to the grading is easy to check as any cut reduces the number of arrows joining nodes with different graphs by the number of cut edges. Finally, for any graded commutative bigebra there is a recursive construction of the antipode, as given in Figueroa and Gracia-Bondia (2001).

Operations on assigned graphs

We now define $\underline{(G,\mu)}$ to be the set of assigned graphs formed by high subgraphs of the assigned graph (G,μ) whose connected components are 1PI and have two or four external edges. An external edge of a subgraph is an edge of G attached to a vertex in g which is not an internal edge of g.

Definition 8.4.10 *The shrinking of a two- or a four-point assigned subgraph (g,ν) inside an assigned graph (G,μ) is defined in the following way. The shrinking of the subgraph g inside the Feynman graph G is done in the usual QFT way, i.e. each connected component of the subgraph is replaced by a vertex (the internal structure of g vanishes); one has the cograph G/g. The scale assignment μ/ν of the cograph G/g is given by the initial scale assignment μ, where we have erased the scale assignment of the internal edges of g (if two external edges are added when shrinking a two-point function, they are assigned a non-dangerous integer). We call the resulting assigned graph $(G/g,\mu/\nu)$ an assigned cograph.*

Remark 8.4.11 *The shrinking operation corresponds to the wave function or mass renormalization, for a two-point subgraph, or to the coupling constant renormalization for a four-point subgraph. In the case of a wave-function renormalization, a decoration indicating the two derivative couplings of the Laplacian must be added to the shrinked two-point vertex to distinguish that renormalization from the mass renormalization.*

Definition 8.4.12 *The gluing data $\circ$ for the insertion of a two- and respectively four-point assigned graph (g,ν) into the propagator and respectively the vertex of an assigned graph (G,μ) is given by a bijection between the external edges of g and the two half-edges of the propagators or respectively the four half-edges of the vertex. It is defined only if the external assignment indices for (g,ν) coincide with the internal indices of the corresponding edges of (G,μ). In that case, the scale assignment of the resulting graph is obtained in the following way. The scale assignment for the internal edges of g are given by ν; the scale assignment for the external edges of g, identified through this operation to internal edges of G are given by their common value in (G,μ) and (g,ν).*

The assigned graph combinatorial Hopf algebra

In this section we define a Hopf algebra on assigned Feynman graphs and we then show the relation between this structure and the combinatorics of multi-scale renormalization.

Consider now the unital associative algebra $\mathcal{H}$ freely generated by the assigned graphs, including the empty assigned graph, which we denote by $1_{\mathcal{H}}$.

The product $m((g_1,\mu_1)(g_2,\mu_2)) = (g,\mu)$ is given by the operation of disjoint union of assigned graphs. This means that the resulting 1PI Feynman graph g is given by the disjoint union of graphs and each disjoint component g_i keeps its scale assignment μ_i $(i = 1,2)$—this gives the resulting scale assignment μ. As in the case of the Connes–Kreimer product, this product is bilinear and commutative.

As we have already mentioned, the integers of the scale assignment μ are bounded by integer cut-off ρ. One has $\mathcal{H}_\rho \subset \mathcal{H}_{\rho+1} \subset \ldots \subset \mathcal{H}_\infty$. Since we do not deal here with distinct cut-offs, we denote $\mathcal{H}_\rho$ by $\mathcal{H}$ in the sequel.

Let us now define the coproduct $\Delta : \mathcal{H} \to \mathcal{H} \otimes \mathcal{H}$ as

$$\Delta(G,\mu) = (G,\mu) \otimes 1_{\mathcal{H}} + 1_{\mathcal{H}} \otimes (G,\mu) + \sum_{(g,\nu)\subset \underline{(G,\mu)}} (g,\nu) \otimes (G/g,\mu/\nu). \tag{8.34}$$

Note that, in this definition, the high subgraphs g are not necessarily connected. The coproduct can be written explicitly as

$$\Delta(G,\mu) = (G,\mu) \otimes 1 + \sum_{\substack{(g_i,\nu_i)\subset(G,\mu)\\ g_i\cap g_i=\emptyset}} \left(\prod_i (g_i,\nu_i) \otimes (G,\mu)\right) \Big/ \prod_i (g_i,\nu_i) + 1 \otimes (G,\mu), \tag{8.35}$$

where the sum runs over divergent and disjoint high 1PI subgraphs, excluding G itself.

In order to illustrate the definition of the let us list all the connected 1PI high superficially divergent subgraphs (i.e. high two- or four-point subgraphs) of the graph of Fig. 8.1,

$$\{1,2,3,4\}, \{7,8,9,10\}, \{3,4\}, \{7,8\}. \tag{8.36}$$

Therefore, the coproduct reads, omitting the explicit expression of the scale assignment,

$$\begin{aligned}\Delta(G) = G \otimes 1_{\mathcal{H}} + 1_{\mathcal{H}} \otimes G &+ \{1,2,3,4,7,8,9,10\} \otimes G/\{1,2,3,4,7,8,9,10\}\\ &+ \{1,2,3,4\} \otimes G/\{1,2,3,4\} + \{7,8,9,10\} \otimes G/\{7,8,9,10\}\\ &+ \{1,2,3,4,7,8\} \otimes G/\{1,2,3,4,7,8\} + \{3,4,7,8,9,10\} \otimes G/\{3,4,7,8,9,10\}\\ &+ \{3,4\} \otimes G/\{3,4\} + \{7,8\} \otimes G/\{7,8\} + \{3,4,7,8\} \otimes G/\{3,4,7,8\}.\end{aligned} \tag{8.37}$$

For example, the reduced graph $G/\{7,8,9,10\}$ is illustrated in Fig. 8.3.

Observe that the vector space $\mathcal{H}$ is graded, as in the usual Connes–Kreimer case, by the number of independent loops, number of edges, or by the number of vertices minus one.

Lemma 8.2.4 holding for the combinatorial Connes–Kreimer Hopf algebra of Feynman graphs naturally generalizes for assigned graphs:

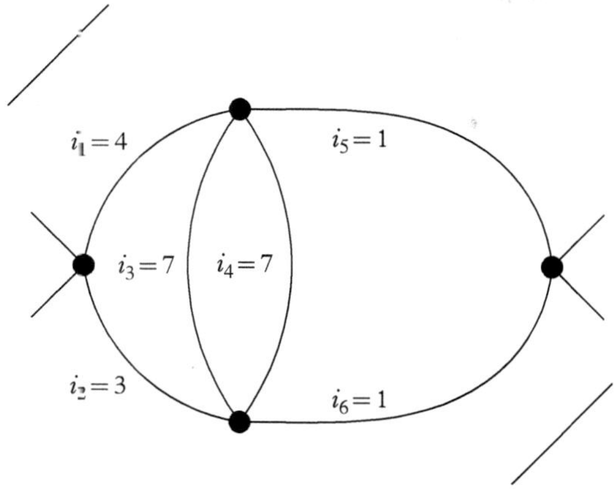

Figure 8.3 *The reduced graph* $G/\{7,8,9,10\}$

Lemma 8.4.13 *Let* (G,μ) *an assigned graph. Provided:*

1. $\forall (g,\nu) \in \underline{(G,\mu)}, \forall (g',\nu') \in \underline{(g,\nu)}$, *the assigned cograph* $(G/G',\nu/\nu')$ *is a two- or four-point high-assigned graph; and*
2. $\forall (g_1,\nu_1),(g_2,\nu_2)$ *such that* (g_1,ν_1) *and* (g_2,ν_2) *are two- or four-point high-assigned graphs, there exists gluing data such that* $(g_1 \circ g_2, \nu_1 \circ \nu_2)$ *is a two- or four-point high-assigned graph,*

the coproduct given by formula (8.34) *is coassociative.*

Proof The proof of Lemma 3.2 of Tanasa and Vignes-Tourneret (2008) generalizes in a straightforward manner.

Furthermore, we define the counit $\varepsilon : \mathcal{H} \to \mathbb{K}$ as

$$\varepsilon(1_{\mathcal{H}}) = 1,\ \varepsilon((G,\mu) = 0,\ \forall (G,\mu) \neq 1_{\mathcal{H}}. \tag{8.38}$$

Finally, the antipode is given recursively by

$$\begin{aligned} S : \mathcal{H} &\to \mathcal{H} \\ (G,\mu) &\mapsto -(G,\mu) - \sum_{(g,\nu)\in \underline{(G,\mu)}} S((g,\nu))(G/g,\mu/nu). \end{aligned} \tag{8.39}$$

This antipode can be computed, as the inverse for the convolution of the identity map. In the case of the Connes–Kreimer Hopf algebra of trees, this was done in Figueroa and Gracia-Bondia (2001). For the graph algebra defined here, an analogous computation leads to

$$S(G,\mu) = \sum_{n=1}^{L} \sum_{i} c_{n-1}^{(i)} (-(G_1^{(i)}, \mu_1^{(i)})) \dots (-(G_n^{(i)}, \mu_n^{(i)})), \tag{8.40}$$

where

$$\Delta'^n(G,\mu) = \sum_i c_n^{(i)} (G_1^{(1)}, \mu_1^{(i)}) \otimes \dots \otimes (G_{n+1}^{(i)}, \mu_{n+1}^{(i)}). \tag{8.41}$$

and $c_n^{(i)}$ are the appropriate combinatorial factors obtained from the explicit coproduct computation. Observe that, from the explicit definition of the unit and of the counit map, one can prove that the sum over n in (8.40) has a finite number of terms, equal to the number of independent cycles of the respective graph, denoted here by L. The non-recursive formula (8.40) is then equivalent to the sum over Zimmermann forests of high superficially divergent graphs.

We can now state the main result of this section:

Theorem 8.4.14 *The quadruple* $(\mathcal{H}, \Delta, \varepsilon, S)$ *is a Hopf algebra.*

Proof We first prove the coassociativity of the coproduct (8.34), using Lemma 8.4.13. Let us first check the first condition of this lemma. The fact that the resulting cograph has two or four external edges (the only thing to check in the usual Connes–Kreimer case) is trivial (since the shrinking does not affect the external structure of g, see Definition 8.4.10).

Let us now check in detail how the situation stands for the scale's assignments. We denote by $i_{g'}$ the minimum of the scale assignments of the edges of g' and $e_{g'}$ the maximum of the scale assignments of the edges of g'. Similarly, we denote by i_g the minimum of the scale assignments of the edges of g and e_g the maximum of the scale assignments of the edges of g. We also denote by $i_{g/g'}$ the minimum of the scale assignments of the edges of the cograph g/g' and by $e_{g/g'}$ the maximum of the scale assignments of the external edges of the cograph g/g'.

Since the external edges of g' are internal edges of g, using Definition 8.4.10, this means that

$$i_{g/g'}(\mu) > e_{g/g'}(\mu) = e_g(\mu), \tag{8.42}$$

because, as mentioned previously, the shrinking does not affect the external structure of g. We have thus checked the first condition of Lemma 8.4.13.

The second condition of Lemma 8.4.13 is checked similarly, using Definition 8.4.12. This concludes the proof of the coassociativity of the coproduct Since $\mathcal{H}$ is graded (see previous example), connected, and from the coassociativity of the coproduct (8.34), the definition (8.39) of the antipode leads to the result.

Let us notice that another way to prove the coassociativity (in a direct way) is by the simple use of the isomorphism with the span of Gallavotti–Nicolò forests (see the following subsection). ■

Just like in the Connes–Kreimer case, one has a straightforward pre-Lie algebra structure, given by the operation of insertion of assigned graphs. Antisymmetrizing this operation leads to a Lie algebra of assigned graphs. Consider now the graded dual of the universal enveloping algebra of this Lie structure. This gives the renormalization Hopf algebra defined in this section.

CHAs morphisms

Let us notice that the Gallavotti–Nicolò tree algebra is isomorphic to the algebra $\mathcal{H}$.

Proposition 8.4.15 *The algebra morphism $\pi : \mathcal{H} \to \mathcal{H}_{\mathrm{GN}}$ defined on the generators by $\pi_{\mathrm{GN}}(G,\mu) = T_{(G,\mu)}$ is a Hopf algebra isomorphism.*

Proof The proof is done by a direct verification.

The relation between the Connes–Kreimer CHA of the previous section and the multi-scale renormalization CHA presented in this section is given by:

Proposition 8.4.16 *For every $\rho \in \mathbb{Z}_+$, the algebra morphism defined on the generators of $\mathcal{H}_{\mathrm{CK}}$ by*

$$\pi^{\rho}_{\mathrm{CK}}(G) = \sum_{|\mu|\leq\rho} (G,\mu) \tag{8.43}$$

extends to a Hopf algebra morphism from $\mathcal{H}_{\mathrm{CK}}$ to $\widetilde{\mathcal{H}}$ where $\widetilde{\mathcal{H}}$ is identical to $\mathcal{H}$ as an algebra but equipped with a coproduct that extracts all assigned graphs with 2 or 4 external edges, not only high subgraphs.

Proof The proof is done by a direct calculation.

For example, for the 'sunset' graph to follow, the morphism formula mentioned earlier leads to:

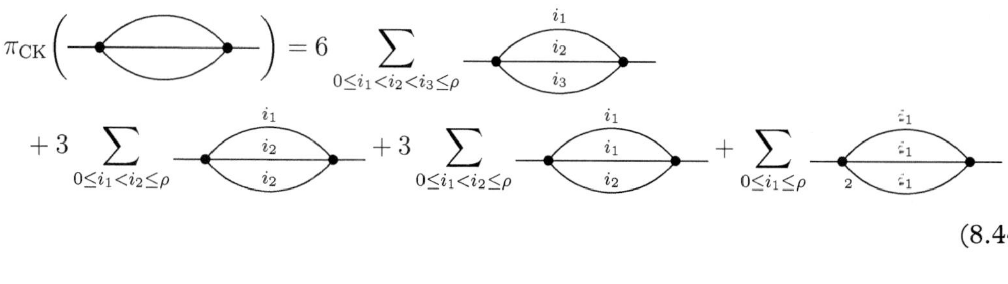

(8.44)

Let us also remark that this morphism allows to evaluate Feynman integrals with the cut-off ρ as

$$A^{\rho}(G) = A \circ \pi^{\rho}_{\mathrm{CK}}(G). \tag{8.45}$$

8.5 Selected further reading

- Karen Yeats. (2016). *A Combinatorial Perspective on Quantum Field Theory*. Springer.
  A book focusing on the algebraic aspects of the combinatorics of QFT (perturbative renormalization and DSE).

9

QFT on the non-commutative Moyal space and combinatorics

Let us begin this subsection by giving some explanations, following the review article Tanasa (2012), on mathematical physicists' interest in QFT on non-commutative spaces.

Non-commutative QFT can be seen as a potential candidate for new physics beyond the celebrated Standard Model of elementary particle physics. Moreover, non-commutativity of space-time is believed to be a possible framework for the quantification of gravity. Thus, it was argued in Doplicher, Fredenhagen, and Roberts (1994) that space-time becomes non-commutative at the Planck scale (physical energy scale where gravitational effects have the same order of magnitude as quantum effects, or the energy scale where gravity and quantum physics can no longer ignore one another). Moreover, non-commutative QFT has been shown to be related to loop quantum gravity as well as GFT (Freidel and Livine 2006; Noui 2009; Baratin and Oriti 2010), which are two of the candidates for a quantified theory of gravity (see Chapter 10). Non-commutative models have also been shown to be effective theories of certain models of strings and of matrix theories, see Connes, Douglas, and Schwarz (1998); Seiberg and Witten (1999). Finally, from another perspective, it was argued that non-commutative models can be better adapted for describing the physics of effective non-local interactions, as is the case for the fractional quantum Hall effect, see Morariu and Polychronakos (2001); Polychronakos (2001). For more details on non-commutative QFT, the interested reader may refer to the review paper Rivasseau (2007).

The chapter is structured as follows. In this chapter we present the Phi? QFT model on the non-commutative Moyal space and the UV/IR mixing issue, which prevents it from being renormalizable. We then present the Grosse–Wulkenhaar Phi? QFT model on the non-commutative Moyal space, which changes the usual propagator of the Phi? model (based on the heat kernel formula) to a Mehler kernel-based propagator. This Grosse–Wulkenhaar model is perturbatively renormalizable but it is not translation-invariant (translation-invariance being a usual property of high energy physics models). We then show how the Mehlin transform technique can be used to express the Feynman integrals of the Grosse–Wulkenhaar model. In the last part of the chapter, we present another Phi? QFT model on the non-commutative Moyal space, which is however both

Combinatorial Physics: Combinatorics, quantum field theory, and quantum gravity models. Adrian Tanasa, Oxford University Press (2021). © Adrian Tanasa. DOI: 10.1093/oso/9780192895493.003.0009

renormalizable and translation-invariant. We show the relation between the parametric representation of this model and the Bollobas–Riordan polynomial. Finally, we show how to define a Connes–Kreimer Hopf algebra for non-commutative renormalization and how to study its Hochschild cohomology in relation to the combinatorial DSE of these QFT models.

9.1 Mathematical setting: Renormalizability

The Moyal space, one of the simplest cases of such a non-commutative setting, can be defined in the following way:

Definition 9.1.1 *The* **Moyal algebra** *is the linear space of smooth and rapidly decreasing functions $\mathcal{S}(\mathbb{R}^D)$ equipped with the* **Moyal product,**

$$(f \star g)(x) = \int \frac{d^D k}{(2\pi)^D} d^D y\, f(x + \frac{1}{2}\Theta \cdot k) g(x+y) e^{ik\cdot y}, \forall x \in \mathbb{R}^D, \tag{9.1}$$

where Θ is a D-dimensional skew-symmetric matrix.

Remark 9.1.2

1. *The Moyal product is a non-local and non-commutative but still associative product.*
2. *Taking the matrix Θ to be vanishing leads to the usual commutative multiplication of functions.*

The Moyal algebra defined previously can be extended by duality, considering the product of a tempered distribution with a Schwartz-class function. The identity, the polynomials, the δ function, and its derivatives can then be shown to belong to this extended Moyal algebra (see Gracia-Bondia and Varilly (1988) for details). One can then prove that

$$[x_\mu, x_\nu]_\star := x_\mu \star x_\nu - x_\nu \star x_\mu = i\Theta^{\mu\nu}, \ \forall \mu, \nu = 1, \ldots, D. \tag{9.2}$$

As in the commutative case (see Chapter 3), we are interested in the case $D = 4$ (the usual four-dimensional space-time). The *non-commutativity matrix* Θ is taken to be

$$\Theta = \begin{pmatrix} \Theta_2 & 0 \\ 0 & \Theta_2 \end{pmatrix}, \quad \Theta_2 = \begin{pmatrix} 0 & -\theta \\ \theta & 0 \end{pmatrix}.$$

For a general even D, the non-commutativity matrix is constructed similarly, in a block-diagonal form, the blocks being given by the two-dimensional square matrix Θ_2. This represents a canonical form of the noncommutativity matrix.

The most common way to implement non-commutative QFT is to replace the usual commutative multiplication of fields with the previously mentioned non-commutative

$\star$-product. Thus, the non-commutative Φ^4 model's action produces:

$$S[\Phi(x)] = \int d^4x \left[\frac{1}{2} \sum_{\mu=1}^{4} (\partial_\mu \Phi \star \partial_\mu \Phi)(x) + \frac{1}{2} m \Phi^{\star 2}(x) + V^{\star}_{\text{int}}[\Phi(x)] \right], \tag{9.3}$$

where the non-commutative interaction potential is:

$$V^{\star}_{\text{int}}[\Phi(x)] = \frac{\lambda}{4} \Phi^{\star 4}(x). \tag{9.4}$$

As in the commutative case, one can Fourier transform this action to momentum space. A first consequence of the use of the Moyal product is that no change is made to the propagation part of the action. This is a direct consequence of the following mathematical property:

Proposition 9.1.3 *One has:*

$$\int d^4x \, (\Phi \star \Psi)(x) = \int d^4x \, \Phi(x) \, \Psi(x), \quad \forall \Phi, \Psi. \tag{9.5}$$

Proof One can prove this identity through a direct check using Definition 9.1.

One can thus conclude that the propagator of the theory has the same form as in the commutative case. Nevertheless, the use of the Moyal product at the level of the interaction terms completely changes the situation: the non-commutative interaction is not equivalent to the commutative one (as it is the case for the propagation). Using Definition 9.1 of the Moyal product again, one can prove the following:

Proposition 9.1.4 *One has:*

$$\begin{aligned} \int d^4x \phi^{\star 4}(x) &= \int \prod_{i=1}^{4} d^4x_i \phi(x_i) \delta(x_1 - x_2 + x_3 - x_4) e^{\imath (x_1 - x_2) \wedge (x_3 - x_4)} \\ &= \int \prod_{i=1}^{4} d^4x_i \phi(x_i) \delta(x_1 - x_2 + x_3 - x_4) e^{\imath (x_1 \wedge x_2 + x_3 \wedge x_4)}, \end{aligned} \tag{9.6}$$

where we have used the notation $x \wedge y = \sum_{\mu\nu} x_\mu \Theta_{\mu\nu} x_\nu$.

One notices that an oscillatory phase (proportional to the area of the parallelogram formed by the points $x_1, \ldots, x_4$) is present in the interaction term. A similar formula can be obtained by Fourier transforming this to momentum space. Note that one uses a cyclic Fourier transform, particularly adapted for Moyal QFT (the interested reader may check for details Vincent Rivasseau's review paper Rivasseau (2007)).

A first consequence of this formula is that the interaction part is manifestly **non-local**. Furthermore, this Moyal interaction no longer preserves the invariance under

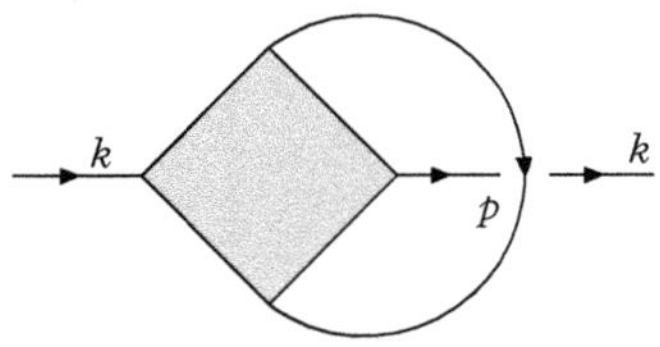

Figure 9.1 *The 'non-planar' tadpole. The external momentum has been denoted by k, while the internal one has been denoted by p*

permutation of external fields. This invariance is restricted to only *cyclic permutations*. Furthermore, there exists a basis—the matrix base—of the Moyal algebra where the Moyal product takes the form of an ordinary matrix product. For these reasons, the associated Feynman graphs are *ribbon graphs*, just like the ones described in the previous subsection. In this book, we will represent these graphs both as ribbon graphs and as usual graphs with non-local vertices depending on the particular character of the Moyal interaction that we need to stress.

Let us also mention that Proposition 9.1.3 and Proposition 9.1.4 hold, in the same form, for a general even D.

Let us now present how the situation stands with respect to the *renormalizability* of QFTs implemented in such a way on non-commutative Moyal space.

Thus, for planar regular ribbon graphs (see Definition 2.2.4), the renormalizability remains identical to that in the commutative case. The situation changes in the planar irregular sector. Thus, if one considers the planar irregular tadpole graph (also called in the literature 'non-planar' tadpole, even though its genus is vanishing) of Fig. 9.1, the oscillating phase appearing in (9.6) leads, see Minwalla, Van Raamsdonk, and Seiberg (2001), to the following Feynman integral (in momentum space):

$$\lambda \int d^4p \frac{e^{ik_\mu \Theta^{\mu\nu} p_\nu}}{p^2+m^2}. \tag{9.7}$$

This integral is, up to an irrelevant constant, a **modified Bessel function of the first type**:

$$\lambda\sqrt{\frac{m^2}{(\Theta\cdot k)^2}}K_1(\sqrt(m^2(\Theta\cdot p)^2)). \tag{9.8}$$

With or without the identification (9.8) given previously, one can prove that in the infrared regime of the external momentum k the Feynman integral behaves like

$$\frac{1}{\theta^2 k^2}. \tag{9.9}$$

This quantity is finite (for non-vanishing k^2), but when inserting this non-planar tadpole graph into some Feynman graph, it needs to be integrated on k (like any other internal momentum, see the momentum space Feynman rules of scalar QFT introduced in Section 3.3). This leads to a divergence in the infrared regime of k (a divergence which arises from the initial integration in the ultraviolet regime of the momentum p). This is a new type of divergence appearing at the level of the two-point function, which is called an *ultraviolet/infrared mixing*. An important characteristic of this type of divergence is that it is *non-local* (this can be seen by a direct inspection of the Fourier transform, in position space, of expression (9.9)).

In Magnen, Rivasseau, and Tanasa (2009) it was proven that this type of phenomenon appears at *any order* in perturbation theory. Moreover, in Tanasa and Vitale (2010), it was proven that replacing the Moyal product with another type of non-commutative product, the Voros product, does not prevent ultraviolet/infrared mixing.

Since the quadratic part of the action has the same type of behaviour as in commutative QFT (see Proposition 9.1.3), such a non-local divergence **cannot** be re-absorbed (that is, renormalized) in a local part of the action. This leads to the non-renormalizability of the Moyal action (9.3).

There exist, in the literature, two types of solution for restoring renormalizability of scalar QFT on non-commutative Moyal space. They are both based on modifications of the propagator of the model:

1. the Grosse–Wulkenhaar model, based on a Mehler-type kernel (see Section 9.2); and
2. a heat-type kernel (see Section 9.6).

For more details on the implementation of QFTs on non-commutative Moyal space and their renormalizability, the interested reader may refer again to Vincent Rivasseau's review paper Rivasseau (2007) and references within.

9.2 The Mehler kernel and the Grosse–Wulkenhaar model

In Grosse and Wulkenhaar (2005) it was shown that the model (9.3), to whom one adds a harmonic oscillator term

$$\delta_{GW} S[\Phi(x)] = \int d^4x \frac{\Omega}{2} \tilde{x}^2 \Phi^2, \tag{9.10}$$

(where Ω is a constant and $\tilde{x} = 2\Theta^{-1}x$) is renormalizable at any order in perturbation theory. This result was then proven using other QFT methods, in Rivasseau, Vignes-Tourneret, and Wulkenhaar (2006); Gurau et al. (2006). Let us also mention here that the parameter Ω introduced previously is free to flow under the renormalization group.

Remark 9.2.1 *The Grosse–Wulkenhaar modification* (9.10) *explicitly breaks the translation invariance of the action of the initial model (see Subsection 3.1).*

The modification (9.10) of the quadratic part of the action leads to a model with a propagator with a Mehler kernel. This propagator is expressed as, up to some irrelevant constants:

$$\int_0^\infty \frac{dt}{\sinh^2(2\Omega t)} e^{-\frac{\Omega}{2}\coth(2\Omega t)(x-y)^2 - \frac{\Omega}{2}\tanh(2\Omega t)(x+y)^2 - m^2 t}. \tag{9.11}$$

Working in position space, one defines the *short and long variables*:

$$u = y - x, \; v = y + x, \tag{9.12}$$

The propagator thus denotes:

$$\int_0^\infty \frac{dt}{\sinh^2(2\Omega t)} e^{-\frac{\Omega}{2}\coth(2\Omega t)u^2 - \frac{\Omega}{2}\tanh(2\Omega t)v^2 - m^2 t}. \tag{9.13}$$

Let us mention that in de Goursac, Tanasa, and Wallet (2008) some non-scalar solutions of the equation of motion (see Chapter 3) for the Grosse–Wulkenhaar model have been obtained and analysed. From a mathematical point of view, the equation of motion is a partial differential equation involving the Moyal product (9.1).

Moreover, the renormalization group of this model has been studied, at any order in perturbation theory, in Disertori et al. (2007) the result proving to be promising for a *constructive proof* of its renormalizability.

A similar modification of the propagator leads again to a perturbatively renormalizable model—the Langman–Szabo–Zarembo model (also extensively studied from the perspective of integrable models Langmann, Szabo, and Zarembo (2004)). Its action is written:

$$S[\Phi] = \int d^4x \frac{1}{2}\sum_{\mu=1}^{4}\left(\frac{\partial}{\partial x_\mu}\Phi - i\Omega\tilde{x}_\mu\Phi\right) \star \left(\frac{\partial}{\partial x_\mu}\bar{\Phi} - i\Omega\tilde{x}_\mu\bar{\Phi}\right) + \frac{\lambda}{4}\Phi\star\bar{\Phi}\star\Phi\star\bar{\Phi}, \tag{9.14}$$

where Φ is now a complex field.

Let us end this section by pointing out that various attempts at generalizing this approach to restoring renormalizability for a gauge theory on non-commutative Moyal space have been made in literature (see de Goursac, Wallet, and Wulkenhaar (2007); Axel de Goursac's PhD thesis (2009), and references within).

9.3 Parametric representation of Grosse–Wulkenhaar-like models

This section follows the review section 2 of Gurau et al. (2007). Let us also mention that a brief review of the parametric representation of Grosse–Wulkenhaar-like models that will now be presented was given in Tanasa (2008).

Let us define the antisymmetric matrix σ as

$$\sigma = \begin{pmatrix} \sigma_2 & 0 \\ 0 & \sigma_2 \end{pmatrix} \text{ with } \sigma_2 = \begin{pmatrix} 0 & -i \\ i & 0 \end{pmatrix}. \tag{9.15}$$

The δ-functions appearing in the vertices can be rewritten as an integral over some new variables p_V called *hypermomenta*, *via* the relation

$$\delta(x_1 - x_2 + x_3 - x_4) = \int \frac{dp'}{(2\pi)^4} e^{ip'(x_1 - x_2 + x_3 - x_4)} = \int \frac{dp}{(2\pi)^4} e^{p\sigma(x_1 - x_2 + x_3 - x_4)}. \tag{9.16}$$

Let Γ be a ribbon graph. Choosing a particular root vertex $\bar{V}$, the parametric representation for the integral of Γ is expressed in terms of $t_\ell = \tanh \alpha_\ell / 2$ (where α_ℓ is the former Schwinger parameter) as

$$\phi_\Gamma = \int_0^1 \prod_\ell dt_\ell (1 - t_\ell^2)^{\frac{D}{2} - 1} \int dx dp e^{-\frac{\Omega}{2} X G X^t}, \tag{9.17}$$

where D is the space-time dimension, X summarizes all positions and hypermomenta, and G is a certain quadratic form. Calling x_e and $p_{\bar{V}}$ the external variables and x_i, p_i the internal ones, we decompose G into an internal quadratic form Q, an external one M and a coupling part P such that

$$X = \begin{pmatrix} x_e & p_{\bar{V}} & x_i & p_i \end{pmatrix}, \; G = \begin{pmatrix} M & P \\ P^t & Q \end{pmatrix}. \tag{9.18}$$

Performing the Gaussian integration over all internal variables one gets the non-commutative parametric representation:

$$\phi_\Gamma = \int \prod_\ell dt_\ell (1 - t_\ell^2)^{\frac{D}{2} - 1} \frac{1}{\sqrt{\det Q}} e^{-\frac{\Omega}{2} \begin{pmatrix} x_e & p_{\bar{V}} \end{pmatrix} \left[M - P Q^{-1} P^t \right] \begin{pmatrix} x_e \\ p_{\bar{V}} \end{pmatrix}}. \tag{9.19}$$

This representation leads to new polynomials $HU_{\Gamma,\bar{V}}$ and $HV_{\Gamma,\bar{V}}$ in the t_ℓ ($\ell = 1, \ldots, L$) variables, analogues of the Symanzik polynomials U and V of the commutative case, through

$$\phi_\Gamma(\{x_e\}, p_{\bar{V}}) = K' \int_0^1 \prod_\ell dt_\ell (1 - t_\ell^2)^{\frac{D}{2} - 1} \frac{e^{-\frac{HV_{\Gamma,\bar{V}}(t, x_e, p_{\bar{V}})}{HU_{\Gamma,\bar{V}}(t)}}}{HU_{\Gamma,\bar{V}}(t)^{\frac{D}{2}}}. \tag{9.20}$$

The main results of this section are:

- The polynomial $HU_{\Gamma,\bar{V}}$ and the real part of $HV_{\Gamma,\bar{V}}$ have a *positivity property*. They are sums of monomials with positive integer coefficients, which are squares of Pfaffians with integer entries.
- Leading terms can be identified in a given Hepp sector, at least for some particular type of graphs (see Gurau et al. (2007)). In $HU_{\Gamma,\bar{V}}$ they correspond to hyper-trees which are the disjoint union of a tree in the direct graph and another tree in the dual graph. Any connected graph has such hypertrees. Similarly, 'hyper-two-trees' govern the leading behaviour of $HV_{\Gamma,\bar{V}}$ in any Hepp sector.

Let us relabel the $2L$ internal positions of Γ as L short and L long variables (see (9.12)). For the Grosse–Wulkenhaar model one has:

$$HU_{\Gamma,\bar{V}}(t) = \sum_{K_U = I\cup J,\ n+|K_U|\ \text{odd}} s^{2g-k_{K_U}}\, n_{K_U}^2 \prod_{\ell\notin I} t_\ell \prod_{\ell'\in J} t_{\ell'}$$
$$= \sum_{K_U} a_{K_U} \prod_\ell t_\ell^{u_{\ell K_U}} \equiv \sum_{K_U} HU_{K_U}. \tag{9.21}$$

where:

- I is a subset of the first L indices (corresponding to the short variables) with $|I|$ elements, and J a subset of the next L indices (corresponding to the long ones) with $|J|$ elements.
- B is the antisymmetric part of the quadratic form Q restricted to these $2L$ short and long variables (i.e. omitting hypermomenta).
- $n_{K_U} = \mathrm{Pf}(B_{\widehat{K_U}})$ is the Pfaffian of the antisymmetric matrix obtained from B by deleting the lines and columns in the set $K_U = I\cup J$.
- k_{K_U} is $|K_U| - L - F + 1$, where F is the number of faces of the graph, g is the genus of Γ, and s is a constant.
- $a_{K_U} = s^{2g-k_{K_U}} n_{K_U}^2$, and

$$u_{\ell K_U} = \begin{cases} 0 & \text{if } \ell\in I \text{ and } \ell\notin J \\ 1 & \text{if } (\ell\notin I \text{ and } \ell\notin J) \text{ or } (\ell\in I \text{ and } \ell\notin J). \\ 2 & \text{if } \ell\notin I \text{ and } \ell\in J \end{cases} \tag{9.22}$$

The second polynomial HV has both a real part $HV^{\mathcal{R}}$ and an imaginary part $HV^{\mathcal{I}}$. We need to introduce beyond I and J, as was done previously, a particular edge $\tau\notin I$ which is the analogue of a two-tree cut. We define $\mathrm{Pf}(B_{\hat{K_V}\hat{\tau}})$ as the Pfaffian of the matrix obtained from B by deleting the lines and columns in the sets I, J, and τ. Moreover, we define $\epsilon_{I,\tau}$ to be the signature of the permutation obtained from $(1,\dots,d)$ by extracting the positions belonging to I and replacing them at the end in the order

$$1,\dots,d \to 1,\dots,\hat{i_1},\dots,\hat{i_{|I|}},\dots,\hat{i_\tau},\dots,d,i_\tau,i_{|I|}\dots,i_1. \tag{9.23}$$

where d is the dimension of the matrix Q. Then:

$$HV^{\mathcal{R}}_{\Gamma,\bar{V}} = \sum_{K_V=I\cup J} \prod_{\ell\notin I} t_\ell \prod_{\ell'\in J} t_{\ell'} \Big[\sum_{e_1} x_{e_1} \sum_{\tau\notin K_V} P_{e_1\tau}\epsilon_{K_V\tau}\mathrm{Pf}(B_{\hat{K_V}\hat{\tau}})\Big]^2 .$$
$$= \sum_{K_V} s^{\mathcal{R}}_{K_V} \left(\prod_{\ell=1}^{L} t_\ell^{v_{\ell K_V}}\right) \equiv \sum_{K_V} HV^{\mathcal{R}}_{K_V} \tag{9.24}$$

where,

$$s^{\mathcal{R}}_{K_V} = \left(\sum_e x_e \sum_{\tau\notin K_V} P_{e\tau}\varepsilon_{K_V\tau}\mathrm{Pf}(B_{\hat{K_V}\hat{\tau}})\right)^2 , \tag{9.25}$$

and $v_{\ell K_V}$ given by the same formula as $u_{\ell K_U}$.

The imaginary part involves *pairs* of lines τ, τ' and corresponding signatures:

$$HV^{\mathcal{I}}_{\Gamma,\bar{V}} = \sum_{K_V=I\cup J} \prod_{\ell\notin I} t_\ell \prod_{\ell'\in J} t_{\ell'}$$
$$\epsilon_{K_V}\mathrm{Pf}(B_{\hat{K_V}}) \left[\sum_{e_1,e_2}\left(\sum_{\tau\tau'} P_{e_1\tau}\epsilon_{K_V\tau\tau'}\mathrm{Pf}(B_{\hat{K_V}\hat{\tau}\hat{\tau'}})P_{e_2\tau'}\right) x_{e_1}\sigma x_{e_2}\right] .$$
$$= \sum_{K_V} s^{\mathcal{I}}_{K_V} \left(\prod_{\ell=1}^{L} t_\ell^{v_{\ell K_V}}\right) \equiv \sum_{K_V} HV^{\mathcal{I}}_{K_V}, \tag{9.26}$$

where,

$$s^{\mathcal{I}}_{K_V} = \epsilon_{K_V}\mathrm{Pf}(B_{\hat{K_V}}) \left(\sum_{e,e'}\left(\sum_{\tau,\tau'} P_{e\tau}\varepsilon_{K_V\tau\tau'}\mathrm{Pf}(B_{\hat{K_V}\hat{\tau}\hat{\tau'}})P_{e'\tau'}\right) x_e\sigma x_{e'}\right) . \tag{9.27}$$

A similar development exists for the Langman–Szabo–Zarembo model (see the original paper Rivasseau and Tanasa (2008)). The first polynomial in this case is written:

$$HU_{\Gamma,\bar{V}}(t) = \sum_{\substack{I\subset\{1\ldots L\},\\ n+|I|\ \mathrm{odd}}} s^{2g-k_I}\, n_I^2 \prod_{l\in I}\frac{1+t_\ell^2}{2t_\ell} \prod_{l'\in\{1,\ldots,L\}} t_{l'}$$
$$= \sum_{K_U=I\cup J} a_{K_U} \prod_{\ell} t_\ell^{u_{\ell K_U}} \equiv \sum_{K_U} HU_{K_U} . \tag{9.28}$$

Again, I runs over the subsets of $\{1,\ldots,L\}$. But J now runs over subsets of I (representing the expansion of $\prod_{\ell\in I}[1+t_\ell^2]$), and one has again,

$$u_{\ell K_U} = \begin{cases} 0 & \text{if } \ell \in I \text{ and } \ell \in J \\ 1 & \text{if } \ell \notin I \\ 2 & \text{if } \ell \in I \text{ and } \ell \notin J \end{cases} \quad , \; a_{K_U} = \frac{1}{2^{|I|}} s^{2g-k_I} n_I^2. \tag{9.29}$$

There are similar expressions for the second polynomial.

To summarize, the main differences between the parametric representation presented in this section and the parametric representation of the commutative case (see Section 3.4) are:

- the presence of the constants a_j in the form (9.21) of HU;
- the presence of an imaginary part $i\,HV^{\mathcal{I}}$ in HV, see (9.26); and
- the fact that the parameters $u_{\ell j}$ and $v_{\ell k}$ in the previous formulas can also have the value 2 (and not only 0 or 1).

Rescaling the non-commutative Schwinger parameters, one also has:

Theorem 9.3.1 (Power counting theorem for the Grosse–Wulkenhaar model). *The superficially divergent ribbon graphs of the Grosse–Wulkenhaar models are the planar regular two- and four-point ribbon graphs.*

9.4 The Mellin transform and the Grosse–Wulkenhaar model

In this section, we present the implementation of the complete Mellin representation of the Feynman integrals of the Grosse–Wulkenhaar model. This survey closely follows the results obtained in Gurau et al. (2007). Let us also mention that a brief review of the parametric representation of Grosse–Wulkenhaar-like models (presented in the previous subsection) and the following Mellin representation were given in Tanasa (2007).

We start with the real part $HV^{\mathcal{R}}$ of the polynomial HV; one uses the identity (7.3):

$$e^{-HV^{\mathcal{R}}_{K_V}/U} = \int_{\tau^{\mathcal{R}}_{K_V}} \Gamma(-y^{\mathcal{R}}_{K_V}) \left(\frac{HV^{\mathcal{R}}_{K_V}}{U}\right)^{y^{\mathcal{R}}_{K_V}}. \tag{9.30}$$

This introduces the set of Mellin parameters $y^{\mathcal{R}}_{K_V}$.

For the imaginary part one cannot apply the same identity, which only holds in the sense of distributions. For $HV^{\mathcal{I}}_{K_V}/U > 0$ and $-1 < \tau^{\mathcal{I}}_{K_V} < 0$, one has (see Appendix A of Gurau et al. (2007) for the proof):

$$e^{-i\,HV^{\mathcal{I}}_{K_V}/U} = \int_{\tau^{\mathcal{I}}_{K_V}} \Gamma(-y^{\mathcal{I}}_{K_V}) \left(\frac{i\,HV^{\mathcal{I}}_{K_V}}{U}\right)^{y^{\mathcal{I}}_{K_V}}. \tag{9.31}$$

Note that this introduces another set of Mellin parameters $y^{\mathcal{I}}_{K_V}$. The distributional sense of formula (9.31) comes from the distributional form of the vertex contribution (the δ function forces the four points to form a parallelogram). This represents a major difference with respect to the commutative case treated in Section 7.1.

For the polynomial HU one can again use formula (7.5) which rewrites here as

$$\Gamma\left(\sum_{K_V} y_{K_V} + \frac{D}{2}\right)(HU)^{-\sum_{K_V}(y^{\mathcal{R}}_{K_V}+y^{\mathcal{I}}_{K_V})-\frac{D}{2}} = \int_\sigma \prod_{K_U} \Gamma(-x_{K_U}) U_{K_U}^{x_{K_U}}. \tag{9.32}$$

Note that the role of the Mellin parameters y_{K_V} of the commutative case is now played by the sum $y^{\mathcal{R}}_{K_V} + y^{\mathcal{I}}_{K_V}$.

As in the commutative case, we now insert the distribution formulas (9.30), (9.31), and (9.32) in the general form of the Feynman integral. This gives

$$\begin{aligned}\mathcal{A}_G = \mathrm{K}' \int_\Delta \frac{\prod_{K_U} a_{K_U}^{x_{K_U}} \Gamma(-x_{K_U})}{\Gamma(-\sum_{K_U} x_{K_U})} \left(\prod_{K_V} (s^{\mathcal{R}}_{K_V})^{y^{\mathcal{R}}_{K_V}} \Gamma(-y^{\mathcal{R}}_{K_V})\right) \\ \left(\prod_{K_V} (s^{\mathcal{I}}_{K_V})^{y^{\mathcal{I}}_{K_V}} \Gamma(-y^{\mathcal{I}}_{K_V})\right) \int_0^1 \prod_{\ell=1}^{L} dt_\ell (1-t_\ell^2)^{\frac{D}{2}-1} t_\ell^{\phi_\ell - 1}\end{aligned} \tag{9.33}$$

where

$$\phi_\ell \equiv \sum_{K_U} u_{\ell K_U} x_{K_U} + \sum_{K_V} (v^{\mathcal{R}}_{\ell K_V} y^{\mathcal{R}}_{K_V} + v^{\mathcal{I}}_{\ell K_V} y^{\mathcal{I}}_{K_V}) + 1. \tag{9.34}$$

Here, $\int_\Delta$ means integration over the variables $\frac{\mathrm{Im}\, x_{K_U}}{2\pi i}$, $\frac{\mathrm{Im}\, y^{\mathcal{R}}_{K_V}}{2\pi i}$, and $\frac{\mathrm{Im}\, y^{\mathcal{I}}_{K_V}}{2\pi i}$, where Δ is the convex domain:

$$\Delta = \left\{ \sigma, \tau^{\mathcal{R}}, \tau^{\mathcal{I}} \;\middle|\; \begin{array}{l} \sigma_{K_U} < 0;\ \tau^{\mathcal{R}}_{K_V} < 0;\ -1 < \tau^{\mathcal{I}}_{K_V} < 0; \\ \sum_{K_U} x_{K_U} + \sum_{K_V} (y^{\mathcal{R}}_{K_V} + y^{\mathcal{I}}_{K_V}) = -\frac{D}{2}; \\ \forall \ell,\ \mathrm{Re}\, \phi_\ell \equiv \sum_{K_U} u_{\ell K_U} \sigma_{K_U} \\ + \sum_{K_V} (v^{\mathcal{R}}_{\ell K_V} \tau^{\mathcal{R}}_{K_V} + v^{\mathcal{I}}_{\ell K_V} \tau^{\mathcal{I}}_{K_V}) + 1 > 0 \end{array} \right\} \tag{9.35}$$

and σ, $\tau^{\mathcal{R}}$ and $\tau^{\mathcal{I}}$ stand for $\mathrm{Re}\, x_{K_U}$, $\mathrm{Re}\, y^{\mathcal{R}}_{K_V}$ and $\mathrm{Re}\, y^{\mathcal{I}}_{K_V}$.

The dt_ℓ integrations in (9.33) may be performed using the representation for the beta function

$$\int_0^1 \prod_{\ell=1}^{L} dt_\ell (1-t_\ell^2)^{\frac{D}{2}-1} t_\ell^{\phi_\ell - 1} = \frac{1}{2} \beta\left(\frac{\phi_\ell}{2}, \frac{D}{2}\right). \tag{9.36}$$

Furthermore, one has

$$\beta(\frac{\phi_\ell}{2}, \frac{D}{2}) = \frac{\Gamma(\frac{\phi_\ell}{2})\Gamma(\frac{D}{2})}{\Gamma(\frac{\phi_\ell+D}{2})}.$$

This representation is convergent for $0 < \Re D < 2$. One has:

Theorem 9.4.1 *(Theorem* 4.1 *of Gurau et al. (2007)). Any Feynman integral of a Grosse–Wulkenhaar graph G is analytic in the strip* $0 < \Re D < 2$, *where it admits the following complete Mellin representation:*

$$\mathcal{A}_G = \mathrm{K}' \int_\Delta \frac{\prod_{K_U} a_{K_U}^{x_{K_U}} \Gamma(-x_{K_U})}{\Gamma(-\sum_{K_U} x_{K_U})} \left(\prod_{K_V} (s_{K_V}^{\mathcal{R}})^{y_{K_V}^{\mathcal{R}}} \Gamma(-y_{K_V}^{\mathcal{R}}) \right)$$
$$\left(\prod_{K_V} (s_{K_V}^{\mathcal{I}})^{y_{K_V}^{\mathcal{I}}} \Gamma(-y_{K_V}^{\mathcal{I}}) \right) \left(\prod_{\ell=1}^{L} \frac{\Gamma(\frac{\phi_\ell}{2})\Gamma(\frac{D}{2})}{2\Gamma(\frac{\phi_\ell+D}{2})} \right). \tag{9.37}$$

This holds as tempered distribution of the external invariants.

Let us first point out that the distributional aspect does not affect the last product, that involving the dimensionality of space-time D. Therefore, one can still prove the meromorphy of the Feynman integral $\phi_\Gamma(D)$ in the space-time dimension D essentially as in the commutative case.

Another difference with respect to the commutative Φ^4 model (see again Section 7.1) is that the dt_ℓ integration brings out β functions (see (9.36)), hence more complicated ratios of Γ functions. In particular singularities appear for negative D which come from the integration near $t_\ell = 1$, hence the 'infrared side'. They have no analogue in the commutative case and are due to the hyperbolic nature of the Mehler kernel. At zero mass these singularities occur for D real negative even integer and, in the massive case (i.e. non vanishing mass), they occur for D real but sufficiently negative depending on the mass.

The complete Mellin representation (9.37) for non-commutative theories allows an immediate proof of the meromorphy of the Feynman integral ϕ_G in the variable D, since D appears only as an argument of (meromorphic) Γ functions.

Theorem 9.4.2 *(Theorem* 5.1 *of Gurau et al. (2007)). A Feynman integral* $\mathcal{A}_G$ *of the Grosse-Wulkenhaar model is a tempered meromorphic distribution in D, that is* $\mathcal{A}_G$ *smeared against any fixed Schwarz-class test function of the external invariants yields a meromorphic function in D in the entire complex plane, with singularities located among a discrete rational set* S_G *which depends only on the ribbon graph G not on the test function. No such singularity can occur in the strip* $0 < \mathrm{Re}D < 2$.

9.5 Dimensional renormalization for the Grosse–Wulkenhaar model

Using the parametric representation of the Grosse–Wulkenhaar model (see Section 9.3), we can implement the steps described in Subsection 3.6.5 for the commutative case. This section follows the original article Gurau and Tanasa (2008).

Let us denote all the leading terms in the first polynomial associated with a graph γ by HU^l_γ. If we rescale all the t_ℓ parameters corresponding to a subgraph γ by ρ^2, HU_Γ becomes a polynomial in ρ. We denote the terms of minimal degree in ρ in this polynomial by $HU^{l(\rho)}_\Gamma$. One can check that for the subgraph γ we have $HU^{l(\rho)}_\gamma = \rho^{2[L(\gamma)-n(\gamma)+1]} HU^l_\gamma|_{\rho=1}$.

One has:

Theorem 9.5.1 *(Theorem 4.1 of Gurau and Tanasa (2008)). Under the rescaling*

$$t_\alpha \mapsto \rho^2 t_\alpha \tag{9.38}$$

of the parameters corresponding to a divergent subgraph γ of some Feynman ribbon graph Γ, one has:

$$HU^{l(\rho)}_{\Gamma,\bar{V}} = HU^{l(\rho)}_{\gamma,\bar{V}_\gamma} HU_{\Gamma/\gamma,\bar{V}}\,. \tag{9.39}$$

The proof is done using a Grassmann manipulation of the respective determinants.

Proposition 9.5.2 *Under the rescaling $t_\alpha \mapsto \rho^2 t_\alpha$ of all the edges of the subgraph S one has:*

$$\left.\frac{HV_\Gamma}{HU_\Gamma}\right|_{\rho=0} = \frac{HV_{\Gamma/\gamma}}{HU_{\Gamma/\gamma}}\,. \tag{9.40}$$

The analytic continuation in the complex plane being trivial (like in the commutative case), one then proves that the Feynman integral $\phi(\Gamma)$ of some ribbon graph Γ is a *meromorphic function* (see again Gurau and Tanasa (2008) for details). One then applies the subtraction operator R in order to obtain the renormalized Feynman integrals. The subtraction operator is defined as in the commutative case (see Section 3.6.4), with the only difference being that the class of superficially divergent graphs is now different (planar regular two- and four-point graphs)—see Theorem 9.3.1.

The main result of this section is:

Theorem 9.5.3 *(Theorem 5.1 of Gurau and Tanasa (2008)). The renormalized integral is an analytic function of the space-time dimension D.*

For the sake of completeness, let us end this section with by mentioning that the Mellin transform technique has also been recently used for simplifying conformal field theoretical correlation functions Nandan, Volovich, and Wen (2012), within the framework of the Anti de Sitter—conformal field theory correspondence (which is a relation between conformal field theory in D dimensions and a gravity theory in $(D+1)$-dimensional Anti de Sitter space, theories which both have a $SO(2,D)$ symmetry group. The case of interest for theoretical physicists is $D=4$, where the conformal field theoretical model is the so-called $N=4$ supersymmetric Yang–Mills theories with gauge group $SU(N)$ and the gravitational theory is *string theory*. For a general reference on conformal field theory, which is a particular type of QFT, admitting the conformal group as a symmetry group, we refer the interested reader to Philippe Di Francesco, Pierre Mathieu, and David Sénéchal (1997)).

9.6 A heat kernel–based renormalizable model

Getting back to Moyal QFT, let us recall from Section 7.1 that a second type of modification of the propagator of the non-commutative Φ^4 model exists in the literature; this modification also transforms the original non-renormalizable model (9.3) into a model renormalizable at any order in perturbation theory.

This modification was proposed in Gurau et al. (2009), and adds to the propagator its own quantum corrections. In momentum space, this action is written as:

$$S=\int d^4p\,[\frac{1}{2}p_\mu\phi\star p^\mu\phi+\frac{1}{2}a\frac{1}{(\Theta\cdot p)^2}\phi\star\phi+\frac{1}{2}m^2\phi\star\phi+\frac{\lambda}{4}\phi\star\phi\star\phi\star\phi]. \quad (9.41)$$

As for the Grosse–Wulkenhaar model, the new parameter a is free to flow under the renormalization group.

Remark 9.6.1 *This modification does not break the translation-invariance of the initial non-commutative model* (9.3).

The details of the renormalization computations can be found in the original paper Gurau et al. (2009). Let us just give the following table here, summarizing the behaviour of the Feynman integrals (and thus the *power counting theorems*) of the three types of non-commutative models exposed in this chapter (the non-renormalizable (9.3) one, the Grosse–Wulkenhaar one, and the (9.41) one):

	the (9.3) *model*		*Grosse–Wulkenhaar model*		*the* (9.41) *model*	
	2P	4P	2P	4P	2P	4P
planar regular	ren.	ren.	ren.	ren.	ren.	ren.
planar irregular	UV/IR	log UV/IR	conv.	conv.	finite ren.	conv.
non-planar	IR div.	IR div.	conv.	conv.	conv.	conv.

The notations used have the following meanings: 'nP' ($n = 2$ or $n = 4$) stands for the $n-$point function, 'ren'. stands for the divergences which can be tamed through renormalization, 'UV/IR' stands for the phenomenon of ultraviolet/infrared mixing (see Section 9.1), 'log UV/IR' stands for a logarithmic ultraviolet/infrared mixing (similar to the previous one, but with a logarithmic behaviour in the external momentum of the graph to be inserted), 'conv'. stands for convergent, 'div'. stands for divergences which cannot be tamed through renormalization (the non-renormalizability of the model (9.3)) and finally 'finite ren'. stands for finite Feynman integrals which lead to a trivial renormalization of the parameters of the action (in this case the parameter a).

Let us also mention that a brief review of the renormalizability of the model (9.41) has been given in Tanasa (2010). Moreover, the issue of the commutative limit of this model has been analysed in Magnen, Rivasseau, and Tanasa (2009).

We end this section by explaining why we refer to the propagator

$$\frac{1}{p^2 + m^2 + \frac{a}{\theta^2 p^2}} \tag{9.42}$$

of this model as a heat kernel propagator (note that here we have used the explicit form (9.3)) of the non-commutativity matrix Θ). Using the formula

$$\frac{1}{A+B} = \frac{1}{A} - \frac{1}{A} B \frac{1}{A+B}, \tag{9.43}$$

with

$$A = p^2 + m^2 \text{ and } B = \frac{a}{\theta^2 p^2}, \tag{9.44}$$

the propagator (9.42) becomes

$$\begin{aligned} &\frac{1}{p^2+m^2} - \frac{1}{p^2+m^2}\frac{a}{\theta^2 p^2(p^2+m^2)+a} \\ &= \frac{1}{p^2+m^2} - \frac{1}{p^2+m^2}\frac{a}{\theta^2(p^2+m_1^2)(p^2+m_2^2)}, \end{aligned} \tag{9.45}$$

where $-m_1^2$ and $-m_2^2$ are the roots of the denominator of the second term in the right-hand side of the previous first line (considered as a second-order equation in p^2, namely $\frac{-\theta^2 m^2 \pm \sqrt{\theta^4 m^4 - 4\theta^2 a}}{2\theta^2} < 0$. Note that one needs to have $\frac{1}{4}\theta^2 m^4 \geq a > 0$ for this to hold.

This allows to have a heat kernel propagator (just like in the commutative case) with a correction which is a product of three such heat kernels. It is this representation of the propagator that was used in Tanasa (2009) for the implementation of the parametric representation (see next section) as well as for the calculation of the flow under the renormalization group of the parameters of the model (see Geloun and Tanasa (2008)). Let us also mention that explicit quantum corrections at 1-loop level—using various analytical properties of Bessel functions—was performed in Blaschke et al. (2008).

For the sake of completeness, we end this section by pointing out that important work is in progress to generalize this idea for restoring the renormalizability of gauge theories on non-commutative Moyal space (see Blaschke et al. 2009, 2010; as well as Tanasa (2008b), for a short survey).

9.7 Parametric representation and the Bollobás–Riordan polynomial

9.7.1 Parametric representation

Let us now present the implementation of the *non-commutative parametric representation*. This can be achieved using the same trick (3.25) as in the commutative case. The difference comes however from the fact that the interaction part is non-trivially dependent on the non-commutativity parameter θ and this dependence will consequently appear in the form of the two polynomials which give the parametric representation of the respective Feynman integral (polynomials which we denote by $U^\star(\alpha,\theta)$ and $V^\star(p_{\text{ext}},\alpha,\theta)$).

Using a Grassmannian development of the Pfaffians obtained through these Gaussian integrations, one can prove (see again Krajewski et al. (2010b)) that the Feynman integral of a generic graph Γ is written as:

$$\phi(\Gamma) = \int_0^\infty \frac{e^{-V^\star(p_{\text{ext}},\alpha,\theta)/U^\star(\alpha)}}{U^\star(\alpha,\theta)^2} \prod_{\ell=1}^{E} (e^{-m^2\alpha_\ell} d\alpha_\ell). \tag{9.46}$$

In order to present the exact form of the polynomials $U^\star_\Gamma(\alpha,\theta)$ and $V^\star_\Gamma(p_{\text{ext}},\alpha,\theta)$, one needs to give the following definition (see Krajewski et al. (2010b) for details):

Definition 9.7.1

1. *A $\star$-tree of a connected ribbon graph is a subribbon graph with one boundary component.*
2. *A two $\star$-tree is a subribbon graph with two boundary components.*

Note that these definitions are non-trivial generalizations of the topological notions of trees and respectively two-trees (see Definition 2.1.2 8 and respectively 9). Let us also mention that $\star$-trees are also known as quasi trees in knot theory literature and as *unicellular maps* in combinatorics literature (see, for example, (Vignes-Tourneret 2009; Champanerkar, Kofman, and Stoltzfus 2011), and references within).

One has

$$U^\star_\Gamma(\alpha,\theta) = \left(\frac{\theta}{2}\right)^b \sum_{\mathcal{T}^\star\ \star-\text{tree}} \prod_{e\notin \mathcal{T}^\star} 2\frac{\alpha_e}{\theta}, \tag{9.47}$$

where $b = F - 1 + 2g$.

The case of the polynomial $V^{\star}(p_{\text{ext}}, \alpha, \theta)$ is more involved, because the non-commutative setting presented here allows for it to be complex. One has the following theorems:

Theorem 9.7.2 *(Theorem 5.2 of Krajewski et al. (2010b)). The real part of the polynomial $V_\Gamma^{\star}(p_{\text{ext}}, \alpha, \theta)$ is written as:*

$$\mathcal{X}_\Gamma^{\star}(p_{\text{ext}}, \alpha, \theta) = \left(\frac{\theta}{2}\right)^{b+1} \sum_{\mathcal{T}_2^{\star}\ 2-\star\ \text{tree}} \prod_{e \notin \mathcal{T}_2^{\star}} 2\frac{\alpha_e}{\theta} (p_{\mathcal{T}_2^{\star}})^2, \tag{9.48}$$

where $p_{\mathcal{T}_2^{\star}}$ is the sum of the momenta entering one of the two faces of the two-$\star$−tree $\mathcal{T}_2^{\star}$.

Note that we say that an external momentum enters a face if the respective external edge is hooked to one of the vertices of the face (we also consider in the previous sum, for each such external momentum, the sign $+$ if the momentum is ingoing, respectively $-$ if it is outgoing). As in the commutative case, the choice of the face in the previous theorem is irrelevant (by momentum conservation).

Theorem 9.7.3 *(Theorem 5.3 of Krajewski et al. (2010b)). The imaginary part of the polynomial is expressed as $V_\Gamma^{\star}(p_{\text{ext}}, \alpha, \theta)$*

$$\mathcal{Y}_\Gamma^{\star}(p_{\text{ext}}, \alpha, \theta) = \left(\frac{\theta}{2}\right)^{b} \sum_{\mathcal{T}^{\star}\ \star\ \text{tree}} \prod_{e \notin \mathcal{T}^{\star}} 2\frac{\alpha_e}{\theta} \psi(p), \tag{9.49}$$

where $\psi(p)$ is the phase obtained by following the momenta entering the face of the $\star$−tree $\mathcal{T}^{\star}$ as if it were a Moyal vertex (see Krajewski et al. (2010b) for details).

Let us give some explanations on the definition of $\psi(p)$. We denote the (unique) face of the $\star$-tree by f. One can then define an incidence matrix $(\tilde{\epsilon}_{f,i})$ between this face and the external edges of the graph (in the same way one defines the usual incidence matrix between vertices and edges). Following the external momenta entering the face, one can define an order relation on these momenta. One then has $\psi(p) = \sum_{i<j} (\tilde{\epsilon}_{f,i} p_i) \wedge (\tilde{\epsilon}_{f,i} p_j)$, where $p_i \wedge p_j = \frac{1}{\theta} p_i^{\mu} \Theta_{\mu\nu} p_j^{\nu}$. This expression generalizes in a straightforward manner the expression which gives, as a function of the incoming/outgoing momenta, the phase corresponding to a Moyal vertex.

9.7.2 Relation between the multi-variate Bollobás–Riordan and the polynomials of the parametric representation

The polynomials $U_\Gamma^{\star}(\alpha, \theta)_\Gamma$ and $V_\Gamma^{\star}(p_{\text{ext}}, \alpha, \theta)$ can also be proven to respect the deletion/contraction relation. One has the following theorem:

Theorem 9.7.4 *(Theorem 5.1 of Krajewski et al. (2010b)). For any semi-regular edge e*

$$U_\Gamma^{\star}(\alpha, \theta) = \alpha_e U_{\Gamma-e}^{\star}(\alpha, \theta) + U_{\Gamma/e}^{\star}(\alpha, \theta). \tag{9.50}$$

One can further prove the following relation between the multi-variate Bollobás–Riordan and the polynomial $U^\star_\Gamma(\alpha,\theta)$ of the parametric representation:

$$U^\star_\Gamma(\alpha,\theta) = (\theta/2)^{\frac{1}{2}(V-E-1)}\Big(\prod_{e\in E}\alpha_e\Big) \times \lim_{w\to 0} w^{-1} Z_\Gamma\Big(1, \tfrac{\theta}{2\alpha_e}, w\Big). \tag{9.51}$$

Note that the commutative limit ($\theta \to 0$) gives the usual Φ^4 model on the commutative $\mathbb{R}^4$. This can be seen as well in the parametric representation, where $U^\star_\Gamma(\alpha,\theta)$ is given by $U_\Gamma(\alpha)$ plus some non-commutative corrections which factor θ^2 (corrections thus vanishing in the commutative limit). This directly applies to equation (9.51) (where one obviously has to take the limits in the appropriate order on the right-hand side):

$$U_\Gamma(\alpha) = \lim_{\theta\to 0} U^\star_\Gamma(\alpha,\theta) = \lim_{\theta\to 0}\left((\theta/2)^{\frac{1}{2}(V-E-1)}\Big(\prod_{e\in E}\alpha_e\Big) \times \lim_{w\to 0} w^{-1} Z_\Gamma\Big(1, \tfrac{\theta}{2\alpha_e}, w\Big)\right). \tag{9.52}$$

The second polynomial $V^\star_\Gamma(\alpha, p_{\text{ext}}, \theta)$ can also be related to the multi-variate Bollobás–Riordan polynomial in a similar way (see Krajewski et al. (2010b) for details).

For the sake of completeness, let us mention that the polynomials of the parametric representation of the Grosse–Wulkenhaar model (see previous notation) do not obey a deletion/contraction relation but they obey some other types of recursive relations, as proved by T. Krajewski et al. in Krajewski, Rivasseau, and Vignes-Tourneret (2011). Algebraic topological properties of the parametric representation of the Grosse–Wulkenhaar model (as well as of the parametric representation of the Φ^4 commutative model presented in Chapter 3) were studied in P. Aluffi and M. Marcolli (2009).

9.8 Combinatorial Connes–Kreimer Hopf algebra and its Hochschild cohomology

In this section, we present, following Tanasa and Vignes-Tourneret (2008); Tanasa and Kreimer (2012), the implementation of the combinatorial Connes–Kreimer Hopf algebra describing renormalizability on the Φ^4 models on Moyal space; we then analyse the Hochschild cohomology of its algebra and its relation to the combinatorics of the DSE of these models. Unless otherwise stated, the results in this section hold for both the Grosse–Wulkenhaar model and the translation-invariant model (9.41).

9.8.1 Combinatorial Connes–Kreimer Hopf algebra

The definition of the Hopf algebra of non-commutative Feynman graphs (driving the combinatorics of Moyal renormalization) is formally identical to the commutative case (see Section 8.2). Nevertheless, the class of superficially divergent graphs (one has to sum over in the definition of the coproduct) explicitly takes into account the topology of the graph. Thus, for the Grosse–Wulkenhaar model, the superficially divergent graphs of the renormalizable non-commutative models are the *planar regular* ribbon graphs with two and four external edges (see previous sections).

We now denote by $\mathcal{H}$ the unital, associative algebra freely generated by 1PI ribbon Feynman graphs (including the empty set, which we denote by $1_{\mathcal{H}}$). The product m is bilinear, commutative and given by the operation of disjoint union. Let the coproduct $\Delta : \mathcal{H} \to \mathcal{H} \otimes \mathcal{H}$ be defined as

$$\Delta\Gamma = \Gamma \otimes 1_{\mathcal{H}} + 1_{\mathcal{H}} \otimes \Gamma + \sum_{\gamma \in \underline{\Gamma}} \gamma \otimes \Gamma/\gamma, \ \forall \Gamma \in \mathcal{H}, \tag{9.53}$$

where, as announced previously, we have denoted by $\underline{\Gamma}$ the set of two and four external edges planar regular ribbon subgraphs of Γ. Furthermore, we define the counit $\varepsilon : \mathcal{H} \to \mathbb{K}$ as:

$$\varepsilon(1_{\mathcal{H}}) = 1, \ \varepsilon(\Gamma) = 0, \ \forall \Gamma \neq 1_{\mathcal{H}}. \tag{9.54}$$

Finally, the antipode is given recursively by

$$\begin{aligned} S : \mathcal{H} &\to \mathcal{H} \\ \Gamma &\mapsto -\Gamma - \sum_{\gamma \in \underline{\Gamma}} S(\gamma)\Gamma/\gamma. \end{aligned} \tag{9.55}$$

The combinatorics of the renormalizability of the Grosse–Wulkenhaar model is set in the following algebraic framework:

Theorem 9.8.1 *(Theorem 4.1 of Tanasa and Vignes-Tourneret (2008)). The quadruple $(\mathcal{H}^\star, \Delta, \varepsilon, S)$ is a Hopf algebra.*

The main difficulty in the proof of this theorem (in comparison with the original Connes–Kreimer one) comes from the necessity to define gluing data which respect the ribbon graph cyclic ordering. Thus, one can see in Fig. 9.2 an example of an insertion of a four external edges graph with a gluing data which respects the cyclic ordering, while the example of Fig. 9.3 does not.

Let us now investigate the case of the heat-kernel renormalizable non-commutative model (9.41). In order to do this, we begin with a short discussion on what the superficially divergent graphs of this model are.

As proven in Gurau et al. (2009), the superficially divergent graphs of the translation-invariant model (9.41) are again the two- and four-point graphs. However, in the case of planar regular graphs, these two- and four-point graphs will lead to the renormalization of the mass, field strength and coupling constant, just like in the case of the commutative ϕ^4 model. A more tricky situation appears for planar irregular graphs. The four-point function graphs are proven to be convergent. The two-point function graphs (which are

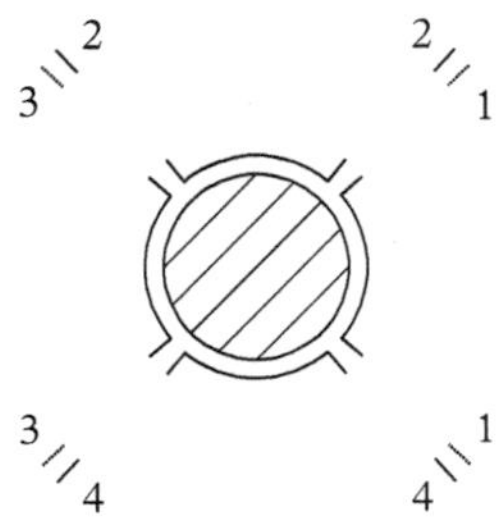

Figure 9.2 *Insertion of a regular ribbon graph with* 4 *external edges. The cyclic ordering is respected by the chosen gluing data*

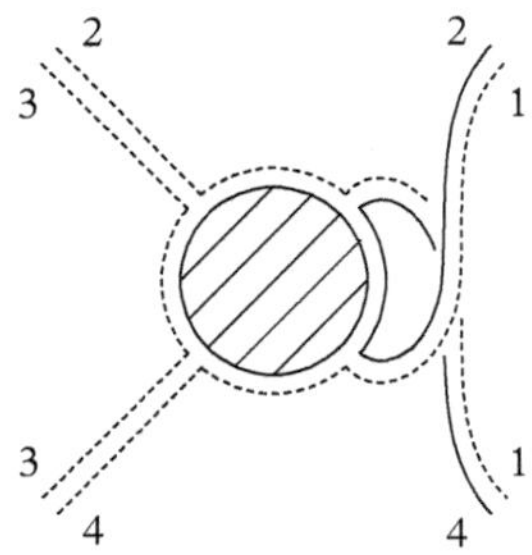

Figure 9.3 *Another example of the insertion of a ribbon graph. The cyclic ordering is not respected by the chosen gluing data*

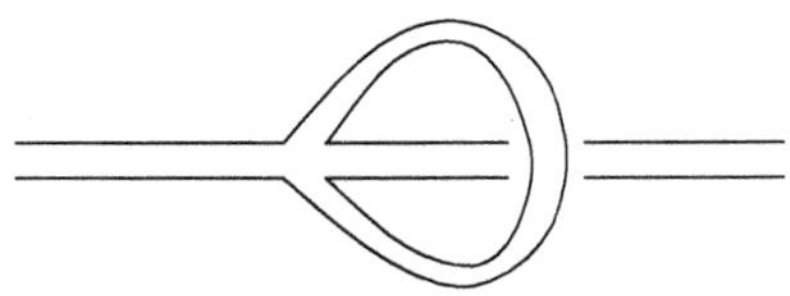

Figure 9.4 *The 'non-planar' tadpole. Note that this ribbon representation corresponds to the same Feynman as the one in Fig. 9.1*

those encoding the ultraviolet/infrared mixing) are again *convergent*. Nevertheless, when placing ourselves in the ultraviolet regime of their internal momenta, they lead a priori to a $1/p^2$ contribution to the respective Feynman integral (p being the external momenta of the respective two-point graph). However, the modification of the propagation given in (9.41) will ensure the renormalizability of the model. These two-point planar irregular graphs will lead to a *finite* renormalization for the coefficient a in the action (9.41).

When inserting these two-point graphs into 'larger' graphs, the latter become non-planar (for example, when inserting the tadpole of Fig. 9.4 in any planar ribbon graph, the resulting ribbon graph is non-planar). As explained in this chapter, in the case of the model (9.3), the Feynman integrals of these graphs are ultraviolet convergent but infrared divergent (because of ultraviolet/infrared mixing). In the case of the model (9.41), these non-planar graphs are also convergent in the infrared regime, thanks to the $1/p^2$ terms in the propagator (see the original proofs of Gurau et al. (2009), where multi-scale analysis was used, or Tanasa (2009), where the parametric representation was used).

Let us conclude by stating that, for the reasons already explained, the primitive divergent graphs of the model (9.41) are taken to be the planar regular two- and four-point graphs.

This justifies the fact that one can describe the combinatorics of the renormalizability of the model (9.41) using the same Hopf algebra as in the case of the Grosse–Wulkenhaar model—Theorem 9.8.1. Moreover, one can prove the following:

Proposition 9.8.2 *(Proposition 3.3 of Tanasa and Kreimer (2012)). The ideal $\mathcal{H}^{\text{pli}}$ generated by the 1PI two-point planar irregular Feynman graphs is a Hopf ideal and coideal in $\mathcal{H}$,*

$$\Delta(\mathcal{H}^{\text{pli}}) \subseteq \mathcal{H}^{\text{pli}} \otimes \mathcal{H} + \mathcal{H} \otimes \mathcal{H}^{\text{pli}},\ \varepsilon(\mathcal{H}^{\text{pli}}) = 0, S(\mathcal{H}^{\text{pli}}) \subseteq \mathcal{H}^{\text{pli}}. \tag{9.56}$$

A direct consequence of this proposition is that one can discard the two-point planar irregular sector by simply taking the respective quotient. Let us also note that this cannot be done for the non-planar sector, because the non-planar sector does not form a Hopf coideal (one can easily find some counterexamples).

Let us now make a short remark on the accompanying (pre-)Lie algebras. The insertion operation is defined as the bilinear map

$$\Gamma_1 \circ \Gamma_2 := \sum_{\Gamma} n(\Gamma_1, \Gamma_2, \Gamma)\Gamma. \tag{9.57}$$

The coefficient $n(\Gamma_1, \Gamma_2, \Gamma)$ counts the number of ways to shrink Γ_2 to its residue in the graph Γ such that Γ_1 is obtained.

Let us look closely at the insertion of a planar irregular four-point ribbon graph. One can insert such a graph in a Moyal vertex in a way that reduces the number of broken faces and does not increase the genus of the resulting graph. This becomes clear in Fig. 9.5. Inserting planar irregular graphs into planar graphs in a way that a planar

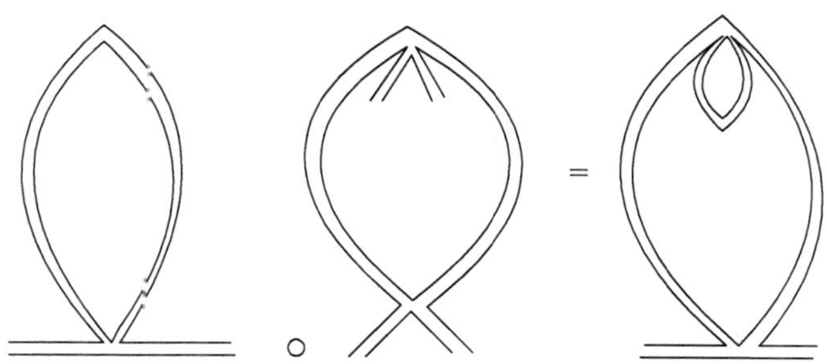

Figure 9.5 *An example of an insertion of a planar irregular graph leading to a planar regular graph. The number of broken faces is thus reduced. We have inserted a planar irregular (thus convergent) graph in a planar regular one. The result is again planar regular*

Figure 9.6 *An example of a four-point graph with four faces broken by external edges. Each face is broken by a single edge. For this reason, when inserting such a graph in a Moyal vertex, the resulting graph is non-planar*

(regular or irregular) graph is obtained is also possible for four-point graphs with three broken faces, in a similar way. This is still possible because one has two edges on a particular face.

Nevertheless, this is no longer possible for four-point planar irregular graphs with four faces broken by external edges (consider Fig. 9.6). This comes from the fact that one no longer has two external edges on the same face, whose presence, by a proper defined gluing data, could have prevented the final graph from being non-planar. One can thus conclude that in any way such a graph is inserted into a Moyal vertex the genus of the resulting graph increases.

When inserting planar regular graphs, one can consider gluing data compatible with the cyclic ordering, which we denote by $\circ_c$ (see Fig. 9.7). One then has a pre-Lie algebra structure. Antisymmetrizing the pre-Lie product gives a Lie bracket

$$[\Gamma_1,\Gamma_2]=\Gamma_1\circ_c\Gamma_2-\Gamma_2\circ_c\Gamma_1, \tag{9.58}$$

which defines a Lie algebra structure L. Consider now the graded dual of the universal enveloping algebra of this Lie structure. This gives the renormalization Hopf algebra introduced previously. The fact that these products are indeed pre-Lie and respectively Lie products can thus be seen as a direct consequence of the existence of the Hopf algebra structure.

The renormalized Feynman integrals are defined analogously as in the commutative case (see the renormalized Feynman integral formula (8.19)).

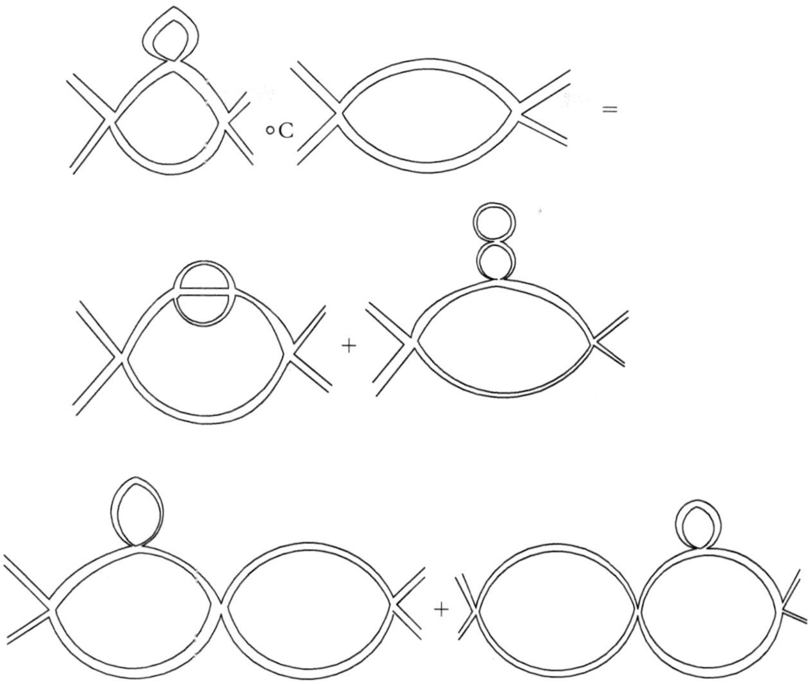

Figure 9.7 *Example of a ribbon graph insertion which respects the cyclic ordering*

9.8.2 Hochschild cohomology and the combinatorial DSE

Let us now investigate how Hochschild cohomology issues and the combinatorial DSE generalize within this Moyal QFT framework. Following Tanasa and Kreimer (2012), we define a map,

$$B_+^{k;r} := \sum_{\substack{|\gamma|=k \\ \mathrm{res}(\gamma)=r}} \frac{1}{\mathbf{sym}(\gamma)} B_+^{\gamma}, \tag{9.59}$$

where

$$B_+^{\gamma}(X) := \sum_{\Gamma \in \mathcal{H}_{\mathrm{lin}}} \frac{\mathbf{bij}(\gamma, X, \Gamma)}{|X|_{\vee}} \frac{1}{\mathrm{maxf}(\Gamma)} \frac{1}{[\gamma|X]} \Gamma, \tag{9.60}$$

We also set:

$$B_+^{\gamma}(1_{\mathcal{H}}) := \gamma. \tag{9.61}$$

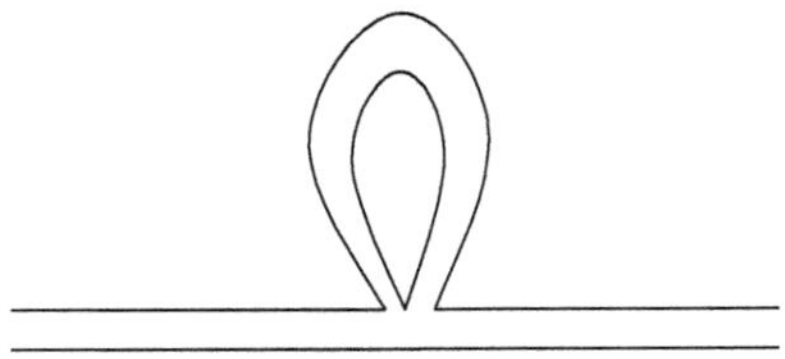

Figure 9.8 *The up tadpole. This graph is planar regular and hence superficially divergent. Its symmetry factor is equal to* 1

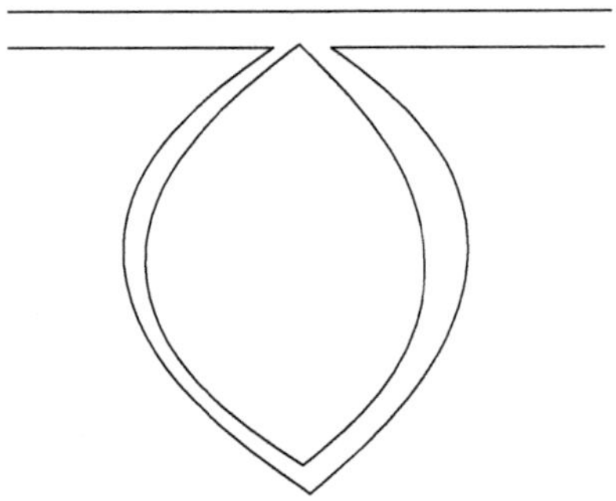

Figure 9.9 *The down tadpole. This graph is planar regular and hence superficially divergent. Its symmetry factor is equal to* 1

Figure 9.10 *An example for* Γ_V

Note that this *formal* definition is identical to Definition 8.25 of the commutative case (see Section 8.3). Nevertheless, the combinatorial factors previously given are *different*.

One has:

Proposition 9.8.3 *(Proposition* 4.1 *of Tanasa and Kreimer (2012)). The symmetry factor of a ribbon graph is equal to* 1.

For example, the up and down ribbon tadpoles of Figs. 9.8 and 9.9 have unital symmetry factors.

Let us now give an example which shows that the number Γ_V of distinct ribbon graphs which are equal upon removal of external edges is again different with respect to the commutative case. When considering Fig. 9.10, a third graph is missing since only the planar regular sector of the theory is relevant here.

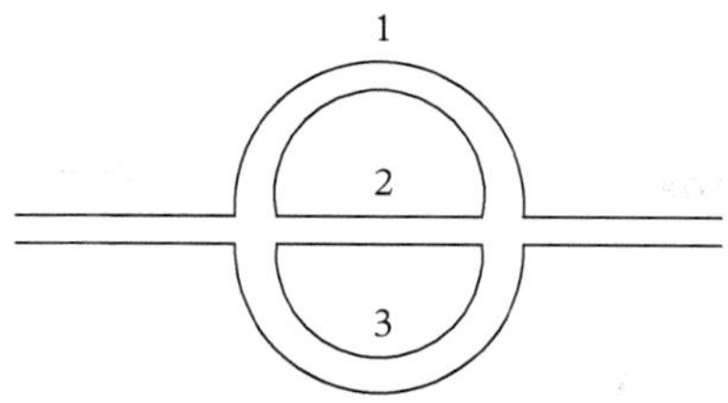

Figure 9.11 *The sunshine ribbon graph, with three internal edges. Its maximal number of forests is equal to* 2

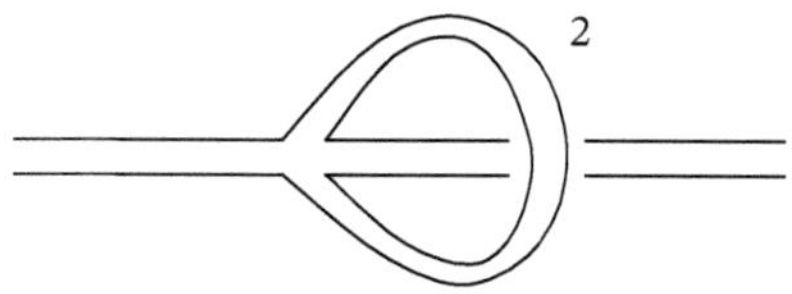

Figure 9.12 *The 'non-planar' tadpole obtained from shrinking the bubble graph formed by edges* 1 *and* 3 *in the sunshine graph of Fig. 9.11*

Finally, let us also give an example showing that the number maxf of maximal forests (see the commutative case for the definition) is also different with respect to the commutative case. The graph of Fig. 9.11 has a maximal number of forests equal to two. Indeed, when the sub-divergence is represented by the bubble graph formed of the internal edges 1 and 2 then the resulting cograph is the down tadpole of Fig. 9.9. If the sub-divergence is taken to be the bubble graph formed of the internal edges 2 and 3 then the resulting cograph is the up tadpole of Fig. 9.8. Finally, if one takes the divergence to be given by the internal edges 1 and 3 then the resulting cograph is the 'non-planar' tadpole of Fig. 9.12, which is not divergent. Hence, maxf is equal to 2. In the case of the commutative ϕ^4 theory, the previous three tadpoles are equivalent and thus the number of maximal forests will be equal to 3.

As in the commutative case, the B_+ operator already defined generalizes in a non-trivial way the insertion operator used to define a pre-Lie algebra structure on the vector space of 1PI ribbon Feynman graphs (see previous subsection).

One can then prove that such an operator is a Hochschild 1−cocycle:

Theorem 9.8.4 *(Theorem 5.4 ii) of Tanasa and Kreimer (2012)). One has:*

$$\Delta B_+^{k;r} = B_+^{k;r} \otimes 1_{\mathcal{H}} + (\mathrm{id} \otimes B_+^{k;r})\Delta. \tag{9.62}$$

Furthermore, one has:

Theorem 9.8.5 *(Theorem 5.4 i) of Tanasa and Kreimer (2012)). Any relevant graph in the perturbative expansion (with the right combinatorial factor) is in the range of a B_+ cocycle.*

As in the commutative case (see again Section 8.3), this theorem can be interpreted as follows: the combinatorial DSE is as a formal construction based on the Hochschild cohomology of the combinatorial Connes–Kreimer Hopf algebra.

Let us end this subsection by defining, as in the commutative case, the sum of graphs of a given residue r and a given loop number k,

$$c_k^r := \sum_{|\Gamma|=k} \Gamma. \tag{9.63}$$

Finally, one has:

Theorem 9.8.6 *(Theorem 5.4 iii) of Tanasa and Kreimer (2012)).*

$$\Delta(c_k^r) = \sum_{j=0}^{k} \mathrm{Pol}_j^r(c) \otimes c_{k-j}^r, \tag{9.64}$$

where $\mathrm{Pol}_j^r(c)$ *is a polynomial in the variables* c_m^r *of degree* j.

The interested reader can refer to Appendix C for some explicit computations, up to two loops in perturbation theory. Computations have been done for all the results given in this subsection.

We end this chapter by mentioning here that a similar combinatorial Hopf algebraic structure, for quantum gravity spin-foam models, and its Hochschild cohomology have been studied in detail in Tanasa (2010a).

9.9 Selected further reading

- Vincent Rivasseau. (2007). *Non-commutative Renormalization.* Xth Poincaré (BourbaPhy) Seminar, 15–95, arXiv:0705.0705.

  A review on the renormalization of the Grosse–Wulkenhaar scalar model on the non-commutative Moyal space.

10

Quantum gravity, group field theory (GFT), and combinatorics

This chapter is the first chapter of the book dedicated to the study of the combinatorics of various quantum gravity approaches.

The chapter is structured as follows. After a brief introductory section to quantum gravity, we shortly mention the main candidates for a quantum theory of gravity: string theory, loop quantum gravity (LQG), and group field theory (GFT), causal dynamical triangulations, matrix models. The next sections introduce some GFT models such as the Boulatov model, the colourable and the multi-orientable model. The saddle point method for some specific GFT Feynman integrals are presented in the fifth section. Finally, some algebraic combinatorics results are presented: definition of an appropriate Connes–Kreimer Hopf algebra describing the combinatorics of the renormalization of a certain tensor GFT model (the so-called Ben Geloun–Rivasseau (BGR) model) and the use of its Hochschild cohomology for the study of the combinatorial DSE of this specific model.

10.1 Quantum gravity

The fundamental interactions of the Universe such as:

1. the electromagnetic interaction, responsible for the interactions such as the ones between elementary particles such as the electrons or the photons, and thus responsible for electromagnetism;
2. the strong interactions, responsible for the interactions within a nucleus (interactions between the quarks forming the protons and the nucleons);
3. the weak interactions, responsible for the interactions of more exotic elementary particles, such as the neutrinos;

can all be compatible with the principles of quantum mechanics. All the three previous fundamental interactions can be described within a QFT framework. This is known

Combinatorial Physics: Combinatorics, quantum field theory, and quantum gravity models. Adrian Tanasa, Oxford University Press (2021). © Adrian Tanasa. DOI: 10.1093/oso/9780192895493.003.0010

under the name of the *Standard Model of elementary particles*. A key notion here is the notion of *quanta*, which is an elementary particle carrying a given fundamental interaction. This means that the respective fundamental interaction can be seen as an exchange of these quantas. For example, the quanta of the electromagnetic force is the photon.

However, there is one last fundamental interaction, the gravitational one (described by Einstein's general relativity), which cannot be quantified as where quantified the three fundamental interactions listed previously. This quantification according to the usual 'recipes' would lead to a non-renormalizable theory (see Chapter 3). This means that Einstein's general relativity, which gives us a classical (i.e. non-quantic theory of gravity, and quantum mechanics are not compatible!

This made the quest for a theory of quantum gravity to be considered as the Holy Grail of modern theoretical physics. This unification is not only of purely 'academic' interest, but it is also motivated by the fact that it exists a certain energy scale, called the Planck scale

$$\ell_{\text{Planck}} \propto 10^{-35}\, m, \tag{10.1}$$

where the effects of general relativity and of quantum mechanics are of the same order of magnitude. Certainly, the Planck scale is very small (and thus corresponds to huge energy units, which are not expected to be reached by current technology for quite a while), but at this scale, general relativity and quantum mechanics cannot ignore each other anymore—an unified theory is needed!

10.2 Main candidates for a theory of quantum gravity: The holographic principle

The most important candidate for a theory of quantum gravity has been over several decades the 'all-mighty' **string theory** (see the books Schwarz, Green, and Witten (2012); Polchinski (1998); or B. Zwiebach (2009)). The main idea of string theory is that the point-like particles of particle physics QFT should be replaced by 1D objects called strings. One of the *many* vibrational states of the string corresponds to the graviton, the quanta of gravity.

One of the many criticisms brought to string theory is that it is a *background depend* theory. This means that space-time (which, by the way, needs a considerable number of supplementary dimensions for string theory to be mathematically consistent!) is some kind of a spectator of the quantification of particles and fields (this being the case in QFT as well, see again Chapter 3, but QFT usually doesn't deal with space-time, as string theory is supposed to).

Another approach for the quantification of gravity is given by **loop quantum gravity** (LQG) (see C. Rovelli (2004); or T. Thiemann (2017)). LQG is a background dependent attempt to provide a quantum theory of gravity. In this approach, space and time are quantized, analogously to the way energy and momentum are quantized in quantum mechanics. The formal equation of quantum gravity, the so-called Wheeler-DeWitt equation, is proved to admit solutions labelled by loops when rewritten in terms of the so-called Ashtekar variables.

A relation between LQG and the QFT formalism exposed in Chapter 3 is given by **GFT**—the interested reader can turn to the review papers Freidel (2005); or Oriti (2006). The following sections of this chapter are dedicated to the presentation of some GFT models, namely the coloured and the multi-orientable ones. It is interesting noticing that these two models, which lie nowadays at the foundation of modern tensor theory (see the following chapters), have been initially proposed within a GFT approach.

Another approach for quantum gravity which developed through the last decades is given by the **causal dynamical triangulations**. Moreover, if one restricts to 2D Quantum Gravity, then **matrix models** represent a successful approach for quantifying gravity (see the Chapter 11).

Let us end this section by mentioning a very interesting principle, the **holographic principle**, which is now expected, within some quantum gravity approaches, to be a property of quantum gravity. Thus, in a theory with gravity one is expected to have a certain correspondence between a volume of space and the boundary enclosing it. The most studied example of holography is the Anti de Sitter/Conformal Field Theory correspondence (sometimes called Maldacena duality or gauge/gravity duality), which was conjectured by Maldacena (1999). In this framework, holographic duality refers to the duality between quantum many-body systems in d spatial dimensions and gravitational theories in $(d+1)$-dimensional asymptotically anti-de Sitter geometries.

A toy model exhibiting this kind of AdS/CFT (although called nearly-AdS/CFT, since it is a $(0+1)$-dimensional model) is the Sachdev–Ye–Kitaev model, see Chapter 15.

10.3 GFT models: the Boulatov and the colourable models

As announced previously, let us now give some more details on the definition of some particular GFT models. GFT is defined as a QFT on *group manifolds* (hence the "group" in its name). Its associated Feynman graphs are, for the D-dimensional case, rank D tensor graphs.

As in the commutative Φ^4 model described in Chapter 3 (or in the non-commutative Φ^4 models described in Chapter 9), one only has one type of field, but this time this field does not depend on space-time (space-time is not the 'scene' of the action, one does not have background dependence, as mentioned in the previous section). From a mathematical point of view, the GFT field ϕ is a function

$$\phi : G^D \to \mathbb{R}, \tag{10.2}$$

where G is some group. In the rest of this chapter we consider $D = 3$.

In order to have a connection with the 3D topological quantum gravity, one needs to use as group the holonomy group $SU(2)$ (one can thus refer to GFT as a theory of holonomies).

The action of this topological 3D model (also known as the Boulatov model Boulatov (1992)) written as:

$$
\begin{aligned}
S[\phi] = &\frac{1}{2}\int dg_1 dg_2 dg_3 \phi(g_1,g_2,g_3)\phi(g_1,g_2,g_3) \\
&+\frac{\lambda}{4!}\int dg_1 \dots dg_6 dg_{1'} \dots dg_{6'} \phi(g_1,g_2,g_3)\phi(g_{3'},g_4,g_5)\phi(g_{5'},g_{2'},g_6)\phi(g_{6'},g_{4'},g_{1'}) \\
&\delta(g_1 g_{1'}^{-1})\dots\delta(g_6 g_{6'}^{-1}),
\end{aligned}
\tag{10.3}
$$

Using the previous group δ-functions to perform some of the integrals, one has

$$
\begin{aligned}
S[\phi] = &\frac{1}{2}\int dg_1 dg_2 dg_3 \phi(g_1,g_2,g_3)\phi(g_1,g_2,g_3) \\
&+\frac{\lambda}{4!}\int dg_1 \dots dg_6 \phi(g_1,g_2,g_3)\phi(g_3,g_4,g_5)\phi(g_5,g_2,g_6)\phi(g_6,g_4,g_1).
\end{aligned}
\tag{10.4}
$$

It is this form of the interaction potential which corresponds to the graphical representation of Fig. 11.3. Note that the integration over the group is done with the invariant Haar measure, both in (10.3) and in (10.4).

Here, the field ϕ in equation (10.2), taken real-valued, is not assumed to have specific symmetry properties under the permutations of its arguments: only the identical permutation is associated with such an edge. This type of model is called *orientable* in GFT literature.

The propagator of such a topological model is written:

$$
C(g_1,g_2,g_3;g_1',g_2',g_3') = \int dh \prod_{i=1}^{3} \delta(g_i h (g_i')^{-1}).
\tag{10.5}
$$

As in any QFT model, in order to write down a Feynman integral, one needs to put together the contributions of the vertices and the propagators (10.5). For a general Feynman integral, one has

$$
A_\Gamma = \int \prod_{\ell \in L_\Gamma} dh_\ell \prod_{f \in F_\Gamma} \delta\left(\vec{\prod}_{\ell \in f} h_\ell^{\eta_{\ell f}}\right),
\tag{10.6}
$$

where L_Γ, F_Γ are the set of edges and faces of Γ, respectively. The oriented product $\vec{\prod}_{l\in f} h^{\eta_{\ell f}}$ means that the product of the variables h_ℓ has to be taken in the cyclic ordering corresponding to the face orientation (starting anywhere on the cycle).

The Feynman integral (10.6) depends neither on the chosen orientation of the edges, nor on those of the faces. One can see this through a direct check $h \to h^{-1}$ (using the parity of the δ function and of the Haar measure under this transformation). Let us emphasize that the formula (10.6) can lead (as for regular QFT) to divergences. For explicit computations of GFT integrals, we refer the interested reader to Appendix D.

The colourable 3D model is defined in the following way. The (real-valued) field (10.2) is replaced by four complex-valued fields ϕ_p; the index $p = 0,\ldots,3$ is a colour index. Moreover, one now has two types of vertices, a ϕ^4 one and a $\bar{\phi}^4$ one. Furthermore, a clockwise cyclic ordering of the four colours adjacent at one of these types of vertices (and an anticlockwise cyclic ordering at the second type of vertex) is also imposed; the action thus writes:

$$S_{\text{col}} = \frac{1}{2}\sum_p \int \bar{\phi}_p \phi_p + \frac{\lambda}{4!}\int \phi_0 \ldots \phi_4 + \frac{\lambda}{4!}\int \bar{\phi}_0 \ldots \bar{\phi}_4. \tag{10.7}$$

Note that, from now on, the integrations over the group are left implicit.

From a graph-theoretical point of view, this means that one imposes a particular four-colouring of the edges and that only bipartite graphs are kept.

10.4 The multi-orientable GFT model

Following the original paper Tanasa (2012a); and the review Tanasa (2012b), let us now introduce a distinct QFT particularization of the model (10.4). As in the coloured case, one has a complex-valued field ϕ. Nevertheless, we do not need more copies of this field. The proposed interaction is restricted to vertices in which each corner has a $+$ or a $-$ label. Furthermore, each vertex has two corners labelled with $+$ and two corners labelled with $-$, which are cyclically ordered as shown in Fig. 10.1. A field propagates from a $-$ to a $+$ corner of some vertex.

This notion of corners of a vertex is natural within the framework of non-local QFT. Moreover, GFT can also be seen in a sense as a non-local QFT on the group manifold

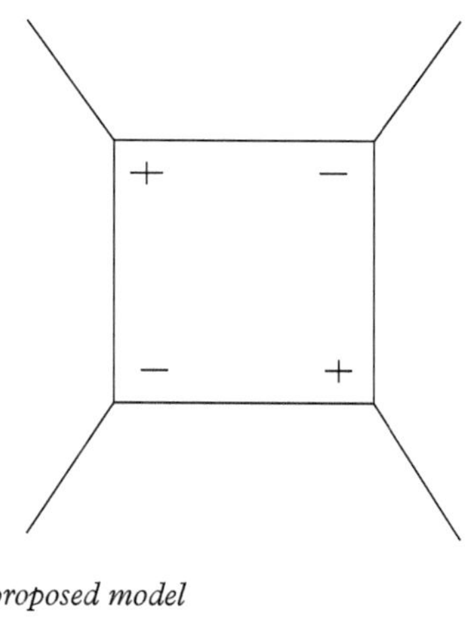

Figure 10.1 *The vertex of the proposed model*

on which it lives. This comes from the fact that the interaction is not defined on some 'group point' g (see (10.3) and (10.4)); this is distinct from the interaction of local QFT models (see (3.3)), where the interaction is $\Phi^4(x)$ and is localized on some point x of the space-time $\mathbb{R}^4$.

Let us make the following comment. We have seen that the notion of locality generalizes to Moyal QFT to some kind of 'Moyality', stating that the counterterms must have the same Moyal form as the terms of the bare action. One hopes to also obtain such a generalized locality for tensor model counterterms; we propose to call this eventual property **traciality**, since one should keep, for the counterterms, the same form given by some appropriate tensor trace (thus generalizing the notion of matrix trace).

The action of the proposed model is written as:

$$S[\phi] = \frac{1}{2}\int \bar{\phi}\phi + \frac{\lambda}{4!}\int \bar{\phi}\phi\bar{\phi}\phi. \tag{10.8}$$

In the original paper Tanasa (2012a), this model was called multi-orientable because one has, on one hand, the orientability of the edges mentioned previously (no twists between the various strands) and on the other hand, the orientability of the vertex. Let us mention that combinatorial maps with four-valence vertices such as the orientable ones of Fig. 10.1 have not been counted in combinatorics Fusy (2011).

As explained in Chapter 3, a QFT action leads, through an appropriate perturbative expansion, to a certain class of graphs. Through this mechanism, colourability discards a highly significant class of graphs, including the so-called 'wrapping singularities', which correspond, amongst others, to graphs containing loops, (the *tadpoles* in QFT terminology, as we have already mentioned several times in this book), like Figs. 10.2 and 10.3 (see the review paper Gurau and Ryan (2012), and references within for more details on these types of singularities).

Figure 10.2 *A tadpole graph which is generated by the QFT action* (10.8). *This graph is not permitted by the colourable GFT. The indices correspond to the group elements associated to the two external edges*

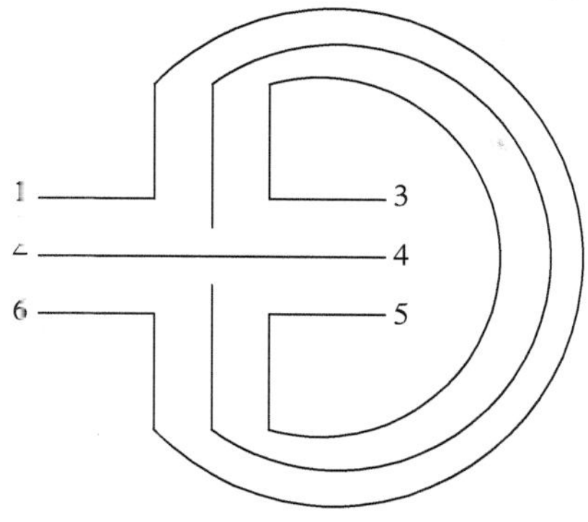

Figure 10.3 *A tadpole graph which is not generated by the QFT action* (10.8). *The indices correspond to the group elements associated with the two external edges*

Figure 10.4 *A non-tadpole example of a GFT graph*

Proposition 10.4.1 *(Proposition* 3.1 *of Tanasa (2012a)). Every GFT graph which is colourable is also generated by the QFT action* (10.8).

The reciprocal statement however is not true. A counterexample is the tadpole graph of Fig. 10.2. Another example, which is not a tadpole graph, is the graph of Fig. 10.4. This can be rephrased as: the proposed model (10.8) discards a less important class of graphs than the colourable model.

10.4.1 Tadpoles and generalized tadpoles

Before investigating the issue of tadpoles within the QFT framework defined by the action (10.8), let us get back in more detail to a remark previously mentioned in this chapter. Since the edges of the models we deal with here do not allow for twists, one can drop the 'middle' strand of any edge and obtain a one-to-one correspondence with some ribbon graph (or combinatorial map). The ribbon graph thus obtained is the *jacket* (introduced in Ben Geloun et al. (2010); and generalized in Gurau (2011b)). One can thus refer to the planarity of the respective tensor GFT graph as the planarity of the associated ribbon graph. Moreover, exactly as for ribbon graphs (see Section 2.2) one can count the number of faces broken by the external edges; we denote this number by B.

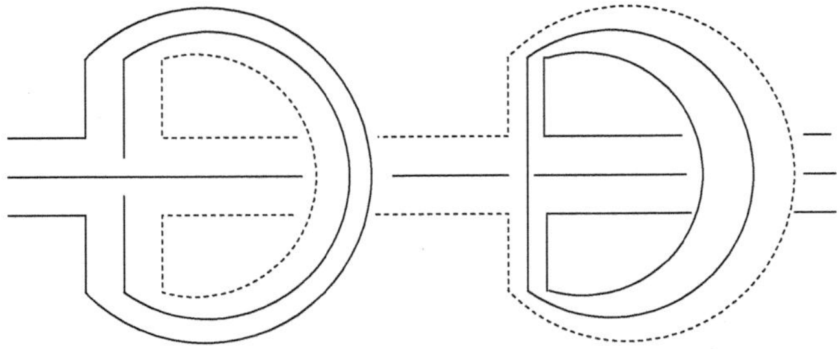

Figure 10.5 *A tadface graph. The tadface is represented by the dashed line. One notices that the respective face goes through the edge connecting the two tadpoles twice, once through one strand and once through another strand of the edge*

If $B > 1$ we call the respective graph *irregular*, just like in Moyal QFT (see previous chapter). The tadpole in Fig. 10.2 is thus referred to as a *planar* tadpole, while the tadpole in Fig. 10.3 is referred to as a *'non-planar'* tadpole (see again the previous chapter). We have seen that planar tadpoles are permitted by the model (10.8), while the 'non-planar' ones are not.

Let us now recall from Ben Geloun, Magnen, and Rivasseau (2010) the following definition:

Definition 10.4.2 *A generalized tadpole is a graph with one external vertex.*

Planar generalized tadpoles are permitted by models like that defined in (10.8). A 'non-planar' generalized tadpole is a graph with two external edges and with $B = 2$. These graphs are not permitted by models of type (10.8).

10.4.2 Tadfaces

We recall from Ben Geloun et al. (2010) the following definition:

Definition 10.4.3 *A tadface is a face that goes through an edge at least twice.*

Let us give further explanation on this. Such a tadface can be obtained if one traces the respective edge a first time through one strand and a second time through another strand. A second order (in perturbation theory) example of such a graph is given in Fig. 10.5. Nevertheless, this graph is not permitted by the model (10.8), since it is made up of two 'non-planar' tadpoles like those in Fig. 10.3 (each of them not being permitted by the model (10.8), as explained previously).

One has the following result:

Theorem 10.4.4 *(Theorem* 3.1 *of Tanasa (2012a)). Tadfaces are not permitted by the multi-orientable tensor model* (10.8).

One can also obtain the following result, as announced in Ben Geloun et al. (2010):

Corollary 10.4.5 *(Corollary 3.1 of Tanasa (2012a)). Tadfaces are not permitted by colourability.*

Furthermore, one has:

Corollary 10.4.6 *(Corollary 3.2 of Tanasa (2012a)). 'Non-planar' generalized tadpoles are not permitted by multi-orientability.*

Finally, one has:

Corollary 10.4.7 *(Corollary 3.3 of Tanasa (2012a)). 'Non-planar' generalized tadpoles are not permitted by colourability.*

We summarize all these results in the following table, which compares the two models:

	colourable	model (10.8)
generalized planar tadpoles	forbidden	permitted
generalized 'non-planar' tadpoles	forbidden	forbidden
tadfaces	forbidden	forbidden

The comparison with the colourable model was advanced further in Tanasa (2012a) through the investigation of aspects related to the renormalizability (see Chapter 3) of these two types of GFT models. Thus, it was exhibited that there exist graphs which, within the colourable framework, are divergent (when one computes the associated Feynman integral) but that do not correspond to terms present in the action (10.7). For the QFT framework defined by the model (10.8), this is cured, in the sense that these graphs are still divergent but correspond to terms present in the action (10.8) (see the original paper Tanasa (2012a) for more details). An example of such a graph is given in Fig. 10.6.

10.5 Saddle point method for GFT Feynman integrals

In order to use the saddle point method, we need to re-express the propagator (10.5) of the topological model (10.3) using $SU(2)$ coherent states (see Appendix E).

One can introduce two distinct $SU(2)$ gauge-averaging variables, u and v, (one for each half-edge of an edge) instead of the single group variable h (to avoid any ambiguity, these two new variables can be introduced using the incidence matrix of the graph); one can then insert the partition of unity (E.1).

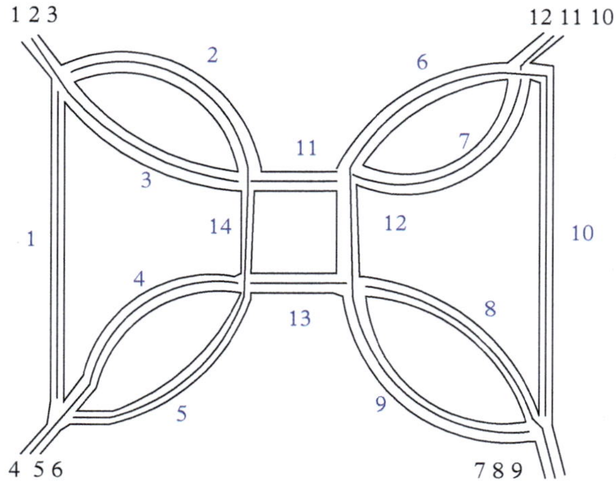

Figure 10.6 *An example of a tensor graph which represents a contribution of a form not present in the bare action. In the QFT framework of the model* (10.8), *it represents a contribution of a form already present in the bare action. The fourteen internal edges and the twelve group elements associated with the four external edges have been labelled*

After some algebra (see Krajewski et al. (2010) for details), one has a general integral of some tensor graph Γ factorized over faces:

$$A_\Gamma = \int \prod_{\ell \in L_\Gamma} \mathrm{d}u_\ell \mathrm{d}v_\ell \prod_{f \in F_\Gamma} \mathcal{A}_f. \tag{10.9}$$

To write down $\mathcal{A}_f$, let us number the vertices and edges in the (anti)-cyclic order along a face f of length p as $\ell_1, v_1 \cdots \ell_p, v_p$, with by definition $\ell_{p+1} = \ell_1$. We then have

$$\mathcal{A}_f = \sum_j d_j^{p+1} \int \prod_{a=1}^{p} dn_{\ell_a f} < j, n_{\ell_a f} | h_{\ell_a, v_a}^{\eta_{\ell_a f}} h_{\ell_{a+1}, v_a}^{\eta_{\ell_{a+1} f}} | j, n_{\ell a+1, f} >, \tag{10.10}$$

where,

$$\begin{aligned} h_{\ell_a, v_a} &= v_{\ell_a} \text{ if } \epsilon_{v_a, \ell_a} = +1 \\ h_{\ell_a, v_a} &= u_{\ell_a} \text{ if } \epsilon_{v_a, \ell_a} = -1. \end{aligned} \tag{10.11}$$

When working on the 4D case, the case of interest for quantum gravity, one needs to use the $SU(2) \times SU(2)$ group (the universal covering group of $SO(4)$). Similarly, one has a coherent state system $|j_+, n_+\rangle \otimes |j_-, n_-\rangle$ and the following resolution of identity formula holds:

$$\mathbf{1}_{j_+} \otimes \mathbf{1}_{j_-} = \mathbf{1}_j = \mathrm{d}_{j_+}\mathrm{d}_{j_-} \int \mathrm{d}n_+ \mathrm{d}n_- |j_+, n_+\rangle \otimes |j_-, n_-\rangle\langle j_+, n_+|\langle j_-, n_-|. \quad (10.12)$$

Using this identity and the notation $u = (u_+, u_-)$ and $v = (v_+, v_-)$ (elements of $SU(2) \times SU(2)$), one can re-express the propagator of the model as:

$$\int \mathrm{d}u\mathrm{d}v \prod_{f=1}^{4} \sum_{j_{f+}, j_{f-}} \mathrm{d}_{j_{f+}} \mathrm{d}_{j_{f-}} \tilde{\rho}_{V_{j_{f+}} \otimes V_{j_{f-}}} \left(u g_f (g'_f)^{-1} v^{-1} \mathbf{1}_{j_f}\right), \quad (10.13)$$

with g_f, g'_f, u and v elements of $SU(2) \times SU(2)$.

Let Γ be a tensor graph, with V vertices, L edges and F faces, usually labelled by letters v, ℓ and f. In the coherent state representation, the GFT integral can in general be written as

$$\phi_\Gamma = \sum_{j_f \leq \Lambda} \mathcal{N} \int \prod dh \prod dn \exp\{\sum_f j_f S_f[h, n]\}, \quad (10.14)$$

where $\mathcal{N}$ is a normalization factor which is a rational function of the spins (or highest weight of the representation). The exact form of the face action and of the number of group variables $h \in \mathrm{SU}(2)$ and unit vectors $n \in S^2$ depends on the choice of model. As explained previously, the sums over the spins j_f may lead to divergences; we thus use a standard QFT recipe and introduce an 'ultraspin' cut-off Λ (this restricts the summation to spins below Λ just like the an usual momentum space cutoff, see Chapter 3). We set $j_f = jk_f$ with $k_f \in [0, 1]$ and use the stationary phase method to derive the large j behaviour of

$$\int \prod dh \prod dn \exp\{j \sum_f k_f S_f[h, n]\}. \quad (10.15)$$

If the action is complex but has a negative real part, the contribution to this integral is given by quadratic fluctuations around zeroes of the real part of S; these are stationary points of its imaginary part, otherwise the integral is exponentially suppressed as $j \to \infty$.

For the topologic Boulatov model, one has to integrate against one group element $h_l \in \mathrm{SU}(2)$ for each edge. In the 4D topologic model, as previously mentioned, one must work with $\mathrm{SU}(2) \times \mathrm{SU}(2)$ instead of $SU(2)$; this leads to two independent copies of the previous Feynman integral. We thus restrict ourselves to SU(2) for simplification; the Feynman integral is given by (10.10). Introducing, as already explained, the variable k_f one can write the Feynman integral under the form (10.14), with (see again Krajewski et al. (2010) for details):

$$S_f[h, n] = k_f \log \mathrm{Tr}\Big[\big(\overrightarrow{\prod_{\ell \in \partial f}} h_\ell^{\eta_{\ell, f}}\big)(1 + \sigma \cdot n)\Big]. \quad (10.16)$$

To perform the saddle point expansion, we expand the group element to the second order as

$$h_\ell = 1 - \frac{A_\ell^2}{2} + \mathrm{i} A_\ell \cdot \sigma + O(A_\ell^3), \tag{10.17}$$

with $A \in su(2) \times su(2)$ a Lie algebra element. By the same token, we expand the unit vectors as

$$n_f = n_f^{(0)} + \xi_f - \frac{\xi_f^2}{2} n_f^{(0)} + O(\xi_f^3), \qquad \text{with} \quad n_f^{(0)} \cdot \xi_f = 0. \tag{10.18}$$

To simplify the notation, we drop the superscript (0). Let us consider a face with edges $\ell_1, \ldots, \ell_p$, then to the second order

$$\prod_{\ell \in \partial f}^{\longrightarrow} h_\ell^{\eta_{\ell,f}} = 1 - \frac{A_f^2}{2} + \mathrm{i}\sigma \cdot A_f - \mathrm{i}\sigma \cdot \Phi_f, \tag{10.19}$$

with

$$A_f = \sum_{1 \leq a \leq p} \eta_{\ell_a, f} A_{\ell_a} \qquad \text{and} \qquad \Phi_f = \sum_{1 \leq a < b \leq p} \eta_{\ell_a, f} \eta_{\ell_b, f} A_{\ell_a} \wedge A_{\ell_b}. \tag{10.20}$$

Inserting these developments in the expression (10.16), one gets

$$S_f[A_\ell, \xi_f] = 2k_f \Big\{ \mathrm{i} n_f \cdot A_f - \frac{A_f^2}{2} + \frac{(n_f \cdot A_f)^2}{2} + \mathrm{i}\xi_f \cdot A_f + \mathrm{i} n_f \cdot \Phi_f \Big\}. \tag{10.21}$$

We thus have to estimate

$$\int \prod_\ell dA_\ell \prod_f d\xi_f \exp 2j \sum_f k_f \Big\{ \mathrm{i} n_f \cdot A_f - \frac{A_f^2}{2} + \frac{(n_f \cdot A_f)^2}{2} + \mathrm{i}\xi_f \cdot A_f + \mathrm{i} n_f \cdot \Phi_f \Big\} \tag{10.22}$$

as $j \to \infty$.

This quadratic saddle point approximation yields an upper bound estimate (see again Krajewski et al. (2010) for details)

$$\phi_\Gamma \lesssim \Lambda^{3F - 3r}, \tag{10.23}$$

with r being the rank of the $L \times F$ incidence matrix $\eta_{\ell,f}$ between the L edges and the F faces of the respective tensor graph. This estimation agrees with general results for topological GFT theory, see also Ben Geloun et al. (2010) and Bonzom and Smerlak (2010).

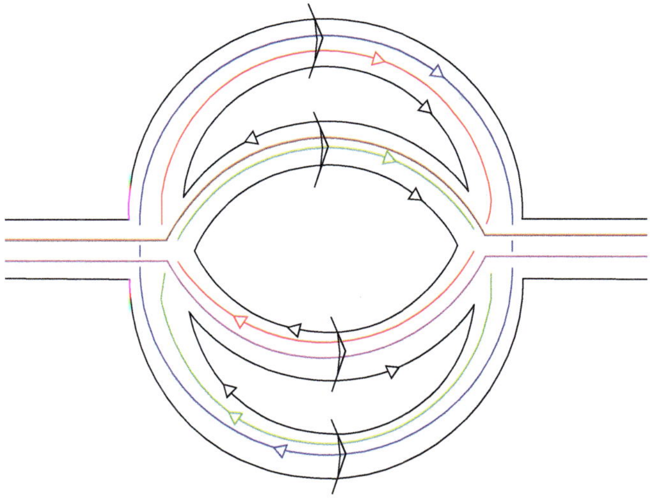

Figure 10.7 *A quantum correction to a tensor model propagator*

This saddle point approximation also proves fruitful for a more involved GFT model—the Engle–Pereira–Rovelli–Livine/Freidel–Krasnov (EPRL/FK, see the original articles Engle, Pereira, and Rovelli (2007); Livine and Speziale (2007); Freidel and Krasnov (2008))—model which aims to provide a non-topological formulation of 4D quantum gravity. The EPRL/FK model introduces a modification of the propagator of the topological Boulatov model (10.4) but keeps the interaction part identical (see Krajewski et al. (2010) for more details on this). One can then use the saddle point method to study the behaviour of the Feynman integral associated with the graph 10.7 (again, see the original article Krajewski et al. (2010)).

10.6 Algebraic combinatorics and tensorial GFT

In this section we present the implementation of algebraic combinatorial techniques (Connes-Kreimer Hopf algebra and its Hochschil cohomology) for the study of various tensorial GFT models. This section is based on the original research articles (Raasakka and Tanasa 2013; Avohou, Rivasseau, and Tanasa 2015).

10.6.1 The Ben Geloun–Rivasseau (BGR) model

In this section we review the – BGR tensorial GFT model Ben Geloun and Rivasseau (2011). The form of the BGR model derives from the combinatorics of 4D simplicial geometry. Let us consider a 4 simplex. Its boundary is given by five 3 simplices, i.e. tetrahedra. Each pair of these boundary tetrahedra shares one boundary triangle. This combinatorial structure can be illustrated by a stranded graph, as in Fig. 10.8. Here, each

Figure 10.8 *The two orientations of the stranded representation of the combinatorics of a 4-simplex, where we consider the colours to be: blue, red, green, grey, and yellow*

individual strand represents a triangle, while the five rectangular boxes, each connected to four strands entering the graph, represent the five boundary tetrahedra. The tetrahedra are considered to be coloured, and we allow for two opposite orientations of the colouring. One considers coloured tetrahedra and only two orientations of the colouring in order to simplify the combinatorics and the topology of this type of models (see, for example, the paper Gurau (2017), where this type of model was first introduced).

One can then build larger stranded graphs from these building blocks by identifying boundary tetrahedra with matching colours. Moreover, we allow only for identification of tetrahedra belonging to 4-simplices of opposite orientations, taken into account by alignment of arrows in the boxes representing tetrahedra. The resulting stranded graph will then be combinatorially dual to a 4D simplicial pseudo-manifold.

A (quantum) statistical description of such simplicial geometries can be formulated as a coloured tensor model. This is a QFT model, whose Feynman graphs have exactly such stranded structure. The model has fields of five different colours (each coloured block in Fig. 10.8 being a field), and each such field is a complex-valued function on $U(1)^{\times 4}$. The two stranded graphs in Fig. 10.8 correspond to the two interaction terms of the model, while the propagator simply identifies the strands of the same colour associated with interaction terms with opposite orientations. Such models are sometimes called *combinatorially non-local* due to the combinatorial nature of the non-locality of their interaction terms.

The Feynman graphs of the BGR model are obtained from such stranded representations of simplicial geometry by removing four of the five fields (namely the blue, the red, the green, and the yellow fields in Fig. 10.8). Thus, one considers interaction terms such as those illustrated in Figs. 10.10 and 10.11. These terms correspond to one-to-one relations between particular coloured graphs and new 'uncoloured' vertices which, by construction, only have grey fields on the boundaries.

These vertices are dual to 4D polytopes, whose boundaries are composed of six or four tetrahedra. Again, the combinatorial connectivity of the strands represents the identifications of boundary triangles of the boundary tetrahedra.

Note that these are not the only possible 'uncoloured' vertices that one can construct. However, these are the only type of vertices considered in the BGR model in order to ensure renormalizability (see again Ben Geloun and Rivasseau (2011), for details).

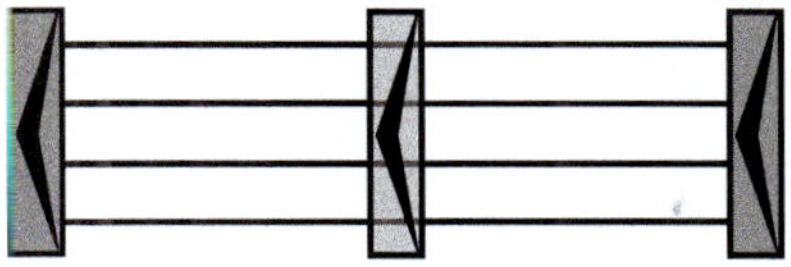

Figure 10.9 *The kinetic term*

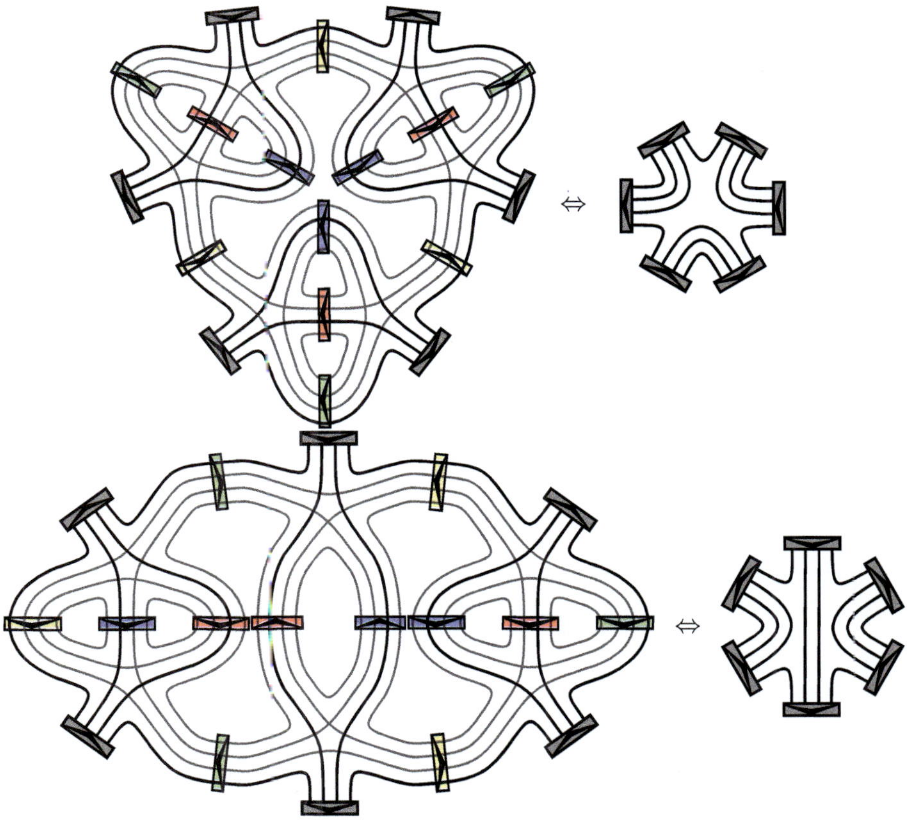

Figure 10.10 *Stranded structures of the vertices $V_{6;1}$ (upper) and $V_{6;2}$ (lower) of the BGR model*

In addition, a non-trivial kinetic term is added in order to induce dynamics, which allows for renormalization (see Fig. 10.9 for a graphical representation). Note that propagators cannot shuffle strands.

In Fig. 10.12 we illustrate a simple example of a Feynman graph of the BGR model, where the propagators are represented by transparent rectangles, while the external edges end in solid rectangles.

Let us describe in detail the action functional of the BGR model. The fields of the model live on a direct product of four copies of the group $U(1)$. The momentum

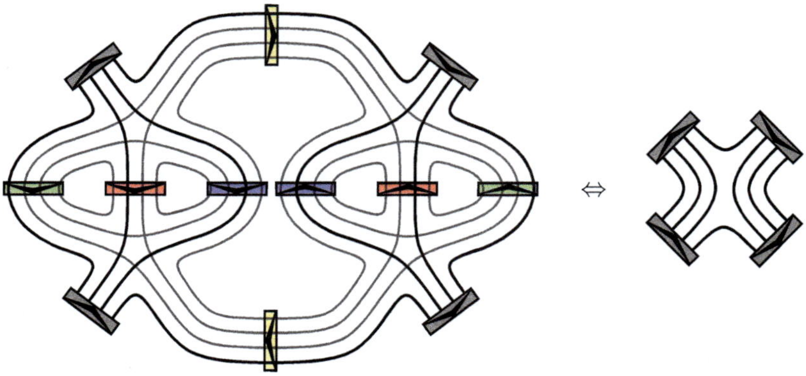

Figure 10.11 *Stranded structure of the vertex $V_{4;1}$ of the BGR model*

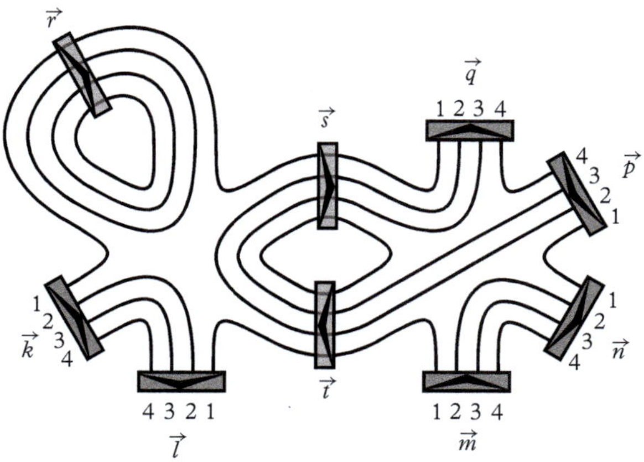

Figure 10.12 *A simple example of a Feynman graph of the BGR model. There is a $V_{6;1}$ vertex on the left and a $V_{6;2}$ vertex on the right. $\vec{k}, \vec{l}, \vec{m}, \vec{n}, \vec{p}, \vec{q} \in \bar{m}Z^4$ label the external momenta, while $\vec{r}, \vec{s}, \vec{t} \in \bar{m}Z^4$ label momenta running through internal propagators. The momentum components $k_i \in \bar{m}Z, i = 1, 2, 3, 4$ are associated with the strands as indicated by the numbers (and similarly for the other momenta)*

representation considered in the following is obtained through harmonic analysis, which yields the momentum space $\bar{m}Z^4$. In the momentum representation, the kinetic term of the action functional reads explicitly

$$S_{kin}[\phi] = \sum_{\vec{p} \in \bar{m}Z^4} \overline{\phi(\vec{p})}(|\vec{p}|^2 + m^2)\phi(\vec{p}),$$

where $|p|^2 = \sum_{i=1}^{4} |p_i|^2$. This leads to the propagator

$$\mathcal{P}(\vec{p},\vec{q}) = \frac{1}{\sum_i |p_i|^2 + m^2} \prod_{i=1}^{4} \delta_{p_i q_i}.$$

In the direct (i.e. $U(1)$) representation, by Fourier transform, the kinetic term reads

$$S_{kin}[\tilde{\phi}] = \int \overline{\tilde{\phi}}_{1,2,3,4} \left(-\sum_{s=1}^{4} \Delta_s + m^2 \right) \tilde{\phi}_{1,2,3,4},$$

where the integral is performed for the Haar measure over the respective group variables, Δ_s denotes the Laplace operator on $U(1)$ acting on the strand index s. Note that $\tilde{\phi}$ denotes the Fourier transform of the function ϕ.

The interaction terms are based on a completely different principle—instead of locality, they are built on the notion of simplicial combinatorics, as described previously. The interaction terms read

$$S^{6;1}_{int}[\phi] = \lambda_{6;1} \sum_{\vec{p},\vec{p}',\vec{p}'' \in \bar{m} Z^4} \sum_{\sigma \in \Sigma_4} \phi_{\sigma_1,\sigma_2,\sigma_3,\sigma_4} \overline{\phi_{\sigma_1',\sigma_2,\sigma_3,\sigma_4}} \phi_{\sigma_1',\sigma_2',\sigma_3',\sigma_4'} \overline{\phi_{\sigma_1'',\sigma_2',\sigma_3',\sigma_4'}}$$
$$\times \phi_{\sigma_1'',\sigma_2'',\sigma_3'',\sigma_4''} \overline{\phi_{\sigma_1,\sigma_2'',\sigma_3'',\sigma_4''}},$$

and

$$S^{6;2}_{int}[\phi] = \lambda_{6;2} \sum_{\vec{p},\vec{p}',\vec{p}'' \in \bar{m} Z^4} \sum_{\sigma \in \Sigma_4} \phi_{\sigma_1,\sigma_2,\sigma_3,\sigma_4} \overline{\phi_{\sigma_1',\sigma_2',\sigma_3',\sigma_4}} \phi_{\sigma_1',\sigma_2',\sigma_3',\sigma_4'} \overline{\phi_{\sigma_1'',\sigma_2,\sigma_3,\sigma_4'}}$$
$$\times \phi_{\sigma_1'',\sigma_2'',\sigma_3'',\sigma_4''} \overline{\phi_{\sigma_1,\sigma_2'',\sigma_3'',\sigma_4''}},$$

where we wrote

$$\phi(p_1,p_2,p_3,p_4) =: \phi_{1,2,3,4}, \; \phi(p_1',p_2,p_3,p_4) =: \phi_{1',2,3,4}, \; \phi(p_1,p_2',p_3,p_4) =: \phi_{2',1,3,4}, \text{ etc.}$$

for simplicity, and we sum over all permutations $\sigma \in \Sigma_4$ of the four colour indices. These interactions lead exactly to the identifications of the field variables represented by the stranded Feynman graphs of the type in Fig. 10.10 modulo permutations of the ordering of the strands.

In addition to the previously mentioned 6-valent interaction vertices, there are also two 4-valent vertices in this model. The first one is shown in Fig. 10.11, while the second one is a disconnected combination of two propagators, which arises from the vertex $V_{6;2}$ through the insertion of a propagator, as in Fig. 10.13. The corresponding interaction terms of the action functional read

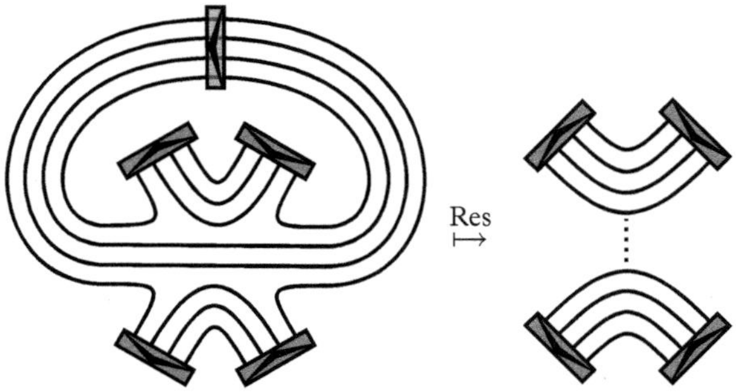

Figure 10.13 *Vertex $V_{4;2}$ of the BGR model obtained through the contraction of a self-loop from the vertex $V_{6;2}$*

$$S^{4;1}_{int}[\phi] = \lambda_{4;1} \sum_{\vec{p},\vec{p}' \in \bar{m}Z^4} \sum_{\sigma\in\Sigma_4} \phi_{\sigma_1,\sigma_2,\sigma_3,\sigma_4} \overline{\phi_{\sigma'_1,\sigma_2,\sigma_3,\sigma_4}} \phi_{\sigma'_1,\sigma'_2,\sigma'_3,\sigma'_4} \overline{\phi_{\sigma_1,\sigma'_2,\sigma'_3,\sigma'_4}},$$

and

$$S^{4;2}_{int}[\phi] = \lambda_{4;2} \sum_{\vec{p},\vec{p}' \in \bar{m}Z^4} \sum_{\sigma\in\Sigma_4} \phi_{\sigma_1,\sigma_2,\sigma_3,\sigma_4} \overline{\phi_{\sigma_1,\sigma_2,\sigma_3,\sigma_4}} \phi_{\sigma'_1,\sigma'_2,\sigma'_3,\sigma'_4} \overline{\phi_{\sigma'_1,\sigma'_2,\sigma'_3,\sigma'_4}}.$$

Notice the peculiar, disconnected structure of the vertex $V_{4;2}$. This can be interpreted geometrically as a four-sphere with two holes of the form of four simplices.

The Feynman amplitude associated with a particular Feynman graph is constructed as follows:

1. Let $\vec{p}_e \in \bar{m}Z^4$ be the momenta labels on the edges $e \in \Gamma^{[1]}$ of the Feynman graph Γ. We construct the product of free propagators

$$\prod_{e\in\Gamma^{[1]}} \frac{1}{|\vec{p}_e|^2 + m^2}.$$

2. For each vertex $v_t \in \Gamma^{[0]}$, multiply by the coupling constant λ_t, where t labels the type of the vertex, and impose identifications of strands (momentum variables) by multiplying the previous product by

$$\prod_{v\in\Gamma^{[0]}} \prod_k \delta(c^{t(v)}_k(\vec{p}_l)),$$

where $\{c^k_{t(v)}(\vec{p}_l) = 0\}$ is the set of constraints imposed on the incoming momenta by the vertex type $t(v)$ of vertex $v \in \Gamma^{[0]}$. Explicitly, these are given by the following:

- $\prod_k \delta(c_k^{6;1}(\vec{p_l})) = \delta_{p_1n_1}\delta_{p_2q_2}\delta_{p_3q_3}\delta_{p_4q_4}\delta_{q_1k_1}\delta_{k_2l_2}\delta_{k_3l_3}\delta_{k_4l_4}\delta_{l_1m_1}\delta_{m_2n_2}\delta_{m_3n_3}\delta_{m_4n_4}$,
- $\prod_k \delta(c_k^{6;2}(\vec{p_l})) = \delta_{p_1n_1}\delta_{p_2l_2}\delta_{p_3l_3}\delta_{p_4q_4}\delta_{q_1k_1}\delta_{q_2k_2}\delta_{q_3k_3}\delta_{k_4l_4}\delta_{l_1m_1}\delta_{m_2n_2}\delta_{m_3n_3}\delta_{m_4n_4}$,
- $\prod_k \delta(c_k^{4;1}(\vec{p_l})) = \delta_{p_1l_1}\delta_{p_2q_2}\delta_{p_3q_3}\delta_{p_4q_4}\delta_{q_1k_1}\delta_{k_2l_2}\delta_{k_3l_3}\delta_{k_4l_4}$,
- $\prod_k \delta(c_k^{4;2}(\vec{p_l})) = \delta_{p_1q_1}\delta_{p_2q_2}\delta_{p_3q_3}\delta_{p_4q_4}\delta_{k_1l_1}\delta_{k_2l_2}\delta_{k_3l_3}\delta_{k_4l_4}$,

where $\vec{p},\vec{k},\vec{m}$ are incoming and $\vec{q},\vec{l},\vec{n}$ outgoing momenta. Notice that the vertex functions are not symmetric with respect to all permutations of the edges, so, in addition to specifying the type of vertices, one must also specify the orientations. (This is naturally taken care of in the stranded representation.)

3. Sum over $\vec{p}_e \in \bar{m}Z^4$ associated with the internal edges $e \in \Gamma^{[1]}_{int}$ of the graph.

The quantity that results from applying these rules to a Feynman graph is the Feynman amplitude of that graph.

Since the momenta are conserved by a simple identification of variables in the propagator and the vertices, as already mentioned, the momentum vector components may be associated with the individual strands $s \in \Gamma^s$ of a Feynman graph in the stranded representation. We emphasize that this combinatorial identification of field variables is, in fact, what allows for the stranded representation in the first place. As a result, the Feynman amplitude of a stranded graph consists of a product over Kronecker deltas identifying the external momentum vector components through the strands, multiplied by the product over internal propagators of the graph, which depend on the momenta associated with the strands that pass through them, summed over the momenta associated with internal (looped) strands. For example, for the Feynman graph in Fig. 10.12, the amplitude reads

$$\phi(\Gamma_{\text{Fig. 10.12}}) := \lambda_{6;1}\lambda_{6;2}\ \delta_{k_1q_1}\delta_{k_2l_2}\delta_{k_3l_3}\delta_{k_4l_4}\delta_{l_1m_1}\delta_{m_2n_2}\delta_{m_3n_3}\delta_{m_4n_4}\delta_{n_1p_1}\delta_{p_2q_2}\delta_{p_3q_3}\delta_{p_4q_4}$$
$$\times \sum_{\substack{r_2,r_3,\\ r_4\in\bar{m}Z}} \left(\frac{1}{k_1^2+r_2^2+r_3^2+r_4^2+m^2}\right) \sum_{s_4\in\bar{m}Z} \left(\frac{1}{l_1^2+p_2^2+p_3^2+s_4^2+m^2}\right)\left(\frac{1}{k_1^2+p_2^2+p_3^2+s_4^2+m^2}\right), \quad (10.24)$$

where $\vec{k},\vec{l},\vec{m},\vec{n},\vec{p},\vec{q},\in\mathbb{Z}^4$ are the momenta associated with the external edges and r_i $(i=2,3,4,)$ and s_4 are the momentum vector components associated with the internal strands. We observe that, for large absolute values of the internal momenta r_2,r_3,r_4,s_4, the sums behave asymptotically as:

$$\phi(\Gamma_{\text{Fig. 10.12}}) \propto \sum_{\substack{r_2,r_3,\\ r_4\in\bar{m}Z}} \frac{1}{r_2^2+r_3^2+r_4^2} \sum_{s_4\in\bar{m}Z} \frac{1}{s_4^4} \approx \int_{\bar{m}R^3} \frac{\mathrm{d}^3 r}{r^2} \int_{\bar{m}R} \frac{\mathrm{d}s}{s^4} \propto \Lambda \times \Lambda^{-3} = \Lambda^{-2},$$

where Λ is a cut-off on the momentum summation. Here, the first factor, leading to a linear divergence in the limit $\Lambda \to \infty$, corresponds to the three strand loops in the upper left part of the graph, and the second convergent factor corresponds to the single strand loop in the middle. Although the first factor leads to a sub-divergence,

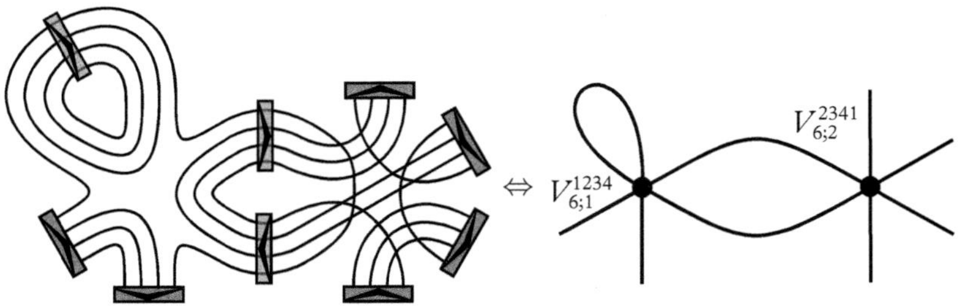

Figure 10.14 *An example of the correspondence between stranded and non-stranded representation of a Feynman graph of the BGR model*

the superficial divergence degree of the total graph $\Gamma_{\text{Fig. 10.12}}$ is -2, and therefore the graph is superficially convergent.

Similarly, the Feynman amplitude for the graph in Fig. 10.14 reads

$$\phi(\Gamma_{\text{Fig. 10.14}}) := \lambda_{6;1}\lambda_{6;2}\ \delta_{k_1l_1}\delta_{k_2l_2}\delta_{k_3l_3}\delta_{k_4l_4}\delta_{m_1n_1}\delta_{m_2n_2}\delta_{m_3n_3}\delta_{m_4q_4}\delta_{n_4p_4}\delta_{p_1q_1}\delta_{p_2q_2}\delta_{p_3q_3}$$
$$\times \sum_{\substack{r_2,r_3,\\ r_4\in \tilde{m}Z}} \left(\frac{1}{k_1^2+r_2^2+r_3^2+r_4^2+m^2}\right)\left(\frac{1}{l_1^2+p_2^2+p_3^2+m_4^2+m^2}\right)\left(\frac{1}{k_1^2+p_2^2+p_3^2+m_4^2+m^2}\right), \tag{10.25}$$

with the same notation for the momenta as in Fig. 10.12. The difference to the previous graph is that now there are only three internal strands corresponding to r_i, $i=2,3,4$. The superficial divergence degree of this graph is -3 despite the sub-divergence, since the propagators contribute negatively to the power counting. In the following we state the power counting more explicitly.

The Feynman amplitude of any other BGR tensor graph may be computed in an analogous way. We leave the calculation of amplitudes for the rest of the graphs of this section as an exercise to the interested reader.

We may also draw the Feynman graphs as ordinary graphs, but then the combinatorics of the internal structure must be taken into account by additional labelling of vertices. We have illustrated the correspondence between the ordinary and the stranded representation of Feynman graphs of the BGR model in Fig. 10.14. The vertices are labelled, for example, as $V^{1234}_{6;1}$, where the lower index indicates the vertex type, and the upper indices indicate the permutation of field variables. Since the vertex is not symmetrical under permutations of edges, the edges meeting at a vertex must be distinguished. This can be accomplished, for example, by placing the vertex label between the first and the last edge, where the edges are ordered in a right-handed orientation around the vertex with respect to the normal pointing upwards from the surface, on which the graph is drawn.

Definition 10.6.1 *A* face *of a stranded Feynman graph is a strand that forms a loop.*

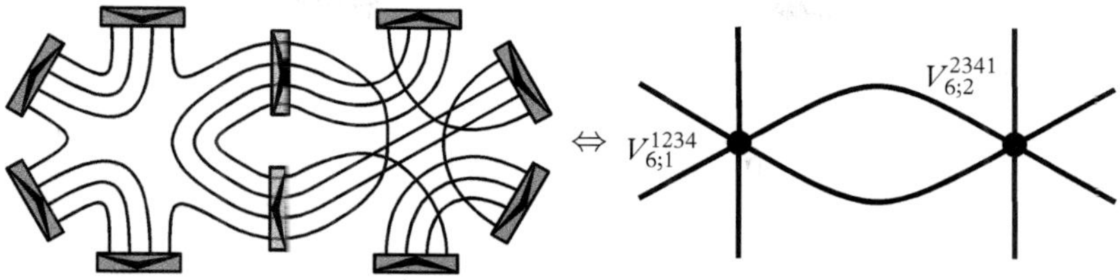

Figure 10.15 *An example of a graph without a face but with a loop in the non-stranded representation*

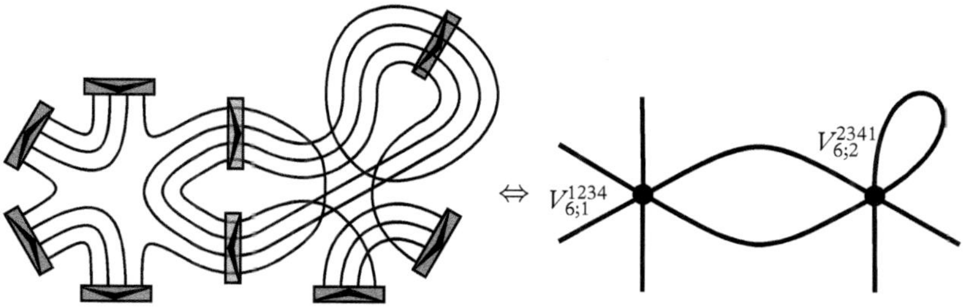

Figure 10.16 *An example of a graph with faces, which extend around several loops of the non-stranded representation*

Note that a face of a stranded graph is dual to the corresponding triangle of the tetrahedra of the 4D triangulation.

The integrations over face momenta may give rise to divergences, as we have already observed. In order for a strand to form a loop, the non-stranded representation must clearly have a loop as well. However, a loop in the non-stranded representation does not imply the existence of a face as can for instance be seen in Fig. 10.15. Moreover, a face may extend around several loops of the non-stranded graph, as in Fig. 10.16. Therefore, the faces can be highly non-local objects on the graph. Fortunately, the power counting restricts the non-locality of divergent faces significantly, as we will see in the following. Notice also that faces cannot extend over bridges of the non-stranded Feynman graph, since due to the colouring, a single strand can pass the same propagator only once. Therefore, 1PI Feynman graphs are still sufficient to resolve the divergence structure of Feynman amplitudes of the BGR model.

We have the following auxiliary result:

Lemma 10.6.2 (Ben Geloun and Rivasseau (2011); Lemma 2) *A superficial divergence degree for the BGR model is given by $\omega(\Gamma) = F_\Gamma - 2P_\Gamma$, where F_Γ and P_Γ are the numbers of faces and propagators in Γ, respectively.*

Thus, we see that not all faces lead to divergences, but there is a balance between the number of propagators and the number of strands. It was further proved that the superficial divergence degree ω for connected graphs, where $V_{4;2}$ is considered disconnected, can be written in terms of the combinatorial and topological structure of the stranded Feynman graphs as follows.

Definition 10.6.3 (Ben Geloun and Rivasseau (2011); Definition 1) *Let Γ be a Feynman graph of the BGR model. Then:*

- *the* coloured extension Γ_{col} *of Γ is the corresponding unique stranded graph obtained by replacing the vertices of Γ by the corresponding graphs of the simplicial coloured model from Figs. 10.10 and 10.11.*
- *a* jacket J *of Γ_{col} is a ribbon graph obtained from Γ_{col} by removing a subset of the strands. Specifically, a jacket J is determined by a permutation $(0abcd)$ of colours (up to cyclic permutations), and the strands that are included in J are the ones that connect the tetrahedra of colours $(0a), (ab), (bc), (cd), (d0)$, where $a, b, c, d, \in \{R, G, B, Y\}$ are all different. Here, we have chosen 0 to represent the grey colour, while the other colours are labelled by their initials;*
- *the* pinched jacket $\tilde{J}$ *is the vacuum ribbon graph obtained from the jacket J by closing all external ribbon edges of J; (See Fig. 10.17 for an example.)*
- *the* boundary graph $\partial\Gamma$ *of Γ is the 3D tensor graph, which corresponds to the boundary of Γ. It is obtained by first removing all internal faces of Γ, closing the external edges by connecting their strands to a vertex per external edge, replacing the strands by bundles of three strands, and finally replacing the vertices of external edges by tensorial simplicial vertices; (See Fig. 10.18 for an example.)*
- *the* boundary jacket J_∂ *is a jacket of the boundary graph $\partial\Gamma$.*

(See Ben Geloun and Rivasseau (2011) and references therein for further details.) Ben Geloun and Rivasseau proved the following result.

Theorem 10.6.4 *The superficial divergence degree ω_{BGR} of a connected Feynman graph Γ can be written as*

$$\omega_{BGR}(\Gamma) = -\frac{1}{3}\left[\sum_J g_{\tilde{J}} - \sum_{J_\partial} g_{J_\partial}\right] - (C_\partial - 1) - V_4 - 2V_2'' - \frac{1}{2}(N_{ext} - 6),$$

where g_J denotes the genus of the ribbon graph J, C_∂ is the number of disconnected components of $\partial\Gamma$, V_4 is the number of vertices of type $V_{4;1}$, $V_2''/2$ is the number of vertices of type $V_{4;2}$, and N_{ext} is the number of external edges of Γ.

Moreover, they classified all superficially divergent Feynman graphs in Table 10.1, which is the departure point for our subsequent analysis.

Note that all superficially divergent graphs have a combinatorial structure that corresponds to one of the vertices of the original model, which facilitates perturbative renormalizability. Ben Geloun and Rivasseau (2011) showed that the BGR model is perturbatively renormalizable.

Figure 10.17 *An example of a pinched jacket*

10.6.2 Cones–Kreimer Hopf algebraic description of the combinatorics of the renormalizability of the BGR model

When implementing a Connes–Kreimer Hopf algebraic description of the combinatorics of the renormalizability of the BGR model, one needs to take into consideration the following:

(i) The constraints on the external parameters of a Feynman graph Γ are more involved. One does not need to impose only momentum conservation. Instead, one needs to identify the variables represented by the stranded structure of the Feynman diagrams.

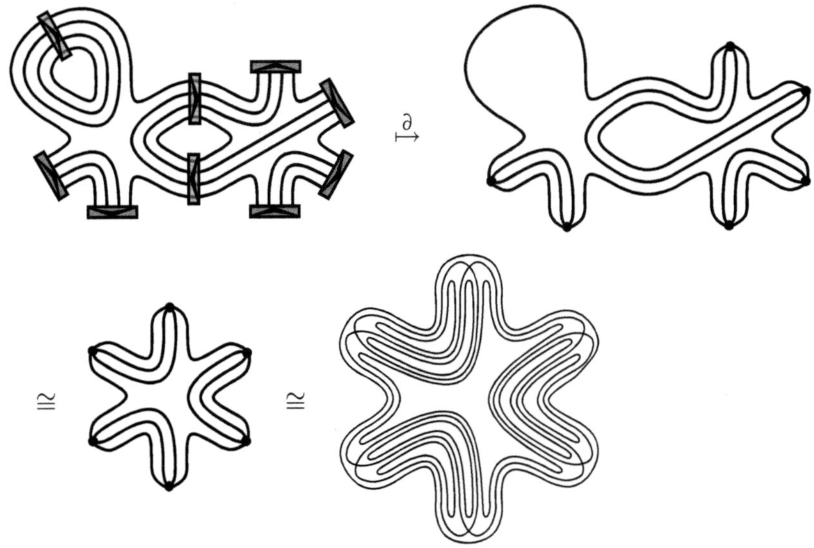

Figure 10.18 *An example of a boundary graph*

Table 10.1 *Classification and residues of superficially divergent Feynman graphs of the BGR model*

class	N_{ext}	V_4	$\sum_{J_\partial} g_{J_\partial}$	$C_\partial - 1$	$\sum_{\tilde{J}} g_{\tilde{J}}$	ω_{BGR}	**Res**
6_1	6	0	0	0	0	0	$V_{6;i}$
4_1	4	0	0	0	0	1	$V^*_{4;1}$
4_2	4	1	0	0	0	0	$V_{4;1}$
4_3	4	0	0	1	0	0	$V_{4;2}$
2_1	2	0	0	0	0	2	$V_{2;1}$
2_2	2	1	0	0	0	1	$V_{2;2}$
2_3	2	2	0	0	0	0	$V_{2;3}$
2_4	2	0	0	0	6	0	$V_{2;3}$

Definition 10.6.5 *A Feynman graph* Γ *of a QFT model is equipped with a set of* external constraints $\{c_k^\Gamma(\vec{p}_i) = 0\}$ *on the external momenta. We denote the space of the external momenta of* Γ *by* P_Γ*. The Feynman amplitudes are then generalized functions on* P_Γ*, as before.*

Remark 10.6.6 *Notice that this generalizes the case of local QFT, where the external constraints impose the momentum conservation. In contrast, in tensor field theories the interactions are combinatorially non-local in the sense that the boundary variables are identified according to tensorial invariance. Thus, the notion of tensorial invariance in TFT constitutes the*

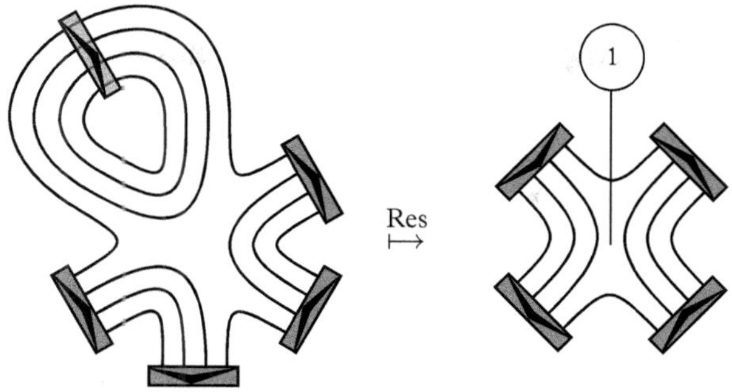

Figure 10.19 *Illustration of the linear 4-valent counter-term vertex*

substitute for the notion of locality in local QFT. A gluing of tensor invariant graphs always yields another tensor invariant graph.

Accordingly, we also need to generalize the notion of residue to tensorial Feynman graphs:

Definition 10.6.7 *A residue* Res(Γ) *of a tensorial Feynman graph* $\Gamma \in \mathcal{G}_{1PI}$ *of the BGR model is the graph obtained, in the non-stranded representation, by shrinking all internal edges of* Γ *to a point, and retaining the external constraints and structure of the graph. In the stranded representation, this corresponds to the removal of all internal faces and internal propagators of* Γ.

(ii) There are vertices of valence four and vertices of valence six. By inserting a self-loop into a 6-valent vertex $V_{6;1}$, we obtain a self-loop Feynman graph with four external edges and divergence degree 1. This leads us to introduce a new 4-valent counter-term vertex $V^{*}_{4;1}$ that has the same structure as $V_{4;1}$ but a linear counter-term attached. This may be illustrated as in Fig. 10.19.

(iii) Similarly, the contraction of superficially divergent subgraphs with two external edges leads us to introduce 2-valent counter-term vertices $V_{2;1}$, $V_{2;2}$, and $V_{2;3}$, with a quadratic, linear, and logarithmic counter-term attached, respectively. This is very different, for example, from the ϕ^4-model, for which we only have quadratically divergent two-point functions. See Fig. 10.20 for illustration. Of the 2-valent counter-term vertices, only the one with a quadratic divergence can be replaced by the original propagator, since only in this case the divergence degree of the graph is preserved.

Let us make here a few more remarks on the renormalizability of the BGR model. When one does the Taylor expansion of the two-point function (see Lemmas 11 and 12 in Ben Geloun and Rivasseau (2011) for the respective analytic details), the sub-leading divergence of the first graph in Fig. 10.20 is logarithmic and renormalizes the wave function. Note that there is no linear sub-leading divergence. Moreover, there are no sub-leading divergences for the rest of the graphs of Fig. 10.20.

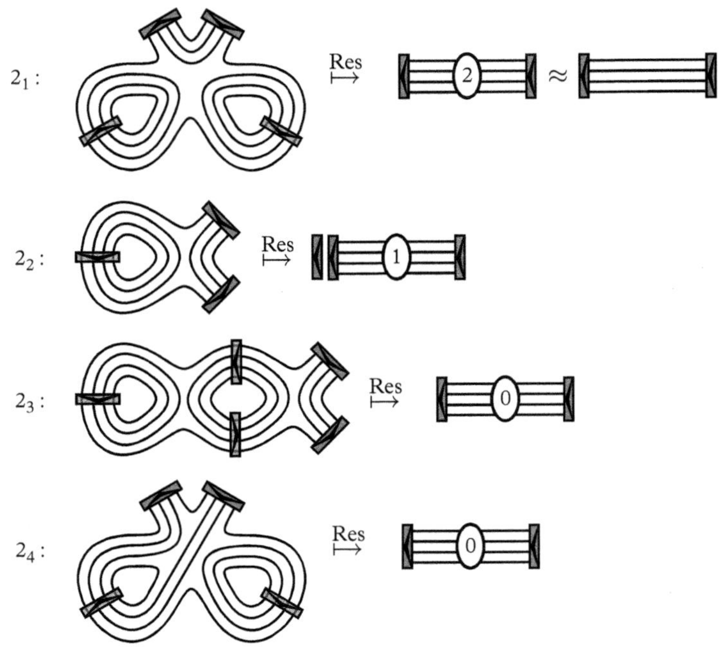

Figure 10.20 *Some examples of the 2-valent counter-term as residues of two-point functions from the classes 2_1, 2_2, 2_3, and 2_4, in the respective order. Only the quadratic counter-term vertex may be inserted into a propagator, since this conserves the divergence degree*

Remark 10.6.8 *Note that a given two-point graph cannot have several counterterms of type* $0, 1$ *and* 2 *(see again Fig. 10.20), since one can identify to which line of Table 10.1 the respective tensor graph belongs (due to the numbers V_4, V_2'', and to the graph topology). If the graph belongs to line 2_1, then it has a counterterm of type* 2 *(which means a quadratic counterterm of mass type and a logarithmic counterterm of wave function-type, see previous example), if it belongs to line 2_2, then it has a counterterm of type* 1*, and so on.*

(iv) The superficial divergence degree does not only depend on the external data of a Feynman graph, such as the number of external edges, but also some internal combinatorial and topological information, in particular, the number of 4-valent vertices and the sum over the genera of pinched jackets.

One has:

Lemma 10.6.9 *Let $\Gamma \in \mathcal{G}, \gamma \in \mathcal{G}_{sd}$ be Feynman graphs of the BGR model such that $\gamma \subseteq \Gamma$. For the sum over the genera of pinched jackets, we have the relation*

$$\sum_J g_{\tilde{J}}(\Gamma) = \sum_J g_{\tilde{J}}(\gamma) + \sum_J g_{\tilde{J}}(\Gamma/\gamma).$$

Proof First note that the genus for a single pinched jacket may be expressed via the Euler characteristic formula as $g_{\tilde{J}} = -\frac{1}{2}(F_{\tilde{J}} - L_{\tilde{J}} + V_{\tilde{J}} - 2C_{\tilde{J}})$, where $F_{\tilde{J}}$, $L_{\tilde{J}}$, $V_{\tilde{J}}$, and $C_{\tilde{J}}$ are the numbers of faces, edges, vertices, and connected components of $\tilde{J}$. (For a disconnected jacket, we define the genus to be given by the sum over the genera of its connected components.) One may further write $F_{\tilde{J}} = F_{\tilde{J},ext} + F_{\tilde{J},int}$, where $F_{\tilde{J},ext}$ is the number of faces formed by pinching the external edges of J, and $F_{\tilde{J},int}$ is the number of (internal) faces of J. Now, one may easily verify that each of $F_{\tilde{J},int}$, $L_{\tilde{J}}$, and $V_{\tilde{J}}$ separately satisfies the relation

$$Q(\Gamma) - Q(\gamma) = Q(\Gamma/\gamma) - Q(\mathrm{Res}(\gamma))$$

with respect to the contraction of subgraphs $\gamma \in \mathcal{G}_{sd}$, as defined for Feynman graphs of the BGR model. (Notice, in particular, that, in general, $F_{\tilde{J},int}(\mathrm{Res}(\gamma)) \neq 0$, $L_{\tilde{J}}(\mathrm{Res}(\gamma)) \neq 0$, and $V_{\tilde{J}}(\mathrm{Res}(\gamma)) \neq 0$, since the vertices of the BGR model correspond to extended subgraphs in the coloured extension of the Feynman graphs, from which the jackets are derived.) For example, it is clear that both sides of

$$Q(\Gamma) - Q(\Gamma/\gamma) = Q(\gamma) - Q(\mathrm{Res}(\gamma))$$

represent the number of internal faces, edges, or vertices, which are removed from $\tilde{J}$ by contracting $\gamma \subseteq \Gamma$, since by definition the subgraph contraction operation replaces γ by $\mathrm{Res}(\gamma)$ inside Γ, which does not affect the numbers of internal faces, edges, or vertices that do not belong to γ. (Notice that some of the internal faces of Γ not internal to γ may still pass through γ, but are not affected by the contraction of γ.) Accordingly, we have

$$Q(\Gamma) - Q(\mathrm{Res}(\gamma)) = Q(\gamma) - Q(\mathrm{Res}(\gamma)) + Q(\Gamma/\gamma) - Q(\mathrm{Res}(\gamma)),$$

so, $G(\Gamma) := Q(\Gamma) - Q(\mathrm{Res}(\gamma))$ gives a grading of jackets with respect to the contraction of γ, in the sense that $G(\Gamma) = G(\gamma) + G(\Gamma/\gamma)$. Moreover, if we have a set of gradings G_i, then it is easy to verify that $\sum_i (-1)^{n_i} G_i$ is also a grading for any choice of $n_i \in \{0,1\}$.

Now, note that we may write

$$\begin{aligned} &g_{\tilde{J}}(\Gamma) + \frac{1}{2}\left(F_{\tilde{J},ext}(\Gamma) + F_{\tilde{J}}(\mathrm{Res}(\gamma)) - L_{\tilde{J}}(\mathrm{Res}(\gamma)) + V_{\tilde{J}}(\mathrm{Res}(\gamma)) - 2C_{\tilde{J}}(\mathrm{Res}(\gamma))\right) \\ &= -\frac{1}{2}\left((F_{\tilde{J},int}(\Gamma) - F_{\tilde{J},int}(\mathrm{Res}(\gamma))) - (L_{\tilde{J}}(\Gamma) - L_{\tilde{J}}(\mathrm{Res}(\gamma))) + (V_{\tilde{J}}(\Gamma) - V_{\tilde{J}}(\mathrm{Res}(\gamma)))\right). \end{aligned}$$

Thus, we find that $g_{\tilde{J}}(\Gamma) + \frac{1}{2}F_{\tilde{J},ext}(\Gamma) + X_{\tilde{J}}(\mathrm{Res}(\gamma))$ is a grading with respect to the contraction of γ, where we wrote

$$X_{\tilde{J}}(\mathrm{Res}(\gamma)) := \frac{1}{2}\left(F_{\tilde{J},int}(\mathrm{Res}(\gamma)) - L_{\tilde{J}}(\mathrm{Res}(\gamma)) + V_{\tilde{J}}(\mathrm{Res}(\gamma)) - 2C_{\tilde{J}}(\mathrm{Res}(\gamma))\right).$$

Accordingly, we find that, as a sum of gradings,

$$\sum_J g_{\bar{J}}(\Gamma) + \frac{1}{2}\sum_J F_{\bar{J},ext}(\Gamma) + \sum_J X_{\bar{J}}(\mathrm{Res}(\gamma))$$

is also a grading with respect to the contraction of γ. Let us write $X(\mathrm{Res}(\gamma)) := \sum_J X_{\bar{J}}(\mathrm{Res}(\gamma))$. Moreover, $\sum_J F_{\bar{J},ext}(\Gamma) = 2F_{\partial\Gamma}$, the number of faces of the boundary $\partial\Gamma$ of Γ. Then, due to the grading property, we have

$$\sum_J g_{\bar{J}}(\Gamma) + F_{\partial\Gamma} = \sum_J g_{\bar{J}}(\gamma) + F_{\partial\gamma} + \sum_J g_{\bar{J}}(\Gamma/\gamma) + F_{\partial(\Gamma/\gamma)} + X(\mathrm{Res}(\gamma)).$$

We may further simplify by noting that $F_{\partial\Gamma} = F_{\partial(\Gamma/\gamma)}$, since a subgraph contraction does not affect the boundary. Hence, we may write

$$\sum_J g_{\bar{J}}(\Gamma) = \sum_J g_{\bar{J}}(\gamma) + \sum_J g_{\bar{J}}(\Gamma/\gamma) + F_{\partial\gamma}(\mathrm{Res}(\gamma)) + X(\mathrm{Res}(\gamma)).$$

But now

$$\begin{aligned} F_{\partial\gamma}(\mathrm{Res}(\gamma)) + X(\mathrm{Res}(\gamma)) &= \frac{1}{2}\sum_J \left(F_{\bar{J}}(\mathrm{Res}(\gamma)) - L_{\bar{J}}(\mathrm{Res}(\gamma)) + V_{\bar{J}}(\mathrm{Res}(\gamma)) - 2C_{\bar{J}}(\mathrm{Res}(\gamma))\right) \\ &= -\sum_J g_{\bar{J}}(\mathrm{Res}(\gamma)), \end{aligned}$$

so, we finally obtain

$$\sum_J g_{\bar{J}}(\Gamma) = \sum_J g_{\bar{J}}(\gamma) + \sum_J g_{\bar{J}}(\Gamma/\gamma) - \sum_J g_{\bar{J}}(\mathrm{Res}(\gamma)).$$

One may then explicitly check that all the genera of pinched jackets vanish for different choices of $\mathrm{Res}(\gamma)$ for $\gamma \in \mathcal{G}_{sd}$, by using the Euler character formula for $g_{\bar{J}}$. For example, for $\mathrm{Res}(\gamma) = V_{6;1}$, we have $V_{\bar{J}}(V_{6;1}) = 6$, $L_{\bar{J}}(V_{6;1}) = 12$, and $C_{\bar{J}}(V_{6;1}) = 1$ for all J, so, $g_{\bar{J}} = -\frac{1}{2}(F_{\bar{J}}(V_{6;1}) - 8)$. We must then count the number of faces for each jacket, which gives $F_{\bar{J}}(V_{6;1}) = 8$ for all J, so $g_{\bar{J}}(V_{6;1}) = 0$ for all J. Similar calculations may be done for the other possible residues of superficially divergent graphs, which result in the previous statement.

With the previous generalizations of the Connes–Kreimer framework, and the power-counting for the model given in Table 10.1, we can now define the combinatorial Hopf algebra of BGR model Feynman graphs in the same way as for local QFT. In particular, we define the coproduct for the 1PI Feynman graphs $\Gamma \in \mathcal{G}_{1PI}$ by

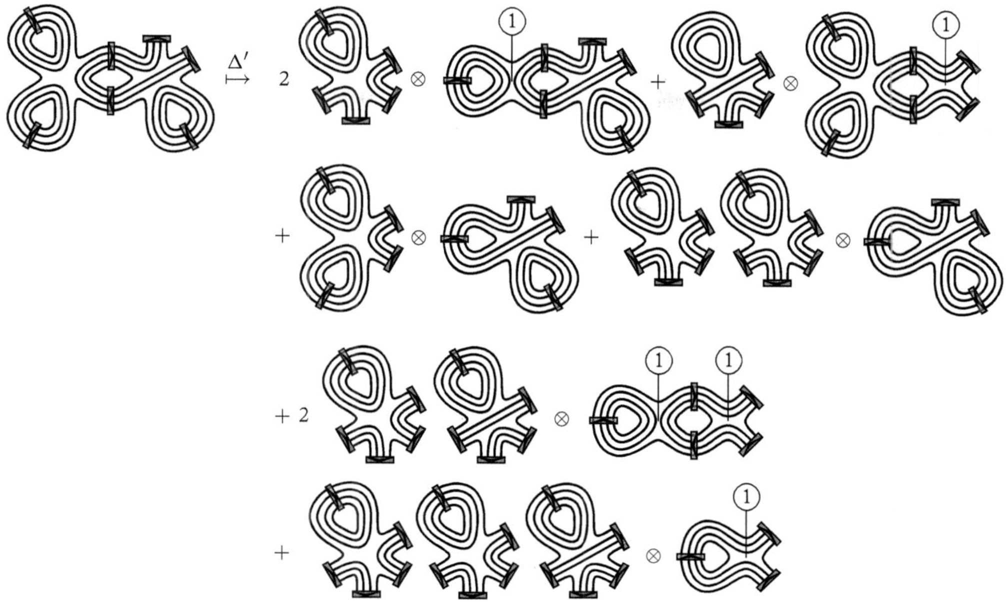

Figure 10.21 *An example of the coproduct* $\Delta(\Gamma) \equiv \Gamma \otimes frm[o] - - + frm[o] - - \otimes \Gamma + \Delta'(\Gamma)$

$$\Delta(\Gamma) = \Gamma \otimes frm[o] - - + frm[o] - - \otimes \Gamma + \sum_{\substack{\gamma \in \bigcup \mathcal{G}^{\omega}_{sd} \\ \gamma \subsetneq \Gamma}} \gamma \otimes \Gamma/\gamma. \tag{10.26}$$

Recall that the summation runs over the set $\bigcup \mathcal{G}^{\omega}_{sd}$ of all disconnected unions of superficially divergent 1PI subgraphs. (See Fig. 10.21 for an example.)

We now have the following auxiliary result:

Lemma 10.6.10 *The set of superficially divergent Feynman diagrams of the BGR model is closed under the operations of subgraph contraction and insertion.*

Proof Let us first consider the subgraph contraction on a class-by-class basis. Notice that a subgraph contraction does not affect the boundary of a graph, and therefore we do not need to consider the boundary properties. Accordingly, the contractions of subgraphs lead to the following, depending on the class to which they belong:

- 6_1: Contracting a subgraph in $6_1, 4_1, 4_3, 2_1$ gives a graph in 6_1. Other classes are not possible, since 6_1 does not contain 4-valent vertices, and the genera of its pinched jackets are zero.
- 4_1: Contracting a subgraph in $6_1, 4_1, 4_3, 2_1$ gives a graph in 4_1. Other classes are not possible for the previously given reasons.
- 4_2: Contracting a subgraph in $6_1, 4_1, 4_2, 4_3, 2_1$ gives a graph in 4_2. Contracting a subgraph in 2_2 gives a graph in 4_1. Other subgraph classes are not possible.

Table 10.2 *Various possibilities of contractions of superficially divergent Feynman graphs of the BGR model*

row/col	6_1	4_1	4_2	4_3	2_1	2_2	2_3	2_4
6_1	6_1	6_1	0	6_1	6_1	0	0	0
4_1	4_1	4_1	0	4_1	4_1	0	0	0
4_2	4_2	4_2	4_2	4_2	4_2	4_1	0	0
4_3	4_3	4_3	0	4_3	4_3	0	0	0
2_1	2_1	2_1	0	2_1	2_1	0	0	0
2_2	2_2	2_2	2_2	2_2	2_2	2_1	0	0
2_3	2_3	2_3	2_3	2_3	2_3	2_2	2_1	0
2_4	2_4	2_4	0	2_4	2_4	0	0	2_1

4_3: Contracting a subgraph in $6_1, 4_1, 4_3, 2_1$ gives a graph in 4_3. Other subgraph classes are not possible.

2_1: Contracting a subgraph in $6_1, 4_1, 4_3, 2_1$ gives a graph in 2_1. Other subgraph classes are not possible.

2_2: Contracting a subgraph in $6_1, 4_1, 4_2, 4_3, 2_1$ gives a graph in 2_2. Contracting a subgraph in 2_2 gives a graph in 2_1. Other subgraph classes are not possible.

2_3: Contracting a subgraph in $6_1, 4_1, 4_2, 4_3, 2_1$ gives a graph in 2_3. Contracting a subgraph in 2_2 gives a graph in 2_2. Contracting a subgraph in 2_3 gives a graph in 2_1. Other subgraph classes are not possible.

2_4: Contracting a subgraph in $6_1, 4_1, 4_3, 2_1$ gives a graph in 2_4. Contracting a subgraph in 2_4 gives a graph in 2_1. Other subgraph classes are not possible.

We may condense the analysis into Table 10.2. We have marked a vanishing entry in the table if a graph from the column class cannot be a subgraph of a graph from the row class.

Similarly, we have to consider the insertions of superficially divergent graphs into other superficially divergent graphs. One has:

6_1: Inserting a graph from class $6_1, 4_3, 2_1$ yields a graph in class 6_1. Insertion of a graph in 4_2 is not possible, since 6_1 does not contain $V_{4;1}$ vertices.

4_1: Inserting a graph from class $6_1, 4_3, 2_1$ yields a graph in class 4_1. Insertion of a graph in 4_2 is not possible, since 4_1 does not contain $V_{4;1}$ vertices.

4_2: Inserting a graph from class $6_1, 4_2, 4_3, 2_1$ yields a graph in class 4_2.

4_3: Inserting a graph from class $6_1, 4_3, 2_1$ yields a graph in class 4_3. Insertion of a graph in 4_2 is not possible, since 4_3 does not contain $V_{4;1}$ vertices.

2_1: Inserting a graph from class $6_1, 4_3, 2_1$ yields a graph in class 2_1. Insertion of a graph in 4_2 is not possible, since 2_1 does not contain $V_{4;1}$ vertices.

2_2: Inserting a graph from class $6_1, 4_2, 4_3, 2_1$ yields a graph in class 2_2.

2_3: Inserting a graph from class $6_1, 4_2, 4_3, 2_1$ yields a graph in class 2_3.

2_4: Inserting a graph from class $6_1, 4_3, 2_1$ yields a graph in class 2_4. Insertion of a graph in 4_2 is not possible, since 2_4 does not contain $V_{4;1}$ vertices.

Accordingly, all the allowed insertions lead to superficially divergent graphs, which concludes the proof.

One now has:

Lemma 10.6.11 *The coproduct defined in* (10.26) *is coassociative, if the set of superficially divergent Feynman diagrams is closed under the operations of subgraph contraction and insertion.*

Proof First, by a direct calculation, we find that

$$(\Delta \otimes \mathrm{id}_{\mathcal{H}}) \circ \Delta - (\mathrm{id}_{\mathcal{H}} \otimes \Delta) \circ \Delta = (\Delta' \otimes \mathrm{id}_{\mathcal{H}}) \circ \Delta' - (\mathrm{id}_{\mathcal{H}} \otimes \Delta') \circ \Delta'$$

for $\Delta(\Gamma) \equiv \Gamma \otimes frm[o] - - + frm[o] - - \otimes \Gamma + \Delta'(\Gamma)$, and therefore

$$(\Delta \otimes \mathrm{id}_{\mathcal{H}}) \circ \Delta = (\mathrm{id}_{\mathcal{H}} \otimes \Delta) \circ \Delta \quad \text{if and only if} \quad (\Delta' \otimes \mathrm{id}_{\mathcal{H}}) \circ \Delta' = (\mathrm{id}_{\mathcal{H}} \otimes \Delta') \circ \Delta'.$$

Moreover, it is enough to show the coassociativity property for 1PI Feynman graphs, which are the generators of $\mathcal{H}$, since Δ' is an algebra homomorphism, i.e. $\Delta(\Gamma \cup \Gamma') = \Delta(\Gamma) \cup \Delta(\Gamma')$, which is also easy to verify by a direct calculation. Now, on the one hand, we have

$$(\Delta' \otimes \mathrm{id}_{\mathcal{H}}) \circ \Delta'\Gamma = \sum_{\substack{\gamma \in \bigcup \mathcal{G}^{\omega}_{sd} \\ \gamma \subsetneq \Gamma}} \Delta'\gamma \otimes \Gamma/\gamma = \sum_{\substack{\gamma \in \bigcup \mathcal{G}^{\omega}_{sd} \\ \gamma \subsetneq \Gamma}} \sum_{\substack{\gamma' \in \bigcup \mathcal{G}^{\omega}_{sd} \\ \gamma' \subsetneq \gamma}} \gamma' \otimes \gamma/\gamma' \otimes \Gamma/\gamma.$$

On the other hand, we have

$$(\mathrm{id}_{\mathcal{H}} \otimes \Delta') \circ \Delta'\Gamma = \sum_{\substack{\gamma' \in \bigcup \mathcal{G}^{\omega}_{sd} \\ \gamma' \subsetneq \Gamma}} \gamma' \otimes \Delta'(\Gamma/\gamma') = \sum_{\substack{\gamma' \in \bigcup \mathcal{G}^{\omega}_{sd} \\ \gamma' \subsetneq \Gamma}} \sum_{\substack{\gamma'' \in \bigcup \mathcal{G}^{\omega}_{sd} \\ \gamma'' \subsetneq \Gamma/\gamma'}} \gamma' \otimes \gamma'' \otimes (\Gamma/\gamma')/\gamma''.$$

For the coassociativity property, we must show that these are equal for all algebra generators $\Gamma \in \mathcal{G}_{1\mathrm{PI}}$. Notice first that in the first sum we may interchange the order of the two summations,

$$(\Delta' \otimes \mathrm{id}_{\mathcal{H}}) \circ \Delta'\Gamma = \sum_{\substack{\gamma' \in \bigcup \mathcal{G}^{\omega}_{sd} \\ \gamma' \subsetneq \Gamma}} \sum_{\substack{\gamma \in \bigcup \mathcal{G}^{\omega}_{sd} \\ \gamma' \subsetneq \gamma \subseteq \Gamma}} \gamma' \otimes \gamma/\gamma' \otimes \Gamma/\gamma.$$

It is therefore enough to show that, for a fixed γ', we have

$$\sum_{\substack{\gamma\in\bigcup\mathcal{G}^{\omega}_{sd}\\ \gamma'\subsetneq\gamma\subsetneq\Gamma}} \gamma/\gamma'\otimes\Gamma/\gamma = \sum_{\substack{\gamma''\in\bigcup\mathcal{G}^{\omega}_{sd}\\ \gamma''\subsetneq\Gamma/\gamma'}} \gamma''\otimes(\Gamma/\gamma')/\gamma''.$$

These two expressions coincide, if we can set $\gamma''=\gamma/\gamma'$ in the summation for all $\gamma,\gamma',\gamma''\in\bigcup\mathcal{G}^{\omega}_{sd}$ such that $\gamma'\subsetneq\gamma\subsetneq\Gamma$ and $\gamma''\subsetneq\Gamma/\gamma'$, since then we have

$$(\Gamma/\gamma')/\gamma''=(\Gamma/\gamma')/(\gamma/\gamma')=\Gamma/\gamma.$$

This amounts to two requirements:

1. Closure of $\mathcal{G}^{\omega}_{sd}$ under contraction: $\gamma/\gamma'\in\mathcal{G}^{\omega}_{sd}$ for all $\gamma,\gamma'\in\mathcal{G}^{\omega}_{sd}$ such that $\gamma'\subsetneq\gamma$, so that we can find $\gamma''=\gamma/\gamma'\subsetneq\Gamma/\gamma'$ in $\bigcup\mathcal{G}^{\omega}_{sd}$.
2. Closure of $\mathcal{G}^{\omega}_{sd}$ under insertion: for all $\gamma',\gamma''\in\mathcal{G}^{\omega}_{sd}$ we have $\gamma'\circ_v\gamma''\in\mathcal{G}^{\omega}_{sd}$ for any allowed insertion, so that we can find $\gamma=\gamma'\circ_v\gamma''\subsetneq\Gamma$ in $\bigcup\mathcal{G}^{\omega}_{sd}$.

As we have assumed these properties to hold, this finishes the proof of the lemma.

Let us now state the main result of this subsection:

Theorem 10.6.12 *Let $\mathcal{H}_{BGR}$ be the unital associative algebra freely generated by the 1PI Feynman graphs of the BGR TGFT model. Then $(\mathcal{H}_{BGR},u,m,\epsilon,\Delta,S)$ is a Hopf algebra, where*

- *the unit $u:\bar{m}C\to\mathcal{H}_{BGR}$ is given by $u(1)=1$, where 1 is the empty graph;*
- *the product $m:\mathcal{H}_{BGR}\otimes\mathcal{H}_{BGR}\to\mathcal{H}_{BGR}$ is given by the disjoint union of graphs;*
- *the counit $\epsilon:\mathcal{H}_{BGR}\to\bar{m}C$ is given by $\epsilon(frm[o]--)=1$ and $\epsilon(\Gamma)=0$ for $\Gamma\neq 1$;*
- *the coproduct $\Delta:\mathcal{H}_{BGR}\to\mathcal{H}_{BGR}\otimes\mathcal{H}_{BGR}$ is given by*

$$\Delta(\Gamma)=\Gamma\otimes 1+1\otimes\Gamma+\sum_{\substack{\gamma\in\bigcup\mathcal{G}^{\omega}_{sd}\\ \gamma\subsetneq\Gamma}}\gamma\otimes\Gamma/\gamma,$$

  which is extended to $\mathcal{H}_{BGR}$ as an algebra homomorphism.
- *the antipode $S:\mathcal{H}_{BGR}\to\mathcal{H}_{BGR}$ is given by the recursive formula*

$$S(\Gamma)=-\Gamma-\sum_{\substack{\gamma\in\mathcal{G}^{\omega}_{sd}\\ \gamma\subsetneq\Gamma}}S(\gamma)\Gamma/\gamma,$$

  with $S(frm[o]--)=1$.

Proof We have already shown that the coproduct is coassociative. All the other Hopf algebra properties follow. In particular, we note that $(\mathcal{H}_{\text{BGR}},u,m,\epsilon,\Delta)$ is a graded connected bialgebra, graded by the number of internal edges which then leads to the formula of the antipode. This concludes the proof.

The six-point function Feynman amplitude of the BGR model is then given by

$$\phi_R(\Gamma) = S_R^{\phi}(\phi(\Gamma)) \star \phi(\Gamma),$$

where S_R^{ϕ} is given recursively by

$$S_R^{\phi}(\phi(\Gamma)) = -R[\phi(\Gamma)] - R\left[\sum_{\substack{\gamma \subsetneq \Gamma \\ \gamma \in \mathcal{G}_{sd}^{\omega}}} S_R^{\phi}(\phi(\gamma))\phi(\Gamma/\gamma)\right].$$

If one uses a dimensional regularization scheme, then the given formula can be identified with the renormalized Feynman amplitude of any graph (recall however that in the original paper Ben Geloun and Rivasseau (2011) the position space multi-scale analysis Taylor expansion was used to prove renormalizability).

10.6.3 Hochschild cohomology and the combinatorial DSE for tensorial GFT

In this section we deepen the analysis of one of the simplest renormalizable tensor GFT models, namely the rank five tensor model with propagator $1/(p^2+m^2)$ and a specific type of interaction, called quartic melonic interaction.

This model is just renormalizable and has been studied in Samary et al. (2015). We give the power counting theorem for this model and we define the Connes–Kreimer combinatorial Hopf algebra for this model and the Hochschild one-cocycles allowing to write the combinatorial DSE. We show the corresponding diagrams at one- and two-loop levels. Furthermore, Hopf subalgebras are exhibited, again at one- and two-loop levels.

The model and its power counting theorem

Consider a pair of conjugate rank-5 tensor fields

$$T_n, \bar{T}_{\bar{n}}, \text{ with } n = \{n_1, n_2, n_3, n_4, n_5\} \in \mathbb{Z}^5,\ \bar{n} = \{\bar{n}_1, \bar{n}_2, \bar{n}_3, \bar{n}_4, \bar{n}_5\} \in \mathbb{Z}^5. \quad (10.27)$$

They belong respectively to the tensor product $\mathcal{H} = \mathcal{H}_1 \otimes \mathcal{H}_2 \otimes \mathcal{H}_3 \otimes \mathcal{H}_4 \otimes \mathcal{H}_5$ and to its dual, where each $\mathcal{H}_c$ is an independent copy of $\ell_2(\mathbb{Z}) = L_2(U(1))$, and the colour or strand index c takes values $c = 1,2,3,4,5$. By Fourier transform the field T can be considered also as an ordinary complex scalar field $T(\theta_1,\theta_2,\theta_3,\theta_4,\theta_5)$ on the 5D torus $\mathbf{T}_5 = U(1)^5$ and $\bar{T}(\bar{\theta}_1,\bar{\theta}_2,\bar{\theta}_3,\bar{\theta}_4,\bar{\theta}_5)$ is simply its complex conjugate Ben Geloun and Rivasseau (2011). The tensor indices n can be identified as the *momenta* associated to the positions θ, and we therefore also call *momentum basis* the canonical basis $e_n = e_{n_1} \otimes \cdots \otimes e_{n_5}$ of $\mathcal{H}$.

If we restrict the momenta n to lie in $[-N,N]^5$ rather than in $\mathbb{Z}^5$ we have a proper (finite dimensional) tensor model. We consider N as the ultraviolet cutoff, and we are interested in performing the ultraviolet limit $N \to \infty$.

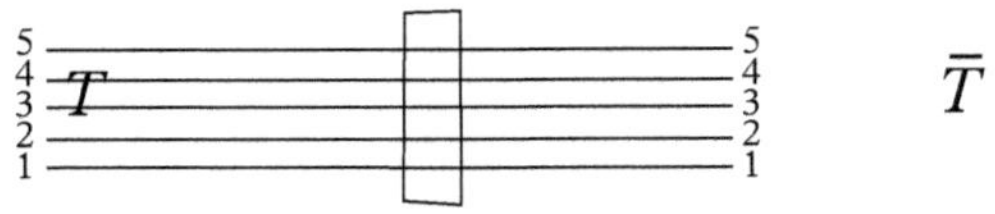

Figure 10.22 *The strands represent the five different colours, and the box suggests the C propagator that links them*

We introduce the normalized Gaussian measure

$$d\mu_C(T,\bar{T}) = \left(\prod_{n,\bar{n}\in[-N,N]^5} \frac{dT_n d\bar{T}_{\bar{n}}}{2i\pi} \right) \mathrm{Det}(C)^{-1}\, e^{-\sum_n T_n C^{-1}(n)\bar{T}_n}, \tag{10.28}$$

where the bare covariance C is, up to a bare field strength parameter Z, the inverse of the Laplacian on $\mathbf{T}_5$ with cut-off N on each momentum plus a bare mass term

$$C(n,\bar{n}) = \delta_{n,\bar{n}} C(n), \quad C(n) = \frac{1}{Z}\frac{1}{n^2 + m_b^2}. \tag{10.29}$$

In such formulas, letters such as $n, m, p, q, \cdots$ will be used for elements in $[-N,N]^5$, $n^2 = \sum_c n_c^2$ and m_b^2 is the square of the bare mass. Fig. 10.22 gives a graphical representation of the propagator of this theory.

We adopt the colour notations of Delepouve, Gurau, and Rivasseau (2016). $\tilde{\rho}$, $\mathbb{I}$ and $<,>$ mean the trace, the identity, and the scalar product on $\mathcal{H}$. $\mathbb{I}_c$ is the identity on $\mathcal{H}_c$, $\tilde{\rho}_c$ is the trace on $\mathcal{H}_c$, and $<,>_c$ the scalar product restricted to $\mathcal{H}_c$. The notation $\hat{c}$ means 'every colour except c'. For instance, $\mathcal{H}_{\hat{c}}$ means $\otimes_{c'\neq c}\mathcal{H}_{c'}$, $\mathbb{I}_{\hat{c}}$ is the identity on the tensor product $\mathcal{H}_{\hat{c}}$, $\tilde{\rho}_{\hat{c}}$ is the partial trace over $\mathcal{H}_{\hat{c}}$ and $<,>_{\hat{c}}$ the scalar product restricted to $\mathcal{H}_{\hat{c}}$.

T and $\bar{T}$ can be considered both as vectors in $\mathcal{H}$ or as diagonal (in the momentum basis) operators acting on $\mathcal{H}$, with eigenvalues T_n and $\bar{T}_n$. An important quantity in tensor models is the partial trace $\tilde{\rho}_{\hat{c}} T\bar{T}$, which we can also identify with the partial product $< T,\bar{T} >_{\hat{c}}$. It is a (in general non-diagonal) operator in $\mathcal{H}_c$ with matrix elements in the momentum basis

$$< T,\bar{T} >_{\hat{c}} (n_c,\bar{n}_c) = (\tilde{\rho}_{\hat{c}} T\bar{T})(n_c,\bar{n}_c) = \sum_{n_{c'},\bar{n}_{c'},c'\neq c}\ \prod_{c'\neq c} \delta_{n_{c'}\bar{n}_{c'}} T_n \bar{T}_{\bar{n}}. \tag{10.30}$$

The main new feature of tensor models compared to ordinary field theories is the non-local form of their interaction, which is chosen invariant under the action of $U(N)^{\otimes 5}$. In this paper we consider only the quartic melonic interaction, which is $\frac{g_b}{2}\sum_c V_c(T,\bar{T})$, where g_b is the bare coupling constant and

$$V_c(T,\bar{T}) = \tilde{\rho}_c[(\tilde{\rho}_{\hat{c}} T\bar{T})^2] = \sum_{n_c,\bar{n}_c,m_c,\bar{m}_c} \left(T_n \bar{T}_{\bar{n}} \prod_{c'\neq c} \delta_{n_{c'}\bar{n}_{c'}} \right) \delta_{n_c \bar{m}_c} \delta_{m_c \bar{n}_c} \left(T_m \bar{T}_{\bar{m}} \prod_{c'\neq c} \delta_{m_{c'}\bar{m}_{c'}} \right) \tag{10.31}$$

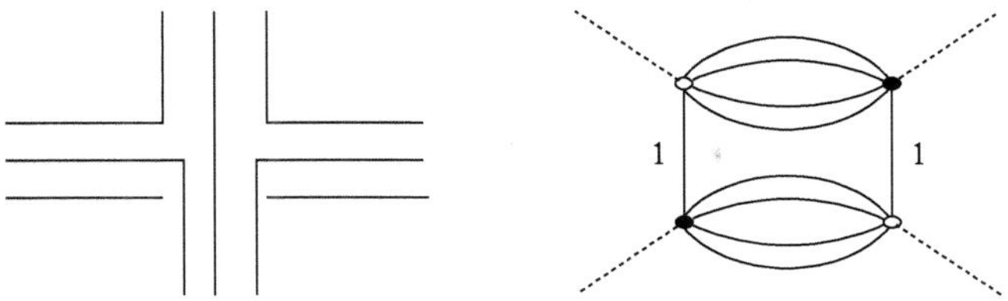

Figure 10.23 *Two equivalent representations of the melonic quartic vertex of type V_1*

for $c = 1, \cdots, 5$ are the five quartic melonic interactions of random tensors at rank five Delepouve, Gurau, and Rivasseau (2016) (see Fig. 10.23).

The generating function for the moments of the model is then

$$\mathcal{Z}(g, J, \bar{J}) = \frac{1}{\mathcal{Z}} \int e^{<\bar{J},T>+<J,\bar{T}>} e^{-g_b \sum_c V_c(T,\bar{T})} d\mu_C(T, \bar{T}), \tag{10.32}$$

where $\mathcal{Z} = \mathcal{Z}(g, J, \bar{J})|_{J=\bar{J}=0}$ is the normalization and the source tensors J and $\bar{J}$ are dual respectively to $\bar{T}$ and T. The main problem in QFT is to compute $\mathcal{W}(g, J, \bar{J}) = \log \mathcal{Z}(g, J, \bar{J})$ which is the generating function for the connected Schwinger functions,

$$S_{2k}(\bar{n}_1, \cdots, \bar{n}_k; n_1, \cdots, n_k) = \frac{\partial}{\partial J_{\bar{n}_1}} \cdots \frac{\partial}{\partial J_{\bar{n}_k}} \frac{\partial}{\partial \bar{J}_{n_1}} \cdots \frac{\partial}{\partial J_{n_k}} \mathcal{W}(g, J, \bar{J})|_{J=\bar{J}=0}. \tag{10.33}$$

This model is globally symmetric under colour permutations and is just renormalizable like ordinary ϕ_4^4. However, the structure of ultraviolet divergences is simpler, as shown in the next subsection.

Let us now turn to the power counting theorem of this model, which comes to the listing of superficial divergent Feynman graphs and of theory superficially divergent degree (see again Subsection 3.6.1).

A way to determine the divergent degree $\omega(\mathcal{G}) = -2L(\mathcal{G}) + F(\mathcal{G})$ of a graph $\mathcal{G}$, where $L(\mathcal{G})$ and $F(\mathcal{G})$ are respectively the number edges and faces of $\mathcal{G}$, is by computing the number of its faces. This quantity is given by

$$F = 4V + 4 - 2E - \frac{1}{12}(\sum_{\mathcal{J}} g_{\tilde{\mathcal{J}}} - \sum_{\mathcal{J}_\partial} g_{\mathcal{J}_\partial}) - (C_{\partial\mathcal{G}} - 1), \tag{10.34}$$

where V and E are respectively the number of vertices and the number of external legs of $\mathcal{G}$. The sum is performed on all jackets $\mathcal{J}$ of $\mathcal{G}_{color}$, $g_{\tilde{\mathcal{J}}}$ is the genus of the pinched jacket $\tilde{\mathcal{J}}$, $g_{\mathcal{J}_\partial}$ is the genus of the jacket $\mathcal{J}_\partial$ and $C_{\partial\mathcal{G}}$ is the number of connected components of the boundary graph $\partial\mathcal{G}$. For precise definitions of all these notions and a proof of (10.34), we refer the interested reader to Ben Geloun and Rivasseau (2011); Raasakka and Tanasa (2013); Samary et al. (2015).

After substituting the combinatorial relation $2L+E=4V_4$, the superficially divergent degree of $\mathcal{G}$ can be written as

$$\omega(\mathcal{G}) = 4 - E - (C_{\partial\mathcal{G}} - 1) - \frac{1}{12}\left(\sum_{\mathcal{J}} g_{\tilde{\mathcal{J}}} - \sum_{\mathcal{J}_\partial} g_{\mathcal{J}_\partial}\right). \tag{10.35}$$

We call 1PI graphs $\mathcal{G}$ with $\omega(\mathcal{G}) \geq 0$ *superficially divergent*, and write $\mathcal{G}^{\omega}_{sd} := \{\mathcal{G} : \omega(\mathcal{G}) \geq 0\}$ for the set of superficially divergent $1PI$ Feynman graphs.

In the following table, we give a list of superficially divergent graphs, all over cases are finite as $N \to \infty$.

E	$\sum_{J_\partial} g_{J_\partial}$	$C_{\partial\mathcal{G}} - 1$	$\sum_{\tilde{J}} g_{\tilde{J}}$	$\omega(\mathcal{G})$
4	0	0	0	0
2	0	0	0	2
0	0	–1	0	5
0	0	–1	36	2
0	0	–1	60	0

List of superficially divergent graphs

Connes-Kreimer combinatorial Hopf algebra

As in the previous subsection, let $\mathcal{H}$ be the vector space freely generated by the 1PI Feynman tensor graphs of our model (including the empty graph, which we denote by $1_{\mathcal{H}}$). We endow this vector space with a product given by the disjoint union and with a unit map $\eta : \mathbb{K} \to \mathcal{H}$ defined by $\eta(1) = 1_{\mathcal{H}}$ and zero elsewhere, where we have denoted by $\mathbb{K}$ a field such as $\mathbb{R}$ or $\mathbb{C}$. We further endow $\mathcal{H}$ with the coproduct $\Delta : \mathcal{H} \to \mathcal{H} \otimes \mathcal{H}$ given by

$$\Delta(\mathcal{G}) := \mathcal{G} \otimes 1_{\mathcal{H}} + 1_{\mathcal{H}} \otimes \mathcal{G} + \sum_{\gamma} \gamma \otimes \mathcal{G}/\gamma, \tag{10.36}$$

where the sum is taken on disjoint unions of superficially divergent subgraphs of $\mathcal{G}$.

Furthermore, we define Δ' to be the non-trivial part of the coproduct:

$$\Delta(\mathcal{G}) = \mathcal{G} \otimes 1_{\mathcal{H}} + 1_{\mathcal{H}} \otimes \mathcal{G} + \Delta'(\mathcal{G}).$$

Note that $\mathcal{H}$ is graded by the loop number. One can then prove that the previous coproduct is coassociative. This follows Raasakka and Tanasa (2013)—one needs to check that the set of 1PI Feynman graphs of the model (see Table 10.1) is closed under tensor subgraph contraction and insertion.

The counit $\epsilon : \mathcal{H} \to \mathbb{K}$ is defined by: $\epsilon(1_{\mathcal{H}}) = 1$ and $\epsilon(\mathcal{G}) = 0$ for all $\mathcal{G} \neq 1_{\mathcal{H}}$. Defining now the antipode in the usual way one concludes that $(\mathcal{H}, \Delta)$ is a Hopf algebra, which is actually the Connes–Kreimer Hopf algebra describing the combinatorics of the renormalization of our 5D quartic tensor models (see again the previous subsection for details).

Combinatorial DSE and Hochschild cohomology

We show here that a primitively divergent graph γ can be used to define a Hochschild $1-$cocycle B_+^γ, where the operator B_+^γ is given by the operation of insertion of tensor graphs into γ. We then use these operators to encode, in a recursive way, the combinatorics of the DSE.

These manipulations follow the ones of Section 3.7 for QFT and the ones of Section 9.8 for QFT on the non-commutative Moyal space.

We denote by $|\mathcal{G}|$ the number of loops of $\mathcal{G}$. We then define the operators $B_+ : \mathcal{H} \to \mathcal{H}$ by

$$B_+^\gamma(X) := \sum_{\mathcal{G}} \frac{bij(\gamma, X, \mathcal{G})}{|X|_V} \frac{1}{maxf(\mathcal{G})} \frac{1}{(\gamma|X)} \mathcal{G}, \quad B_+^{k;r} = \sum_{|\gamma|=k,\, res(\gamma)=r} B_+^\gamma, \tag{10.37}$$

where $maxf(\mathcal{G})$ is the number of ways to shrink subdivergencies such that the resulting graph is primitive and $|X|_V$ is the number of graphs which are equal upon removal of external edges. The quantity $bij(\gamma, X, \mathcal{G})$ is the number of bijections between the external edges of X and adjacent edges of places in γ, such that $\mathcal{G}$ is obtained. Finally, the notation $\gamma|X$ stands for the number of insertion places of the graph X in γ.

We further set

$$B_+^\gamma(1_{\mathcal{H}}) = \gamma, \tag{10.38}$$

and

$$h_k^r := \sum_{|\mathcal{G}|=k, res(\mathcal{G})=r} \mathcal{G}, \tag{10.39}$$

where the sum runs on the graphs of a given loop number and residue. Finally, let M_r be the set of graphs such that $res(\mathcal{G}) = r$.

In commutative or Moyal non-commutative scalar QFT, (see again Sections 3.7 and 9.8) for any Feynman graph $X_{k,r}$ with k loops and residue r, one has the following identities:

$$\mathcal{G}^r \equiv 1 + \sum_{\mathcal{G} \in M_r} \mathcal{G} = 1 + \sum_{k=1}^{\infty} B_+^{k,r}(X_{k,r}), \tag{10.40}$$

$$\Delta(B_+^{k;r}(X_{k,r})) = B_+^{k;r}(X_{k,r}) \otimes \mathbb{I} + (id_{\mathcal{H}} \otimes B_+^{k;r}) \Delta(X_{k,r}), \tag{10.41}$$

$$\Delta(h_k^r) = \sum_{j=0}^{k} Pol_j^r(c) \otimes h_{k-j}^r, \tag{10.42}$$

where $Pol_j^r(c)$ is a polynomial in the variables h_m^r of total degree j.

Let us now recall the meaning of each of these three identities. The first one encodes, in a recursive way, the combinatorics of the DSE—the so-called combinatorial DSE. The second one points out the fact that the operator B_+ is, from a purely algebraic point of view, a Hochschild cocycle of the Hopf algebra of the previous subsection. Finally, the last equation shows that one can define Hopf subalgebras at any loop order.

In the sequel, we will explicitly check under what conditions these algebraic identities hold at the one- and two-loop level, in the case of our 5D quartic tensor model. We set for simplicity, 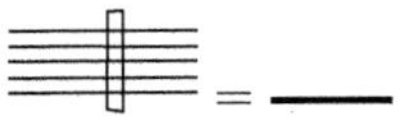 and 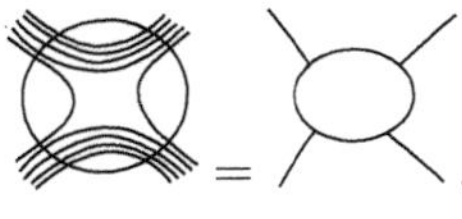.

• **One-loop analysis:**
We have

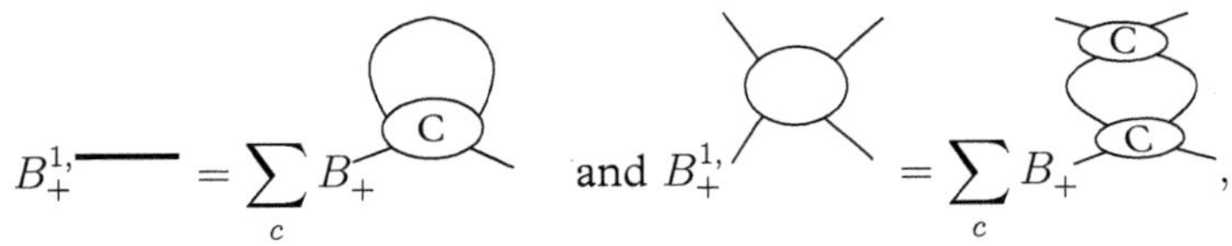

(10.43)

where the sum $\sum_c$ runs over all five colours, and by 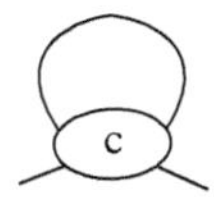 one means 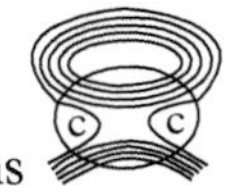 the quartic vertex V_c with special thread of colour c (see Fig. 10.23 for the case $c = 1$). Respectively by 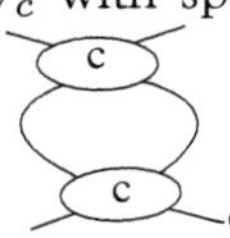 one means 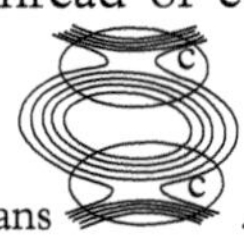.

One then has:

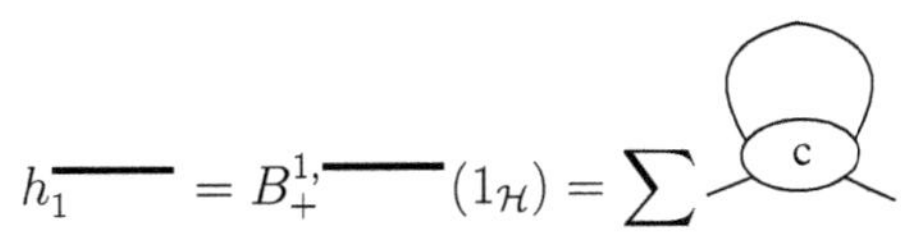

(10.44)

and

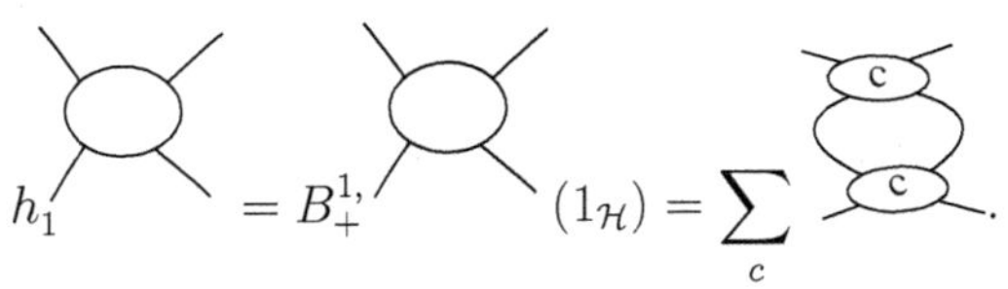

(10.45)

• **Two-loop analysis:**
As in the previous example, one computes

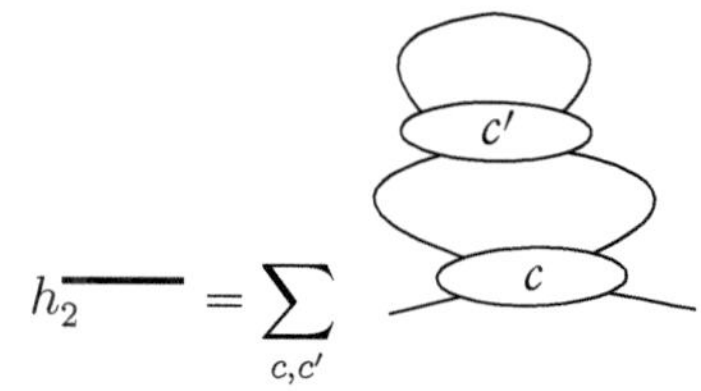

(10.46)

where the sum over c and c' of 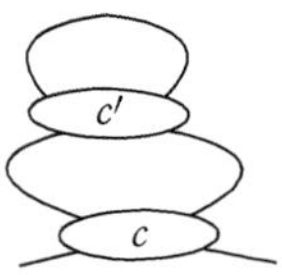 now corresponds to two different types of stranded diagrams. If $c = c'$ we get 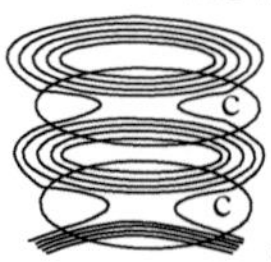. If $c \neq c'$, we get a different diagram namely . Both types of graphs are quadratically divergent, since they have both eight inner closed faces and three propagators. Notice, however, that they do not have the same amplitude: the coefficients in front of the quadratic divergence are different.

The relation in (10.46) coincides with the diagrammatic expression obtained directly from definition (10.39). Let us now apply Δ on each of the graphs in the right-hand side of (10.46). For each of these tensor graphs, the non-trivial part of the coproduct writes:

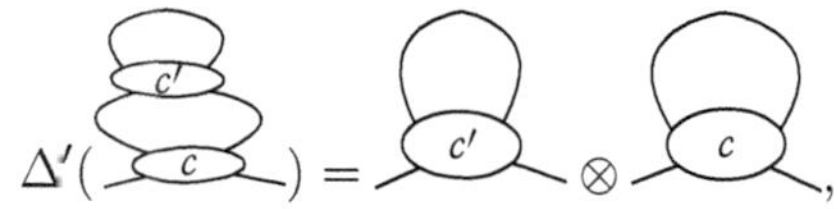

(10.47)

when $(c \neq c')$ and

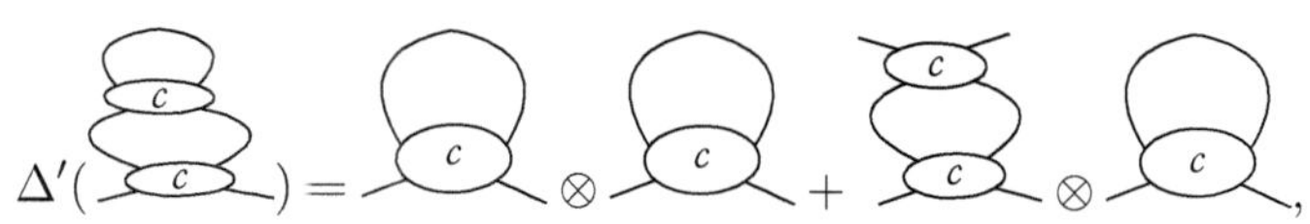

(10.48)

else. As an example, we have:

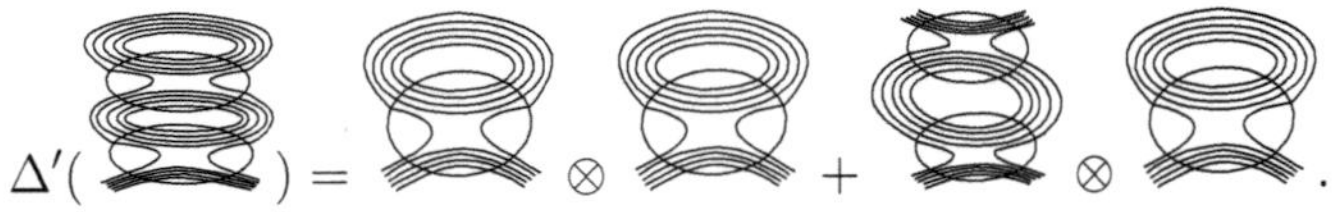

(10.49)

Adding up all these contributions, we obtain

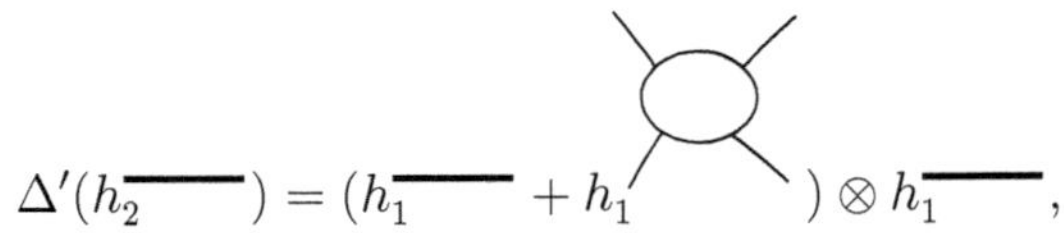

(10.50)

which thus checks relation (10.42) at two-loop level.

For the four-point function, we have

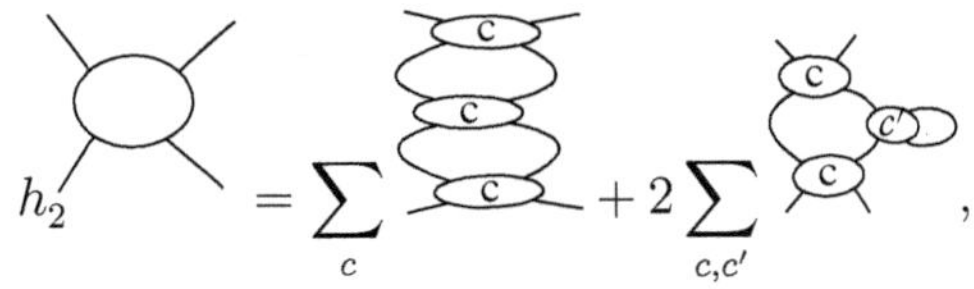

(10.51)

where 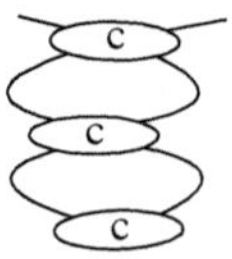 corresponds to 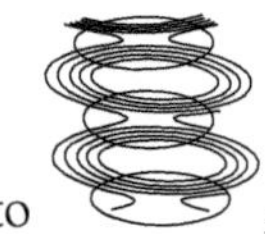, and by 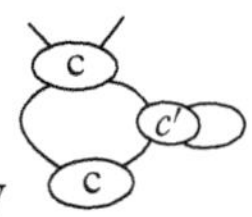 one means 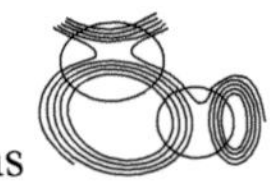 (for $c = c'$) or 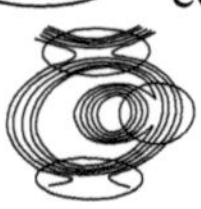 for $c \neq c'$.

Once again, we apply the coproduct on each of these tensor graphs and we obtain for the non-trivial part

(10.52)

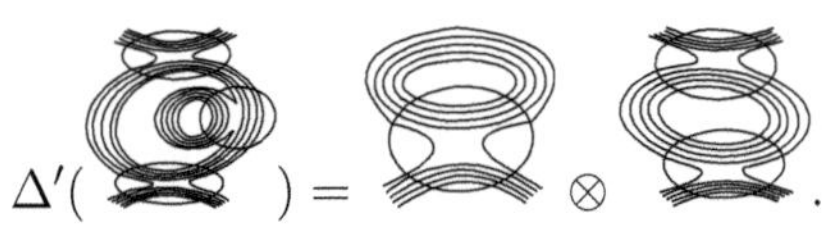

(10.53)

Putting all this together leads to

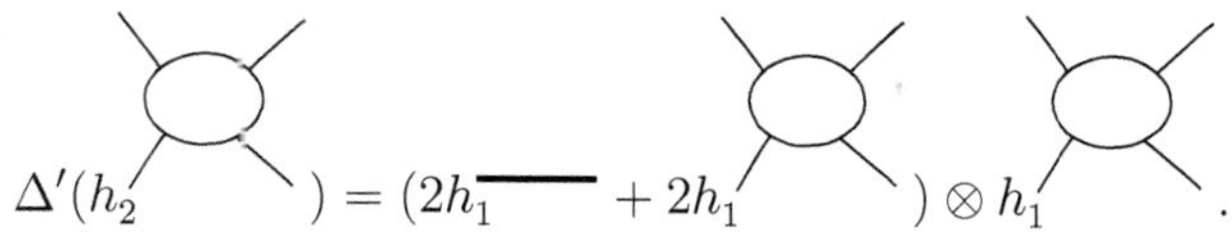

(10.54)

As was the case for (10.50), equation (10.54) illustrates identity (10.42) at two-loop level.

In addition, each of these graphs is in the image of appropriate Hochschild one-cocycles. Indeed, one has

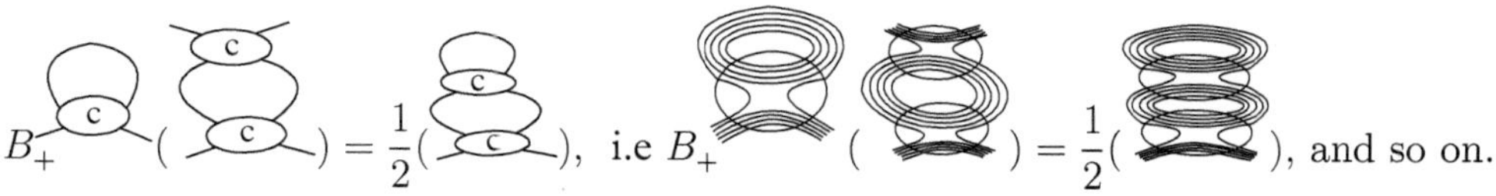

(10.55)

Note that the combinatorial factors $\frac{1}{2}$ come from the definition (10.37).

One can check that putting all these expressions together does *not* give the two-loop quantum corrections of the propagator.

Nevertheless, this issue is fixed by taking into consideration the equivalent, in the tensorial setting, of the so-called planar irregular sector of Moyal QFT. In our current setting, the corresponding graphs have external legs breaking faces of different colours. We call them the two-colour-breaking sectors. The action of the B_+ operator on these graphs is given by:

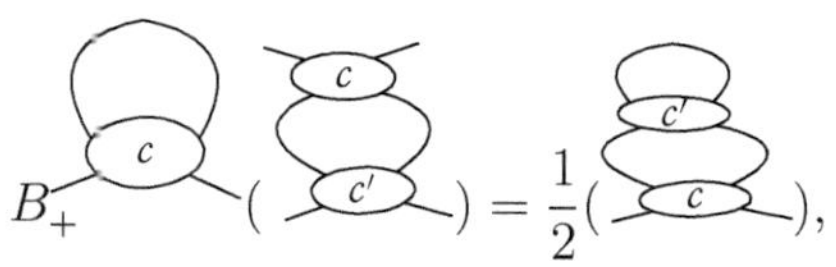

(10.56)

with $c \neq c'$ i.e. explicitly showing strand structure

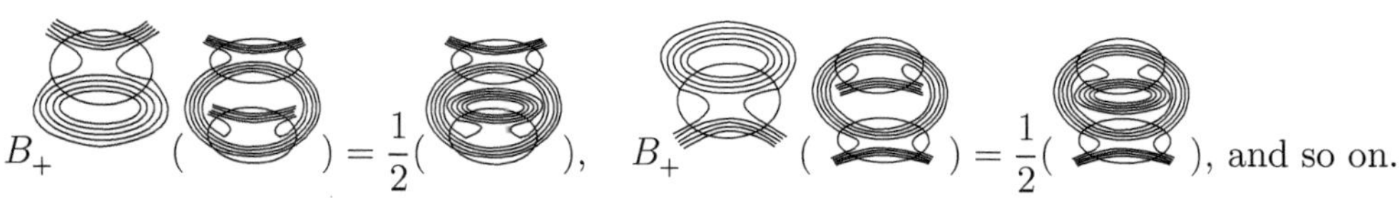

(10.57)

Adding the diagrams in (10.55) to the ones in (10.55) (with the appropriate combinatorial factors) leads to identity (10.40), at two-loop level.

Finally, let us also check the identity (10.41) at two-loop level:

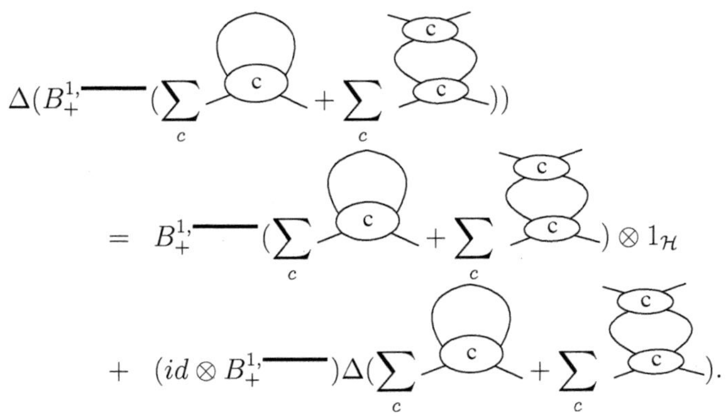

(10.58)

We calculate the left-hand side of this relation. One has:

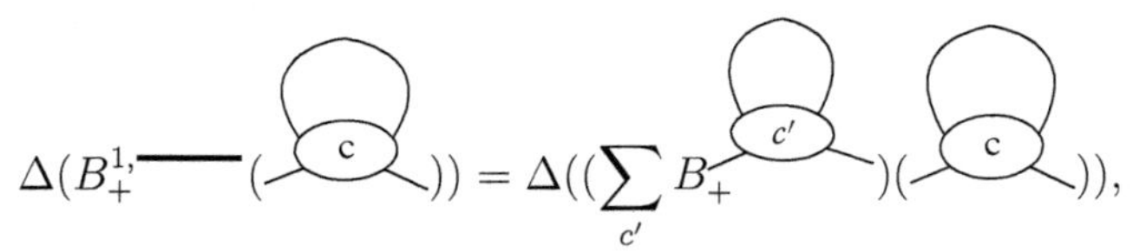

(10.59)

and

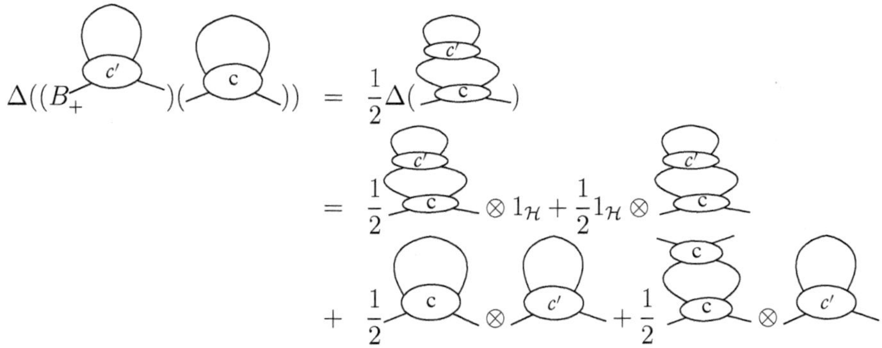

(10.60)

The expression in (10.60) helps to compute (10.59). Analogous expressions are computed for the other tensor Feynman graphs of equation (10.58).

In order to compute the right-hand side of (10.58), we have

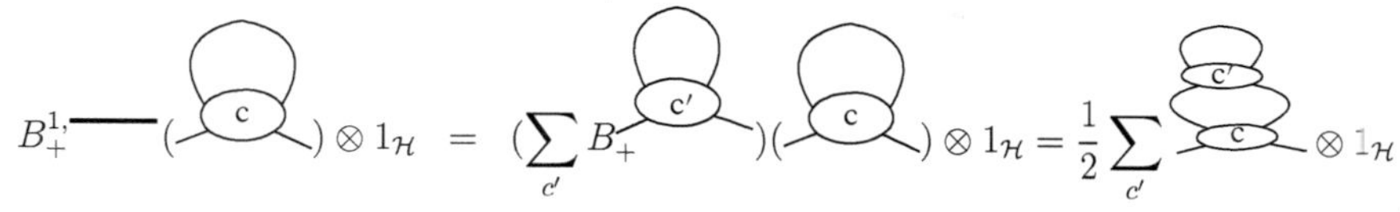

(10.61)

One further has:

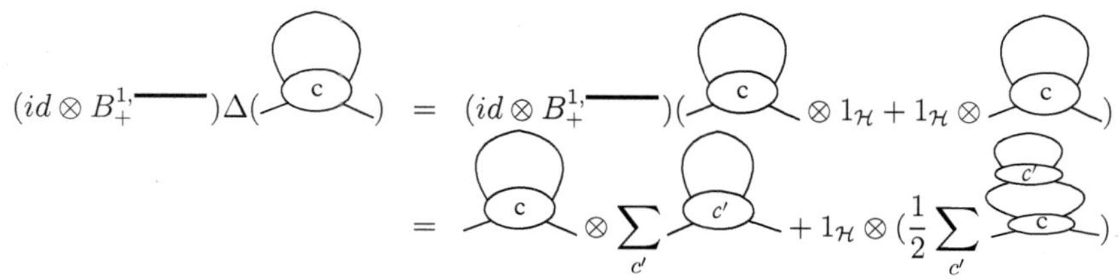

(10.62)

Summing all the contributions of type (10.61) and (10.62), we obtain the right-hand side of (10.58). Nevertheless, one checks that this is not equal to the left-hand side. Once again, the two-colour-breaking sector saves the day. Let us compute the left-hand side of (10.58) for this two-breaking face sector. One has for ($c \neq c'$):

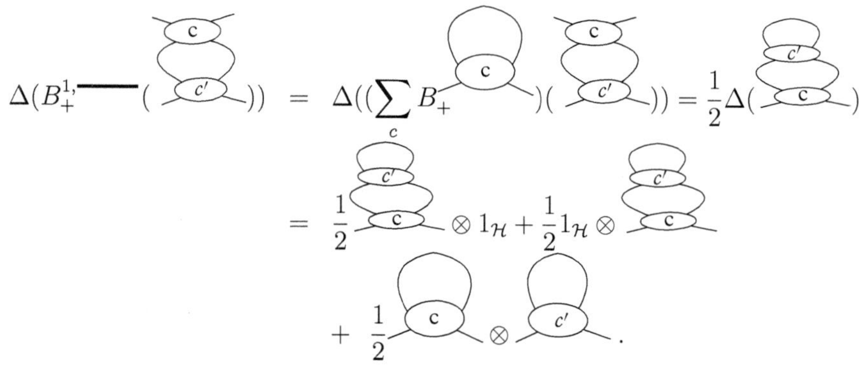

(10.63)

Let us now compute the right-hand side of (10.58) for the two-colour-breaking sector. One has

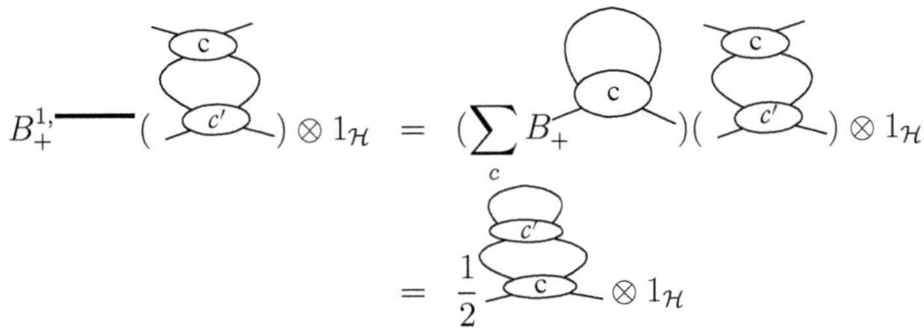

(10.64)

and

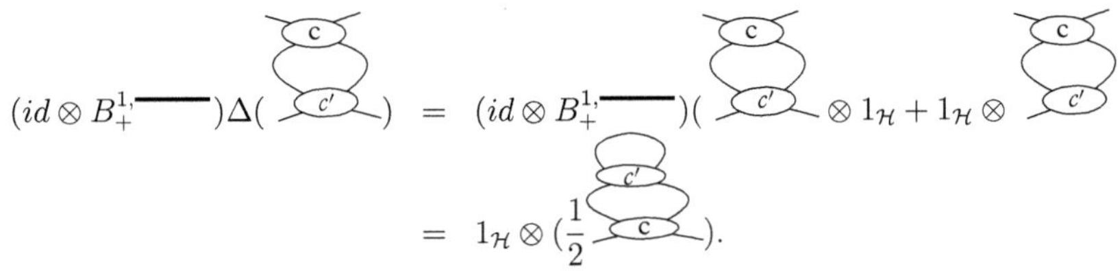

(10.65)

In a similar way, one can compute the corresponding contributions of all such two-colour-breaking Feynman graphs.

Adding all such contributions shows that identity (10.41) holds.

The case of sixth order interaction tensor model

We now show that the algebraic results of the previous subsection do not hold for sixth order interaction QFT models. We focus here on the BGR model.

In this (renormalizable) model, one has both sixth order and fourth order interactions (see again the previous subsection). Thus, one has superficially divergent graphs such as

$$\mathcal{G} =$$  .

A sub-divergence of this graph is given by

$$\gamma =$$ 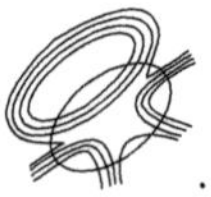 .

The contraction of γ in $\mathcal{G}$ gives

$$\mathcal{G}/\gamma =$$ 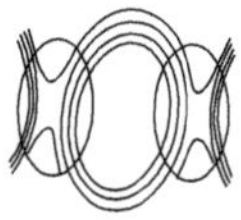 . (10.66)

Nevertheless, the tensor graph $\mathcal{G}/\gamma$ is not superficially divergent, in the bare theory. Since the tensor graph previous $\mathcal{G}$ belongs to 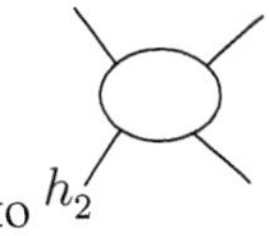 and the graph (10.66) does not belong to 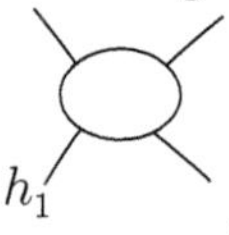, one can obviously not obtain identities of type (10.42). The same type of problem appears for identities (10.40) and (10.41) illustrated in the previous subsection.

Let us end this chapter by emphasizing that this is not related in any way to the tensor character of the models discussed in this paper—the exact same behaviour appears in the case of a standard ϕ^6 theory where the one-loop correction to the ϕ^4 term is not superficially divergent in the bare theory.

10.7 Selected further reading

- D. Oriti. (2011). 'The microscopic dynamics of quantum space as a group field theory', Proceedings, Foundations of Space and Time: Reflections on Quantum Gravity: Cape Town, South Africa', 2011, 257–320.
- D. Oriti, The Group field theory approach to quantum gravity, 2006, expanded version of a contribution to Approaches to Quantum Gravity—toward a new understanding of space, time, and matter D. Oriti forthcoming, Cambridge: Cambridge University Press, gr-qc/0607032.
- L. Freidel, Group field theory: An overview, International 44, 1769–1783, 2005, hep-th/0505016.

  Introductory texts to GFT.

11

From random matrices to random tensors

Let us start this chapter with the following definition:

Definition 11.0.1 *A random matrix is a matrix of given type and size whose entries consist of random numbers from some specified distribution.*

For extensive studies of the particularly vast field of random matrices, we refer the interested reader to the books Guionnet, Anderson, and Zeitouni (2009); Mehta (2004).

In mathematics, random matrices have not been only used per se, but they also led to various developments in other domains, such as Voiculescu's non-commutative probabilities, to name only one. Moreover, Witten's (1990) conjecture on a rigorous approach to the moduli space of punctured Riemann surfaces was solved by Kontsevitch through the use of an appropriate matrix model Kontsevich (1992).

In combinatorics, matrix models have been used extensively for proving map counting theorems, in a different way than the standard ways known by combinatorists. This application of matrix models in combinatorics lies outside the purpose of this book, but we are directing the interested reader to A. Zvonkine's pedagogical review article on this topic Zvonkin (1997).

On the theoretical physics side, the study of the spectra of random matrices was proposed in nuclear physics in relation to the study of spectra of heavy atoms by Wigner in 1955, see the original article Wigner (1955). This comes from the fact that one can use random matrices to study the Hamiltonian of such a physical system. In 1974, G. 't Hooft proposed matrix models to study quantum chromodynamics (the QFT model studying the strong force, one of the four fundamental forces of the Universe, a force which for example is responsible for interactions within the nucleus of an atom) with a large number of colours (the colours being the colours associated to the quarks, the quantas of the strong force), see the original article t' Hooft (1974). His goal was to use this approach in order to explain with the celebrated quark confinement. Unfortunately, matrix models have not yet brought an explanation for the quark confinement, but later on matrix models were proved to relate to 2D quantum gravity and string theory, as we will briefly explain in the sequel.

Combinatorial Physics: Combinatorics, quantum field theory, and quantum gravity models. Adrian Tanasa, Oxford University Press (2021). © Adrian Tanasa. DOI: 10.1093/oso/9780192895493.003.0011

An important fact to emphasize here is that, in mathematical physics, matrix Feynman integrals describe the QFT behaviour of random matrix models, seen as quantum field theoretical models.

Let us now give some detail on matrix models (we refer the interested reader to Bertrand Eynard (2016); or to one of the various review articles on this topic, (Francesco 2004; Di Francesco, Ginsparg, and Zinn-Justin 1995), or David (1985), to name only a few).

Let M be an $N \times N$ matrix, which we take here to be Hermitian. The partition function of such a matrix model is written

$$Z = \int dM e^{-\frac{1}{2}\mathrm{Tr}M^2 + \frac{\lambda}{\sqrt{N}}\mathrm{Tr}M^3}, \qquad (11.1)$$

where the measure is defined to be

$$dM = \prod_i dM_{ii} \prod_{i<j} d\mathrm{Re}M_{ij}\, \mathrm{Im}M_{ij}. \qquad (11.2)$$

Note that we have considered in (11.1), a cubic interaction term $\mathrm{Tr}M^3$, this will lead to the generation of random triangulations, as we will now explain. If one considers, for example, a quartic interaction, this will lead to random quadrangulations.

The diagrammatic expansion of such a matrix model leads to Feynman ribbon graphs, which are dual to random triangulations.

Let us give some further explanations regarding this crucial point. The building block of such a triangularization is of course a triangle. One can now take a dual point of view and encode the information coming from the triangularization in a ribbon graph picture (see the 2D triangularization of Fig. 11.1). As mentioned previously, the appropriate valence of the associated ribbon graph vertex is thus three: the vertex (see Fig. 11.2) corresponds to the triangle (the building block of the triangularization), and the three incoming/outgoing edges correspond to the three edges of the original triangle. The two strands of each edge correspond to the two vertices joined by the respective edge in the initial triangle. The graph obtained in this way is actually a Feynman graph obtained through perturbative expansion of the matrix model (11.1).

Thus, a matrix model naturally leads to a sum over random triangulations. This corresponds to a discretized analogue of an integral over all possible geometries. The connection with 2D quantum gravity comes thus from the fact that matrix models describe a certain statistical ensemble over discrete geometries.

Let us also mention that the so-called 0-dimensional string theory (a pure theory of surfaces with no coupling to matter on the string world sheet) is also described by matrix models (see again David (1985); or Di Francesco, Ginsparg, and Zinn-Justin (1995)).

Two important mathematical physics techniques in the study of matrix models are:

1. the large N expansion; and
2. the double-scaling limit.

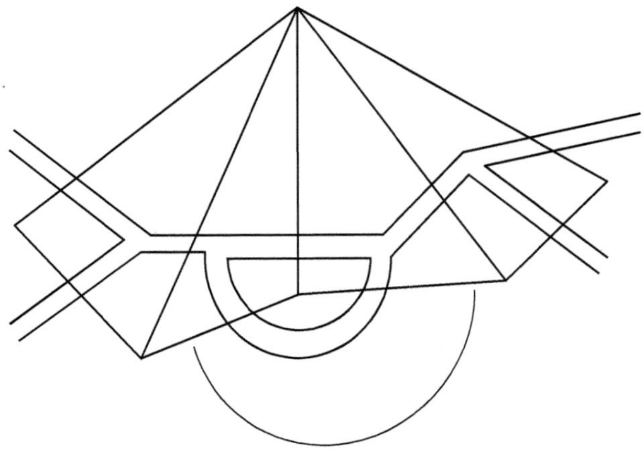

Figure 11.1 *Some triangularization of a 2D surface and the dual point of view leading to valence three ribbon graphs*

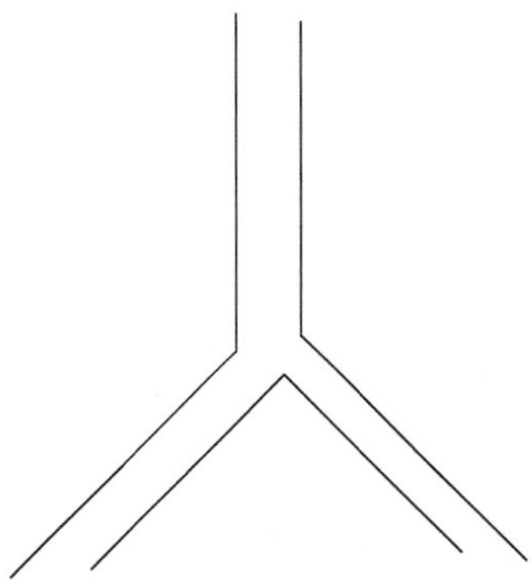

Figure 11.2 *A valence three vertex of a 2D model*

The **large N expansion**, which is from a combinatorial point of view an asymptotic expansion, (since one is taking the limit $N \to \infty$) is controlled by the genus of the ribbon Feynman graphs. We will show in the sequel that the leading order contribution to the partition function is given by planar graphs (which pave the 2D sphere $\mathcal{S}^2$).

By the simultaneous scaling of N and the coupling constant, the **double-scaling limit** allows to define a continuum limit, where topologies contribute, which is proven to be connected to 2D gravity (see again Di Francesco, Ginsparg, and Zinn-Justin (1995)).

In the next sections we will give a few details on these two crucial tools in the study of matrix models, the large N limit and the double-scaling limit. These sections follow the review articles Di Francesco (2004); or Di Francesco, Ginsparg, and Zinn-Justin (1995).

11.1 The large *N* limit

The large N limit of the matrix model can be implemented because the matrix Feynman amplitude can be combinatorially computed—in terms of number of vertices (V), edges, and faces (F) of the graph.

One first needs to do a change of variables:

$$M \to M\sqrt{N}. \tag{11.3}$$

This change of variables makes it easier to count powers of N. One can prove that the Feynman amplitude of a vacuum graph is written:

$$\mathcal{A} = \lambda^V N^{-\frac{1}{2}V+F} = \lambda^V N^{2-2g}. \tag{11.4}$$

Re-expressing the amplitude using only the genus, as previously demonstrated, is possible because of the relation:

$$E = \frac{3}{2}V. \tag{11.5}$$

This then leads to the fact that the partition function supports the $1/N$ expansion or large N expansion:

$$Z = N^2 Z_0(\lambda) + Z_1(\lambda) + \ldots = \sum_g N^{2-2g} Z_g(\lambda), \tag{11.6}$$

where Z_g gives the contribution from surfaces of genus g.

In the large N limit, only **planar surfaces** survive. The planar graphs are thus the **dominant graphs** (recall that planar graphs correspond to *triangulations of the sphere* $\mathcal{S}^2$).

11.2 The double-scaling limit

The successive coefficient functions $Z_g(\lambda)$ diverging at the same critical value of the coupling $\lambda = \lambda_c$.

One can prove that the leading singular piece of Z_g behaves as:

$$Z_g(\lambda) \propto f_g(\lambda_c - \lambda)^{(2-\gamma_{\text{str}})\chi/2} \text{ with } \gamma_{\text{str}} = -\frac{1}{2}. \tag{11.7}$$

Thus, one notices that contributions from higher genera ($\chi < 0$) are enhanced as $\lambda \to \lambda_c$.

Let us define the following parameter:

$$\kappa^{-1} := N(\lambda - \lambda_c)^{(2-\gamma_{\text{str}})/2}. \tag{11.8}$$

One can prove that the partition function is written

$$Z = \sum_g \kappa^{2g-2} f_g \tag{11.9}$$

The **double-scaling limit** corresponds to the limits $N \to \infty$, $\lambda \to \lambda_c$ while holding fixed κ. One can finally prove that this double-scaling limit leads to *coherent contribution from all genus surfaces* (see again Di Francesco (2004); or Di Francesco, Ginsparg, and Zinn-Justin (1995) for details).

In the following chapters, we will show that these two important techniques, the large N limit and the double-scaling limits, can also be implemented for various tensor model (N being in that case the size of the tensor). But, let us first give the general idea of extending matrix models to tensor models.

11.3 From matrices to tensors

Roughly speaking, the main idea is that if matrix models can be seen as dual to triangulations of 2D surfaces, this can then be extended - using tensor models instead of matrix models—to 3D, 4D spaces, and so on.

It is worth emphasizing that in this context, the *dimension* doesn't refer to the dimension of the QFT model (see again Chapter 3), since from a QFT point of view, matrix models and tensor models are 0-dimensional QFT models. Dimension here refers to the dimension of the space the Feynman graphs which are dual to. Thus, the ribbon graphs of matrix models are naturally dual to 2D surfaces (see previous discussion), and this can be generalized to an arbitrary dimension D.

The elementary cells that, when glued together form the space itself, are the D-simplices (a 2-simplex being a triangle, a 3-simplex being a tetrahedron, and so on). Since a D-simplex has $(D+1)$ facets on its boundary, the backbone of tensor models in D-dimension is some abstract $\Phi^{(D+1)}$ interaction on rank D-tensor fields Φ (see Fig. 11.2, Fig. 11.3; and respectively Fig. 11.4 which represent, for $D=2$, $D=3$, and $D=4$ vertices of these Feynman quantum gravity tensor graphs).

In order to define such a scalar Feynman graph, one needs a propagator (associated with the edge of the graph) and an interaction (associated with the graph vertex).

Let us now focus on the $D=3$ graph. The edge is simply given by Fig. 11.5. Each such tensor graph edge has three strands.

The valence four vertex corresponds to a tetrahedron, the simplest building block of a 3D space. The four triangles constructing the tetrahedron correspond to the four edges intersecting at the respective vertex (see, for example, Fig. 11.6). The three strands of such an edge correspond to the three edges of the respective triangle (face of the tetrahedron), thus generalizing the image of the 2D case.

One can use several conventions for drawing the vertex of such a theory. In the 3D case, one has the 'standard' convention given in Fig. 11.3. Nevertheless, one can also use that of Freidel, Gurau, and Oriti (2009), given in Fig. 11.7.

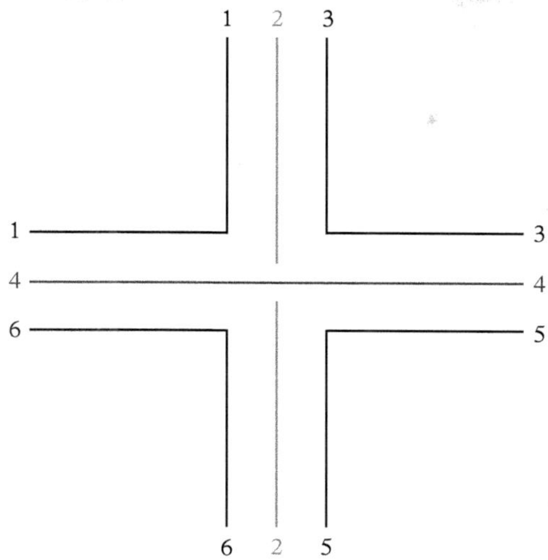

Figure 11.3 *A vertex of a 3D tensor model*

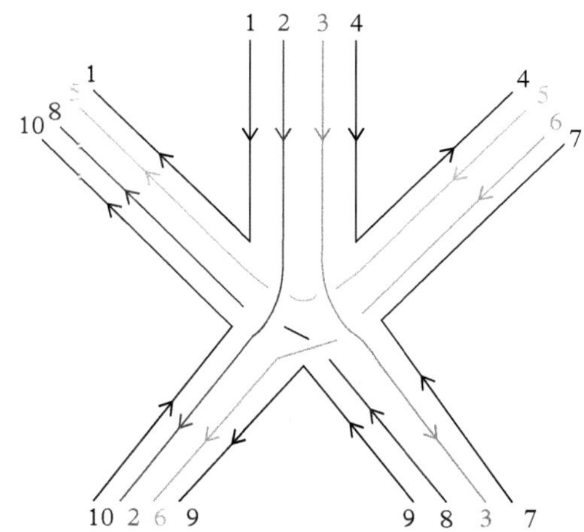

Figure 11.4 *A vertex of a 4D tensor model*

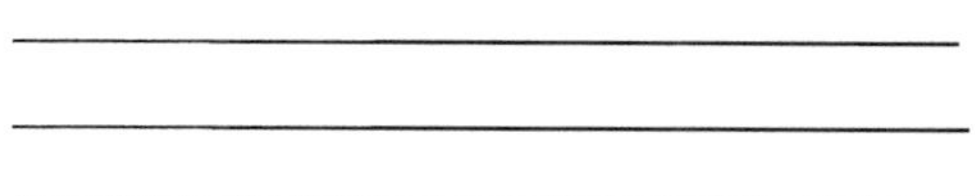

Figure 11.5 *An edge of a 3D GFT tensor graph*

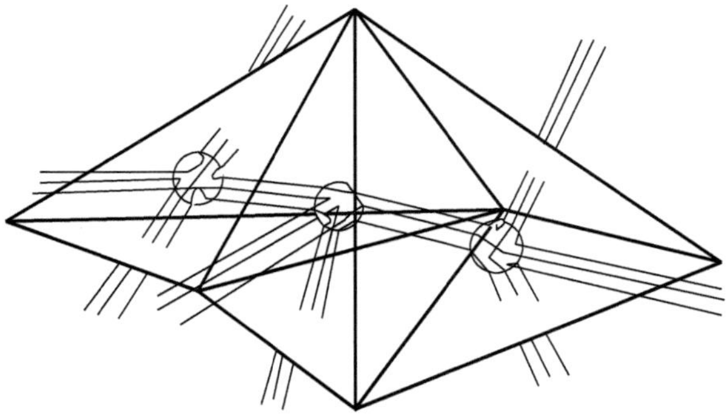

Figure 11.6 *Some triangularization of a 3D surface and the dual point of view leading to valence four tensor graphs*

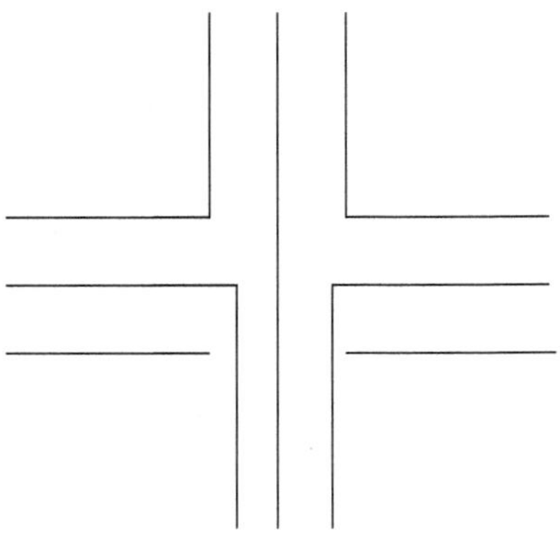

Figure 11.7 *A different convention for picturing the vertex of a 3D tensor graphs*

Assembling together these two constituents, the edge and the vertex, one is now able to build up tensor graphs. An example of such a tensor graph is given in Fig. 11.8.

Let us emphasize that the strand structure described here makes these theories extremely rich from a combinatorial and topological point of view.

One can define *faces* or closed circuits of these tensor graphs. This is a natural generalization of the definition of faces of ribbon graphs (see Definition 2.2.2). One can further generalize this notion in the case of tensor graphs, by defining a **bubble** as a closed 3D region of the graph.

As a consequence, a bubble can be represented as a ribbon graph, and thus, from a topological point of view, it is natural to define its Euler characteristic χ (and hence its genus). As explained in Section 2.2, this relation is a mere translation of the fact that such a ribbon graph can be drawn on a 2D manifold (each graph defining the surface on which it is drawn), g thus being the genus of the respective manifold. One thus speaks of *planar* or *non-planar bubbles* (for more details on this notion of planarity, the interested reader may again refer to Section 2.2).

For the tensor graph of Fig. 11.8, the bubbles are given in Figs. 11.9 and 11.10. The first of them is non-planar (having a single face and thus, is in accordance with the Euler characteristic relation genus 1), while the second is planar.

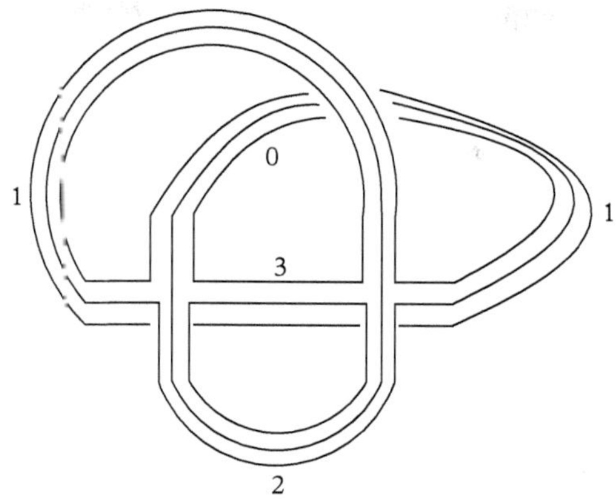

Figure 11.8 *An example of a 3D tensor graph*

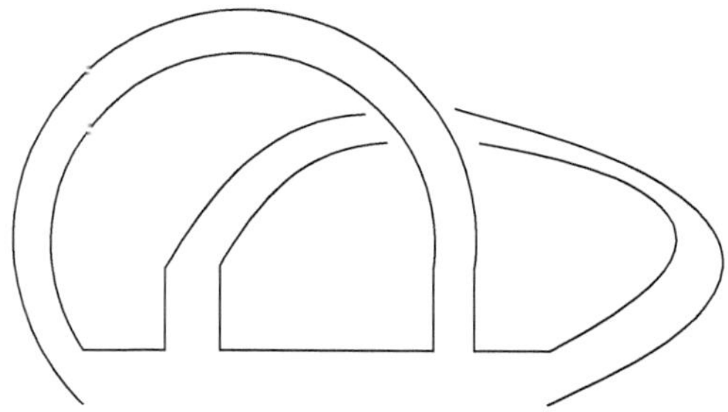

Figure 11.9 *One of the bubbles of the graph of Fig. 11.8. It has two vertices, three edges and 1 face and thus genus 1; it is non-planar*

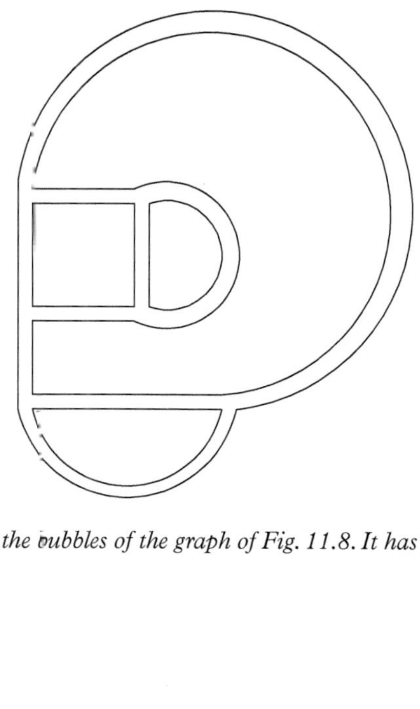

Figure 11.10 *The second of the bubbles of the graph of Fig. 11.8. It has genus 0, being hence planar*

Note that in Freidel, Gurau, and Oriti (2009) an algorithm for characterizing these bubbles was given for an arbitrary graph of this type. A cohomological definition of this notion of bubble was given in Bonzom and Smerlak (2010).

Another important result which we will exploit later on, is the following. As mentioned previously, 3D tensor graphs correspond to triangularizations of 3D spaces. If all the bubbles of the graph are planar, the respective graph is dual to a *manifold.*

11.4 Tensor graph polynomials—a generalization of the Bollobás–Riordan polynomial

In this section, following the original paper Tanasa (2011), we present a generalization of the Bollobás–Riordan polynomial which partially captures the increase in topologic information of 3D tensor graphs.

In order to do so, we need to first introduce a way of contracting edges in a tensor graph, contraction introduced in Gurau (2010) for a particular class of graphs, the colourable ones. Nevertheless, in Tanasa (2011), it was proposed to extend this definition to non-colourable tensor graphs.

The necessity of introducing a special way of contracting edges comes from the following arguments. When contracting an edge of some graphs, the valence of their vertices can increase. This is not a problem for a ribbon graph (faces can still be defined in a straightforward manner even for such a contracted ribbon graph), but for the tensor graphs presented in this section, one has to define a way of matching the middle strands of the respective edges. Furthermore, it is not clear whether the notion of bubble can still be defined for such contracted graphs.

The definition of the contraction of an edge given in Gurau (2010) is the following. When an edge is contracted, it becomes *passive.* The set of edges of the graph G is now formed of the disjoint reunion of two subsets, the active ones, denoted by $\mathcal{L}_1(G)$, and the passive ones, denoted by $\mathcal{L}_2(G)$. Hence, the tensor graph contraction transforms an active edge into a passive one (instead of identifying its two vertices, as it happens in the Tutte or the Bollobás–Riordan cases). The deletion (which is defined in the usual manner) and the contraction operations are only allowed for active edges.

Let us now define the announced generalization of the Bollobás–Riordan polynomial for 3D orientable graphs.

The main idea behind this proposition is, as stated before, to exploit the importance of the sum of genera of the bubbles of the graph. As mentioned previously, if this sum is vanishing, then the respective graph is dual to a manifold. Otherwise, it is dual to a pseudo-manifold or to an even more singular surface.

Let us recall that the Bollobás–Riordan polynomial generalized the Tutte polynomial by adding a supplementary variable z in order to keep track of the increase of complexity of the structure one deals with (see Section 2.2). We thus propose to keep the same scheme by adding one more variable, t, to the set of variables of the Bollobás–Riordan polynomial. This variable will keep track of this sum of genera of the bubbles, telling thus, in a natural manner, whether or not the respective graph is dual to a manifold.

To the tensor graphs studied here one can associate ribbon 'jacket' graphs in a unique way. This is done by getting rid of the 'middle strand' (this operation is not mathematically ambiguous, because no twists are allowed between the various strands of an edge)—we will get back to this point in the following subsections. This allows to define the rank, nullity, genus, number of faces, etc. of the respective graph in a straightforward manner.

One has:

Definition 11.4.1 *If G is a tensor graph, then*

$$\mathcal{T}_G(x,y,z,t) = \sum_{H\subset G} (x-1)^{r(G)-r(H)} y^{n(H)} z^{k(H)-F(H)+n(H)} t^{2\sum_{\mathcal{B}(H)} g_{\mathcal{B}(H)}}, \quad (11.10)$$

where we have denoted by $\mathcal{B}(H)$ the bubbles of the tensor subgraph H.

As in the definition of the Bollobás–Riordan polynomial (see Definition 2.2.6), the previous notation $F(H)$ stands for the number of components of the boundary of the respective subgraph H (seen as a ribbon jacket graph).

The factor two in the last exponent of (11.10) is chosen for similitude with the Euler characteristics formula.

One has:

$$\begin{aligned} \mathcal{T}_G(x,y,z,1) &= R_G(x,y,z), \\ \mathcal{T}_G(x,y,1,1) &= T_G(x,y), \end{aligned} \quad (11.11)$$

where, by an abuse of notation, we have denoted again with G the (ribbon) graph associated with the tensor graph G (from the left-hand side of the equations).

Furthermore, this polynomial satisfies the deletion/contraction relation.

Theorem 11.4.2 *(Theorem 5.1 of Tanasa (2011)) If G is a tensor graph, then*

$$\mathcal{T}_G(x,y,z,t) = \mathcal{T}_{G-e}(x,y,z,t) + \mathcal{T}_{G/e}(x,y,z,t), \quad (11.12)$$

if e is an active regular edge.

We now define a multi-variate version of the polynomial (11.4.1), generalizing in a natural manner the multi-variate Bollobás–Riordan polynomial.

Definition 11.4.3 *If G is a tensor graph, then,*

$$\mathcal{T}_G(x,\{\beta\},z,t) = \sum_{H\subset E} x^{k(H)} \left(\prod_{e\in H} \beta_e\right) z^{F(H)} t^{2\sum_{\mathcal{B}(H)} g_{\mathcal{B}(H)}}, \quad (11.13)$$

where we have denoted by $\mathcal{B}(H)$ the bubbles of the tensor subgraph H.

Furthermore, the polynomial (11.13) can be further generalized to a *hypervariate version*, if we make use of a set of variables $\{\gamma_b\}$, one for each bubble of the graph, instead of a single variable t:

Definition 11.4.4 *If G is a tensor graph, then,*

$$\mathcal{T}_G(x, \{\beta\}, z, \{\gamma\}) = \sum_{H \subset G} x^{k(H)} \left(\prod_{e \in H} \beta_e \right) z^{F(H)} \prod_{b(H)} \gamma_b^{2g_{b(H)}}. \tag{11.14}$$

The multi-variate and hypervariate versions defined previously also satisfy the deletion/contraction relation.

Let us now comment on the universality of the polynomial introduced here. For the Bollobás–Riordan polynomial, this was proven in Bollobás and Riordan (1999), or, for the multi-variate case in Krushkal and Renardy (2010). A restatement of this universality property was proved in a different form in Ellis-Monaghan and Sarmiento (2011).

Such a proof is mainly made possible through a classification of the terminal forms of the ribbon deletion/contraction procedure (referred to as *chord diagrams* in Bollobás and Riordan (1999); or Ellis-Monaghan and Sarmiento (2011); or as *rosettes* in the non-commutative QFT literature (see (Krajewski et al. 1010b; Tanasa 2000; Magnen, Rivasseau, and Tanasa 2009); Gurau et al. (2009); Krajewski, Rivasseau, and Vignes-Tourneret (2011), and references within).

For the graph polynomial described in this section, the situation is similar to the one described in Gurau (2010), in the sense that, in order to have such a property, one needs to start with a classification of 3D piecewise linear manifolds and pseudo-manifolds. Such a result is nowadays an open question in algebraic topology (see again Gurau (2010) for a more detailed discussion).

Before ending this chapter, we must stress that these definitions and properties also hold for graphs with external edges. As mentioned in Section 2.1, from a graph theoretical point of view, these external edges can be seen as flags attached to the vertices (taking care that the valence of the vertices always remains equal to four). Each such flag is attached to a single vertex.

11.5 Selected further reading

- Ph. Di Francesco. (2006). 2D quantum gravity, matrix models and graph combinatorics. 33–88, arXiv:math-ph/0406013, NATO Science Series vol. 221.

  Very well written physicist summer school notes.
- Di Francesco, P. and Ginsparg, Paul H. and Zinn-Justin, Jean. (1995). 2D gravity and random matrices. *Phys. Rept.* 254, 1–133.

  A review article for physicists, presenting in detail matrix models and their relation to 2D quantum gravity.

- Bertrand Eynard. (2016). *Counting Surfaces*. Springer.

  A book presenting matrix models and the celebrated topological recursion.
- A. Zvonkin. (1997). Matrix integrals and map enumeration: An accessible introduction. Computers and Mathematics with Applications. *Mathematical and Computer Modelling*, 26: 281–304.

  A very well-written introductory article presenting some of the rich relations between Feynman integrals of matrix models and map counting, interplay between combinatorics, and QFT which lies outside the aims of this book.

12

Random tensor models—the $U(N)^D$-invariant model

As we have seen in Chapter 11, random tensor models represent a natural generalization, to dimensions greater than two, of the celebrated random matrix models.

The recent revival of interest in the study of tensor models came from the definition of coloured tensor models Gurau (2011a) and, shortly thereafter, from the implementation of a $1/N$ expansion within these models (Gurau 2011; Gurau 2011b; Gurau 2012; Bonzom, Gurau, and Rivasseau 2012).

In the case of matrix models, the $1/N$ expansion is controlled by the genus of the corresponding Feynman ribbon graphs. In dimensions greater than two, the role of the genus is played by the so-called degree of the tensor graphs. The degree of a tensor graph is defined to be the sum of the genera of certain well-chosen ribbon subgraphs (called the jackets of the tensor graph (Ryan 2012)).

Recall that at large N, matrix models are dominated by planar graphs (tiling the 2D sphere $\mathcal{S}^2$), while the dominant graphs for tensor models are the so-called melonic graphs, which we will describe in the sequel. Let us note here that in D dimensions, these melonic graphs correspond to particular triangulations of the 2D sphere $\mathcal{S}^D$. Moreover, the continuum limit of melonic graphs turns out to be the continuous random branched polymer Gurau and Ryan (2014).

The coloured model double-scaling limit was obtained in Gurau and Schaeffer (2016), using a purely combinatorial approach; the double-scaling limit of the $U(N)$-invariant model that we will analyse in this chapter was obtained in Dartois, Gurau, and Rivasseau (2013), using an appropriate intermediate field method (see Chapter 3 for a description of this method). Both results, therefore rely on a thorough analysis of each term of the series (in the coupling constant) associated to each order in the initial large N expansion.

However, the method of Dartois, Gurau, and Rivasseau (2013) is not known to be applicable beyond the case of quartic interactions. To improve these results and possibly go beyond, it is reasonable to look for a more effective way to reach the double-scaling regime, which we argue here could be the DSE.

Combinatorial Physics: Combinatorics, quantum field theory, and quantum gravity models. Adrian Tanasa,
Oxford University Press (2021). © Adrian Tanasa. DOI: 10.1093/oso/9780192895493.003.0012

Taking some inspiration from matrix models (and later, string field theory), while there exist various ways to solve them, one interesting technique is the use of their (DSE, often known as loop equations in this context) which have unravelled some fascinating structures (such as integrability) that become transparent in the celebrated topological recursion (see Eynard (2016). Let us emphasize here that topological recursion was initially developed as an intrinsic method to solve the DSE. It is therefore natural to ask whether the DSE could also be used to solve tensor models.

In this chapter we derive the double-scaling limit of the $U(N)$-invariant random tensor model using the DSE. It is worth emphasizing that our method allows to obtain this double-scaling limit not only for tensor models with quartic interactions, thereby reproducing the result of Dartois, Gurau, and Rivasseau (2013), but also for tensor models with a general melonic interactions. This is a result which seem to us rather intricate to derive without the use of the DSE. Indeed, the method of Dartois, Gurau, and Rivasseau (2013) relies on a bijection that is only known to exist in the case of quartic interactions.

This chapter follows the original research article Bonzom (2014). The organization of the chapter is the following. In the next section we briefly present the $U(N)$-invariant random tensor models and of the two relevant DSE which will be used. Section 12.2 is then dedicated to our analysis of the DSE beyond the $1/N$ limit LO, NLO and even another sub-leading-order (SLO) in the case of the quartic model). Those results are essential to implement the double-scaling limit mechanism of the DSE in Section 12.3.1. The main result of this chapter is the doubly scaled two-point function for a model with generic melonic interactions. However, several assumptions on the large N scaling of cumulants are made along the way. They are proved using combinatorial methods in Section 12.3.2 and in Appendix . The reader mostly interested in solving the DSE may skip those parts and simply consider their conclusions as ansatz which allow to extract particular solutions of the DSE.

12.1 Definition of the model and its DSE

12.1.1 $U(N)^D$-invariant bubble interactions

Observables in random tensor theory are generalizations of trace invariants in matrix models (see again the book Gurau (2017b)). They are generated by polynomials in the tensor entries $\mathbb{T}_{a_1\cdots a_D}$ and $\overline{\mathbb{T}}_{a_1\cdots a_D}$ $(a_1,\dots,a_D = 1,\dots,N)$, which are invariant under transformations of the following form:

$$\mathbb{T}'_{a_1\cdots a_D} = \sum_{b_1,\dots,b_D} U^{(1)}_{a_1b_1}\cdots U^{(D)}_{a_Db_D}\,\mathbb{T}_{b_1\cdots b_D}, \tag{12.1}$$

where $(U^{(1)},\dots,U^{(D)})$ is a D-uple of independent unitary matrices. The complex conjugated tensor $\overline{\mathbb{T}}$ transforms in a similar way. Since different unitary transformations

Table 12.1

Polynomials	Bubbles
$\mathbb{T}_{a_1 \cdots a_D}$	White vertex 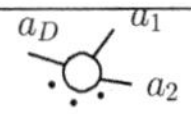
$\overline{\mathbb{T}}_{a_1 \cdots a_D}$	Black vertex 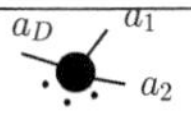
Contraction	Edge with colour label i
$\sum_{a_i} \mathbb{T}_{a_1 \cdots a_i \cdots a_D} \overline{\mathbb{T}}_{b_1 \cdots a_i \cdots b_D}$	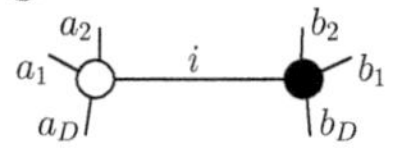

are applied to the different indices, invariants can be obtained by *contracting* (that is, by identifying and summing) indices pairwise only when they have the same position between 1 and D. Invariance also requires all indices to be contracted.

This leads to the fact that the generating polynomials can be labelled by connected, regular, bipartite graphs of degree D, whose edges have a colour label drawn from $\{1,\dots,D\}$ such that the D edges incident to a vertex have distinct colours. In this context, such graphs are called *bubbles* (note that this is a different notion of the GFT bubble notion presented in the Chapter 11). The correspondence between polynomials and bubbles is tabulated in Table 12.1.

Let us denote bubbles by $\mathcal{B}$, the vertex set of $\mathcal{B}$ by $\mathcal{V}(\mathcal{B})$ of cardinality $|\mathcal{V}(\mathcal{B})| = 2p(\mathcal{B})$ and the set of edges of colour c of $\mathcal{B}$ by $\mathcal{E}_c(\mathcal{B})$, of cardinality $|\mathcal{E}_c(\mathcal{B})| = p(\mathcal{B})$. The vertex set of $\mathcal{B}$ is bipartite. We generically denote a white vertex of $\mathcal{B}$ by v, and a black vertex by $\bar{v}$ and an edge of colour c by e_c. We furthermore use the shorthand notation $\vec{a}$ for the D-uple of indices $(a_1,\dots,a_D)$ (hence $\vec{a}^v = (a_1^v,\dots,a_D^v)$) with $a_i = 1,\dots,N$ for $i \in \{1,\dots,D\}$.

The invariant polynomial associated to $\mathcal{B}$ is denoted $\tilde{p}_{\mathcal{B}}(\overline{\mathbb{T}},\mathbb{T})$ and it is written:

$$
\begin{aligned}
\tilde{p}_{\mathcal{B}}(\overline{\mathbb{T}},\mathbb{T}) &= \sum_{\{\vec{a}^v,\vec{b}^{\bar{v}}\}_{v,\bar{v}}} \left(\prod_{\bar{v}\in\mathcal{V}(\mathcal{B})} \overline{\mathbb{T}}_{\vec{b}^{\bar{v}}} \right) \left(\prod_{v\in\mathcal{V}(\mathcal{B})} \mathbb{T}_{\vec{a}^v} \right) \delta^{\mathcal{B}}_{\vec{a}^v,\vec{b}^{\bar{v}}}, \\
\delta^{\mathcal{B}}_{\vec{a}^v,\vec{b}^{\bar{v}}} &= \prod_{c=1}^{D} \prod_{e^c=(x,\bar{y})\in\mathcal{E}^c(\mathcal{B})} \delta_{a_c^x b_c^{\bar{y}}},
\end{aligned}
\tag{12.2}
$$

that is for each white vertex v (resp. black vertex $\bar{v}$) of $\mathcal{B}$ we take a tensor $\mathbb{T}_{\vec{a}^v}$ (resp. $\overline{\mathbb{T}}_{\vec{b}^{\bar{v}}}$), and for each edge of colour c we contract the indices of colour c of the tensors associated to its end vertices. The operator $\delta^{\mathcal{B}}_{\vec{a}^v,\vec{b}^{\bar{v}}}$ is called the *trace-invariant operator* associated to $\mathcal{B}$.

There is the unique 2-vertex bubble $\mathcal{B}_2$, displayed in Fig. 12.1. Secondly, there are several 4−vertex bubble interactions, which are illustrated in Fig. 12.2. They are 1−particle-irreducible, yet 2-particle reducible. They are labeled $\mathcal{B}_{4,\{c\}}$ where $c \in \{1, \ldots, D\}$ is the colour of the edges that, when cut, disconnect the graph. Another example of a bubble interaction is given in Fig. 12.3.

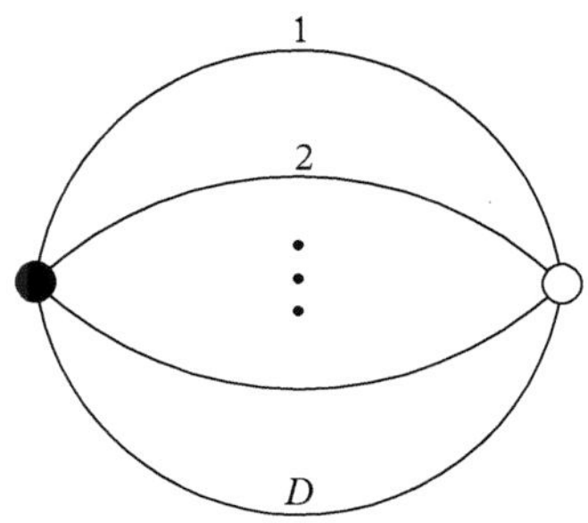

Figure 12.1 *The two vertex bubble interaction* $\mathcal{B}_2$

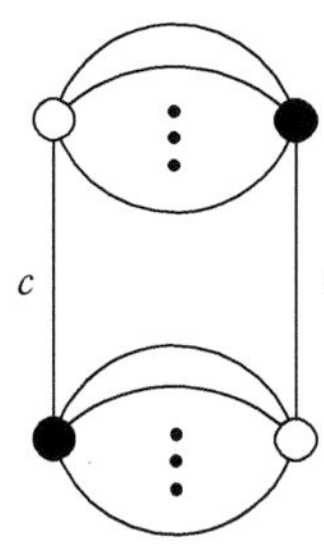

Figure 12.2 *The 4-vertex bubble interactions* $\mathcal{B}_{4,\{c\}}$

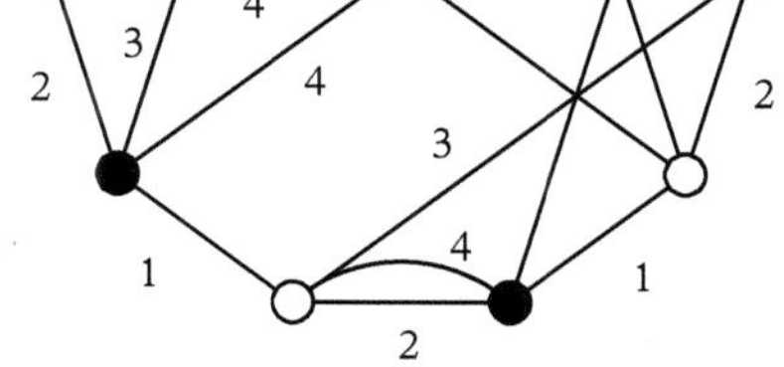

Figure 12.3 *A 10-vertex bubble interaction at* $D = 4$

Let I be a finite set and $\{\mathcal{B}_i\}_{i\in I}$ be a set of bubbles such that $\mathcal{B}_i$ has $p_i \geq 2$ black vertices. We shall denote the set of corresponding invariant polynomials by $\{\tilde{\rho}_{\mathcal{B}_i}\}_{i\in I}$. The *action* is an invariant function of $\mathbb{T}$ and $\overline{\mathbb{T}}$:

$$S(\mathbb{T},\overline{\mathbb{T}}) = \tilde{\rho}_{\mathcal{B}_2}(\mathbb{T},\overline{\mathbb{T}}) - \sum_{i\in I} \frac{z^{p_i-1}}{p_i} t_i\, \tilde{\rho}_{\mathcal{B}_i}(\mathbb{T},\overline{\mathbb{T}}), \tag{12.3}$$

where the parameters z and $\{t_i\}_{i\in I}$ are called the couplings. The partition function of a generic 'single trace' tensor model is:

$$Z(N,z,\{t_i\}) = \int [\mathrm{d}\overline{\mathbb{T}}\mathrm{d}\mathbb{T}]\; e^{-N^{D-1}S}, \qquad [\mathrm{d}\overline{\mathbb{T}}\mathrm{d}\mathbb{T}] = \left(\prod_{\vec{a}} N^{D-1}\frac{\mathrm{d}\mathbb{T}_{\vec{a}}\mathrm{d}\overline{\mathbb{T}}_{\vec{a}}}{2\pi i}\right). \tag{12.4}$$

12.1.2 Bubble observables

Among all the observables one can build out of a tensor and its complex conjugate, the invariant observables labeled by bubbles $\mathcal{B}$ play a central role. Their expectations are:

$$\langle \tilde{\rho}_{\mathcal{B}}(\mathbb{T},\overline{\mathbb{T}})\rangle = \frac{1}{Z(N,\{t_i\})}\int d\mathbb{T}\, d\overline{\mathbb{T}}\; e^{-N^{D-1}S}\; \tilde{\rho}_{\mathcal{B}}(\mathbb{T},\overline{\mathbb{T}}), \tag{12.5}$$

and are functions of N, z and $\{t_i\}_{i\in I}$.

The previous integrals can be computed via their QFT Feynman expansions (see Chapter 3). Feynman graphs result from first Taylor expanding the exponentials $e^{N^{D-1}\frac{z^{p_i-1}}{p_i}t_i\tilde{\rho}_{\mathcal{B}_i}(\mathbb{T},\overline{\mathbb{T}})}$, and commuting the sums with the remaining Gaussian integral. We thus obtain a sum over terms, each term being a Gaussian integral of a product of trace invariants. Graphically, a pairing connects a black vertex ($\mathbb{T}$) to a white vertex ($\overline{\mathbb{T}}$) via an edge to which the fictitious colour 0 is attributed.

The Feynman graphs of the partition function are therefore regular bipartite edge-coloured graphs of degree $D+1$ (the colours of the bubbles plus the colour 0). An example is shown in Fig. 12.4. The Feynman graphs contributing to the expectation of $\tilde{\rho}_{\mathcal{B}}(\mathbb{T},\overline{\mathbb{T}})$ for a connected bubble $\mathcal{B}$ are the connected regular bipartite edge-coloured graphs of degree $D+1$ built from the set of bubbles $\{\mathcal{B}_i\}_{i\in I}$ and containing the bubble $\mathcal{B}$ as a marked sub-graph.

Let us compute the Feynman amplitude associated to a closed connected Feynman graph $\mathcal{G}$ with no marked bubble is (up to some symmetry factor). Each bubble $\mathcal{B}_i$, $i\in I$, contained in $\mathcal{G}$ brings a factor $z^{p_i-1}N^{D-1}$ and its associated trace invariant. Each edge of colour 0 brings $N^{-(D-1)}$ times $\prod_{c=1}^{D}\delta_{a_c^v b_c^{\bar{v}}}$, which identifies the index a_c^v on the white vertex adjacent to the edge with the index $b_c^{\bar{v}}$ on the black vertex adjacent to the edge. A *face with colours cd* is defined as a connected component of the graph obtained from $\mathcal{G}$ by removing all edges of colours different of c and d. Contracting the Kronecker deltas of the propagators with the trace-invariants coming from the bubbles, it follows that the indices are identified along the faces of colour $0c$ of $\mathcal{G}$. One gets a free sum per face,

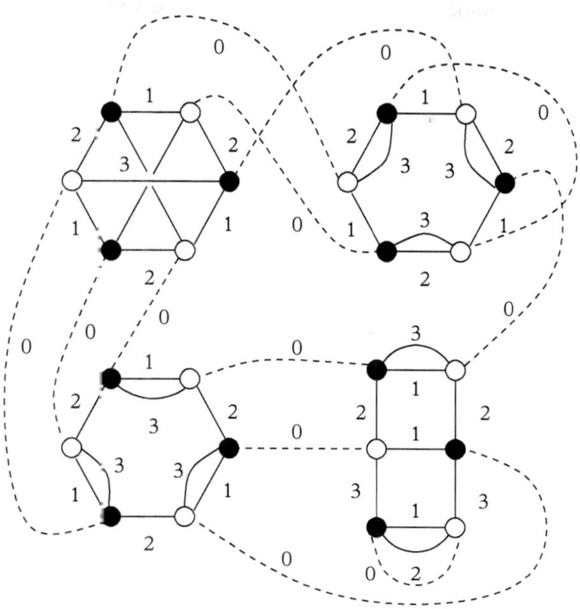

Figure 12.4 *An example of a $(D+1)$–coloured Feynman graph (with $D=4$). The bubbles (solid edges) have colours 1, 2 and 3, while the propagators (dashed edges) are assigned the colour 0*

hence a factor N. We denote b_i the number of bubbles of type $\mathcal{B}_i$ of $\mathcal{G}$, and $b=\sum_{i\in I} b_i$ is the total number of bubbles of $\mathcal{G}$. Moreover, $p=\sum_{i\in I} p_i b_i$ is the total number of black vertices of $\mathcal{G}$, and it also counts the edges of colour 0 of $\mathcal{G}$. The amplitude then reads:

$$A_{\mathcal{G}} = N^{\sum_{c=1}^{D} f_{0c}-(D-1)(v-b)}\, z^{v-b} \prod_{i\in I} (t_i)^{b_i}. \tag{12.6}$$

At $D=2$, the $(D+1)$-coloured graphs $\mathcal{G}$ are also ribbon graphs and the exponent of N in the amplitude reduces to $2-2g$ where g is the genus of the ribbon graph.

The amplitude of a Feynman graph in the expansion of the expectation of $\tilde{\rho}_{\mathcal{B}}$ is similar, expect that the marked sub-graph $\mathcal{B}$ does not bring any power of z and N. The exponent of N for the graphs entering an expectation is bounded and leads to a $1/N$ expansion of expectations of the following form:

$$\frac{1}{N}\langle \tilde{\rho}_{\mathcal{B}}(\mathbb{T},\overline{\mathbb{T}})\rangle = \sum_{\omega\geq 0} N^{-\omega} A_{\omega,\mathcal{B}}(z,\{t_i\}). \tag{12.7}$$

Furthermore, the large N limit is Gaussian:

$$\frac{1}{N}\langle \tilde{\rho}_{\mathcal{B}}(\mathbb{T},\overline{\mathbb{T}})\rangle = N^{-\omega^*_{\mathcal{B}}}\, \alpha_{\mathcal{B}}\, [T(z,\{t_i\})]^{v} + \mathcal{O}(N^{-\omega^*_{\mathcal{B}}}). \tag{12.8}$$

Here, $\omega^*_{\mathcal{B}}$ is the minimal value of ω for which $A_{\omega,\mathcal{B}}(z,\{t_i\}) \neq 0$, $\alpha_{\mathcal{B}}$ counts the number of LO Wick contractions on $\mathcal{B}$, v is the number of black vertices of $\mathcal{B}$ and $T(z,\{t_i\})$ is the large N limit of the two-point function:

$$T(z,\{t_i\}) \equiv \lim_{N\to\infty} \frac{1}{N} \langle \tilde{\rho}_{\mathcal{B}_2}(\mathbb{T},\overline{\mathbb{T}})\rangle = \lim_{N\to\infty} K\Big(\mathcal{B}_2; N, z, \{t_i\}\Big) = K\Big(\mathcal{B}_2; z, \{t_i\}\Big) \,. \quad (12.9)$$

(Here we have introduced the notation $K\Big(\mathcal{B}_2; N, z, \{t_i\}\Big)$ which refers to the two-point cumulant and is equivalent to the full two-point function. It will be generalized to cumulants of higher orders in the Section 12.2.2.)

The equations (12.8) and (12.9) constitute an illustration of the universality theorem for large random tensors equipped with joint distributions that are invariant under (12.1). Details on this theorem can be found in Gurau (2014), where it was proven using mostly the combinatorics of cumulants. In the context of random tensor models defined by an action such as (12.3), it is also possible to get to this universal behaviour through the DSE.

On the one hand, it is an involved task to find $\omega^*_{\mathcal{B}}$ and $\alpha_{\mathcal{B}}$ in general. On the other hand, there is one important class for which they are known: the polynomials $\tilde{\rho}_{\mathcal{B}}$ for which $\omega^*_{\mathcal{B}} = 0$. Members of this class are called *melonic* polynomials and correspond to *melonic* bubbles. The structure of melonic bubbles is defined recursively. The basic building blocks are the so–called $(D-1)$-dipoles. A $(D-1)$*–dipole of colour* c is comprized of two vertices, connected by $D-1$ edges *not* carrying the colour c, and two open edges of colour c:

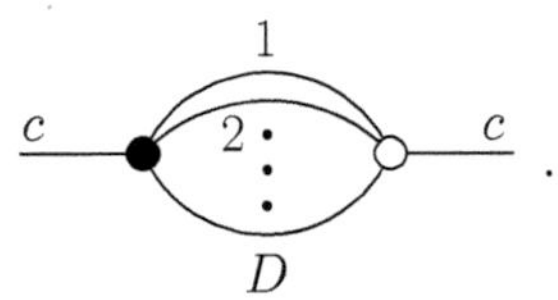

One starts with the bubble $\mathcal{B}_2$ and performs a $(D-1)$-*dipole insertion*, where the edge of colour c is removed and replaced by the $(D-1)$-dipole of the same colour. This results in a bubble with 4 vertices, which is actually $\mathcal{B}_{4,\{c\}}$ (Fig. 12.2). Such a $(D-1)$-dipole insertion can then be performed on an arbitrary edge of the new graph. In this manner, the bubble grows according to a tree structure that records the history of dipole insertions. Melonic bubbles have the following properties:

- They are exactly the bubbles for which $\omega^*_{\mathcal{B}} = 0$.
- The combinatorial coefficient is $\alpha_{\mathcal{B}} = 1$. Indeed, vertices in a melonic bubble come in canonical pairs (the vertices of a pair are those associated to a dipole insertion). $\alpha_{\mathcal{B}} = 1$ means that there is a single family of contributions coming from Wick contracting the two vertices of each pair.

- The large N two-point function T is non-trivial, that is $T \neq 1$, if and only if the set of bubbles $\{\mathcal{B}_i\}_{i\in I}$ contains a melonic bubble (different from $\mathcal{B}_2$). The function T can then be found by means of the DSE, as we will now explain.

12.1.3 The DSE for the model

As already mentioned in Section 3.7, DSE are the quantum equations of motion, governing the behaviour of the expectations. They can be solved at LO to recover the Universality Theorem, as well as to find the equation satisfied by the large N two-point function. Here, we only need a couple of DSE and not the full tower of equations. The first equation we use is the simplest one, coming from the identity

$$\sum_{a_1,\dots,a_D} \frac{1}{Z(N,\{t_i\})} \int d\mathbb{T}\, d\overline{\mathbb{T}}\, \frac{\partial}{\partial \mathbb{T}_{a_1\cdots a_D}} \left(\mathbb{T}_{a_1\cdots a_D}\, e^{-N^{D-1}S} \right) = 0. \tag{12.10}$$

Performing the derivatives explicitly leads to:

$$N - \langle \tilde{\rho}_{\mathcal{B}_2}(\mathbb{T},\overline{\mathbb{T}}) \rangle + \sum_{i\in I} z^{p_i-1} t_i \langle \tilde{\rho}_{\mathcal{B}_i}(\mathbb{T},\overline{\mathbb{T}}) \rangle = 0. \tag{12.11}$$

Here p_i once again denotes the number of black vertices of $\mathcal{B}_i$. The second equation comes from the identity:[1]

$$\sum_{a_1,\dots,a_D} \frac{1}{Z(N,\{t_i\})} \int d\mathbb{T}\, d\overline{\mathbb{T}}\, \frac{\partial}{\partial \mathbb{T}_{a_1\cdots a_D}} \left(\frac{1}{2} \frac{\partial \tilde{\rho}_{\mathcal{B}_{4,\{c\}}}(\mathbb{T},\overline{\mathbb{T}})}{\partial \overline{\mathbb{T}}_{a_1\cdots a_D}}\, e^{-N^{D-1}S} \right) = 0. \tag{12.12}$$

When acting on the first factor the derivative with reference to $\mathbb{T}$ gives:

$$\begin{aligned} &\sum_{a_1,\dots,a_D} \frac{1}{Z(N,\{t_i\})} \int d\mathbb{T}\, d\overline{\mathbb{T}}\, \frac{1}{2} \frac{\partial^2 \tilde{\rho}_{\mathcal{B}_{4,\{c\}}}(\mathbb{T},\overline{\mathbb{T}})}{\partial \mathbb{T}_{a_1\cdots a_D}\, \partial \overline{\mathbb{T}}_{a_1\cdots a_D}}\, e^{-N^{D-1}S} \\ &\quad = N^{D-1} \langle \tilde{\rho}_{\mathcal{B}_2}(\mathbb{T},\overline{\mathbb{T}}) \rangle + N \langle \tilde{\rho}_{\mathcal{B}_2}(\mathbb{T},\overline{\mathbb{T}}) \rangle. \end{aligned} \tag{12.13}$$

We now split its effect on the second factor into two parts. Operating on the quadratic part of the action produces, using the identity $\frac{\partial \tilde{\rho}_{\mathcal{B}_2}(\mathbb{T},\overline{\mathbb{T}})}{\partial \mathbb{T}_{a_1\cdots a_D}} = \overline{\mathbb{T}}_{a_1\cdots a_D}$:

[1] The overall factor $1/2$ has been introduced to counter the fact that the action of the derivative (with reference to $\overline{T}$) on $\tilde{\rho}_{\mathcal{B}_{4,\{c\}}}(\mathbb{T},\overline{\mathbb{T}})$ results in two identical terms.

$$\sum_{a_1,\ldots,a_D} \frac{1}{Z(N,\{t_i\})} \int d\mathbb{T}\, d\overline{\mathbb{T}}\, \frac{1}{2} \frac{\partial\, \tilde{\rho}_{\mathcal{B}_{4,\{c\}}}(\mathbb{T},\overline{\mathbb{T}})}{\partial \overline{\mathbb{T}}_{a_1\cdots a_D}} \left(-\frac{\partial\, \tilde{\rho}_{\mathcal{B}_2}(\mathbb{T},\overline{\mathbb{T}})}{\partial \mathbb{T}_{a_1\cdots a_D}} \right) e^{-N^{D-1}S}$$
$$= -N^{D-1} \langle \tilde{\rho}_{\mathcal{B}_{4,\{c\}}}(\mathbb{T},\overline{\mathbb{T}}) \rangle, \tag{12.14}$$

leads to

$$\left(1+\frac{1}{N^{D-2}}\right) \langle \tilde{\rho}_{\mathcal{B}_2}(\mathbb{T},\overline{\mathbb{T}}) \rangle - \langle \tilde{\rho}_{\mathcal{B}_{4,\{c\}}}(\mathbb{T},\overline{\mathbb{T}}) \rangle$$
$$+ \sum_{a_1,\ldots,a_D} \int \frac{d\mathbb{T}\, d\overline{\mathbb{T}}}{Z(N,\{t_i\})} \frac{1}{2} \frac{\partial\, \tilde{\rho}_{\mathcal{B}_{4,\{c\}}}(\mathbb{T},\overline{\mathbb{T}})}{\partial \overline{\mathbb{T}}_{a_1\cdots a_D}} \frac{\partial}{\partial \mathbb{T}_{a_1\cdots a_D}} \left(\sum_{i\in I} \frac{z^{p_i-1}}{p_i} t_i\, \tilde{\rho}_{\mathcal{B}_i}(\mathbb{T},\overline{\mathbb{T}}) \right) e^{-N^{D-1}S} = 0. \tag{12.15}$$

In order to conclude we must evaluate the final contribution in equation (12.15). Let us explain some of its components in detail. The first factor, $\partial\, \tilde{\rho}_{\mathcal{B}_{4,\{c\}}}(\mathbb{T},\overline{\mathbb{T}})/\partial \overline{\mathbb{T}}_{a_1\cdots a_D}$ is of order two in $\mathbb{T}$ and one in $\overline{\mathbb{T}}$. The graphical rules of Table 12.1 still apply in this context, meaning that the resulting (non-invariant) polynomial has a graphical representative:

$$\frac{1}{2} \frac{\partial\, \tilde{\rho}_{\mathcal{B}_{4,\{c\}}}(\mathbb{T},\overline{\mathbb{T}})}{\partial \overline{\mathbb{T}}_{a_1\cdots a_D}} =$$
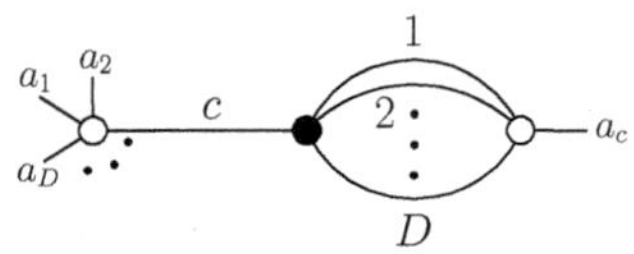

. (12.16)

As expected, the graph is open, with all open edges emerging from white vertices. This means that the polynomial transforms like $\mathbb{T}$ in (12.1).

The factor $\partial\, \tilde{\rho}_{\mathcal{B}_i}(\mathbb{T},\overline{\mathbb{T}})/\partial \mathbb{T}_{a_1\cdots a_D}$ is a polynomial (which transforms like $\overline{\mathbb{T}}$) obtained from $\tilde{\rho}_{\mathcal{B}_i}(\mathbb{T},\overline{\mathbb{T}})$ by summing over all the ways to remove a $\mathbb{T}$. Graphically, this gives to a sum over a collection of open graphs, each of which corresponds to a distinct way to excise a white vertex from $\mathcal{B}_i$.

Now notice that if one removes a white vertex V from an arbitrary bubble $\mathcal{B}$ and inserts the right hand side of (12.16), one produces a new bubble that is $\mathcal{B}$ with a $(D-1)$-dipole inserted on the edge of colour c incident to V. This is precisely the process operating in the last term of (12.15). When it is contracted with the left-hand side of (12.16), it means that the right hand side of (12.16) is glued back instead of the removed white vertex. That yields a sum over all the ways to insert a $(D-1)$-dipole on an edge e_c of colour c in $\mathcal{B}_i$. When the insertion occurs on the edge e_c, we denote the resulting bubble $\mathcal{B}_i + (e_c)$. The DSE (12.15) therefore reads:

$$\left(1+\frac{1}{N^{D-2}}\right) \langle \tilde{\rho}_{\mathcal{B}_2}(\mathbb{T},\overline{\mathbb{T}}) \rangle - \langle \tilde{\rho}_{\mathcal{B}_{4,\{c\}}}(\mathbb{T},\overline{\mathbb{T}}) \rangle + \sum_{i\in I} \frac{z^{p_i-1}}{p_i} t_i \sum_{e_c\in\mathcal{B}_i} \langle \tilde{\rho}_{\mathcal{B}_i+(e_c)}(\mathbb{T},\overline{\mathbb{T}}) \rangle = 0. \tag{12.17}$$

There are D such equations, one for each value of the colour $c \in \{1,\ldots,D\}$.

12.1.4 Navigating the following sections of the chapter

In light of the technical nature of Sections 12.2 and 12.3, the main points of the argument are presented here.

- To begin Section 12.2, we present a quick recapitulation of the large N limit. In particular, combining its Gaussian universality (12.8) with the DSE allows on to solve for the LO contribution $T(z, \{t_i\})$, to the $1/N$-expansion of the two-point function. Only the melonic sub-sector survives and, in the quartic model, one finds:

$$T(z) = \frac{1 - \sqrt{1 - 4Dz}}{2Dz} .$$

- Beyond LO, non-Gaussian contributions creep into the mix. Having catalogued the pertinent examples, we examine the NLO in detail. At NLO, one needs only to utilize the two DSE (12.11) and (12.17) to obtain a pair of coupled equations, linear in $K_2^{\mathrm{NLO}}(z, \{t_i\})$ and $K_{4,\bullet}^{\mathrm{LO}}(z, \{t_i\})$. Respectively, these are the NLO contribution to the two-point function and the LO contribution to the connected four-point function, based on the graphs of species $\mathcal{B}_{4,\{c\}}$. One solves these equations to arrive at the result (in the case of the quartic model):

$$K_2^{\mathrm{NLO}}(z) = \frac{1}{\sqrt{1-4Dz}} \frac{Dz\,T(z)^2}{1 - zT(z)^2},$$
$$K_{4,\bullet}^{\mathrm{LO}}(z) = \frac{zT(z)^4}{1 - zT(z)^2} ,$$

- Specializing to the quartic model allows one to most easily probe deeper into the DSE (12.11) and (12.17), and retrieve information about $K_2^{\mathrm{NNLO}}(z)$ and $K_{4,\emptyset}^{\mathrm{LO}}(z)$. The latter is the LO contribution to the connected four-point function based on the graph $\mathcal{B}_{4,\emptyset}$ (as described next to Fig. 12.6). Again, solving the associated coupled equations yields our next result:

$$K_2^{\mathrm{NNLO}}(z) = \frac{1}{1-4Dz} \frac{D(D-1)\,z^2\,[T(z)]^3}{1 - z\,[T(z)]^2},$$
$$K_{4,\emptyset}^{\mathrm{LO}}(z) = \frac{1}{\sqrt{1-4Dz}} \frac{D(D-1)\,z^2\,[T(z)]^5}{1 - z\,[T(z)]^2}.$$

- In Section 12.3.1, we change the parameter set from $(N, z, \{t_i\})$ to $(N, x, \{t_i\})$ where $x = N^{D-2}(z - \frac{1}{4D})$, dubbed the double-scaling parameter. While this choice for x is, at the outset, an ansatz, its validity is unequivocally confirmed by subsequent analysis. In principle, one can now send $N \to \infty$ and $z \to \frac{1}{4D}$, keeping x fixed. This is the double-scaling limit. Moreover, it allows for a new expansion of the cumulants around the melonic sector evaluated at criticality. Once substituted into the DSE (12.11), one obtains an equation for $K^{DS}(x, \{t_i\})$, the dominant contribution to the two-point function in this new expansion, *provided* i) certain higher order correction terms truly remain sub-dominant as one tunes to criticality in z and *ii*) the contributing series are actually summable (which is the case for $D < 6$). In this instance, one arrives at our main result, the limiting double-scaled behaviour of the two-point function. In the quartic case, this takes the form:

$$K^{\mathrm{DS}}(x) = 4\sqrt{D}\,\sqrt{x - \frac{1}{4(D-1)}}\,.$$

- In Section 12.3.2, we prove that the assumption of sub-dominance, vital for the results of the preceding section, does in fact hold. In Section 12.3.2, we present an argument for the universality of such a double-scaling limit within the subclass of tensor models with melonic interaction terms.

12.2 The DSE beyond the large N limit

12.2.1 The LO

Plugging the result (12.8) of the Universality Theorem into the SDE enables one to obtain a closed the system of equations, since all expectations factorize as products of the large N two-point function $T(z,\{t_i\})$. Only the melonic expectations survive and all SDE ultimately reduce to the same algebraic equation Bonzom (2013):

$$1 - T(z,\{t_i\}) + \sum_{i\in I} z^{p_i-1} t_i \,[T(z,\{t_i\})]^{p_i} = 0. \tag{12.18}$$

Together with the initial condition $T_{|z=0} = 1$, this equation determines T as long as the derivative of (12.18) with respect to T does not vanish. Examining the equation (12.18), one sees that for generic couplings $\{t_i\}_{i\in I}, T$ has a square-root singularity at a finite value of z called the critical coupling z_c:

$$T(z,\{t_i\}) \simeq T_c + T_c'\left(1 - z/z_c\right)^{1/2}, \tag{12.19}$$

where T_c, z_c and T_c' are functions of $\{t_i\}$. The critical values T_c and z_c are determined by equation (12.18) supplemented with the criticality condition that (minus) the derivative of (12.18) with respect to T:

$$C(z,\{t_i\}) \equiv 1 - \sum_{i\in I} t_i p_i \,[zT(z,\{t_i\})]^{p_i-1}, \tag{12.20}$$

vanishes. Notice that $C(z,\{t_i\})$ not only controls how far we are from criticality, but it is also singular at criticality. Indeed, plugging the expansion (12.19) into $C(z,\{t_i\})$, we see that around z_c:

$$C(z,\{t_i\}) = -\Big(\sum_{i\in I} t_i p_i (p_i-1) z_c^{p_i-1} T_c^{p_i-2}\Big) T_c' \left(1-z/z_c\right)^{1/2} = c(\{t_i\}) \left(1-z/z_c\right)^{1/2}. \tag{12.21}$$

Quartic case. The previous scenario is easily illustrated when $I = \{1,\ldots,D\}$ and the interaction part of the action is defined as the set of quartic bubbles $\{\mathcal{B}_i\}_{i\in I} = \{\mathcal{B}_{4,\{c\}}\}_{c=1,\ldots,D}$ with the same global coupling z, i.e. $t_i = 1$ for all $i \in I$. Thus, the model is symmetric with respect to the colours $\{1,\ldots,D\}$, the number of black vertices for each interaction bubble satisfies $p_i = 2$ $(i \in I)$, and the sums over $i \in I$ reduce to multiplication by the factor D. The equations (12.18) specialize to:

$$1 - T(z) + DzT(z)^2 = 0, \tag{12.22}$$

whose physical solution is:

$$T(z) = \frac{1-\sqrt{1-4Dz}}{2Dz}\ . \tag{12.23}$$

The critical point is $z_c = 1/(4D), T_c = 2$, which satisfies the criticality condition, $1 - 2Dz_cT_c = 0$. Moreover, close to criticality:

$$C(z) = 1 - 2DzT = \sqrt{1-4Dz}\,, \tag{12.24}$$

is singular as expected from (12.20).

12.2.2 Moments and Cumulants

The expectations of arbitrary polynomials in the tensor entries can be computed as derivatives of the moment generating function

$$Z(\mathbb{J},\bar{\mathbb{J}}) = \int \left(\prod_{\vec{a}} N^{D-1} \frac{d\mathbb{T}_{\vec{a}} d\bar{\mathbb{T}}_{\vec{a}}}{2\pi i}\right) e^{-N^{D-1} S(\mathbb{T},\bar{\mathbb{T}}) + \tilde{p}_{\mathcal{B}_2}(\mathbb{T},\bar{\mathbb{J}}) + \tilde{p}_{\mathcal{B}_2}(\mathbb{J},\bar{\mathbb{T}})}. \tag{12.25}$$

It turns out that it is more convenient to work with the generating function of connected moments, or *cumulants*:

$$\begin{aligned} W(\mathbb{J},\bar{\mathbb{J}}) &= \ln Z(\mathbb{J},\bar{\mathbb{J}}) \ \text{ with } \ W(\mathbb{J},\bar{\mathbb{J}}) \\ &= \sum_{p=0}^{\infty} \frac{1}{(p!)^2} W^{(2p)}_{\vec{a}^1\ldots\vec{a}^p,\vec{b}^1\ldots\vec{b}^p}\left(N,z,\{t_i\}\right) \mathbb{J}_{\vec{a}^1}\cdots\mathbb{J}_{\vec{a}^p}\bar{\mathbb{J}}_{\vec{b}^1}\cdots\bar{\mathbb{J}}_{\vec{b}^p}. \end{aligned} \tag{12.26}$$

The normalization factor $1/(p!)^2$ is conventional and accounts for the invariance of the cumulant of order $2p$ (or $2p$-point cumulant) $W^{(2p)}_{\vec{a}^1\ldots\vec{a}^p,\vec{b}^1\ldots\vec{b}^p}\left(N,z,\{t_i\}\right)$ under independent permutations σ and τ of its indices:

$$W^{(2p)}_{\vec{a}^1\dots\vec{a}^p,\vec{b}^1\dots\vec{b}^p}\Big(N,z,\{t_i\}\Big) = W^{(2p)}_{\vec{a}^{\sigma(1)}\dots\vec{a}^{\sigma(p)},\vec{b}^{\tau(1)}\dots\vec{b}^{\tau(p)}}\Big(N,z,\{t_i\}\Big)\,. \tag{12.27}$$

The generating function $W(\mathbb{J},\bar{\mathbb{J}})$ is itself invariant under the unitary transformations (12.1). It follows that it admits an expansion in invariants labelled by (not necessarily connected) bubbles. Denoting $p(\mathcal{B})$ the number of black vertices of the bubble $\mathcal{B}$, the $2p$-point cumulant admits an expansion:

$$W^{(2p)}_{\vec{a}^1\dots\vec{a}^p,\vec{b}^1\dots\vec{b}^p}\Big(N,z,\{t_i\}\Big) = \sum_{\mathcal{B},p(\mathcal{B})=p} \bar{\delta}^{\mathcal{B}}_{\vec{a}^1\dots\vec{a}^p,\vec{b}^1\dots\vec{b}^p}\, W\Big(\mathcal{B};N,z,\{t_i\}\Big)\,, \tag{12.28}$$

where $\bar{\delta}^{\mathcal{B}}_{\vec{a}^1\dots\vec{a}^p,\vec{b}^1\dots\vec{b}^p}$ is the trace-invariant operator associated to $\mathcal{B}$, symmetrized over its indices.

The graphical interpretation is as follows. The $2p$-point cumulants are sums over connected Feynman graphs $\mathcal{G}$ with $2p$ external lines, i.e. $(D+1)$ edge-coloured graphs with $2p$ edges of colour 0 hooked to univalent external vertices, and such that all their subgraphs with colours $\{1,\dots,D\}$ are from the set $\{\mathcal{B}_i\}_{i\in I}$. A typical example is presented in Fig. 12.5.

The *faces* of colour $0c$ of a Feynman graph (i.e. the subgraphs with colours $0,c$) fall in two categories. Either they are *internal faces*, that is closed cycles of edges of colours 0 and c, or they are *external faces*, that is chains of edges of colours 0 and c. The external faces start and end at external vertices and there is a single face of colour $0c$ which start/end at an external vertex (there are p external faces of each colour $c\in\{1,\dots,D\}$ in $\mathcal{G}$). The univalent external vertices have D-uples of indices, $\vec{a}^v$ for white vertices and $\vec{b}^{\bar{v}}$ for the black ones. The index of colour c at an external white vertex is identified along the corresponding external face with the index of the same colour at the black vertex on the other end of the chain.

Due to the unitary invariance, the pattern of index identification along the external faces of $\mathcal{G}$ can be encoded in a D-coloured graph called the *boundary graph* of $\mathcal{G}$ and denoted $\partial\mathcal{G}$. It only depends on the external faces and is obtained by only keeping the univalent external vertices of $\mathcal{G}$ and drawing an edge of colour c for every external face of colours $0c$ connecting two external vertices. For instance, Fig. 12.5 shows a four-point graph whose boundary graph is $\mathcal{B}_{4,\{c\}}$. From the boundary graph $\partial\mathcal{G}$, one gets the trace-invariant operator $\bar{\delta}^{\partial\mathcal{G}}_{\vec{a}^1\dots\vec{a}^p,\vec{b}^1\dots\vec{b}^p}$ which contains the whole index dependence of the

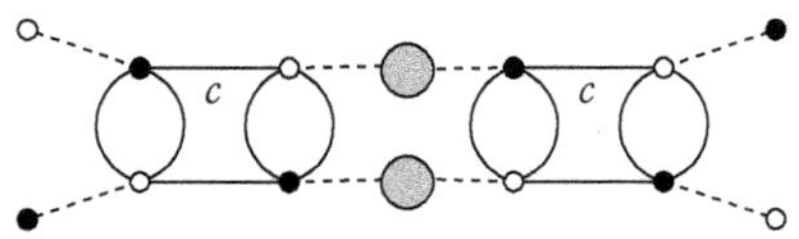

Figure 12.5 *This shows a four-point graph where the grey blobs represent arbitrary two-point insertions, which do not affect the boundary graph*

Feynman graph. The function $W\left(\mathcal{B};N,z,\{t_i\}\right)$ is then the sum of the amplitudes of all graphs $\mathcal{G}$ contributing with the fixed boundary graph $\partial\mathcal{G}=\mathcal{B}$.

The scaling of each of the terms with N in the previous sum has been studied in Gurau (2014). Let us define the rescaled contribution of an invariant:

$$K\left(\mathcal{B};N,z,\{t_i\}\right)\equiv\frac{W\left(\mathcal{B};N,z,\{t_i\}\right)}{N^{D-2(D-1)p(\mathcal{B})-\rho(\mathcal{B})}}\,, \tag{12.29}$$

where $\rho(\mathcal{B})$ denotes the number of connected components of $\mathcal{B}$. In the sense of perturbation theory, for any tensor model, these quantities are bounded for all N, and admit a finite large N limit Gurau (2014):

$$\lim_{N\to\infty}K\left(\mathcal{B};N,z,\{t_i\}\right)=K\left(\mathcal{B},z,\{t_i\}\right). \tag{12.30}$$

In the sequel the two- and four-point cumulants $W^{(2)}_{\vec{a}^1,\vec{b}^1}$ and $W^{(4)}_{\vec{a}^1\vec{a}^2,\vec{b}^1\vec{b}^2}$ will play a distinguished role. $W^{(2)}_{\vec{a}^1,\vec{b}^1}$ is proportional to the trace-invariant $\delta^{\mathcal{B}_2}$ since there is a single bubble on two vertices, $\mathcal{B}_2$.

At order 4, we have several bubbles $\mathcal{B}_{4,\mathcal{C}}$, one for each choice of a subset $\mathcal{C}\subset\{1,\dots,D\}$ of cardinality $0\le|\mathcal{C}|\le D/2$ (it is important to keep in mind that the expansion includes non-connected bubbles, here the union of two copies of $\mathcal{B}_2$). The four vertices of $\mathcal{B}_{4,\mathcal{C}}$ are divided into two pairs. The two vertices in a pair are connected by edges of colours $\{1,\dots,D\}\setminus\mathcal{C}$, while the edges with colour in $\mathcal{C}$ connect the pairs in between them. An example is presented in Fig. 12.6.

Remember that $\mathcal{B}_{4,\emptyset}$ represents two copies of $\mathcal{B}_2$, while the bubbles $\mathcal{B}_{4,\{c\}}$ have already been presented in Fig. 12.2. The trace-invariant operators associated to $\mathcal{B}_2$ and $\mathcal{B}_{4,\mathcal{C}}$ are

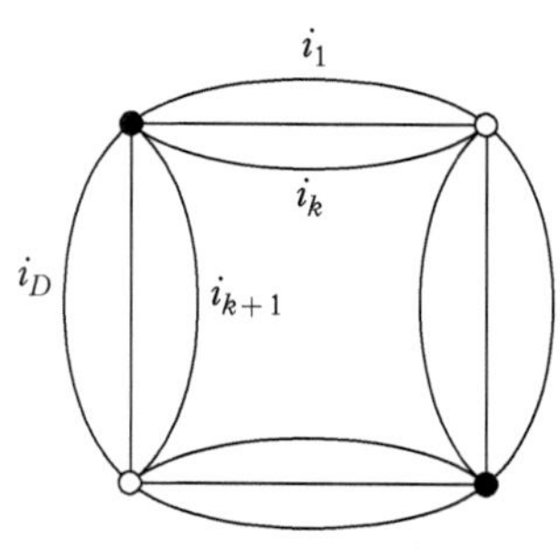

Figure 12.6 *The bubble graph* $\mathcal{B}_{4,\{c_{k+1},\dots,c_D\}}$

$$\bar{\delta}^{\mathcal{B}_2}_{\vec{a}^1,\vec{b}^1} = \prod_{c=1}^{D} \delta_{a_c^1 b_c^1}\,, \tag{12.31}$$

$$\begin{aligned}\bar{\delta}^{\mathcal{B}_{4,\mathcal{C}}}_{\vec{a}^1\vec{a}^2,\vec{b}^1\vec{b}^2} = &\Big(\prod_{c\notin\mathcal{C}} \delta_{a_c^1 b_c^1}\Big)\Big(\prod_{c\notin\mathcal{C}} \delta_{a_c^2 b_c^2}\Big)\Big(\prod_{c\in\mathcal{C}} \delta_{a_c^1 b_c^2}\delta_{a_c^2 b_c^1}\Big)\\ &+\Big(\prod_{c\notin\mathcal{C}} \delta_{a_c^2 b_c^1}\Big)\Big(\prod_{c\notin\mathcal{C}} \delta_{a_c^1 b_c^2}\Big)\Big(\prod_{c\in\mathcal{C}} \delta_{a_c^2 b_c^2}\delta_{a_c^1 b_c^1}\Big),\end{aligned} \tag{12.32}$$

where, according to our previous discussion we use the symmetrized trace-invariant operator for $\mathcal{B}_{4,\mathcal{C}}$. The two- and four-point cumulants are expressed therefore as

$$W^{(2)}_{\vec{a}_1,\vec{b}_1} = \delta^{\mathcal{B}_2}_{\vec{a}^1\vec{b}^1}\, W\Big(\mathcal{B}_2; N, z, \{t_i\}\Big), \tag{12.33}$$

$$W^{(4)}_{\vec{a}_1\vec{a}_2,\vec{b}_1\vec{b}_2} = \sum_{\mathcal{C}\subset\{1,\dots,D\}} \delta^{\mathcal{B}_{4,\mathcal{C}}}_{\vec{a}^1\vec{a}^2;\vec{b}^1\vec{b}^2}\, W\Big(\mathcal{B}_{4,\mathcal{C}}; N, z, \{t_i\}\Big). \tag{12.34}$$

12.2.3 Gaussian and non-Gaussian contributions

The Universality Theorem asserts that the large N limit of a random tensor model is Gaussian with covariance $T(z,\{t_i\})$. At NLO, non-Gaussian contributions eventually show up. Loosely speaking, a non-Gaussian contribution to an expectation $\langle\tilde{\rho}_{\mathcal{B}}(\mathbb{T},\overline{\mathbb{T}})\rangle$ is any Feynman graph $\mathcal{G}$ in the expansion of the expectation which would not appear in a Gaussian distribution whatever the covariance.

Let us denote π a partition of the set of vertices $\mathcal{V}(\mathcal{B})$ of the bubble $\mathcal{B}$ into (disjoint) bipartite subsets, $\mathcal{V}(\mathcal{B}) = \bigcup_\alpha \mathcal{V}_\alpha$, with $\mathcal{V}_\alpha = \{v_1^{(\alpha)},\dots,v_{p_\alpha}^{(\alpha)},\bar{v}_1^{(\alpha)},\dots,\bar{v}_{p_\alpha}^{(\alpha)}\}$ of cardinality $2p_\alpha$ (hence $\sum_\alpha p_\alpha = p(\mathcal{B})$). The expectation of a bubble observable expands in cumulants as:

$$\langle\tilde{\rho}_{\mathcal{B}}(\mathbb{T},\overline{\mathbb{T}})\rangle = \sum_{\{\vec{a}^v,\vec{b}^{\bar{v}}\}_{v,\bar{v}}} \delta^{\mathcal{B}}_{\vec{a}^v,\vec{b}^{\bar{v}}} \Big\langle\Big(\prod_{\bar{v}\in\mathcal{V}(\mathcal{B})} \overline{\mathbb{T}}_{\vec{b}^{\bar{v}}}\Big)\Big(\prod_{v\in\mathcal{V}(\mathcal{B})} \mathbb{T}_{\vec{a}^v}\Big)\Big\rangle \tag{12.35}$$

$$= \sum_{\{\vec{a}^v,\vec{b}^{\bar{v}}\}_{v,\bar{v}}} \delta^{\mathcal{B}}_{\vec{a}^v,\vec{b}^{\bar{v}}} \Big(\sum_{\pi}\prod_{\mathcal{V}_\alpha\in\pi} W^{(2p_\alpha)}_{\vec{a}^{v_1^{(\alpha)}}\dots\vec{a}^{v_{p_\alpha}^{(\alpha)}},\vec{b}^{\bar{v}_1^{(\alpha)}}\dots\vec{b}^{\bar{v}_{p_\alpha}^{(\alpha)}}}\Big(N,z,\{t_i\}\Big)\Big). \tag{12.36}$$

This means that we can classify the Feynman graphs contributing to the expectation using partitions of the vertex set $\mathcal{V}(\mathcal{B})$. Each partition π gives rise to a family of graphs $\mathcal{G}$, such that upon cutting off the edges of colour 0 which connect the marked sub-graph $\mathcal{B}$ to the rest of $\mathcal{G}$, one gets as many connected components as parts in π and these connected components have $2p_\alpha$ external edges which were connected to the vertices of $\mathcal{V}_\alpha$ in $\mathcal{G}$.

We call a *Gaussian contribution* a partition π such that $p_\alpha = 1, \forall\alpha$, and a *non-Gaussian contribution* any other partition. This means that a Gaussian contribution is exactly a graph obtained by attaching to $\mathcal{B}$ only two-point graphs.

The case $p(\mathcal{B}) = 1$ is quite trivial, by definition. Indeed, there is a single partition π which is the vertex set of $\mathcal{B}_2$ and $\langle\tilde{\rho}_{\mathcal{B}_2}(\mathbb{T},\overline{\mathbb{T}})\rangle/N = K(\mathcal{B}_2; N, z, \{t_i\})$. Recall that $K\Big(\mathcal{B}_2; N, z, \{t_i\}\Big)$ is bounded for all N and has a finite limit, namely $T(z,\{t_i\}) \equiv K\Big(\mathcal{B}_2; z, \{t_i\}\Big)$, when N goes to infinity. We can therefore start the $1/N$ expansion of the two-point function as

$$\frac{1}{N}\langle\tilde{\rho}_{\mathcal{B}_2}(\mathbb{T},\overline{\mathbb{T}})\rangle = K\Big(\mathcal{B}_2; N, z, \{t_i\}\Big) = T(z,\{t_i\}) + \frac{1}{N^{D-2}}\Big(K^{\mathrm{NLO}}\Big(\mathcal{B}_2, z, \{t_i\}\Big) + \mathcal{O}(1/N)\Big). \tag{12.37}$$

Next, consider the case $p(\mathcal{B}) = 2$. Labelling the four vertices $v_1, v_2, \bar{v}_1, \bar{v}_2$, three partitions into bipartite subsets are obtained. The partitions $\{\{v_1,\bar{v}_1\},\{v_2,\bar{v}_2\}\}$ and $\{\{v_1,\bar{v}_2\},\{v_2,\bar{v}_1\}\}$ are both Gaussian, since the parts have cardinality two, meaning only two-point functions are attached to $\mathcal{B}$. The third partition is the vertex set itself, corresponding to attaching to $\mathcal{B}$ a four-point cumulant. Graphically, this leads to the following exact expansion on $\mathcal{B}_{4,\{c\}}$,

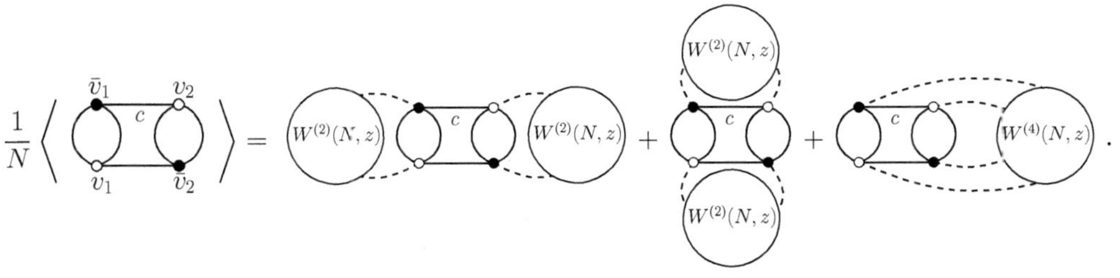

Clearly, the parts of a partition corresponding to a graph can be visualized by cutting the edges of colour 0 around the bubble $\mathcal{B}_{4,\{c\}}$.

Let us consider in more detail the expectation of $\mathcal{B}_{4,\{c\}}$. We have

$$\frac{1}{N}\langle\tilde{\rho}_{\mathcal{B}_{4,\{c\}}}(\mathbb{T},\overline{\mathbb{T}})\rangle = \frac{1}{N}\sum_{\substack{\vec{a}^1,\vec{a}^2\\ \vec{b}^1,\vec{b}^2}}\left(\delta_{a^1_c b^2_c}\delta_{a^2_c b^1_c}\prod_{c_1\neq c}\delta_{a^1_{c_1} b^1_{c_1}}\delta_{a^2_{c_1} b^2_{c_1}}\right)\langle\mathbb{T}_{\vec{a}^1}\mathbb{T}_{\vec{a}^2}\overline{\mathbb{T}}_{\vec{b}^1}\overline{\mathbb{T}}_{\vec{b}^2}\rangle \tag{12.38}$$

$$= \frac{1}{N}\sum_{\substack{\vec{a}^1,\vec{a}^2\\ \vec{b}^1,\vec{b}^2}}\left(\delta_{a^1_c b^2_c}\delta_{a^2_c b^1_c}\prod_{c_1\neq c}\delta_{a^1_{c_1} b^1_{c_1}}\delta_{a^2_{c_1} b^2_{c_1}}\right)\Bigg[W^{(2)}_{\vec{a}^1,\vec{b}^1}\Big(N,z,\{t_i\}\Big)W^{(2)}_{\vec{a}^2,\vec{b}^2}\Big(N,z,\{t_i\}\Big)$$

$$+ W^{(2)}_{\vec{a}^1,\vec{b}^2}\Big(N,z,\{t_i\}\Big)W^{(2)}_{\vec{a}^2,\vec{b}^1}\Big(N,z,\{t_i\}\Big) + W^{(4)}_{\vec{a}^1\vec{a}^2,\vec{b}^1\vec{b}^2}\Big(N,z,\{t_i\}\Big)\Bigg]. \tag{12.39}$$

The two products of two-point cumulants are the two Gaussian contributions.

We treat the previous Gaussian and non-Gaussian contributions separately. Using equation (12.33), the Gaussian contributions is written:

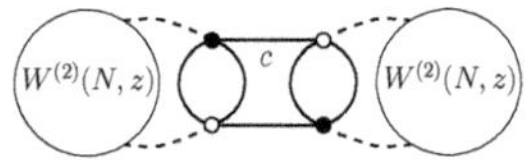

$+$
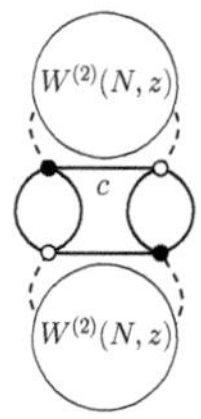

$$= \frac{1}{N}\left(N^{2D-1} + N^{D+1}\right)\left[N^{-(D-1)}K\left(\mathcal{B}_2; N, z, \{t_i\}\right)\right]^2$$

$$= \left(1 + \frac{1}{N^{D-2}}\right)\left[K\left(\mathcal{B}_2; N, z, \{t_i\}\right)\right]^2 =$$

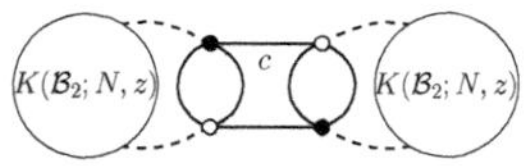

$+ \frac{1}{N^{D-2}}$
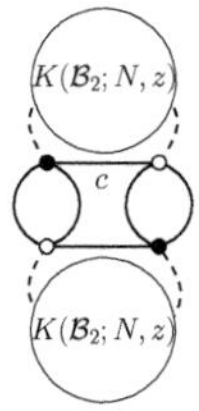

.

(12.40)

The non-Gaussian contributions

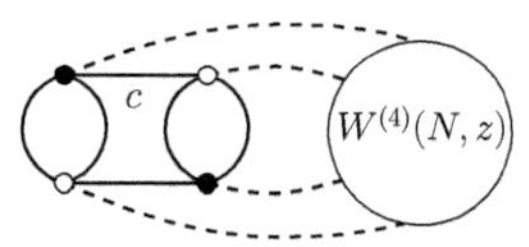

$$= \frac{1}{N}\sum_{a,b}\left(\delta_{a_c^1 b_c^2}\delta_{a_c^2 b_c^1}\prod_{c_1 \neq c}\delta_{a_{c_1}^1 b_{c_1}^1}\delta_{a_{c_1}^2 b_{c_1}^2}\right)W^{(4)}_{\vec{a}^1\vec{a}^2,\vec{b}^1\vec{b}^2}\left(N, z, \{t_i\}\right),$$

(12.41)

yields:

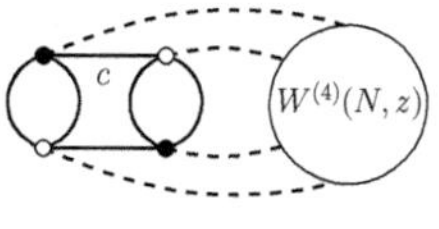

$$= \left(\frac{1}{N^{2(D-1)}} + \frac{1}{N^D}\right)K\left(\mathcal{B}_{4,\emptyset}; N, z, \{t_i\}\right)$$

$$+ \sum_{\mathcal{C}\in\{1,\ldots,D\}, c\in\mathcal{C}, |\mathcal{C}|\leq D/2}\left(\frac{1}{N^{D-2}}N^{1-|\mathcal{C}|} + \frac{1}{N^{D-2}}N^{-D+|\mathcal{C}|-1}\right)K\left(\mathcal{B}_{4,\mathcal{C}}; N, z, \{t_i\}\right)$$

$$+ \sum_{\mathcal{C}\in\{1,\ldots,D\}, c\notin\mathcal{C}, |\mathcal{C}|\leq D/2}\left(\frac{1}{N^{D-2}}N^{-D+1+|\mathcal{C}|} + \frac{1}{N^{D-2}}N^{-1-|\mathcal{C}|}\right)K\left(\mathcal{B}_{4,\mathcal{C}}; N, z, \{t_i\}\right).$$

(12.42)

We conclude that these terms contribute at most at order N^{D-2}. Furthermore, at order N^{D-2}, only the term with $\mathcal{C} = \{c\}$ represented in Fig. 12.7 contributes.

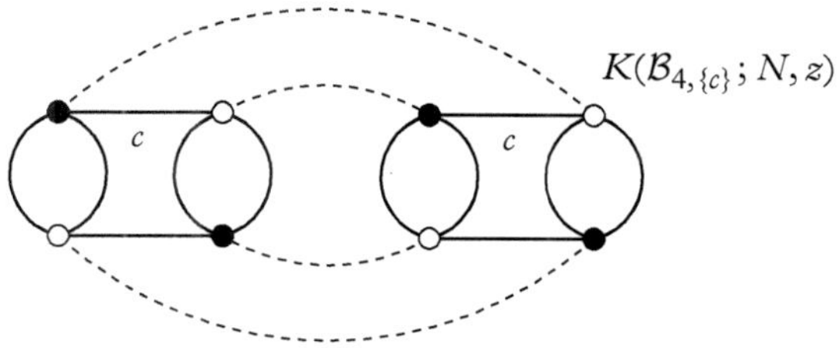

Figure 12.7 *The dominant non-Gaussian contribution to the expectation of $\mathcal{B}_{4,\{c\}}$, where the rightmost bubble represents the boundary graph of $K(\mathcal{B}_{4,\{c\}}; N, z, \{t_i\})$. It is necessary to explicitly draw this boundary graph in order to describe how it is connected to the observable. There are indeed two ways to connect it, corresponding to the two terms proportional to $K(\mathcal{B}_{4,\{c\}}; N, z, \{t_i\})$ in equation (12.43): the dominant one, which comes with a factor $1/N^{D-2}$, and a sub-dominant one, which comes with a scaling N^{-D-1}/N^{D-2}*

Gathering the previously mentioned pieces brings up the following cumulant expansion:

$$\begin{aligned}\frac{1}{N}\langle\tilde{\rho}_{\mathcal{B}_{4,\{c\}}}(\mathbb{T},\overline{\mathbb{T}})\rangle &= \Big(1+\frac{1}{N^{D-2}}\Big)\Big[K\Big(\mathcal{B}_2;N,z,\{t_i\}\Big)\Big]^2 + \Big(\frac{1}{N^{D-2}} \\ &+\frac{1}{N^{2D-1}}\Big)K\Big(\mathcal{B}_{4,\{c\}};N,z,\{t_i\}\Big)+\Big(\frac{1}{N^{2(D-1)}}+\frac{1}{N^{D}}\Big)K\Big(\mathcal{B}_{4,\emptyset};N,z,\{t_i\}\Big) \\ &+\sum_{\mathcal{C}\in\{1,\dots,D\},c\in\mathcal{C},1<|\mathcal{C}|\le D/2}\Big(\frac{1}{N^{D-2}}N^{1-|\mathcal{C}|}+\frac{1}{N^{D-2}}N^{-D+|\mathcal{C}|-1}\Big)K\Big(\mathcal{B}_{4,\mathcal{C}};N,z,\{t_i\}\Big) \\ &+\sum_{\mathcal{C}\in\{1,\dots,D\},c\notin\mathcal{C},1\le|\mathcal{C}|\le D/2}\Big(\frac{1}{N^{D-2}}N^{-D+1+|\mathcal{C}|}+\frac{1}{N^{D-2}}N^{-1-|\mathcal{C}|}\Big)K\Big(\mathcal{B}_{4,\mathcal{C}};N,z,\{t_i\}\Big).\end{aligned} \tag{12.43}$$

This equation is exact. Since we are only interested in the LO and NLO of the expectation, we only keep the first line in the form:

$$\begin{aligned}\frac{1}{N}\langle\tilde{\rho}_{\mathcal{B}_{4,\{c\}}}(\mathbb{T},\overline{\mathbb{T}})\rangle &= \Big(1+\frac{1}{N^{D-2}}\Big)\Big[K\Big(\mathcal{B}_2;N,z,\{t_i\}\Big)\Big]^2 \\ &+\frac{1}{N^{D-2}}K\Big(\mathcal{B}_{4,\{c\}};N,z,\{t_i\}\Big)+\mathcal{O}(1/N^{D-1}).\end{aligned} \tag{12.44}$$

Truncating at order $\frac{1}{N^{D-2}}$ and using (12.37), we obtain:

$$\begin{aligned}\frac{1}{N}\langle\tilde{\rho}_{\mathcal{B}_{4,\{c\}}}(\mathbb{T},\overline{\mathbb{T}})\rangle &= T^2+\frac{1}{N^{D-2}}\Big(2T\,K^{\mathrm{NLO}}\Big(\mathcal{B}_2,z,\{t_i\}\Big)+T^2 \\ &+K\Big(\mathcal{B}_{4,\{c\}};z,\{t_i\}\Big)+\mathcal{O}(1/N)\Big).\end{aligned} \tag{12.45}$$

We will use from now on the shorthand notations $K^{\mathrm{NLO}}\Big(\mathcal{B}_2, z, \{t_i\}\Big) \equiv K_2^{\mathrm{NLO}}$, and $K\Big(\mathcal{B}_{4,\{c\}}; z, \{t_i\}\Big) \equiv K_{4,\{c\}}^{\mathrm{LO}}$. The same type of expansion can be performed on the expectations of the polynomials associated to arbitrary melonic bubbles (see the original article Bonzom (2014) for details)

$$\frac{1}{N}\langle \tilde{\rho}_{\mathcal{B}_i}(\mathbb{T},\overline{\mathbb{T}})\rangle = \Big(1+\frac{\alpha_i}{N^{D-2}}\Big)\Big[K\Big(\mathcal{B}_2; N, z, \{t_i\}\Big)\Big]^{p_i} + \frac{\alpha_i'}{N^{D-2}}\Big[K\Big(\mathcal{B}_2; N, z, \{t_i\}\Big)\Big]^{p_i-2} K\Big(\mathcal{B}_{4,\{c\}}; N, z, \{t_i\}\Big) + \mathcal{O}(1/N^{D-1}), \tag{12.46}$$

reducing at NLO to

$$\frac{1}{N}\langle \tilde{\rho}_{\mathcal{B}_i}(\mathbb{T},\overline{\mathbb{T}})\rangle = T^{p_i} + \frac{1}{N^{D-2}}\left(p_i T^{p_i-1} K_2^{\mathrm{NLO}} + \alpha_i T^{p_i} + T^{p_i-2}\sum_{c=1,\dots,D} \alpha_{i,c}' K_{4,\{c\}}^{\mathrm{LO}} + \mathcal{O}(1/N)\right), \tag{12.47}$$

where α_i and $\alpha_{i,c}'$ are some coefficients which depend on $\mathcal{B}_i$.

From now on, to simplify the analysis, we shall assume that the set of interaction bubbles $\{\mathcal{B}_i\}_{i\in I}$ *(as well as the set of corresponding couplings* $\{t_i\}_{i\in I}$*) is invariant under colour relabelling.* As a consequence, there is a single function $K_{4,\bullet}^{\mathrm{LO}} \equiv K_{4,\{c\}}^{\mathrm{LO}}$ for all $c \in \{1,\dots,D\}$. We denote in (12.47) $\alpha_i' \equiv \sum \alpha_{i,c}'$.

The same kind of expansion holds for $\langle \tilde{\rho}_{\mathcal{B}_i+(e_c)}(\mathbb{T},\overline{\mathbb{T}})\rangle$, albeit with different numerical constants, α_{i,e_c} and α_{i,e_c}'. It is convenient to sum directly over $e \in \mathcal{B}_i$, that is, the edges of colour c, upon which one may insert a $(D-1)$=dipole:

$$\sum_{e_c\in\mathcal{B}_i}\frac{1}{N}\langle \tilde{\rho}_{\mathcal{B}_i+(e_c)}(\mathbb{T},\overline{\mathbb{T}})\rangle = p_i T^{p_i+1} + \frac{1}{N^{D-2}}\big(p_i(p_i+1)T^{p_i} K_2^{\mathrm{NLO}} + \beta_{i,c} T^{p_i+1} + \beta_{i,c}' T^{p_i-1} K_{4,\bullet}^{\mathrm{LO}} + \mathcal{O}(1/N)\big), \tag{12.48}$$

with $\beta_{i,c} = \sum_e \alpha_{i,e_c}$ and $\beta_{i,c}' = \sum_e \alpha_{i,e_c}'$.

six-point bubbles. In the quartic model of Dartois, Gurau, and Rivasseau (2013), the DSE involve the bubbles $\mathcal{B}_{4,\{c'\}} + (e_c)$, that is, $\mathcal{B}_{4,\{c'\}}$ with a melonic insertion of colour c, which have six vertices and are of two types: either $c = c'$ or $c \neq c'$. If $c = c'$, weget a bubble with three $(D-1)$-dipole insertions of the same colour. Setting for convenience $c' = 1$, and $c = 1$, or $c = 2$ (depending on the case), we can write:

$\frac{1}{N}\Big\langle$ 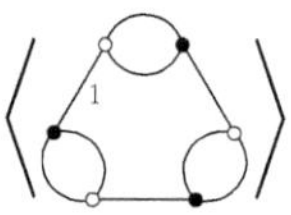 $\Big\rangle = \left(1+\frac{3}{N^{D-2}}\right)\left[K\left(\mathcal{B}_2;N,z,\{t_i\}\right)\right]^3$

$$+\frac{3}{N^{D-2}}\left[K\left(\mathcal{B}_2;N,z,\{t_i\}\right)K\left(\mathcal{B}_{4,\bullet};N,z,\{t_i\}\right)+\mathcal{O}(1/N)\right], \tag{12.49}$$

$\frac{1}{N}\Big\langle$ 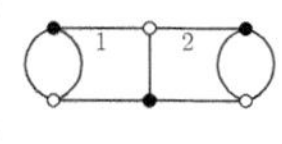 $\Big\rangle = \left(1+\frac{2}{N^{D-2}}\right)\left[K\left(\mathcal{B}_2;N,z,\{t_i\}\right)\right]^3$

$$+\frac{2}{N^{D-2}}\left[K\left(\mathcal{B}_2;N,z,\{t_i\}\right)K\left(\mathcal{B}_{4,\bullet};N,z,\{t_i\}\right)+\mathcal{O}(1/N)\right]. \tag{12.50}$$

In both equations, the $K(\mathcal{B}_2;N,z,\{t_i\})^3$ term is the Gaussian contribution. Its coefficient takes into account the single melonic contraction, as well as the α NLO Gaussian contractions. The second term is the non-Gaussian contribution, where there are α' ways to introduce a four-point contribution with the appropriate $1/N^{D-2}$ scaling. For the first bubble $\alpha = \alpha' = 3$, since there are three contributions of the types:[2]

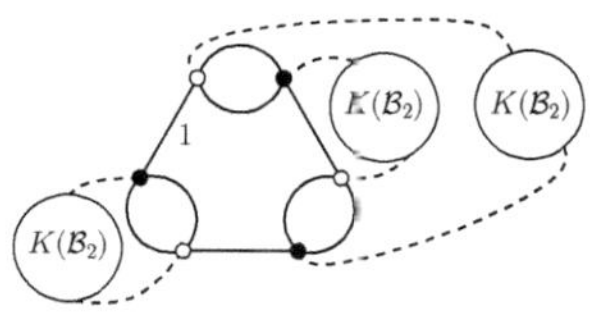

,
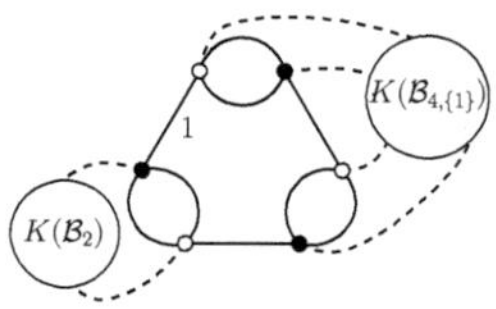

.

For the second bubble $\alpha = 2$, which corresponds to the two terms:

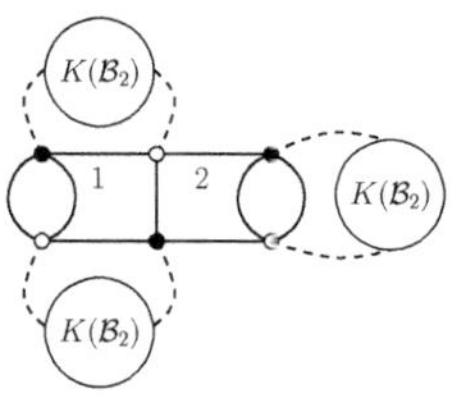

,
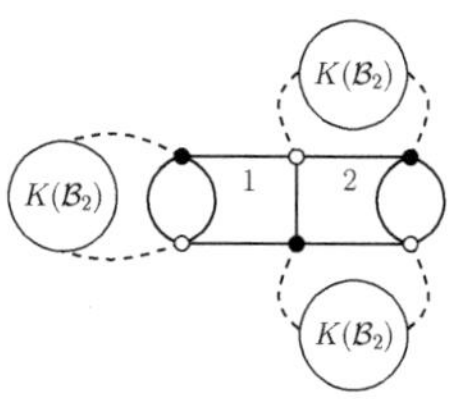

,

and $\alpha' = 2$, corresponding to the following two terms,

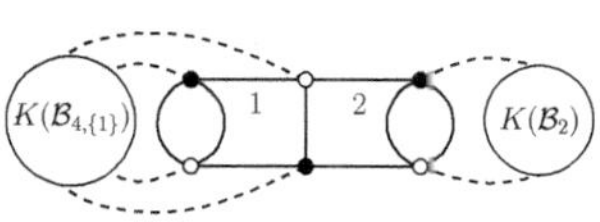

,
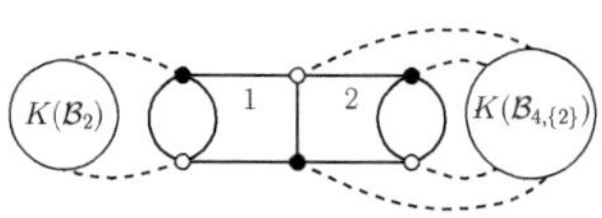

.

[2] Repeatedly rotating the graphs by $2\pi/3$ produces another two graphs of each type.

12.2.4 The DSE at NLO

We are now in position to solve the SD equations at NLO to extract $K_2^{\rm NLO}, K_{4,\bullet}^{\rm LO}$. Plugging the expansions (12.37), (12.45), (12.47), (12.48) into the DSE (12.11) and (12.17) yields a *linear* system on $K_2^{\rm NLO}, K_{4,\bullet}^{\rm LO}$. The first DSE (12.11) reads:

$$\left(-1+\sum_{i\in I} t_i p_i [zT]^{p_i-1}\right) K_2^{\rm NLO} + \left(\sum_{i\in I} t_i \alpha_i' z^{p_i-1} T^{p_i-2}\right) K_{4,\bullet}^{\rm LO} = -\sum_{i\in I} t_i \alpha_i z^{p_i-1} T^{p_i}. \tag{12.51}$$

Notice that the coefficient of $K_2^{\rm NLO}$ is the singular function $C(z,\{t_i\})$ defined in (12.20), which vanishes at criticality.

The second DSE, (12.17), is:

$$\begin{aligned}\left(1-2T+\sum_{i\in I} t_i (p_i+1) z^{p_i-1} T^{p_i}\right) K_2^{\rm NLO} + \left(-1+\sum_{i\in I} t_i \frac{\beta_{i,c}'}{p_i} [zT]^{p_i-1}\right) K_{4,\bullet}^{\rm LO} \\ = -T\left(1-T+\sum_{i\in I} t_i \frac{\beta_{i,c}}{p_i} z^{p_i-1} T^{p_i}\right)\end{aligned} \tag{12.52}$$

It should be noted that, although the numerical coefficients $\beta_{i,c}$ and $\beta_{i,c}'$ depend on both i and c, the sum over $i \in I$ ensures that everything is symmetric with respect to colour relabelling, so that the equation is the same for all values of $c \in \{1,\dots,D\}$. We then set $c=1$ and denote $\beta_i \equiv \beta_{i,c=1}$ and $\beta_i' \equiv \beta_{i,c=1}'$.

To simplify this equation, we use $1-T+\sum_{i\in I} t_i z^{p_i-1} T^{p_i} = 0$, first to observe that the coefficient of $K_2^{\rm NLO}$ is again proportional to the singular function C:

$$1-2T+\sum_{i\in I} t_i (p_i+1) z^{p_i-1} T^{p_i} = -TC, \tag{12.53}$$

and second to reduce the right-hand side:

$$1-T+\sum_{i\in I} t_i \frac{\beta_i}{p_i} z^{p_i-1} T^{p_i} = -\sum_{i\in I} t_i \left(1-\frac{\beta_i}{p_i}\right) z^{p_i-1} T^{p_i}. \tag{12.54}$$

After dividing the equation (12.52) by T, (12.51) and (12.52) can be cast into a linear system:

$$\begin{pmatrix} -C & \sum_{i\in I} t_i \alpha_i' z^{p_i-1} T^{p_i-2} \\ -C & \frac{1}{T}\left(-1+\sum_{i\in I} t_i \frac{\beta_i'}{p_i} [zT]^{p_i-1}\right) \end{pmatrix} \begin{pmatrix} K_2^{\rm NLO} \\ K_{4,\bullet}^{\rm LO} \end{pmatrix} = \begin{pmatrix} -\sum_{i\in I} t_i \alpha_i z^{p_i-1} T^{p_i} \\ \sum_{i\in I} t_i \frac{p_i-\beta_i}{p_i} z^{p_i-1} T^{p_i} \end{pmatrix}, \tag{12.55}$$

whose determinant is:

$$\det\begin{pmatrix} -C & \sum_{i\in I} t_i \alpha_i' z^{p_i-1} T^{p_i-2} \\ -C & \frac{1}{T}\left(-1+\sum_{i\in I} t_i \frac{\beta_i'}{p_i}[zT]^{p_i-1}\right)\end{pmatrix} = \frac{C}{T}\left(1-\sum_{i\in I} t_i\left(\frac{\beta_i'}{p_i}-\alpha_i'\right)[zT]^{p_i-1}\right)$$
$$= \frac{C}{T^2}\left(1-\sum_{i\in I} t_i\left(\frac{\beta_i'}{p_i}-(\alpha_i'+1)\right)z^{p_i-1}T^{p_i}\right). \tag{12.56}$$

Inversion is then straightforward and leads to

$$K_2^{\mathrm{NLO}} = \frac{T}{C}\frac{\left(1-\sum_i t_i\beta_i'[zT]^{p_i-1}/p_i\right)\sum_i t_i\alpha_i z^{p_i-1}T^{p_i} - \sum_{i,j\in I} t_i t_j \alpha_i'(v_j-\beta_j) z^{p_i+v_j-2}T^{p_i+v_j-1}/v_j}{1-\sum_{i\in I} t_i\left(\beta_i'/p_i-(\alpha_i'+1)\right)z^{p_i-1}T^{p_i}}, \tag{12.57}$$

$$K_{4,\bullet}^{\mathrm{LO}} = \frac{\sum_i t_i\left(\beta_i/p_i-(\alpha_i+1)\right)z^{p_i-1}T^{p_i+2}}{1-\sum_{i\in I} t_i\left(\beta_i'/p_i-(\alpha_i'+1)\right)z^{p_i-1}T^{p_i}}. \tag{12.58}$$

The quartic case. For $I=\{1,\dots,D\}$ and $\{\mathcal{B}_i\}_{i\in I}=\{\mathcal{B}_{4,\{c\}}\}_{c=1,\dots,D}$, the combinatorial coefficients are $\alpha=\alpha'=1$ and $\beta_c=\beta_c'$ with $\beta_1=6$ and $\beta_c=4$ for $c\neq 1$. Specializing the previous formula simply gives

$$K_2^{\mathrm{NLO}}(z) = \frac{1}{\sqrt{1-4Dz}}\frac{DzT(z)^2}{1-zT(z)^2},$$
$$K_{4,\bullet}^{\mathrm{LO}}(z) = \frac{zT(z)^4}{1-zT(z)^2}, \tag{12.59}$$

where we have used the explicit solution $T(z)=(1-\sqrt{1-4Dz})/(2Dz)$, $C=1-2DzT=\sqrt{1-4Dz}$.

The function $K_{4,\bullet}^{\mathrm{LO}}$ has a simple combinatorial interpretation as the chain of quartic bubbles transmitting a single colour:

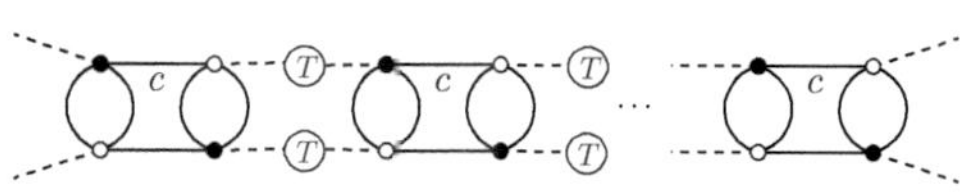

$$= \frac{zT^4}{1-zT^2} = K_{4,\bullet}^{\mathrm{LO}}(z).$$

This makes the link with the results of (Dartois, Gurau, and Rivasseau 2013; Gurau and Schaeffer 2016), where these chains play a prominent role.

12.2.5 The order $1/N^D$ in the quartic model

In the context of the quartic model, it is easy enough to go beyond the order $1/N^{D-2}$. Here, we solve the DSE at the order $1/N^D$ to get the LO of $K(\mathcal{B}_{4,\emptyset};N,z)$. This result will be useful to derive the double-scaling limit (and will actually be also derived in Appendix F, using combinatorics arguments).

Notice that $1/N^D$ is the NLO when $D \geq 5$. For $D=3$, the NLO is however $1/N^{2(D-2)}$, while those two orders coincide at $D=4$. Nonetheless, the DSE hold order by order in the $1/N$ expansion, so $1/N^D$ not being the NLO is not an issue. The only critical case is $D=4$, for which it would seem that looking at terms of order $1/N^D$ is not enough. This is, not true. The terms which scale like $1/N^4$ at $D=4$ can be unambiguously split into contributions which generalize to $D \neq 4$ with scaling $1/N^D$ and others with scaling $1/N^{2(D-2)}$. If the DSE are satisfied by some functions of D at the order $1/N^D$ for $D \neq 4$, they will still be satisfied at this order when $D=4$ by the same functions evaluated on $D=4$.

We here need to come back to the equation (12.43), i.e. the cumulant expansion of the expectation $\langle \tilde{\rho}_{\mathcal{B}_{4,\{c\}}}(\mathbb{T},\overline{\mathbb{T}})\rangle/N$ (fixed colour c), so as to identify the terms which pop up at the order $1/N^D$. We find directly the term $K(\mathcal{B}_{4,\emptyset};N,z)/N^D$ which shows that the disconnected boundary graph is now involved. We also find the term $\frac{N^{-1-|\mathcal{C}|}}{N^{D-2}}K(\mathcal{B}_{4,\mathcal{C}};N,z)$ with $c \notin \mathcal{C}$, which for $|\mathcal{C}|=1$ reduces to $\frac{1}{N^D}K(\mathcal{B}_{4,\{c'\}};N,z)$ with $\mathcal{C}=\{c'\}$. We can now extend the explicit terms of (12.44) to

$$\frac{1}{N}\langle \tilde{\rho}_{\mathcal{B}_{4,\{c\}}}(\mathbb{T},\overline{\mathbb{T}})\rangle = \Big(1+\frac{1}{N^{D-2}}\Big)\Big[K\Big(\mathcal{B}_2;N,z,\{t_i\}\Big)\Big]^2 + \frac{1}{N^{D-2}}K\Big(\mathcal{B}_{4,\{c\}};N,z,\{t_i\}\Big)$$
$$+\frac{1}{N^D}\Big[K\Big(\mathcal{B}_{4,\emptyset};N,z\Big)+\sum_{c'\neq c}K\Big(\mathcal{B}_{4,\{c'\}};N,z\Big)+\mathcal{O}(1/N)\Big]. \quad (12.60)$$

Due to the colour symmetry, all the terms $K\Big(\mathcal{B}_{4,\{c'\}};N,z\Big)$ are equal and their LO is therefore $K^{\mathrm{LO}}_{4,\bullet}(z)$. We introduce $K^{\mathrm{NNLO}}_2(z)$ the restriction of the two-point function at the order $1/N^D$, and $K^{\mathrm{LO}}_{4,\emptyset}(z)$ the LO of $K\Big(\mathcal{B}_{4,\emptyset};N,z\Big)$. Identifying those terms at order $1/N^D$, we find:

$$\frac{1}{N}\langle \tilde{\rho}_{\mathcal{B}_{4,\{c\}}}(\mathbb{T},\overline{\mathbb{T}})\rangle|_{1/N^D} = 2T(z)\,K^{\mathrm{NNLO}}_2(z) + (D-1)K^{\mathrm{LO}}_{4,\bullet}(z) + K^{\mathrm{LO}}_{4,\emptyset}(z). \quad (12.61)$$

Similar cumulant expansions exist for larger melonic bubbles. For the quartic model, we need to extend the expansions (12.49), (12.50) of the two six-point bubbles. This leads to:

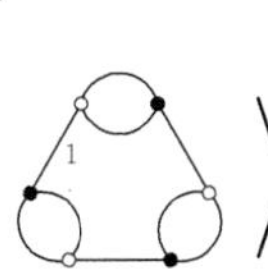

$$\frac{1}{N}\Big\langle \quad \Big\rangle = \Big(1+\frac{3}{N^{D-2}}\Big)\Big[K\Big(\mathcal{B}_2;N,z,\{t_i\}\Big)\Big]^3 + \frac{3}{N^{D-2}}K\Big(\mathcal{B}_2;N,z,\{t_i\}\Big)$$
$$\times K\Big(\mathcal{B}_{4,\bullet};N,z,\{t_i\}\Big) + \frac{3}{N^D}K\Big(\mathcal{B}_2;N,z,\{t_i\}\Big)\Big[(D-1)\,K\Big(\mathcal{B}_{4,\bullet};N,z,\{t_i\}\Big)$$
$$+K\Big(\mathcal{B}_{4,\emptyset};N,z,\{t_i\}\Big)+\mathcal{O}(1/N)\Big], \quad (12.62)$$

and

$$\frac{1}{N}\Big\langle$$ 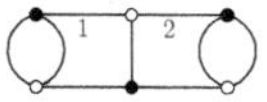 $$\Big\rangle = \left(1+\frac{2}{N^{D-2}}+\frac{1}{N^D}\right)\left[K\left(\mathcal{B}_2;N,z,\{t_i\}\right)\right]^3$$
$$+\frac{2}{N^{D-2}}K\left(\mathcal{B}_2;N,z,\{t_i\}\right)K\left(\mathcal{B}_{4,\bullet};N,z,\{t_i\}\right)+\frac{1}{N^D}K\left(\mathcal{B}_2;N,z,\{t_i\}\right)$$
$$\left[\left(2(D-1)+D\right)K\left(\mathcal{B}_{4,\bullet};N,z,\{t_i\}\right)+3K\left(\mathcal{B}_{4,\emptyset};N,z,\{t_i\}\right)+\mathcal{O}(1/N)\right]. \quad (12.63)$$

This way, the DSE (12.11) gives, at order $1/N^D$,

$$\left(-1+2DzT(z)\right)K_2^{\mathrm{NNLO}}(z)+DzK_{4,\emptyset}^{\mathrm{LO}}(z)=-D(D-1)zK_{4,\bullet}^{\mathrm{LO}}(z), \quad (12.64)$$

while the DSE (12.17) gives

$$\begin{aligned}&\left(1-2T(z)+3DzT(z)^2\right)K_2^{\mathrm{NNLO}}(z)+\left(-1+3DzT(z)\right)K_{4,\emptyset}^{\mathrm{LO}}(z)\\&=(D-1)\left(1-(3D+1)zT(z)\right)K_{4,\bullet}^{\mathrm{LO}}(z)-(D-1)zT(z)^3.\end{aligned} \quad (12.65)$$

Subtracting 2 times the first equation to the second, we get the linear system

$$\begin{pmatrix}-1+2DzT & Dz\\ 1-DzT^2 & -1+DzT\end{pmatrix}\begin{pmatrix}K_2^{\mathrm{NNLO}}(z)\\ K_{4,\emptyset}^{\mathrm{LO}}(z)\end{pmatrix}$$
$$=\begin{pmatrix}-D(D-1)zK_{4,\bullet}^{\mathrm{LO}}(z)\\ (D-1)K_{4,\bullet}^{\mathrm{LO}}(z)-(D+1)(D-1)zTK_{4,\bullet}^{\mathrm{LO}}(z)-(D-1)zT^3\end{pmatrix} \quad (12.66)$$

where $K_{4,\bullet}^{\mathrm{LO}}$ acts as a source. The determinant is found to be

$$\det\begin{pmatrix}-1+2DzT(z) & Dz\\ 1-DzT(z)^2 & -1+DzT(z)\end{pmatrix}=\left[\frac{1-DzT(z)^2}{T(z)}\right]^2=1-4Dz=1-z/z_c, \quad (12.67)$$

which leads to

$$K_2^{\mathrm{NNLO}}(z)=\frac{D(D-1)\,z^2\,[T(z)]^5}{(1-Dz\,[T(z)]^2)^2\,(1-z\,[T(z)]^2)}=\frac{1}{1-4Dz}\,\frac{D(D-1)\,z^2\,[T(z)]^3}{1-z\,[T(z)]^2}, \quad (12.68)$$

$$K_{4,\emptyset}^{\mathrm{LO}}(z)=\frac{D(D-1)\,z^2\,[T(z)]^6}{(1-Dz\,[T(z)]^2)\,(1-z\,[T(z)]^2)}=\frac{1}{\sqrt{1-4Dz}}\,\frac{D(D-1)\,z^2\,[T(z)]^5}{1-z\,[T(z)]^2}. \quad (12.69)$$

Whereas $K_{4,\bullet}^{\mathrm{LO}}(z)$ was found to be the (sum of all) monocloured chains, $K_{4,\emptyset}^{\mathrm{LO}}(z)$ instead corresponds to the (sum of all) strictly multicoloured chains, i.e. the chains of quartic bubbles where there is at least one change of transmitted colour. Indeed, by considering all chains with arbitrary colours and then subtracting the D mono-coloured chains, we obtain:

$$\frac{DzT^4}{1-DzT^2} - \frac{DzT^4}{1-zT^2} = \frac{D(D-1)z^2T^6}{(1-DzT^2)(1-zT^2)} = K^{\mathrm{LO}}_{4,\emptyset}(z). \tag{12.70}$$

All those combinatorial interpretations within the quartic model will be clear in the F which makes use of a powerful combinatorial representation (which, however, only works for the quartic model).

12.3 The double-scaling limit

12.3.1 Double-scaling limit in the DSE

We now look for a different scaling limit, which can be reached by sending both $N \to \infty$ and z to its critical value z_c, at a certain rate such that a combination (which we denote x), of $z_c - z$ and N (to be determined) is held fixed. The system of DSE that we have solved so far is linear in the $1/N$ perturbations: only linear functions arise beyond the LO. To find an interesting limit, the DSE must become *non-linear* in the perturbations, at least in the $1/N$ corrections to the two-point function.

The plan is to expand the two-point function in a different manner, after which we shall test the viability of this ansatz using the simplest DSE (12.11):

$$0 = 1 - \frac{1}{N}\langle \tilde{\rho}_{\mathcal{B}_2}(\mathbb{T},\overline{\mathbb{T}})\rangle + \frac{1}{N}\sum_{i\in I} z^{p_i-1} t_i \langle \tilde{\rho}_{\mathcal{B}_i}(\mathbb{T},\overline{\mathbb{T}})\rangle.$$

A sign that non-linear perturbations are possible stems from the fact that the expectation $\langle \tilde{\rho}_{\mathcal{B}_i}(\mathbb{T},\overline{\mathbb{T}})\rangle$ is of order p_i in the two-point function:

$$\frac{1}{N}\langle \tilde{\rho}_{\mathcal{B}_i}(\mathbb{T},\overline{\mathbb{T}})\rangle - \text{non-Gaussian parts} = \left(1 + \frac{\alpha_i}{N^{D-2}} + \mathcal{O}(1/N^D)\right)\left[K\left(\mathcal{B}_2; N, z, \{t_i\}\right)\right]^{p_i}. \tag{12.71}$$

Therefore, if we make the following generic ansatz for the double-scaled two-point function:

$$K\left(\mathcal{B}_2; N, x, \{t_i\}\right) = T(z_c, \{t_i\}) + \frac{1}{N^a} K^{\mathrm{DS}}(x, \{t_i\}) + \cdots \tag{12.72}$$

when $N \to \infty, z \to z_c$, we see that a quadratic term in $K^{\mathrm{DS}}(x, \{t_i\})$ arises from the expansion:

$$\begin{aligned}
&\frac{1}{N}\langle \tilde{\rho}_{\mathcal{B}_i}(\mathbb{T},\overline{\mathbb{T}})\rangle - \text{non-Gaussian parts} \\
&\quad = T_c^{p_i} + \frac{1}{N^a} p_i T_c^{p_i-1} K^{\mathrm{DS}}(x) + \frac{1}{N^{D-2}}\alpha_i T_c^{p_i} \\
&\qquad + \frac{1}{N^{2a}}\frac{p_i(p_i-1)}{2} T_c^{p_i-2}\left[K^{\mathrm{DS}}(x)\right]^2 + \cdots,
\end{aligned} \tag{12.73}$$

where $T_c \equiv T(z_c, \{t_i\})$, at order $1/N^{2a}$. It follows that a nontrivial double-scaling equation can be obtained for $a = \frac{D-2}{2}$ (when $N^{-2a} = N^{-(D-2)}$) since then the expectation *does* have other contributions at this order, which can provide a source for the quadratic term $[K^{\mathrm{DS}}(x)]^2$.

A reasonable ansatz is thus to start with (12.72) with $a = (D-2)/2$, completed up to the order $1/N^{D-2}$ to which we want to solve the DSE. We introduce the function $K^{\mathrm{DS,NLO}}(N, x, \{t_i\})$ which contains all the orders between $1/N^{\frac{D-2}{2}}$ and $1/N^{D-2}$ and write

$$
\begin{aligned}
K\Big(\mathcal{B}_2; N, x, \{t_i\}\Big) &= T(z_c, \{t_i\}) + \frac{K^{\mathrm{DS}}(x, \{t_i\})}{N^{\frac{D-2}{2}}} + K^{\mathrm{DS,NLO}}(N, x, \{t_i\}) \\
&+ \frac{K^{\mathrm{DS},(D-2)}(x, \{t_i\})}{N^{D-2}} + \mathcal{O}(N^{-D+1}),
\end{aligned}
\tag{12.74}
$$

and we expect $K^{\mathrm{DS}}(x)$ to be determined by the DSE (12.11) at order $1/N^{D-2}$.

In order to determine the fixed coupling x as a function of $z_c - z$ and N, the ansatz (12.74) must be compared with the expansion of the two-point function to NLO, in the vicinity of the critical coupling z_c:

$$
K\Big(\mathcal{B}_2; N, x, \{t_i\}\Big) \underset{z\sim z_c}{=} T(z, \{t_i\}) + \frac{1}{N^{D-2}\sqrt{1 - z/z_c}} (\cdots) + \mathcal{O}(N^{-D+1}),
$$

where the ellipses represent the currently irrelevant contributions to the function K_2^{NLO} given in (12.57) which take a finite value when $z \to z_c$. The singular factor $1/C$ of K_2^{NLO}, which behaves as $1/\sqrt{1 - z/z_c}$, has been explicitly factorized. While $1/N^{D-2}$ goes to zero as $N \to \infty$, the singularity of $1/C$ as z approaches its critical value makes it possible to find a double-scaling limit. Comparing with equation (12.74), it appears that the only way to reach $a = (D-2)/2$ while $z \to z_c$ is to choose:

$$
z = z_c - \frac{x}{N^{D-2}},
\tag{12.75}
$$

where x is held fixed in the double-scaling limit $z \to z_c$, $N \to \infty$.

Substituting the ansatz (12.74) into the expansion (12.47) of the expectation of $\tilde{\rho}_{\mathcal{B}_i}$ leads to:

$$
\begin{aligned}
\frac{1}{N} \langle \tilde{\rho}_{\mathcal{B}_i}(\mathbb{T}, \overline{\mathbb{T}}) \rangle &= \Big(1 + \frac{\alpha_i}{N^{D-2}}\Big) \Bigg[T(z_c, \{t_i\}) + \frac{K^{\mathrm{DS}}(x, \{t_i\})}{N^{\frac{D-2}{2}}} \\
&\quad + K^{\mathrm{DS,NLO}}(N, x, \{t_i\}) + \frac{K^{\mathrm{DS},(D-2)}(x, \{t_i\})}{N^{D-2}} \Bigg]^{p_i} \\
&\quad + \frac{1}{N^{D-2}} \big[T(z_c, \{t_i\})\big]^{p_i - 2} K_{4,\bullet}^{\mathrm{LO}}\Big|_{z \to z_c} + \cdots
\end{aligned}
\tag{12.76}
$$

Importantly, $K^{\mathrm{LO}}_{4,\bullet}$ is finite at criticality, since it does not have the singular factor $1/C$. Therefore, it can be safely evaluated at z_c (it already comes at order $1/N^{D-2}$ so there is no need to expand around z_c which would generate $1/N$ corrections). At this stage, however, it is not clear that the ellipses in the previous equation contain only terms that can be neglected in the double-scaling limit. It could indeed be the case that corrections to the four-point function, as well as non-Gaussian contributions of higher orders, diverge at criticality, and furthermore, that they come with sufficiently many powers of $1/\sqrt{1-z/z_c}$ to counter-balance their $1/N$ suppression. In this way, those terms could contribute in the double-scaling limit. Solving this issue requires a precise analysis that will be performed in the next section 12.3.2.

For the time being, we simply give the result of the analysis, in the context of the *quartic* model. We need to know the behaviour of the various four-point cumulants in the equation (12.43). In the standard $1/N$ expansion:

$$\begin{aligned}
K\Big(\mathcal{B}_{4,\{c\}};N,z\Big) &= \frac{zT(z)^4}{1-zT(z)^2} + \mathcal{O}\Big(\frac{1}{N^{D-2}}\Big), \\
K\Big(\mathcal{B}_{4,\emptyset};N,z\Big) &= \frac{1}{\sqrt{1-4Dz}}\,\frac{D(D-1)\,z^2\,[T(z)]^5}{1-z\,[T(z)]^2} + \mathcal{O}\Big(\frac{1}{N^{D-2}}\Big), \\
K\Big(\mathcal{B}_{4,\mathcal{C}};N,z\Big)\Big|_{|\mathcal{C}|\geq 2} &= \mathcal{O}\Big(\frac{1}{N^{D-2}}\Big).
\end{aligned} \tag{12.77}$$

Both the first and second lines have been obtained by solving the DSE in the Section 12.2 (the estimates on their rests follow combinatorial arguments). The third line is a scaling argument. In the following section 12.3.3, those results will actually be re-derived on the way to showing that the cumulants in the double-scaling limit become:

$$\begin{aligned}
K\Big(\mathcal{B}_{4,\{c\}};N,x\Big) &= f^{(0)}(\mathcal{B}_{4,\{c\}};N,x) + N^{-\frac{D-2}{2}} f^{(-\frac{D-2}{2})}(\mathcal{B}_{4,\{c\}};N,x), \\
K\Big(\mathcal{B}_{4,\emptyset};N,x\Big) &= N^{\frac{D-2}{2}} f^{(\frac{D-2}{2})}(\mathcal{B}_{4,\emptyset};N,x), \\
K\Big(\mathcal{B}_{4,\mathcal{C}};N,x\Big) &= \mathcal{O}\Big(\frac{1}{N^{D-2}}\Big).
\end{aligned} \tag{12.78}$$

Importantly, the functions $f^{(q)}(\mathcal{B};N,x)$ are *bounded for all* N (have a finite value as $N\to\infty$). Note that the sub-leading corrections to $K\Big(\mathcal{B}_{4,\{c\}};N,x\Big)$ contribute at order $N^{-\frac{D-2}{2}}$, that is sooner than expected in the $1/N$ expansion.

As a result, $K\Big(\mathcal{B}_{4,\emptyset};N,x\Big)$ is enhanced in the double-scaling regime by a factor $N^{\frac{D-2}{2}}$. This can be understood already from the $1/N$ expansion (12.77). Its LO indeed contains the singular factor $1/C(z)=1/\sqrt{1-4Dz}$ which as a function of x and N becomes

$$\frac{1}{C(z(N,x))} = \frac{N^{\frac{D-2}{2}}}{\sqrt{x}}\,.$$

The divergence of the LO of $K\left(\mathcal{B}_{4,\emptyset};N,x\right)$ as z approaches z_c is traded for a large N enhancement.

Therefore the non-Gaussian contributions to the expectation of $\tilde{\rho}_{\mathcal{B}_{4,\{c\}}}$ reads:

$$\begin{aligned}&\frac{1}{N^{D-2}}K\left(\mathcal{B}_{4,\{c\}};N,x\right)+\frac{1}{N^{D}}K\left(\mathcal{B}_{4,\emptyset};N,x\right)+\mathcal{O}(N^{-D-1})\\&=\frac{1}{N^{D-2}}\Big(\underbrace{f^{(0)}(\mathcal{B}_{4,\{c\}};N,x)}_{\mathcal{O}(1)}+\underbrace{\frac{N^{D-2}N^{\frac{D-2}{2}}}{N^{D}}f^{(\frac{D-2}{2})}(\mathcal{B}_{4,\emptyset};N,x)+\mathcal{O}(1/N)}_{\mathcal{O}\left(N^{\frac{D-6}{2}}\right)}\Big).\end{aligned}\tag{12.79}$$

From this analysis, we conclude that in the double-scaling limit, the dominant four-point cumulant in the quartic model is dominated by the LO of:

- $f^{(0)}(\mathcal{B}_{4,\{c\}};N,x)$ when $D<6$,
- both $f^{(0)}(\mathcal{B}_{4,\{c\}};N,x)$ and $f^{(\frac{D-2}{2})}(\mathcal{B}_{4,\emptyset};N,x)$ when $D=6$,
- $f^{(\frac{D-2}{2})}(\mathcal{B}_{4,\emptyset};N,x)$ when $D>6$.

This way, we recover the $D=6$ special case found in (Dartois, Gurau, and Rivasseau 2013; Gurau and Schaeffer 2016). Since the doubly scaled four-point function is not summable for $D\geq 6$, we hereafter focus on the case $D<6$. Moreover, we will argue in Section 12.3.2 that this conclusion holds for any model with melonic bubbles in the action (symmetrized on the colours) and not only the quartic model.

Plugging the expansion (12.74) for the two-point function into the DSE (12.11) gives (some terms contributing beyond $1/N^{D-2}$ have already been neglected):

$$\begin{aligned}&1-\left(T_c+\frac{K^{\mathrm{DS}}(x,\{t_i\})}{N^{\frac{D-2}{2}}}+K^{\mathrm{DS,NLO}}(N,x,\{t_i\})+\frac{K^{\mathrm{DS},(D-2)}(x,\{t_i\})}{N^{D-2}}\right)\\&+\sum_{i\in I}t_i\left(z_c-\frac{x}{N^{D-2}}\right)^{p_i-1}\left(T_c^{p_i}+\frac{1}{N^{\frac{D-2}{2}}}p_iT_c^{p_i-1}K^{\mathrm{DS}}(x,\{t_i\})\right.\\&+p_iT_c^{p_i-1}K^{\mathrm{DS,NLO}}(N,x,\{t_i\})+\frac{1}{N^{D-2}}\Big(\alpha_iT_c^{p_i}+p_iT_c^{p_i-1}K^{\mathrm{DS},(D-2)}(x,\{t_i\})\\&\left.+\frac{p_i(p_i-1)}{2}T_c^{p_i-2}\left[K^{\mathrm{DS}}(x,\{t_i\})\right]^2+\alpha_i'T_c^{p_i-2}K_{4,\bullet}^{\mathrm{LO}}(z_c,\{t_i\})\Big)\right)=0,\end{aligned}\tag{12.80}$$

which we shall evaluate order-by-order in the $1/N$ expansion up to $1/N^{D-2}$. The LO equation is trivially satisfied as the evaluation of $1-T+\sum_i t_iz^{p_i-1}T^{p_i}$ at criticality. At order $N^{-\frac{D-2}{2}}$, we get

$$\left(-1+\sum_{i\in I}t_ip_i[z_cT_c]^{p_i-1}\right)K^{\mathrm{DS}}(x,\{t_i\})=0.\tag{12.81}$$

The quantity into brackets is the function C evaluated at criticality, and so vanishes. Thus, it provides no new information. Then we need to take care of all the terms between the order $1/N^{\frac{D-2}{2}}$ and $1/N^{D-2}$. Because they are *strictly bounded* by $N^{-(D-2)/2}$, the functions $K^{\mathrm{DS,NLO}}$ and $K^{\mathrm{DS},(D-2)}$ only appear *linearly* in the previous expansion (their square would only contribute beyond $1/N^{D-2}$). Therefore, they both come in factor of the first derivative of the LO equation, i.e. the function $C(z,\{t_i\})$, evaluated at criticality which is zero.

Remarkably, we are left at order $1/N^{D-2}$ with an equation on the function K^{DS} only:

$$\left(\sum_{i\in I} t_i \frac{p_i(p_i-1)}{2} z_c^{p_i-1} T_c^{p_i-2}\right) [K^{\mathrm{DS}}(x,\{t_i\})]^2 - x\left(\sum_{i\in I} t_i(p_i-1) z_c^{p_i-2} T_c^{p_i}\right) + \sum_{i\in I} t_i z_c^{p_i-1}\left(\alpha_i T_c^{p_i} + \alpha_i' T_c^{p_i-2} K_{4,\bullet}^{\mathrm{LO}}(z_c,\{t_i\})\right) = 0, \tag{12.82}$$

which directly gives its expression as a function of $x,\{t_i\}$. This is the main result of the article.

The quartic case. Specializing this calculation to the quartic case directly reproduces the doubly scaled two-point function found in (Dartois, Gurau, and Rivasseau 2013):

$$[K^{\mathrm{DS}}(x)]^2 = \left(\frac{x}{z_c} - 1\right) T_c^2 - K_{4,\bullet}^{\mathrm{LO}}(z_c). \tag{12.83}$$

Using $z_c = 1/4D$, we find $T_c = 2$ and $K_{4,\bullet}^{\mathrm{LO}}(z_c) = 4/(D-1)$ and finally:

$$K^{\mathrm{DS}}(x) = 4\sqrt{D}\sqrt{x - \frac{1}{4(D-1)}}\,. \tag{12.84}$$

The proof of the double-scaling limit, i.e. to establish the equation (12.78) is done in Appendix F.

12.3.2 From the quartic model to a generic model

Although the last proof is restricted to the quartic model, its result can be easily extended to a generic model whose action has bubbles $\{\mathcal{B}_i\}_{i\in I}$ that are melonic (and symmetrized on their colourings). This is actually expected from universality (in the sense used in statistical mechanics). Indeed, universality means that changing the details of a model (here the building blocks of the graphs) does not change its critical properties.

Let us see to what extent the quartic model and its generic extension coincide:

Graph structure The first thing we shall show is that all the graphs of a generic model can be mapped to graphs of the quartic model. The apparent difference between those two types of graph is that they are built from different bubbles (subgraphs with colours $1, \ldots, D$). However, any melonic bubble can be obtained as the boundary graph of a gluing of quartic bubbles via propagators.

The main idea behind this fact is that, given a bubble $\mathcal{B}$, it is easy to add a $(D-1)$ dipole onto any of its edges, using the propagator and an appropriate quartic bubble. Indeed, say we want to add a $(D-1)$ dipole onto the edge of colour c that is incident to the white vertex v in $\mathcal{B}$. To do so, a propagator is introduced to connect v to a black vertex $\overline{v}$ of $\mathcal{B}_{4,\{c\}}$. The boundary graph is obtained by cutting out v and $\overline{v}$ as well as the propagator, and reconnecting then the open edges of the same colours,

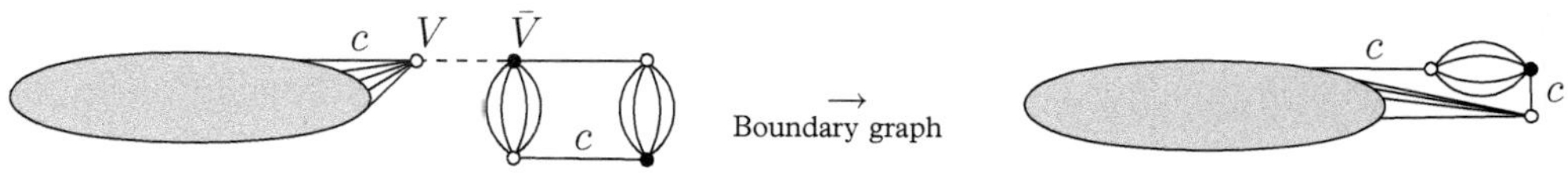

Note that this is the same as removing v and gluing instead the right hand side of (12.16). Importantly, this map can be interpreted the other way around: each $(D-1)$-dipole of external colour c can be replaced by a regular edge of colour c and a propagator connecting the white vertex to the quartic bubble $\mathcal{B}_{4,\{c\}}$.

Since any melonic bubble on $p+1$ black vertices can be obtained through the insertion of a $(D-1)$-dipole on a melonic bubble on p black vertices, a simple inductive argument proves the claim: any melonic bubble with p black vertices is thus represented by a gluing of $p-1$ quartic bubbles. This way the Feynman graphs of an arbitrary model are realized as a sub-family of graphs within the quartic model. (Obviously, the quartic model generates arbitrary melonic bubbles as boundary graphs.)

Scaling in N If $\mathcal{G}$ is a Feynman graph in the generic model, we denote a representative in the quartic model by $Q(\mathcal{G})$. Then, $\mathcal{G}$ and $Q(\mathcal{G})$ turn out to have the same scaling. This stems from the fact that the weight of a melonic bubble represented as a gluing of quartic ones is the same as if the bubble were in the action. Indeed, in the latter case, each bubble $\mathcal{B}_i$, $i \in I$, brings a factor N^{D-1}. On the other hand, we need $p_i - 1$ quartic bubbles to represent it, each of them also coming up with a factor N^{D-1}. There are however $p_i - 2$ propagators, each of them with the weight $N^{-(D-1)}$, which eventually leaves us with an overall N^{D-1}.

z–dependence Furthermore, the amplitudes of $\mathcal{G}$ and $Q(\mathcal{G})$ have the same functional dependence on the coupling z. A bubble $\mathcal{B}_i$, $i \in I$, among those of the action comes with a weight z^{p_i-1}. Since $p_i - 1$ quartic bubbles are required to represented it, each of them bringing a single z, we get z^{p_i-1} as expected for the representant of $\mathcal{B}_i$.

Combinatorial factors The only departure between the two families $\{\mathcal{G}\}$ and $\{Q(\mathcal{G})\}$ are combinatorial factors that weight the graphs when they are summed as contributions to some expectation. Those combinatorial factors come from the Feynman expansion (that is, from the expansion of the exponential in the original integral and the counting the number of times a given graph is obtained). But this is where universality intervenes. Such a microscopic change coming from the use of different building blocks (in other words, the bubbles $\mathcal{B}_i$ are given an internal structure made up of quartic bubbles) may affect the value of the critical coupling but do not change the scaling exponents. Therefore, a family of Feynman graphs of the form $\{Q(\mathcal{G})\}$ which is irrelevant (i.e. suppressed as some powers of $1/N$) in the double-scaling limit of the quartic model cannot be a relevant family $\{\mathcal{G}\}$ in the generic model.

Let us end this chapter by resuming that we have used the DSE to both re-derive the double-scaling limit of the quartic model and also, with a succinct universality argument, to extend it to models with generic melonic interaction bubbles. More precisely, we have obtained the doubly scaled two-point function in the equation (12.82). In addition to this result, we have used the following strategy which combine the DSE with combinatorial scale arguments:

- expand the expectations onto sums of products of cumulants,
- study the (simple- or double-) scaling of the cumulants to identify the relevant boundary graphs at each order,
- solve a linear (in the regular $1/N$ expansion case) or quadratic (in the double-scaling case, at least for the two-point function) system of DSE on those relevant cumulants.

We have thus seen at play the hierarchy of cumulants beyond the LO. The latter is indeed Gaussian, then at NLO, one observes four-point cumulants and so on. Remarkably, the double-scaling of the two-point function only requires to go up to four-point cumulants, and this is why only two DSE were needed instead of the full tower. Finally, let us mention here that the doubly-scaled series is summable for D lower than six.

12.4 Selected further reading

- Razvan Gurau. (2017). *Random Tensors*. Oxford: Oxford University Press.
  A very good book treating in detail the $U(N)^D$-invariant tensor model: perturbative expansion and the large N limit, DSE, the double-scaling limit and so on.

13

Random tensor models—the multi-orientable (MO) model

This chapter presents the multi-orientable (MO) tensor model, and it follows the review article Tanasa (2016).

This rank three model, having $O(N) \times U(N) \times O(N)$ symmetry, can be seen as an intermediate step between the $U(N)^3$-invariant model presented in Chapter 12, and the $O(N)^3$-invariant model presented in Chapter 14.

The class of Feynman graph generated by perturbative expansion of MO model is strictly larger than the class of Feynman graphs of the $U(N)^3$-invariant model and strictly smaller than the one of the $O(N)^3$-invariant model.

This chapter is organized as follows. In the following first section, we present the definition of the MO model. The 1/N expansion and the large N limit of the model are exposed in the second section. In the third section, a thorough enumerative combinatorial analysis of the general term of the 1/N expansion is presented. The implementation of the double-scaling mechanism is then exhibited in the fourth section.

13.1 Definition of the model

In this section, we give the definition of the MO random tensor model. This follows Dartois, Rivasseau, and Tanasa (2014).

Let $\phi_{ijk}, i,j,k = 1,\ldots,N$, be the components of a three-index complex tensor ϕ. The action of the MO tensor model is written:

$$S[\phi] = \sum_{ijk} \phi_{ijk}\bar{\phi}_{ijk} - \frac{\lambda}{2} \sum_{\substack{ijk \\ lmn}} \phi_{ijk}\bar{\phi}_{mlk}\phi_{mjn}\bar{\phi}_{iln}. \tag{13.1}$$

Note that the $1/2$ factor multiplying the coupling constant λ takes into consideration the symmetries of the vertex.

One can check that this tensor model has an

$$O(N) \times U(N) \times O(N) \tag{13.2}$$

symmetry (see Dartois, Rivasseau, and Tanasa (2014) for details).

Combinatorial Physics: Combinatorics, quantum field theory, and quantum gravity models. Adrian Tanasa, Oxford University Press (2021). © Adrian Tanasa. DOI: 10.1093/oso/9780192895493.003.0013

The partition function of the model

$$Z = \int \mathcal{D}[\phi] e^{-S[\phi]} \,, \ \mathcal{D}[\phi] = \prod_{ijk} \frac{\mathrm{d}\phi_{ijk}\, \mathrm{d}\bar{\phi}_{ijk}}{2\pi\imath} \,, \tag{13.3}$$

writes in perturbation theory as a sum over Feynman MO tensor graphs. From a combinatorial point of view, the partition function is a generating function of graphs to whom one associates a certain weight (which is actually the following Feynman amplitude).

The vertices of the MO graphs are four valents, the edges are oriented from ϕ to $\bar{\phi}$ and the orientations alternate around a vertex. The edges can be represented as three parallel strands (one for each index of the tensor) and the vertices as the intersection of four half edges such that every pair of half edges share a strand. This orientation can also be seen in the following way. One labels the four corners of the vertex with alternating signs '+' and '-', see Fig. 13.1. An edge of the graph then connects a '+' to a '-' sign. In Fig. 13.1, an example of a vacuum MO tensor graph is given.

One can then prove that the MO action (13.1) leads to a set of Feynman graphs which is strictly larger than the set of Feynman graphs of the coloured model of section 10.3 (for more details, see Tanasa (2012a)), or the $U(N)^3$-invariant model of Chapter 12. Let us recall here that coloured model allow only for Feynman graphs where each edge has a colour, usually labelled in 3D, from 0 to 3. At each vertex, one has a half-edge of each of these four colours—the graph is bipartite. Moreover, a face of the graph is formed by edges of exactly two colours. The MO model allows for all these Feynman graphs. Moreover, its perturbative expansion allows as well for the following supplementary classes of Feynman graphs:

1. MO graphs which are not edge colourable (with at most four colours) but are not allowed by the coloured model action,
2. MO graphs which are edge colourable (with at most four colours)but are not allowed by the coloured model action, and
3. MO graphs which are not bipartite.

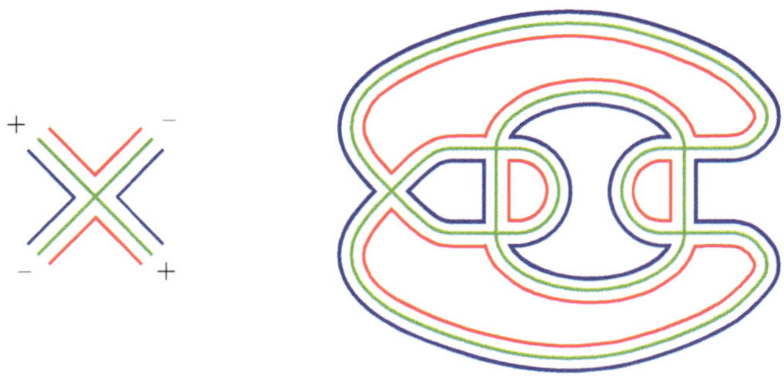

Figure 13.1 *The MO vertex and an MO vacuum Feynman graph*

Note that edge colourability implies bipartiteness—this is a standard result of graph theory. Examples of such MO tensor graphs are given in Fig. 13.2. The left one has a double-tadpole graph, which is an MO graph which is not edge colourable. The right one has a graph which is edge colourable, using four colours, $0,\ldots,3$, but does this not occur in colourable models.

The strands are divided into three classes: the ones in the middle of the edges, called the S (for straight) strands, the ones on the right (with respect to the $\phi \to \bar{\phi}$ orientation) of the edges, called the R (for right) strands and the ones on the left of the edges, called the L (for left) strands. At a vertex the S (R or L) strands only connect to S (R or L) strands.

Note that MO tensor graphs are in one-to-one correspondence with maps. Thus, one can use strandless graphical representations of MO tensor graphs (such as the one given in Section 13.3).

To any MO graph one can associate three canonical ribbon graphs, called the **jackets**, obtained by erasing throughout the graph all the strands in the same class (S, L, or R). For the graph of Fig. 13.1, this leads to the three jackets represented in Fig. 13.3.

The fact that this algorithm always leads, for an MO graph, to a ribbon graph was proved in Prop. 4.1 of Dartois, Rivasseau, and Tanasa (2014). Let us also emphasize that this does not hold for general non-MO graphs (see Fig. 13.4 for such a counterexample).

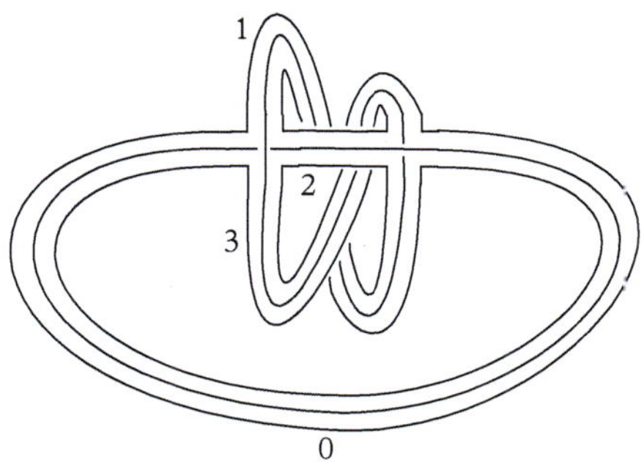

Figure 13.2 *Examples of MO tensor graphs which do not occur in the colourable framework*

Figure 13.3 *The three jackets associated to the graph of Fig. 13.1*

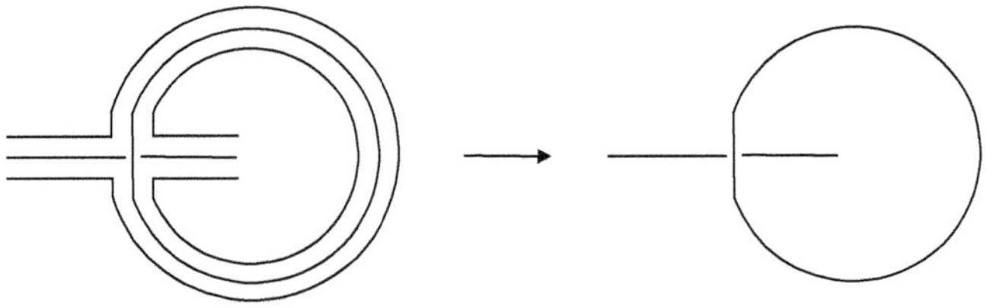

Figure 13.4 *Deletion a pair of opposite corner strands in a non-MO graph*

We further call the **canonical genus** of an MO graph the genus of the jacket obtained by erasing the straight faces (we also call this particular jacket the **canonical jacket**). Thus, a planar MO tensor graph is an MO graph with a vanishing canonical genus.

Let us emphasize here that the jackets of the MO tensor graph can be non-orientable. This is a fundamental difference between the MO model and the coloured-like models.

Definition 13.1.1 *The* **degree** $\delta(\mathcal{G})$ *of a connected MO graph* $\mathcal{G}$ *is the half sum of the non orientable genera of its three jackets:*

$$\delta(\mathcal{G}) = \frac{1}{2}\left(k_{\mathcal{J}_1} + k_{\mathcal{J}_2} + k_{\mathcal{J}_3}\right), \tag{13.4}$$

where the parameters $k_{\mathcal{J}_i}$, $(i = 1, 2, 3)$ *are the non-orientable genera of the three jackets of the graph.*

As a direct consequence of this definition, one concludes that the degree is a positive half integer.

13.2 The $1/N$ expansion and the large N limit

In this section we implement the $1/N$ expansion, and we study the large N limit of the model. In the first two subsections, we analyse the expression of the Feynman amplitudes and the leading order of the $1/N$ expansion. This follows the original article Dartois, Rivasseau, and Tanasa (2014). In the third subsection, we analyse the NLO of the $1/N$ expansion. This follows the original article Raasakka and Tanasa (2015). In the last subsection, we study the leading and next-to-leading order series of the model. This follows again the original article Raasakka and Tanasa (2015).

13.2.1 Feynman amplitudes; the $1/N$ expansion

Let us now organize the free energy series according to powers of N. Since each face corresponds to a closed cycle of Kronecker δ functions, each face contributes with a factor N. The Feynman amplitude of an MO graph is then written

$$A = \lambda^V (k_N)^{-V} N^F, \tag{13.5}$$

where V is the number of vertices of the graph, F is the number of faces of the graph and k_N is a rescaling constant. We choose this rescaling k_N to get the same divergence degree for the leading graphs at any order. We first count the faces of a graph using the previously defined jackets $\mathcal{J}$. Using the Euler characteristic formula, one has

$$f_{\mathcal{J}} = e_{\mathcal{J}} - v_{\mathcal{J}} - k_{\mathcal{J}} + 2, \tag{13.6}$$

where $k_{\mathcal{J}}$ is the non-orientable genus of the jacket $\mathcal{J}$ (see previously illustrated). Since each jacket is a connected vacuum ribbon graph, one has: $e_{\mathcal{J}} = 2v_{\mathcal{J}}$. Let us recall here that the numbers of vertices (in respect of edges) of a jacket $\mathcal{J}$ of a graph is the same than the numbers of vertices (in respect of edges) of the graph. Since each graph has three jackets and each face of a graph occurs in two jackets, summing (13.6) over all the jackets of a graph leads to

$$F = \frac{3}{2}V + 3 - \sum_{\mathcal{J}} \frac{k_{\mathcal{J}}}{2} = \frac{3}{2}V + 3 - \delta. \tag{13.7}$$

The amplitude rewrites as

$$A = \lambda^V (k_N)^{-V} N^{\frac{3}{2}V+3-\delta}. \tag{13.8}$$

To get the same divergence degree for the leading graphs at any order, we choose the scaling constant k_N to be equal to $N^{\frac{3}{2}}$. The amplitude finally can be written as

$$A(\mathcal{G}) = \lambda^V N^{3-\delta}. \tag{13.9}$$

Note that, as in the case of random matrices, random tensor models (such as the MO model described here), have a purely combinatorial expression. This comes from the fact that the propagator in the action (13.1) is, from a QFT point of view, trivial: no Laplacian operator is present in the quadratic part. As already mentioned, adding an appropriate Laplacian operator to the action allows for a more involved expression of the associated Feynman amplitudes and for renormalizability studies (see, for example, the thesis Carrozza (2013)).

Using now the expression (13.9) for the Feynman amplitude we can rewrite the free energy E as a formal series in $1/N$:

$$E = \sum_{\delta \in \mathbb{N}/2} C^{[\delta]}(\lambda) N^{3-\delta}, \tag{13.10}$$

where

$$C^{[\delta]}(\lambda) = \sum_{\mathcal{G}, \delta(\mathcal{G})=\delta} \frac{1}{s(\mathcal{G})} \lambda^{v_{\mathcal{G}}}. \tag{13.11}$$

13.2.2 The large N limit—the LO (melonic graphs)

Having implemented the $1/N$ expansion, one easily sees that the LO in the large N limit (i.e. the limit $N \to \infty$) is given by the tensor graphs satisfying the condition:

$$\delta = 0. \tag{13.12}$$

In this subsection we identify these tensor graphs, and we show that they are the so-called **melonic graphs**. These graphs are obtained by insertions of a fundamental two-point melonic subgraphs in the so-called *elementary melon* (see Fig. 13.5). Another example of a graph obtained in this way is given in Fig. 13.6 (an example where we have used the stranded representation of graph).

Note that one can prove the following crucial fact: melonic insertions preserve the degree. This can be checked by carefully counting the number of faces and vertices and by using the expression (13.7) for the degree.

One can first prove that condition (13.12) cannot be satisfied if the graph is non-bipartite:

Proposition 13.2.1 *A non-bipartite MO graph has at least one non-orientable jacket and thus its degree satisfies the inequality:* $\delta \geq \frac{1}{2}$.

In order to identity the dominant graphs among the bipartite MO graphs, one can prove the following results:

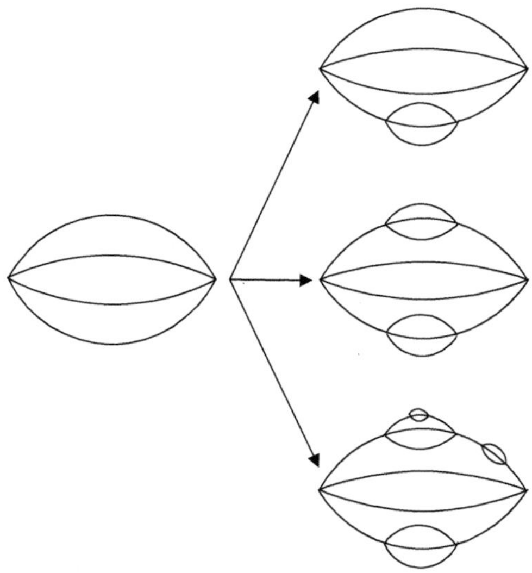

Figure 13.5 *Insertions of melonic subgraphs in the elementary melon graph*

Figure 13.6 *An example of a melonic tensor graph (using the stranded representation)*

Proposition 13.2.2 *If* $\mathcal{G}$ *is a bipartite, vacuum graph of degree zero, then* $\mathcal{G}$ *has a face with two vertices.*

Proposition 13.2.3 *If* $\mathcal{G}$ *is null degree bipartite vacuum graph, then it contains a three-edge coloured subgraph with exactly two vertices.*

Using these two propositions, one can prove that the only graphs satisfying condition (13.12) are the melonic graphs (see Dartois, Rivasseau, and Tanasa (2014) for details). This leads to:

Theorem 13.2.4 *The LO graphs of the* $1/N$ *expansion of the MO model* (13.1) *are the melonic ones.*

Let us end this subsection by stating two important properties of melonic graphs:

1. Melonic graphs maximize the number of faces for a given number of vertices.
2. Melonic graphs correspond to a particular class of triangulations of the three-dimensional sphere S^3.

Let us note that, from a graph theoretical point of view, melonic graphs are a particular type of series-parallel graph. Finally, from a probabilistic point of view, these graphs were proven to correspond to branched polymers (that is, they possess Hausdorff dimension two and spectral dimension $4/3$) Gurau and Ryan (2014).

13.2.3 The large N limit—the NLO

Using again the $1/N$ expansion of Subsection 13.2.1, one easily sees that the NLO in the large N limit is given by the tensor graphs satisfying the condition:

$$\delta = \frac{1}{2}. \tag{13.13}$$

In this subsection we identify these tensor graphs and we prove that they are the so-called **infinity graphs**. These graphs are obtained by melonic insertions in the double-tadpole graph of Fig. 13.2.

One can then prove (see Raasakka and Tanasa (2015); or Fusy and Tanasa (2014) for details) the following result:

Theorem 13.2.5 *The NLO vacuum graphs of the* $1/N$ *expansion of the MO model* (13.1) *are the double tadpole graphs.*

13.2.4 Leading and NLO series

Following Raasakka and Tanasa (2015), we compute in this subsection the radiuses of convergence and the susceptibility exponents of the leading and NLO series of the MO tensor model.

LO series

The analysis of the LO series of the MO model is identical to the one of the coloured model (since one is interested to the same class of graphs, the melonic graphs, and these graphs have the same Feynman amplitude in the two cases).

We denote by $\lambda_{c,LO}$ the LO critical value of the coupling constant λ. From a mathematical point, $\lambda_{c,LO}$ is the radius of convergence of the LO series in λ.

The LO free energy is obtained by summing over the amplitudes of melonic vacuum graphs. We are interested in the asymptotic behaviour of this LO series around $\lambda_{c,LO}$.

The LO connected two-point function can be written as

$$G_{LO}(\lambda) \sim \text{const.} + \left(1 - \frac{\lambda^2}{\lambda_{c,LO}^2}\right)^{\frac{1}{2}} \tag{13.14}$$

around the critical value $\lambda_{c,LO}$ of the coupling constant in the LO. This is related to the behaviour of the LO free energy, for which we obtain

$$E_{LO}(\lambda) \sim \left(1 - \frac{\lambda^2}{\lambda_{c,LO}^2}\right)^{2-\gamma_{LO}}, \tag{13.15}$$

with the susceptibility exponent (or the critical exponent), being

$$\gamma_{LO} = \frac{1}{2}. \tag{13.16}$$

NLO series

In order to study the behaviour of the NLO series, we need to study the connected NLO two-point function. The graphs contributing to the connected NLO two-point function can be obtained from the NLO vacuum graphs by cutting any one of the internal lines of an NLO vacuum graph. Thus, we can in a straightforward manner import the classification of NLO vacuum graphs obtained in the previous section to the case of connected NLO two-point graphs. We express the NLO two-point function in terms of the LO two-point function through algebraic identities relating the LO and NLO two-point functions. More specifically, any two-point function is of the form: bare propagator multiplied by a specific function. We will denote this function associated to the connected LO two-point function as G_{LO}, the function associated to the connected NLO two-point function as G_{NLO}, and the function associated to the one-particle-irreducible (1PI) NLO two-point function as Σ_{NLO}.

The first identity for the two-point functions illustrated in Fig. 13.7 states that the connected NLO two-point function G_{NLO} (on the left-hand side of the figure) is obtained by gluing connected LO two-point functions G_{LO} on both sides of the 1PI NLO two-point function Σ_{NLO} (on the right-hand side of the figure).

This writes:

$$G_{NLO} = G_{LO}^2 \Sigma_{NLO}. \tag{13.17}$$

The second identity comes from the two different combinatorial ways of obtaining 1PI NLO two-point graphs from connected LO and NLO two-point graphs, and is expressed as:

$$\Sigma_{NLO} = \lambda G_{LO} + 3\lambda^2 G_{LO}^2 G_{NLO}, \tag{13.18}$$

where the combinatorial factor three arises from the three different internal lines of the elementary melon, on which the NLO two-point function can be inserted. The two previously mentioned terms exhaust all contributions to the 1PI NLO two-point function, since any such graph contains only one tadpole due to Theorem 13.2.5, and thus either all melonic subgraphs are inside the tadpole (the first term) or the tadpole is a subgraph of a melonic graph (the second term).

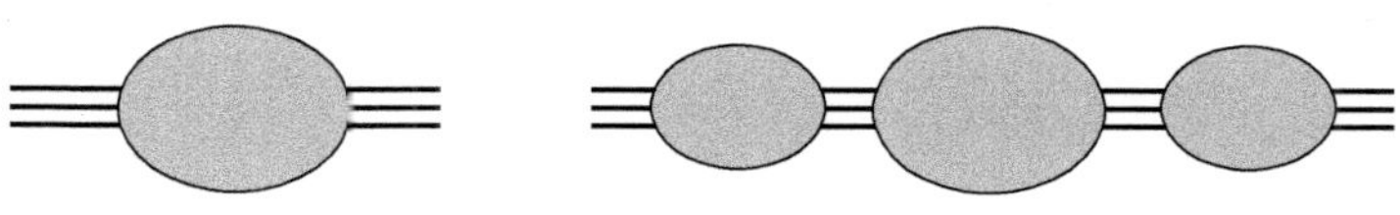

Figure 13.7 *The connected NLO two-point function (left-hand side of the figure) is obtained by gluing connected LO two-point functions on both sides of the 1PI NLO two-point function (right-hand side of the figure)*

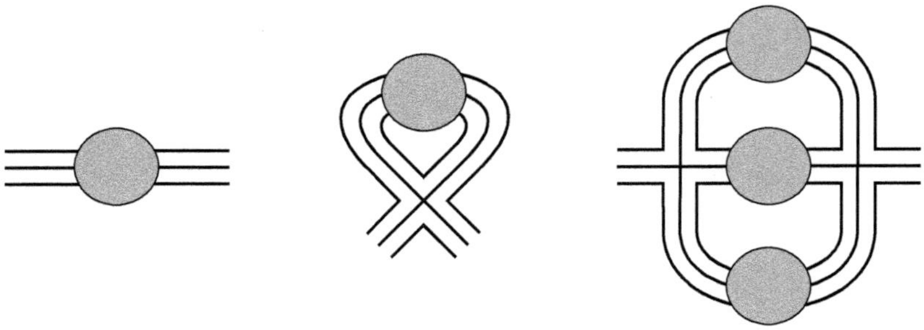

Figure 13.8 *Illustration of* (13.18). *The 1PI NLO two-point function* Σ_{NLO} *is represented on the left-hand side of the figure; the remaining two drawings represent the two terms on the right-hand side of* (13.18)

This is illustrated in Fig. 13.8, where the 1PI NLO two-point function Σ_{NLO} is represented on the left-hand side of the figure. The remaining two drawings represent the two terms on the right-hand side of (13.18).

Now, putting these two identities together, we obtain first the identity

$$G_{LO}^{-2} G_{NLO} = \lambda G_{LO} + 3\lambda^2 G_{LO}^2 G_{NLO}, \tag{13.19}$$

from which we can solve for G_{NLO} in terms of the connected LO two-point function

$$G_{NLO} = \frac{\lambda G_{LO}^3}{1 - 3\lambda^2 G_{LO}^4}. \tag{13.20}$$

On the other hand, differentiating the LO two-point function relation $G_{LO} = 1 + \lambda^2 G_{LO}^4$ we get

$$\frac{\partial}{\partial \lambda^2} G_{LO} = \frac{G_{LO}^4}{1 - 4\lambda^2 G_{LO}^3} = \frac{G_{LO}^5}{1 - 3\lambda^2 G_{LO}^4}, \tag{13.21}$$

where, for the last equality, we used the LO two-point function identity again. Thus, we get the expression

$$G_{NLO} = \frac{\lambda}{G_{LO}^2} \frac{\partial}{\partial \lambda^2} G_{LO}, \tag{13.22}$$

which implies, together with $G_{LO} \sim \text{const.} + (1 - (\lambda^2/\lambda_c^2))^{1/2}$,

$$G_{NLO} \sim \left(1 - \frac{\lambda^2}{\lambda_c^2}\right)^{-1/2}. \tag{13.23}$$

Finally, we use the following DSE:

$$\mathrm{C} = \int \mathrm{d}\bar{\phi}\, \mathrm{d}\phi \, \frac{\delta}{\delta\phi_{ijk}} \left(\phi_{i'j'k'} e^{-S[\phi,\bar{\phi}]} \right) \tag{13.24}$$

We thus obtain the relation

$$G_{NLO} = 1 - 4\lambda^2 \frac{\partial}{\partial\lambda^2} E_{NLO} \tag{13.25}$$

relating the connected two-point function G_{NLO} to the free energy E_{NLO}. Accordingly, we have $E_{NLO} \sim (1-(\lambda/\lambda_c)^2)^{1/2}$ from (13.25), and thus find the same critical value of the coupling constant (i.e. the radius of convergence) for the NLO series (as series in the coupling constant λ) as for the LO series. Nevertheless, one has a distinct value for the NLO susceptibility exponent:

$$\gamma_{NLO} = \frac{3}{2}. \tag{13.26}$$

13.3 Combinatorial analysis of the general term of the large N expansion

In this section we perform a thorough combinatorial analysis of the general term of the $1/N$ expansion of the previous section. We perform their enumeration, and we establish which are the dominant configurations of a given degree. This section follows the original article Fusy and Tanasa (2014).

Note that the '+' and '-' signs canonically induce an orientation of the edges. One can thus represent MO tensor graphs as a particular class of *oriented four regular maps*—see Fig. 13.9 for the representation of the vertex of the MO tensor graph of Fig. 13.1 in such a way. Moreover, we work in this section which rooted maps, i.e. a connected map with a marked edge. This allows us to better handle symmetries issues when counting such combinatorial objects. In physics language marking an edge transforms a vacuum graph into a two-point Feynman graph. One can further see this root as a fake vertex of valence two placed on some root edge.

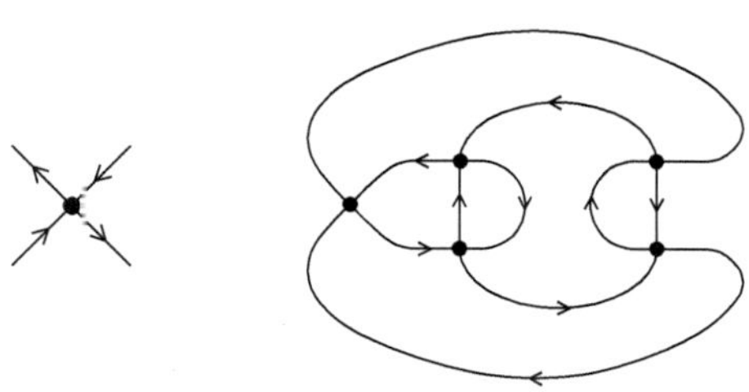

Figure 13.9 *Representation of MO tensor graphs as oriented four regular maps*

Figure 13.10 *Melon removal*

Figure 13.11 *Removing a tadpole in an MO graph*

We then define the **cycle graph** as an oriented self-loop carrying no vertex. The cycle graph is connected, has $V = 0$, $F = 3$ (one face in each type), hence has degree 0. In its rooted version, the (rooted) *cycle graph* is made of an oriented tadpole incident to the root vertex.

In this context, define the *removal* of a melon as the operation represented in Fig. 13.10 (where possibly $u = v$, and possibly u or v might be the root if the MO graph is rooted):

The reverse operation (where e is allowed to be a tadpole and is allowed to be incident to the root vertex if the MO graph is rooted) is called the *insertion* of a melon at an edge.

Analogously, one can define a tadpole (or loop, in graph theoretical language) removal—see Fig. 13.11.

One then has:

Lemma 13.3.1 *Let $\mathcal{G}$ be an MO graph of degree δ, let ℓ be a tadpole of $\mathcal{G}$, and let $\mathcal{G}'$ be the MO graph obtained from $\mathcal{G}$ by erasing the tadpole and its incident vertex, as shown in Fig. 13.11. Then $\mathcal{G}'$ has degree $\delta - 1/2$. Hence, $\mathcal{G}$ has at most 2δ tadpoles.*

Proof *The resulting graph $\mathcal{G}'$ has the same number of connected components as $\mathcal{G}$, has one vertex less, and has one face less (which has length one).*

A melonic graph (previously discussed) can then be seen as an MO graph which can be reduced to the rooted cycle graph by successive removal of melons.

Getting back to melonic insertions (previously illustrated), one needs to stress that *there is an infinite number of melon-free graphs of a given degree.* Nevertheless, as we will explain the sequel, some particular types of subgraphs can be repeated without increasing the degree!

13.3.1 Dipoles, chains, schemes, and all that

Let us now give the following definition:

Definition 13.3.2 *A* **dipole** *is a subgraph of a (possibly rooted) MO graph formed by a couple of vertices connected by two parallel edges containing a face of length two incident to these distinct vertices, and not passing by the root if the graph is rooted.*

Accordingly, one has three distinct types of dipoles: L, R, or S, see Fig. 13.13 (left part). As the figure shows, each dipole has two exterior half-edges on one side and two exterior half-edges on the other side. Notice also that a double edge does not necessarily delimit a dipole, as shown in Fig. 13.13 (right part).

Let us also notice that a face of length two is always incident to two distinct vertices, except in the MO graphs that are made of one vertex and two tadpoles—the infinity graphs. They have no dipole but have two faces of length two.

Let us now give the following definitions:

Definition 13.3.3 *In an MO graph $\mathcal{G}$, a* **chain** *is a sequence of dipoles $d_1 \ldots, d_p$ (not passing by the root if the graph is rooted) such that for each $1 \leq i < p$, d_i and d_{i+1} are connected by two edges involving two half-edges on the same side of d_i (left or right) and two half-edges on the same side (left or right) of d_{i+1}.*

For an example of a chain and of a sequence of dipoles which is not a chain, see Fig. 13.14.

Let us now give the following definitions:

Definition 13.3.4 *An* **unbroken chain** *is a chain for whom all the dipoles are of the same type (L, R, or S).*

Definition 13.3.5 *A* **broken chain** *is a chain which is not unbroken.*

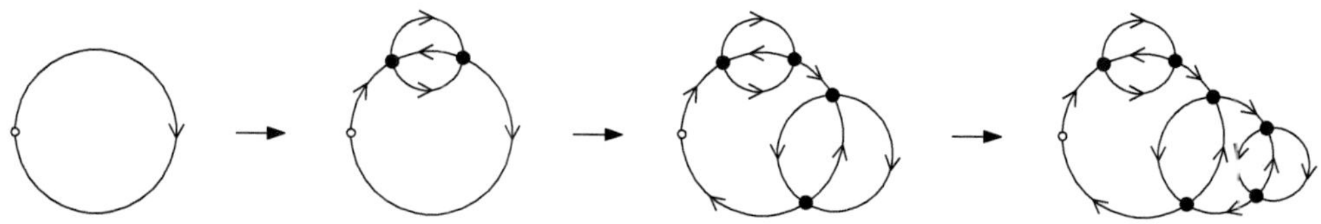

Figure 13.12 *A rooted melonic graph can be built from the cycle graph, by melonic insertions*

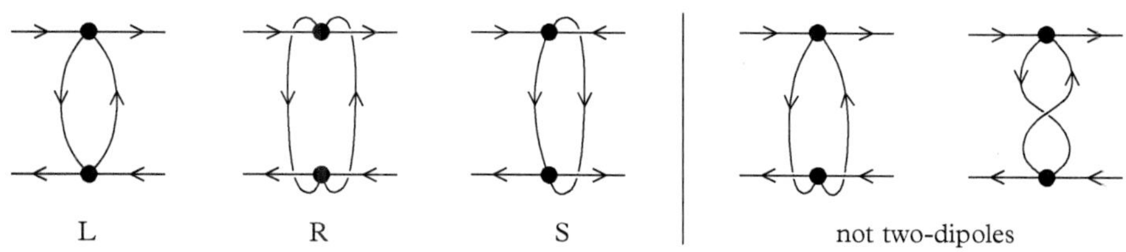

Figure 13.13 *Left: the three types of dipole. Right: examples of double edges that do not form a dipole*

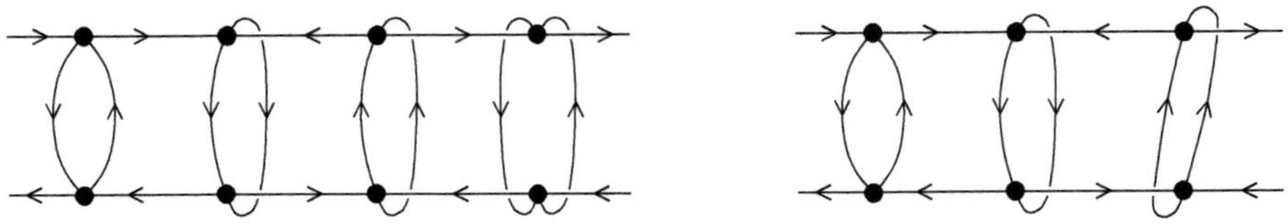

Figure 13.14 *Left: a chain of four dipoles. Right: a sequence of three dipoles that does not form a chain*

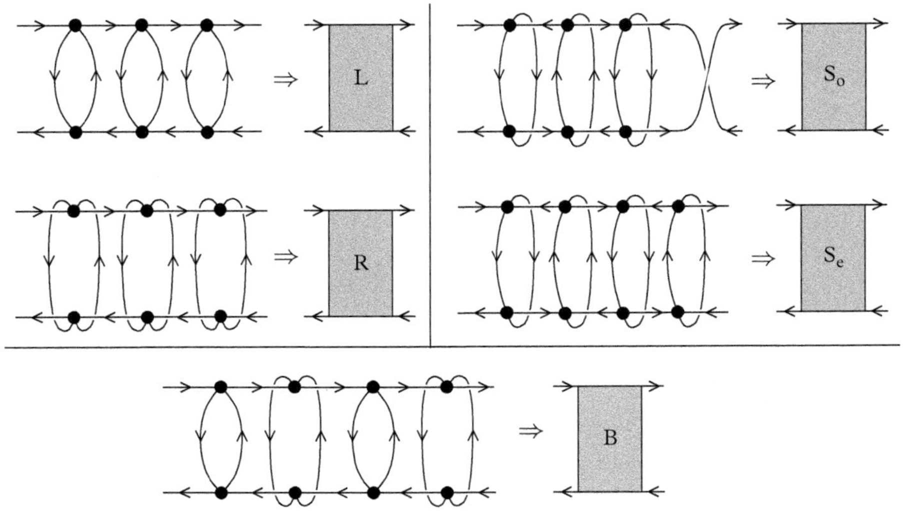

Figure 13.15 *Examples of the types of unbroken proper chains; example of a broken proper chain*

Definition 13.3.6 *A **proper chain** is a chain of at least two dipoles.*

Definition 13.3.7 *A **maximal proper chain** is a proper chain for whom there is no larger chain in the given map that the respective maximal proper chain is part of.*

In a graph, one can then replace a proper chain by a **chain vertex**. In Fig. 13.15, we give examples of the four possible types of unbroken proper chains, replaced by corresponding chain vertices, marked with an appropriate graphical symbol. Note that one has two distinct cases, if the number of dipoles of type S is even or odd. The bottom part of the figure gives an example of a broken proper chain replaced by the corresponding chain vertex.

The corresponding configurations of strands are shown in Fig. 13.16. Note that these are all the possible strand configurations of the MO setting.

We can now give the following definition:

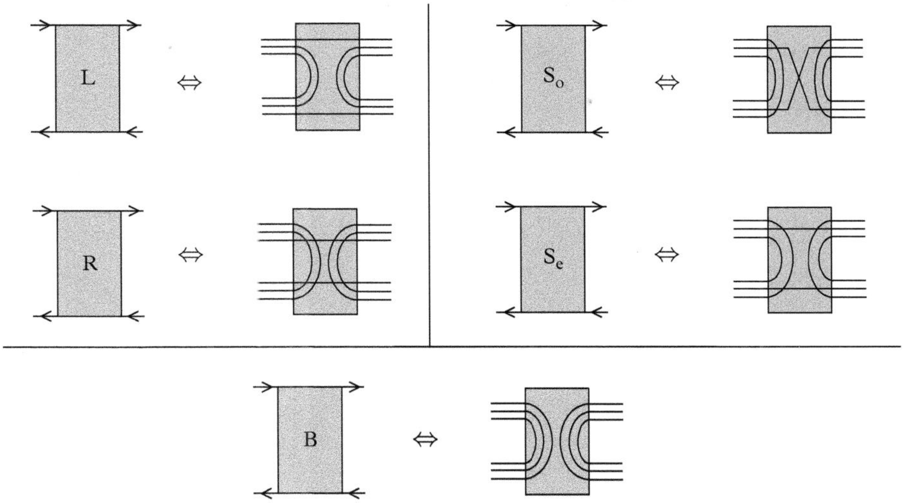

Figure 13.16 *The configurations of the strands for each type of chain vertex*

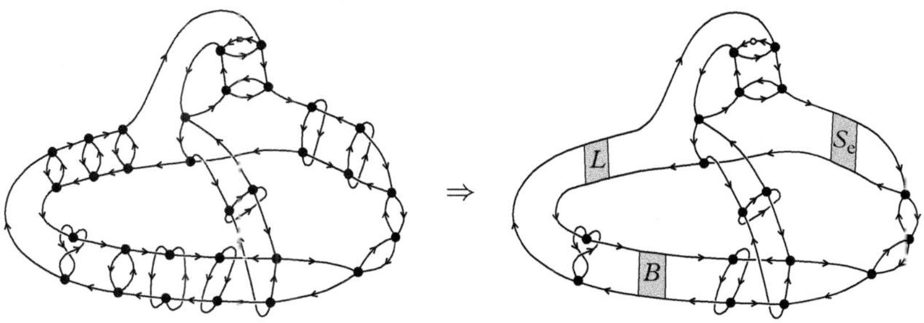

Figure 13.17 *Left: a rooted melon-free MO graph. Right: the associated scheme*

Definition 13.3.8 *Let $\mathcal{G}$ be a rooted melon-free MO graph. The* **scheme** *of $\mathcal{G}$ is the graph obtained by simultaneously replacing any maximal proper chain of $\mathcal{G}$ by the corresponding chain vertex.*

One can see how such a scheme is obtained from a melon-free graph from the example of Fig. 13.17.

Finally, one can give the following definition:

Definition 13.3.9 *A* ***reduced scheme*** *is a rooted melon-free MO graph with chain vertices and with no proper chain.*

Note that, by construction, the scheme of a rooted melon-free MO graph (with no chain vertices) is a reduced scheme.

The following result ensures that the degree definition for MO graphs with chain vertices is consistent with the replacement of chains by chain vertices:

Proposition 13.3.10 *Let $\mathcal{G}$ be an MO graph with chain vertices and let $\mathcal{G}'$ be an MO graph with chain vertices obtained from $\mathcal{G}$ by consistently substituting a chain-vertex by a chain of dipoles. Then the degrees of $\mathcal{G}$ and $\mathcal{G}'$ are equal.*

One can then proof the following finiteness result:

Theorem 13.3.11 *The set of reduced schemes of a given degree δ is finite.*

The proof (see Fusy and Tanasa (2014) for details) relies on the following lemmas:

Lemma 13.3.12 *For each reduced scheme of degree δ, the sum $N(G)$ of the numbers of dipoles and chain vertices satisfies the bound: $N(G) \leq 7\delta - 1$.*

Lemma 13.3.13 *For $k \geq 1$ and a given degree δ, there is a constant $n_{k,\delta}$ such that any connected un-rooted MO graph (without chain vertices) of degree δ with at most k dipoles has at most $n_{k,\delta}$ vertices.*

In order to prove Theorem 13.3.11 using these two lemmas, one can explicitly construct an injective map $S \mapsto G(S)$ which associates to a reduced scheme S a rooted MO graph with no chain vertices. This construction can be done using the choice of canonical substitution of chain vertices by proper chains given in Fig. 13.18. This substitution preserves the strand structure (hence the degree) and yields an injective mapping from MO graphs with chain-vertices to MO graphs without chain vertices.

One then needs to analyse how the degree of an MO graph with chain vertices (not necessarily a reduced scheme, possibly with melons) evolves when removing

1. a chain vertex; and
2. a melon.

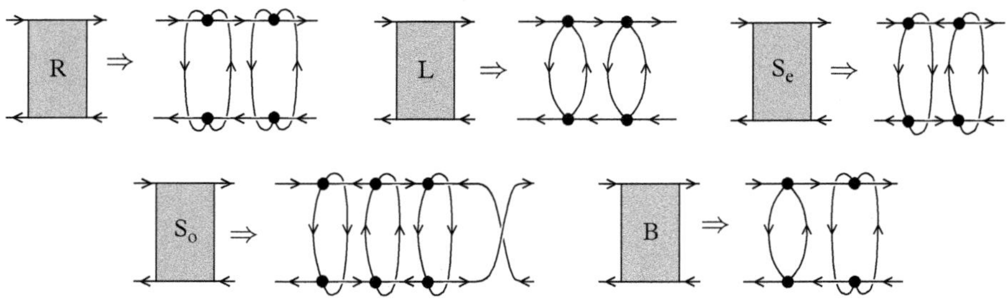

Figure 13.18 *A choice of canonical substitution leading to the map $S \mapsto G(S)$*

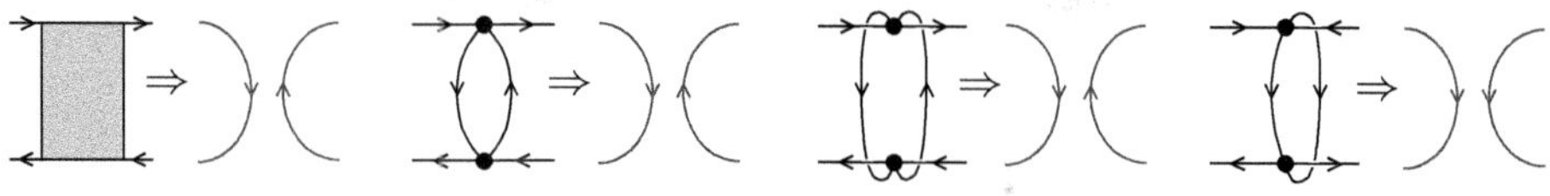

Figure 13.19 *Removal of a chain vertex (whatever its type) and of type L,R, and S*

Let $\mathcal{G}$ be an MO graph with chain vertices, and let m be a chain vertex of $\mathcal{G}$. The *removal* of m consists of the following operations:

(i) delete m from $\mathcal{G}$; and

(ii) on each side of m, connect together the two detached legs (without creating a new vertex).

Such a removal of a chain vertex is represented in the left part of Fig. 13.19. For the removal of a dipole (of types L, R, and S), see the right part of Fig. 13.19.

The chain vertex m is said to be

1. **non-separating** if $\mathcal{G}'$ has the same number of connected components as $\mathcal{G}$; and
2. **separating** otherwise (in which case $\mathcal{G}'$ has one more connected component).

By carefully counting the modifications made to the number of faces, vertices and connected components, one can prove the following lemma (see again Fusy and Tanasa (2014) for details):

Lemma 13.3.14

- *The degree is unchanged when removing a separating chain vertex or a separating dipole (the degree is distributed among the resulting components).*
- *The degree decreases by three when removing a non-separating broken chain vertex.*
- *The degree decreases by one or three when removing a non-separating unbroken chain vertex or a non-separating dipole.*

In order to prove Lemma 13.3.12, one needs to analyse the process of iterative removal of dipoles and chain vertices (first the non-separating ones and then the separating ones)—see again Fusy and Tanasa (2014) for details. One can show that this process leads to a decorated tree of components.

An illustration of this process transforming a scheme into some decorated tree is given in Fig. 13.20. The first drawing shows a reduced scheme $\mathcal{G}$. Then, iteratively, one removes at each step a non-separating dipole or a non-separating chain vertex (at each step the non-separating dipole or chain vertex to be removed next is surrounded). Let $\mathcal{G}'$ be the MO graph with chain vertices thus obtained (where coloured edges are drawn bolder). As the last two drawings show, the removal of uncoloured dipoles and chain vertices (which are all separating) of $\mathcal{G}'$ yields a tree of components (a tree edge is labelled c_x if it comes from a chain vertex of type x and is labelled d_x if it comes from an uncoloured dipole of type x).

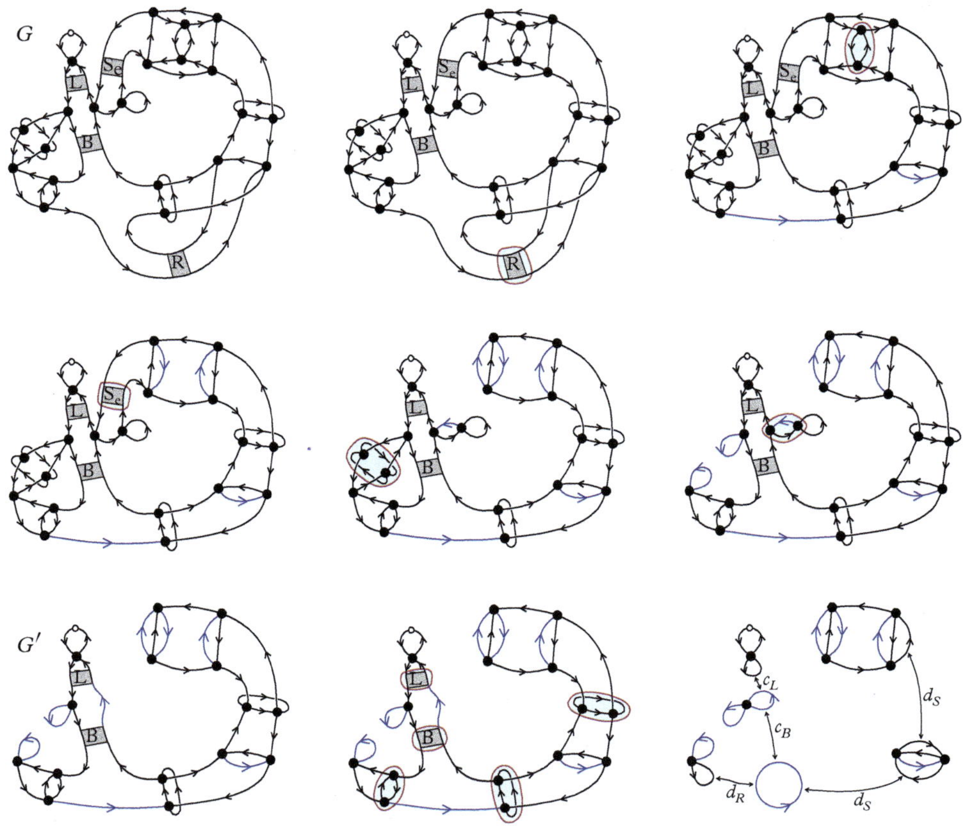

Figure 13.20 *An illustration of the process transforming a scheme into some decorated tree of components*

13.3.2 Generating functions, asymptotic enumeration, and dominant schemes

Let us start this subsection by recalling that the generating function of rooted melonic graphs is written as:

$$T(\lambda^2) = 1 + \lambda^2 (T(\lambda^2))^4 \tag{13.27}$$

(see Fig. 13.5).

Let now $\mathcal{S}_\delta$ be the (finite) set of reduced schemes of degree δ. For each $S \in \mathcal{S}_\delta$, let $G_S^{(\delta)}(u)$ be the generating function of rooted melon-free MO graphs of reduced scheme S.

Let p be half the number of non-root standard vertices of S, b the number of broken chain vertices, a the number of unbroken chain vertices of type L or R, s_e the number of even straight-chain vertices, and s_o the number of odd straight-chain vertices. The generating functions for

- unbroken chains of type L (resp. R) is $(\lambda^2)^2/(1-\lambda^2)$;
- the one for even straight chains is $(\lambda^2)^2/(1-(\lambda^2)^2)$;
- the one for odd straight chains is $(\lambda^2)^3/(1-(\lambda^2)^2)$;
- and the one for broken chains is $(3\lambda^2)^2/(1-3(\lambda^2))-3(\lambda^2)^2/(1-\lambda^2)=6(\lambda^2)^2/((1-3\lambda^2)(1-\lambda^2))$.

Putting all this together leads to

$$G_S^{(\delta)}(\lambda^2)=(\lambda^2)^p\frac{(\lambda^2)^{2a}}{(1-\lambda^2)^a}\frac{(\lambda^2)^{2s_e}}{(1-(\lambda^2)^2)^{s_e}}\frac{(\lambda^2)^{3s_o}}{(1-(\lambda^2)^2)^{s_o}}\frac{6^b(\lambda^2)^{2b}}{(1-3\lambda^2)^b(1-\lambda^2)^b}. \tag{13.28}$$

Denoting by c the total number of chain vertices and by $s=s_e+s_o$ the total number of straight-chain vertices, this expression simplifies to

$$G_S^{(\delta)}(\lambda^2)=\frac{6^b(\lambda^2)^{p+2c+s_o}}{(1-\lambda^2)^{c-s}(1-(\lambda^2)^2)^s(1-3\lambda^2)^b}. \tag{13.29}$$

Now, in order to take melons into considerations (see equation (13.27), recall that a rooted melon-free MO graph with $2p$ non-root vertices has $4p+1$ edges (since the root edge is split into two edges, see the previous subsection) where one can insert a rooted melonic subgraph.

Let us now define:

$$U(\lambda^2):=\lambda^2T(\lambda^2)^4=T(\lambda^2)-1. \tag{13.30}$$

The generating function $F_S^{(\delta)}(\lambda^2)$ of rooted MO graphs of reduced scheme S is then given by

$$F_S^{(\delta)}(\lambda^2)=T(\lambda^2)\frac{6^bU(\lambda^2)^{p+2c+s_o}}{(1-U(\lambda^2))^{c-s}(1-U(\lambda^2)^2)^s(1-3U(\lambda^2))^b}. \tag{13.31}$$

Finally, the generating function $F^{(\delta)}(\lambda^2)$ of rooted MO graphs of degree δ is simply given by

$$F^{(\delta)}(\lambda^2)=\sum_{S\in\mathcal{S}_\delta}F_S^{(\delta)}(\lambda^2) \tag{13.32}$$

(see again Fusy and Tanasa (2014) for details).

The melon generating function has its main singularity at

$$\lambda_c^2 = \frac{3^3}{2^8}. \tag{13.33}$$

Moreover, $T(\lambda_c^2) = \frac{4}{3}$, and

$$(1 - 3U(\lambda^2))^{-b} \sim_{\lambda\to\lambda_c} (2^{3/2}3^{-1/2})^{-b}(1 - \lambda^2/\lambda_c^2)^{-b/2}. \tag{13.34}$$

Therefore, using the expression (13.31), one concludes that **the dominant schemes are those for which the number of broken chains b is maximized** (the larger b, the larger the singularity order).

Using now an appropriate algorithm of iterative removal of broken chains (see again Fusy and Tanasa (2014) for details), one obtains some tree of components. This allows us to prove the following bound:

$$b \leq 4\delta - 1, \tag{13.35}$$

on the number of broken chains. If this bound is saturated then:

- all broken chains are separating;
- the component containing the root has degree zero;
- all the components of positive degree and the component containing the root are leaves of the tree and the remaining components (of degree zero) have three neighbours in the tree; and
- **all positive degree components have degree** $1/2$.

One can then prove (see again Fusy and Tanasa (2014) for details) the following correspondence between these dominant schemes and rooted binary trees:

Theorem 13.3.15 *The dominant schemes of degree δ arise from rooted binary trees with*

- $2\delta + 1$ *leaves;*
- $2\delta - 1$ *inner nodes;*
- $4\delta - 1$ *edges;*

where

- *the root-leaf is occupied by the rooted cycle-graph;*
- *the 2δ leaves are occupied by a infinity graph;*
- *the $2\delta - 1$ inner nodes are occupied by the cycle graph or the quadruple edge graph;*
- *the $4\delta - 1$ edges are occupied by separating broken chain vertices.*

Moreover, each rooted binary tree with $2\delta + 1$ leaves yields $2^{6\delta-2}$ dominant schemes.

In Fig. 13.21, one has an illustration of this correspondence for the case $\delta = 2$.

Using Corollary VI.1 of Flajolet and Sedgewick (2009), one has:

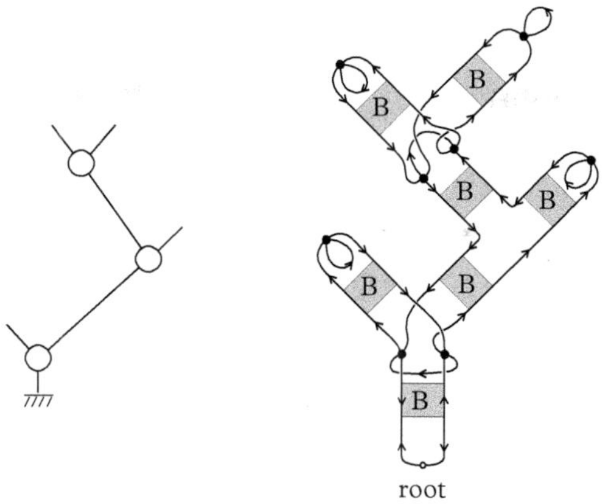

(a) (b)

Figure 13.21 *(a) A rooted binary tree with 5 leaves (b) One of the 2^{10} dominant schemes arising from it*

Theorem 13.3.16 *For δ and n in $\frac{1}{2}\mathbb{Z}_+$, let $a_n^{(\delta)}$ be the number of rooted MO graphs with $2n$ vertices and degree δ. Then, δ being fixed, for $n \in \delta + \mathbb{Z}$ (and $\Gamma(\cdot)$ denoting the Euler gamma function), one has:*

$$a_n^{(\delta)} \sim \mathrm{Cat}_{2\delta-1} \cdot \frac{3^{\delta-3/2}}{2^{2\delta-5/2}} \cdot \frac{n^{2\delta-3/2}}{\Gamma(2\delta-1/2)} \cdot (2^8/3^3)^n \text{ as } n \to \infty. \tag{13.36}$$

and $a^{\delta}_{n+1/2} = O(a^{\delta}_n/\sqrt{n})$ as $n \to \infty$.

Recall that one can define a tadpole erasing mechanism which keeps the degree constant (as peviously mentioned). Using this mechanism for the 2δ tadpoles of a dominant scheme, one can prove the following planarity result:

Theorem 13.3.17 *Each rooted melon-free MO graph whose reduced scheme is dominant is planar.*

An illustration of this planarity result is shown in Fig. 13.22.
Moreover, one has:

Corollary 13.3.18 *For each fixed $\delta \in \frac{1}{2}\mathbb{Z}_+$ and for $n \in \delta + \mathbb{Z}$, the probability that a rooted MO graph of degree δ with $2n$ vertices is planar tends to 1 as $n \to \infty$.*

This is a direct consequence of the fact that edge substitution by melonic components preserves planarity, since it is these graphs which dominate the asymptotic expansion.

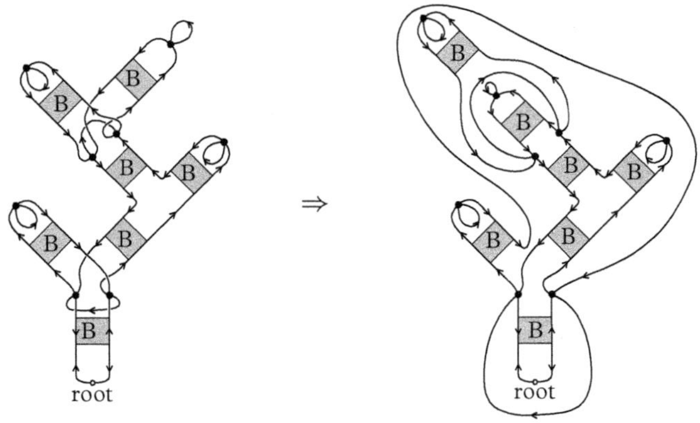

Figure 13.22 *Planar redrawing of the scheme of Fig. 13.21*

In the planar case, rooted MO graphs correspond (bijectively) to rooted 4 regular maps and the straight faces of the MO graph identify to the knot components of the map, and the number of straight faces is equal to

$$V/2+1-\delta, \tag{13.37}$$

with V the number of vertices and δ the degree. Let us recall that straight faces are the closed circuits given by the straight stands of our tensor graphs.

One can then prove:

Proposition 13.3.19 *For $n \in \frac{1}{2}\mathbb{Z}_+$ and $k \in \mathbb{Z}_+$, let $b_n^{(k)}$ be the number of rooted 4 regular planar maps with $2n$ vertices and k knot components. Then $b_n^{(k)} = 0$ for $k > n+1$. In addition, for each fixed $\delta \in \frac{1}{2}\mathbb{Z}_+$, and for $n \in \delta + \mathbb{Z}$, $b_n^{(n+1-\delta)}$ has the same asymptotic estimate as $a_n^{(\delta)}$, given by* (13.36).

13.4 The double-scaling limit

Using the results of the previous section, we implement here the double-scaling limit of the MO model. This section follows the original article Gurau, Tanasa, and Youmans (2015).

The contribution to the $2r$-point functions is given by the dominant schemes with r root edges (see Gurau, Tanasa, and Youmans (2015)). Recall here that all the chains of these schemes are broken chains.

Generalizing the results of the previous section, one can show (see again Gurau, Tanasa, and Youmans (2015) for details) that the set of dominant schemes with degree δ and r roots, $\mathcal{S}_{\delta,r}^{\mathrm{dom}}$, consists in binary trees with r univalent root vertices and another 2δ univalent vertices. Such trees have $2\delta+r-2$ three valent internal vertices and $4\delta+2r-3$ edges. The leading singular contribution to the $2r$-point function is then:

$$
\mathfrak{K}_{2r}^{(1)\text{sing}} = N^{3(1-r)} \sum_{\delta \in \mathbb{N}/2} N^{-\delta} \sum_{S \in \mathcal{S}^{\text{dom}}_{\delta,r}} T(\lambda^2)^r [2^2 U(\lambda^2)]^{\delta} [1+3U(\lambda^2)]^{2\delta+r-2} \times
$$

$$
\times \left(\frac{6[U(\lambda^2)]^2}{[1-U(\lambda^2)][1-3U(\lambda^2)]} \right)^{4\delta+2r-3} , \tag{13.38}
$$

where for the two-point function ($r = 1$) one needs to add the contribution $T(\lambda^2)$ of the degenerate dominant scheme consisting in a unique root vertex.

In the **double-scaling limit**, one compensates the $1/N$ suppression of the higher-order terms in the series in equation (13.38) by the enhancement at criticality due to the $[1-3U(\lambda^2)]^{-1}$ factors. This is achieved by sending at the same time N to infinity and λ to criticality while keeping the double-scaling parameter

$$
N^{\frac{1}{2}} \left(1 - \frac{\lambda^2}{\lambda_c^2} \right) \equiv \kappa^{-1} ,
$$

fixed. In the double-scaling regime, we get:

$$
T(\lambda^2) \sim \frac{4}{3} \left(1 - \sqrt{\frac{1}{6\kappa\sqrt{N}}} \right) , 1 - 3U(\lambda^2) \sim \sqrt{\frac{8}{3\kappa\sqrt{N}}} .
$$

13.4.1 The two-point function

The dominant schemes are rooted binary trees which are counted by the Catalan numbers: there are Cat_{n-1} such trees with $n-1$ three valent vertices, hence with n non-root univalent vertices. The degree of the scheme is $\delta = n/2$ and taking into account the contribution of the degenerate scheme consisting in a unique root vertex we obtain:

$$
\begin{aligned}
\mathfrak{K}_2^{\text{sing}} &= T(\lambda^2) + \sum_{n \geq 1} \text{Cat}_{n-1} \frac{1}{N^{\frac{n}{2}}} T(\lambda^2) [2^2 U(\lambda^2)]^{\frac{n}{2}} [1+3U(\lambda^2)]^{n-1} \\
&\quad \times \left(\frac{6[U(\lambda^2)]^2}{[1-U(\lambda^2)][1-3U(\lambda^2)]} \right)^{2n-1} \\
&= T(\lambda^2) \left(1 + \frac{1}{N^{1/2}} \frac{12 \cdot U(\lambda^2)^{5/2}}{[1-U(\lambda^2)][1-3U(\lambda^2)]} \times \right. \\
&\quad \left. \times \sum_{n \geq 0} \text{Cat}_n \left(\frac{1}{N^{1/2}} \frac{72 \cdot U(\lambda^2)^{9/2} [1+3U(\lambda^2)]}{[1-U(\lambda^2)]^2 [1-3U(\lambda)]^2} \right)^n \right) ,
\end{aligned}
$$

which, in the double-scaling limit, becomes:

$$\mathfrak{K}_2^{DS} = \frac{4}{3}\left(1 - \sqrt{\frac{1}{6\kappa\sqrt{N}}}\right) + \frac{4}{3}\frac{1}{N^{\frac{1}{4}}}\sqrt{\frac{\kappa}{2}}\sum_{n\geq 0}\mathrm{Cat}_n\left(\kappa\frac{\sqrt{3}}{2}\right)^n$$
$$= \frac{4}{3}\left(1 - \frac{\sqrt{1-2\sqrt{3}\kappa}}{N^{\frac{1}{4}}\sqrt{6\kappa}}\right),$$

where the sum over n converges for $\kappa < 2^{-1}3^{-\frac{1}{2}}$.

13.4.2 The four-point function

The dominant schemes are binary trees with two roots. There are again Cat_{n-1} such trees with $n-1$ three valent vertices, but this time they have $n-1$ non-root univalent vertices and degree $\delta = \frac{n-1}{2}$. We obtain:

$$\mathfrak{K}_4^{(1)\mathrm{sing}} = N^{-3}\sum_{n\geq 1}\mathrm{Cat}_{n-1}\frac{1}{N^{\frac{n-1}{2}}}T(\lambda^2)^2[2^2U(\lambda^2)]^{\frac{n-1}{2}}[1+3U(\lambda^2)]^{n-1}\times$$
$$\times\left(\frac{6[U(\lambda^2)]^2}{[1-U(\lambda^2)][1-3U(\lambda^2)]}\right)^{2(n-1)+1}$$
$$= N^{-3}T(\lambda^2)^2\left(\frac{6[U(\lambda^2)]^2}{[1-U(\lambda^2)][1-3U(\lambda^2)]}\right)\times$$
$$\times\sum_{n\geq 0}\mathrm{Cat}_n\left(\frac{1}{N^{1/2}}\frac{72\cdot U(\lambda^2)^{9/2}[1+3U(\lambda^2)]}{[1-U(\lambda^2)]^2[1-3U(\lambda)]^2}\right)^n. \tag{13.39}$$

In the double-scaling limit this becomes

$$\mathfrak{K}_4^{(1)DS} = N^{-3+\frac{1}{4}}\sqrt{\kappa}\frac{8\sqrt{3}}{9\sqrt{2}}\left(\frac{1-\sqrt{1-2\sqrt{3}\kappa}}{\sqrt{3}\kappa}\right). \tag{13.40}$$

Note that $\mathfrak{K}_4^{(1)DS}$ is *enhanced* by a factor $N^{\frac{1}{4}}$ with respect to the natural N scaling of $\mathfrak{K}_4^{(1)}$. This is a consequence of the fact that the singularity of the generating function of dominant schemes boosts this four-point function in double scaling.

13.4.3 The 2r-point function

The dominant schemes are binary trees with r roots and the double-scaling limit of $\mathfrak{K}_{2r}^{(1)}$ is

$$\mathfrak{K}_{2r}^{(1)DS} \sim N^{3(1-r)}N^{\frac{1}{4}(2r-3)}f_{2r}(\kappa), \tag{13.41}$$

for some function f_{2r} depending only on the double-scaling parameter κ. As was the case for the four-point function, the higher-point functions are also boosted in double scaling with respect to their natural scaling in N.

Let us emphasize the fact that the functions $f_{2r}(\kappa)$ are convergent for $2\sqrt{3}\kappa < 1$ and exhibit a square-root singularity at the critical double-scaling coupling $\kappa_c = 2^{-1}3^{-\frac{1}{2}}$.

13.5 Selected further reading

- Adrian Tanasa. (2016). The multi-orientable random tensor model, a review. SIGMA, 12: 056.

  Invited review paper which constitutes the backbone of the present chapter.

14

Random tensor models—the $O(N)^3$-invariant model

We define, in this chapter, a class of tensor models, endowed with $O(N)^3$ invariance This allows to generate, via the usual QFT perturbative expansion, a class of Feynman tensor graphs which is strictly larger than the class of Feynman graphs of both the MO model (and hence, of the coloured model) of Chapter 13 and the $U(N)^3$-invariant models of Chapter 12. We first exhibit the existence of a large N expansion for such a model with general interactions (not-necessary quartic). We then focus on the quartic model and we identify the LO and NLO Feynman graphs of the large N expansion. Finally, we prove the existence of a critical regime and we compute the critical exponents, both at LO and at NLO. This is achieved through the use of various analytic combinatorics techniques that one can find in standard combinatorics reference such as the book Flajolet and Sedgewick (2008). This chapter follows the original research article Carrozza and Tanasa (2016).

14.1 General model and large N expansion

In this section, we define the tensor model we are interested in and we implement its large N expansion. We consider a real tensor $T_{i_1 i_2 i_3}$ with three indices ranging from 1 to $N \in \mathbb{N}^*$. The group $\mathrm{O}(N)^3$ acts on T as

$$T_{i_1 i_2 i_3} \to O^{(1)}_{i_1 j_1} O^{(2)}_{i_2 j_2} O^{(3)}_{i_3 j_3} T_{j_1 j_2 j_3}, \tag{14.1}$$

where $O^{(k)}$ are three independent orthogonal matrices. We further define the partition function

$$Z_N = \int [\mathrm{d}T] \, \varepsilon^{-N^{3/2} S_N(T)}, \tag{14.2}$$

where the action $S_N(T)$ is required to be invariant under $\mathrm{O}(N)^{\otimes 3}$, and $[\mathrm{d}T]$ is the product of the Lebesgue measures associated to the entries $T_{i_1 i_2 i_3}$. The action S_N decomposes as a sum over tensor invariants I_b, weighted by coupling constants t_b:

Combinatorial Physics: Combinatorics, quantum field theory, and quantum gravity models. Adrian Tanasa, Oxford University Press (2021). © Adrian Tanasa. DOI: 10.1093/oso/9780192895493.003.0014

$$S_N(T) = \sum_{b \in \mathcal{B}} t_b \, N^{-\rho(b)} \, I_b(T), \tag{14.3}$$

where we have denoted by $\mathcal{B}$ the set of $O(N)$ tensor invariants, and introduced a free scaling parameter $\rho(b)$ for each $b \in \mathcal{B}$. These tensor invariants are in one-to-one correspondence with 3-coloured graphs[1] but *not necessarily bipartite.* This is the only difference with respect to $\mathrm{U}(N)$-invariant models. The mapping between graphs and invariants is as follows: each node of the graph is associated to a tensor, and each line of colour ℓ represents the contraction of the ℓth indices of two tensors.

Let us recall that, in the tensor model framework, one calls these invariants *bubbles* and they are vertices of the QFT model to be considered (see Chapter 12).

We further restrict to connected invariants, that is to those invariants which are labelled by connected coloured graphs. In Fig. 14.1, we have depicted the two- and four-point bubbles (up to a permutation of the colour labels). The second bubble in Fig. 14.1 has the same combinatorial structure as the tetrahedron in spin foam (Andandrov, Geiller, and Noui 2012; Perez 2013) and GFT models (Freidel 2005; Oriti 2011) for 3D quantum gravity. Note nonetheless that as a coloured graph it encodes the topology of the projective plane Vince (1993). The third bubble in Fig. 14.1 represents the so-called pillow term (see (Freidel and Louapre 2003; Magnen 2009), and like the first bubble of Fig. 14.1 is associated to a spherical triangulation. The first and the third bubbles can be given a bipartite structure such that they become vertices of the complex model. However, the second bubble is not allowed by the $\mathrm{U}(N)^3$ invariance; it is a new ingredient of the $\mathrm{O}(N)^3$ model.

The leftmost bubble of Fig. 14.1—call it b_2—encodes the quadratic part of the action. Since we will evaluate the model in perturbation around the Gaussian contribution, we fix the normalisation of the invariant associated to b_2 to: $t_{b_2} = 1$ and $\rho(b_2) = 0$. The weight $N^{3/2}$ in front of the action is the natural one if one wants the second bubble of Fig. 14.1 to contribute at LO (assuming that $\rho = 0$ for this bubble), as already observed in coloured Gurau (2012) and MO Dartois, Rivasseau, and Tanasa (2014) models. The function ρ will be adjusted below after analysis of the amplitudes. Note that the $U(N)^3$-invariant models of Chapter 12 come with a global weight N^2 instead, and individual

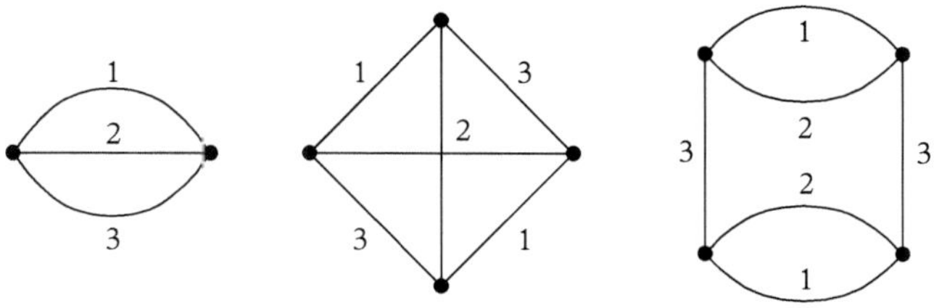

Figure 14.1 *The two- and four-valent bubbles/vertices of the model*

[1] We refer here to edge-colouring, with colour labels $\ell \in \{1, 2, 3\}$.

weights $N^{-2\omega(b)}$ (where $\omega(b)$ is the *degree* of the coloured graph b) in front of the coupling constants t_b.

As usually in QFT, one splits the action between the quadratic part, which provides a notion of propagation, and the higher-order terms which are Taylor expanded (see Chapter 3). As in $U(N)^3$-invariant models of Chapter 12, the resulting Feynman amplitudes $\mathcal{A}_{\mathcal{G}}$ are labeled by graphs $\mathcal{G}$, made out of bubble vertices and propagator lines, which are represented as dashed lines in our figures. Feynman diagrams are in particular 4-coloured graphs, a fourth colour 0 being attributed to the dashed lines. Three simple examples are given in Figs. 14.3 and 14.4 (we will leave some of the colour labels implicit in our pictures).

As already mentioned in the introduction, the model we propose here generates a larger class of Feynman graphs than both the MO model (and hence than the coloured one), and the U(N)-invariant models. One has:

Proposition 14.1.1 *The sets of Feynman graphs generated by the MO action or by an* U$(N)^3$*-invariant action are strict subsets of the set of Feynman graphs generated by the action* (14.3).

Proof One can easily check that the interaction given by the second bubble in Fig. 14.1 already generates, by perturbative expansion, a strictly larger class of graphs than the class of MO tensor graphs. Indeed, MO graphs may be generated by the vertex of Fig. 14.2, with the additional condition that propagator lines must connect black nodes to white nodes. Hence all MO graphs may be realized in our O(N) model, while for instance the Feynman graph of Fig. 14.3 cannot be given the structure of a MO graph. Finally, the graphs generated by the U$(N)^3$ models coincides with the subclass of O$(N)^3$ graphs which can be given a bipartite structure. It is a strict subclass since the graph of Fig. 14.3 does not admit any bipartite colouring of its nodes.

One may perturbatively expand the partition function around the Gaussian, leading to

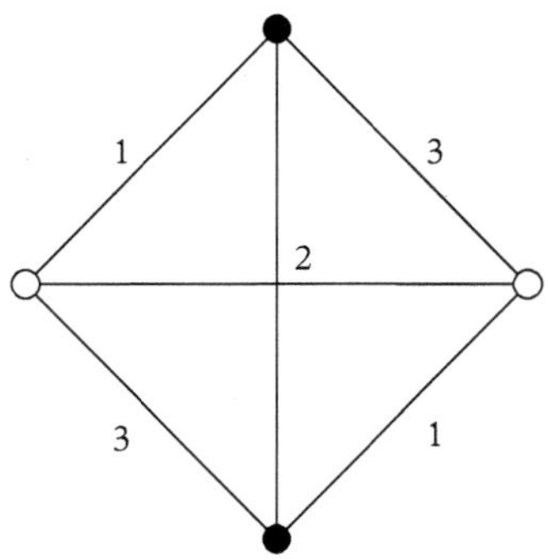

Figure 14.2 *The unique MO interaction vertex*

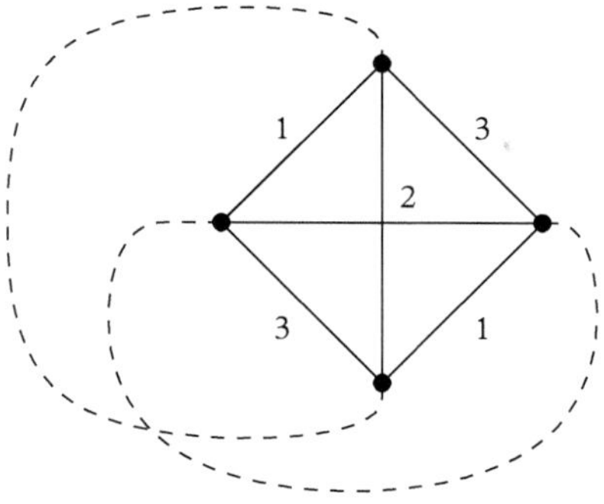

Figure 14.3 *A Feynman graph of the* $\mathrm{O}(N)^3$ *model which is neither MO nor bipartite*

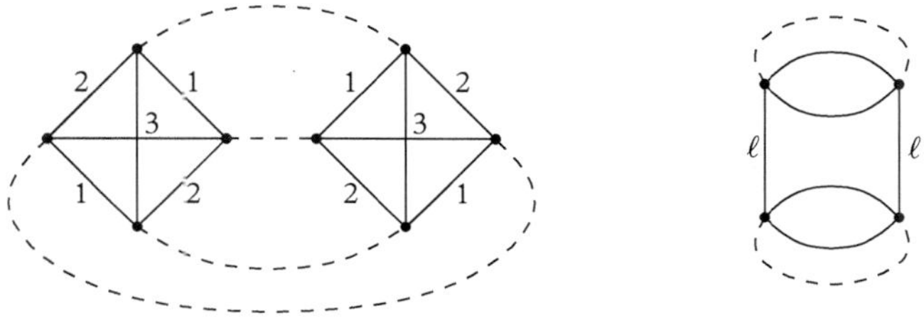

Figure 14.4 *Two-vacuum (and melonic) graphs*

$$Z_N = \sum_{\mathcal{G}} \prod_{b \in \mathcal{B} | N_b > 2} (-t_b N^{-\rho(b)})^{n_b(\mathcal{G})} \mathcal{A}_{\mathcal{G}}, \tag{14.4}$$

where $\mathcal{A}_{\mathcal{G}}$ is the amplitude associated to a graph $\mathcal{G}$, and $n_b(\mathcal{G})$ is the number of bubble vertices of type b in $\mathcal{G}$. A standard calculation (see, for example, Gurau (2011b)) shows that the amplitude $\mathcal{A}_{\mathcal{G}}$ of a closed graph $\mathcal{G}$ is

$$\mathcal{A}_{\mathcal{G}} \propto N^{3-\omega(\mathcal{G})}, \tag{14.5}$$

where we have defined the degree $\omega(\mathcal{G})$ as

$$\omega(\mathcal{G}) := 3 + \frac{3}{2} L(\mathcal{G}) - \sum_{b | N_b > 2} \left(\frac{3}{2} - \rho(b) \right) n_b(\mathcal{G}) - F(\mathcal{G}). \tag{14.6}$$

In the previous definition, we have denoted by L and F respectively the number of lines and faces[2] of $\mathcal{G}$ (these are notations we will stick with here).

[2] In this context, recall that a *face* of colour ℓ is a cycle of alternating colour-0 lines and colour-ℓ edges. We also define the *length* of a face as its number of colour-0 lines.

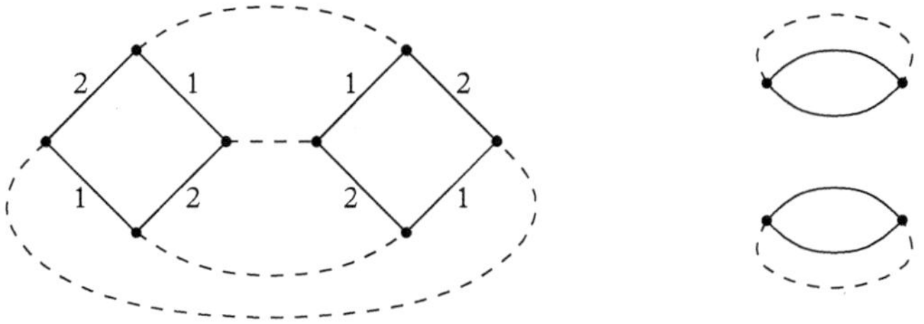

Figure 14.5 *Jackets associated to the tensor graphs of Fig. 14.4. On the right-hand side, two connected components are generated by the removal of the lines of colour ℓ*

This form of the Feynman amplitudes entails the existence of a $1/N$ expansion, provided that ω is bounded from below. We will show that this can be achieved. We will furthermore choose ρ as small as possible, so that the class of LO graphs in N is as large as possible.

In order to conveniently count the number of faces F, we introduce the notion of *jacket*, which is defined similarly as in the rank-3 coloured framework (Gurau 2011b; Ryan 2012).

Definition 14.1.2 *For any graph $\mathcal{G}$ and any $\ell \in \{1,2,3\}$, the* jacket *$J_\ell(\mathcal{G})$ is the 3-coloured graph obtained from $\mathcal{G}$ after deletion of all it colour-ℓ edges.*

Equivalently, each J_ℓ is obtained by deletion of all the faces of colour ℓ, and hence represents a ribbon graph with faces of colours in $\{1,2,3\}\setminus\{\ell\}$. A jacket therefore represents a closed and possibly non-orientable surface. Unlike in the cases studied in (Gurau 2011b; Ryan 2012) the jacket of a connected graph is not necessarily connected—see Fig. 14.5 for examples of such jackets, associated to the tensor graphs of Fig. 14.4.

Note that, for graphs which are MO, the notion of jackets introduced previously coincides with the one of MO jackets (see Chapter 13).

Faces can be counted as follows:

$$2F = \sum_{\ell=1}^{3} f(J_\ell) = \sum_{\ell;i} f(J_\ell^{(i)}), \tag{14.7}$$

where $f(J_\ell^{(i)})$ is the number of faces of the ith connected component of J_ℓ. This number can be expressed in terms of the non-orientable genus k as

$$f(J_\ell^{(i)}) = 2 - v(J_\ell^{(i)}) + e(J_\ell^{(i)}) - k(J_\ell^{(i)}), \tag{14.8}$$

where e (resp. v) is the number of edges (resp. vertices) of the jackets.

To facilitate the analysis in the case of multiple connected components, let us introduce the quantity

$$\delta_\ell := |J_\ell| - 1\,, \tag{14.9}$$

where $|J_\ell|$ denotes the number of connected components of J_ℓ. Likewise, we define

$$\delta_\ell^{(b)} := C_b^\ell - 1\,, \tag{14.10}$$

where C_b^ℓ is the number of connected components of the 2-coloured graph obtained from b after deletion of all the colour-ℓ lines. Notice that

$$\sum_i e(J_\ell^{(i)}) = L(\mathcal{G}) = \sum_{b|N_b>2} n_b \frac{N_b}{2}\,, \tag{14.11}$$

$$\sum_i v(J_\ell^{(i)}) = \sum_{b|N_b>2} n_b \left(1 + \delta_\ell^{(b)}\right), \tag{14.12}$$

where N_b is the number of nodes of the bubble b (i.e. the valence of the tensor interaction). In terms of these quantities, ω can be re-expressed in the form

$$\omega(\mathcal{G}) = \frac{1}{2}\sum_{\ell;i} k(J_\ell^{(i)}) + \sum_{b|N_b>2} n_b \left(\rho(b) + \frac{1}{2}\sum_\ell \delta_\ell^{(b)}\right) - \sum_\ell \delta_\ell. \tag{14.13}$$

The following is the key simple observation leading to the definition of a $1/N$ expansion.

Lemma 14.1.3 *For any $\ell \in \{1,2,3\}$, one has*

$$\sum_{b|N_b>2} n_b\, \delta_\ell^{(b)} \geq \delta_\ell\,. \tag{14.14}$$

Moreover, the number n_b of vertices of each type being given, one can always construct a graph $\mathcal{G}$ on these vertices which saturates the bound.

Proof By definition, each connected component of J_ℓ is supported on at least one connected component of the coloured subgraph with colours $\{1,2,3\}\setminus\{\ell\}$, hence the inequality. And one obtains an equality by pairing half-lines to half-lines in the same connected component of the colour $\{1,2,3\}\setminus\{\ell\}$ sub-graph.

This allows us to conclude that the optimal choice for ρ is

$$\boxed{\rho(b) := \frac{1}{2}\sum_{\ell} \delta_{\ell}^{(b)}} \tag{14.15}$$

which we assume in the sequel. Note that, in terms of the number of coloured faces[3] F_b of b, this definition coincides with

$$\rho(b) = \frac{F_b - 3}{2}. \tag{14.16}$$

Proposition 14.1.4 *Assuming a scaling of the coupling constants as defined by (14.15) and (14.3), the amplitude of a graph $\mathcal{A}_{\mathcal{G}}$ is proportional to $N^{3-\omega(\mathcal{G})}$, where the degree can be expressed as:*

$$\omega = \frac{1}{2}\sum_{\ell;i} k(J_{\ell}^{(i)}) + \sum_{b|N_b>2} n_b \sum_{\ell} \delta_{\ell}^{(b)} - \sum_{\ell} \delta_{\ell}. \tag{14.17}$$

Furthermore, $\omega(\mathcal{G}) \in \frac{\mathbb{N}}{2}$.

Proof The expression of ω is a simple consequence of the relations defining ρ and ω and the other combinatorial quantities. Equation (14.17), together with the fact that the demigenus k is an integer[4], implies that $\omega \in \frac{\mathbb{Z}}{2}$. Furthermore, Lemma 14.1.3 and the positivity of k imply that ω is itself positive.

Notice that the choice of weight (14.15) ensures that

$$\rho(b_1 \sharp b_2) = \rho(b_1) + \rho(b_2), \quad \forall b_1, b_2 \in \mathcal{B}, \tag{14.18}$$

where we loosely denote by $b_1 \sharp b_2$ any connected sum of b_1 and b_2, that is any bubble obtained by first connecting b_1 and b_2 with a propagator and then contracting this propagator. Such a contraction is called a 1-*dipole* contraction in the literature. A posteriori, an independent motivation for our choice of scaling function ρ, is that it renders the degree ω-invariant under these 1-dipole contractions.

Proposition 14.1.5 *The LO graphs are characterized by:*

$$\forall (\ell, i), \qquad k(J_{\ell}^{(i)}) = 0, \tag{14.19}$$

$$\forall \ell, \qquad \delta_{\ell} = \sum_{b|N_b>2} n_b \delta_{\ell}^{(b)}. \tag{14.20}$$

[3] Where a *coloured face* of b is a cycle of alternating colours i and $j \in \{1,2,3\}$ in b.

[4] The demigenus (or half-genus) k is usually invoked to label the topology of non-orientable surfaces, while for orientable surfaces, the genus $g = \frac{k}{2}$ is in general preferred. In the sequel, we will work with the demigenus, irrespectively of the orientability of the surface. This means in particular that k will be even in the case of an orientable surface, and odd in the case of a non-orientable one.

Proof The two terms appearing in the expression of the degree (14.17) are positive or 0. Hence, they must both be 0 when $\omega = 0$, leading to equations (14.19) and (14.20). Furthermore, it is easy to check that the class of degree-0 graphs is non-empty: for instance, the two graphs shown in Fig. 14.4 have 0 degrees.

At least melonic graphs in the sense of the $U(N)^3$-invariant models are dominant. We refer to these types of melons as respectively of type I and II. It is also easy to generate LO graphs which are genuinely new, for example by contraction of an arbitrary number of tree lines in a φ^4 LO graph of type I (this will create non-bipartite bubbles of valency higher than 4 without changing the degree).

Classifying LO graphs of this general model lies outside the purpose of this chapter. In the remaining sections, we focus instead on the quartic interaction model.

14.2 Quartic model, large *N* expansion

The action of the quartic model is written:

$$S_N = \frac{1}{2} I_{\ominus} + \frac{\lambda_1}{4} I_1 + \frac{\lambda_2}{12\sqrt{N}} I_2 , \tag{14.21}$$

where $\lambda_1, \lambda_2 \in \mathbb{R}$ and

$$I_1 := I_{\oplus}, \qquad I_2 := \sum_{\ell=1}^{3} I_{\ominus^{\ell}} . \tag{14.22}$$

The interactions we allow are therefore: the tetrahedral interaction term, for which $\rho = 0$; and the three pillow invariants which have $\rho = \frac{1}{2}$, hence the scaling of $N^{-1/2}$ in front of I_2. We will denote by n_1 the number of tetrahedral interactions in a given graph, and by n_2 the number of pillow vertices. Note that we have introduced normalization factors which take into account the symmetries of the respective interactions (this will of course facilitate the enumeration of graphs in the sequel). We took into account:

- the number of automorphisms[5] of each bubble (four in both cases), and
- an additional $1/3$ factor for the pillow interactions (this comes from the fact that we also require invariance under permutation of the colour labels, and therefore use a single coupling constant for the three pillow interactions).

The invariant I_2 is an explicitly positive interaction, but I_1 is not. It is therefore likely that S_N itself is unbounded from below, making the definition of the path integral

[5] That is, graph automorphisms which also preserve the colouring.

questionable.[6] Our calculations should be interpreted as formal manipulations of power series in λ_1, λ_2, and N. This being said, we will prove that the LO graphs of this model are exactly the melonic graphs of coloured and MO tensor models. The method of proof that we will use is itself a generalization of these cases.

14.2.1 Large N expansion: LO

In this model, we define two types of melonic moves, called type I (resp. type II) contractions or insertions—see Fig. 14.6 and resp. type II. Note that, which are nothing but the melonic moves already relevant in MO (resp. $U(N)^3$-invariant) tensor models, see Chapters 12 and 13. This moves allow us to formalize the notion of melonic graph in the context of $O(N)^3$-invariant Feynman tensor graphs in the following way:

Definition 14.2.1 *The family of vacuum* melonic graphs *is the set of graphs generated by the two graphs shown in Fig. 14.4, and the melonic insertion operations of type I and II.*

One can directly check that melonic moves conserve the degree.

Lemma 14.2.2 *The degree ω is invariant under the melonic moves of type I and II.*

Proof A melonic move of type I changes L to $L-4$, n_1 to n_1-2, F to $F-3$, and does not change n_2. Hence, the new degree is:

$$\omega - \frac{3}{2}4 + 2\left(\frac{3}{2} - 0\right) + 3 = \omega\,. \tag{14.23}$$

Similarly, a melonic move of type II changes L to $L-2$, n_2 to n_2-1, F to $F-2$, and does not change n_1. Hence, the new degree is:

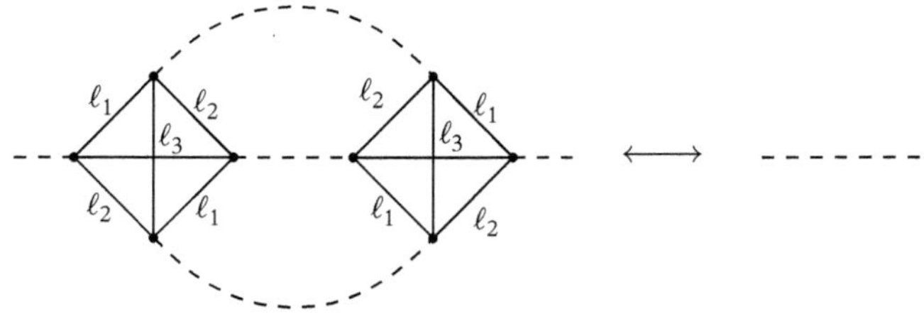

Figure 14.6 *Melonic moves of type I*

[6] We refer to Freidel and Louapre (2003) in which this question is explored at length, though in a slightly different context.

$$\omega - \frac{3}{2}2 + \left(\frac{3}{2} - \frac{1}{2}\right) + 2 = \omega\,. \tag{14.24}$$

This concludes the proof.

Hence, we have already proven that vacuum melonic graphs have degree 0 and are therefore LO. We will now prove that they are the only ones. Let us first give a list of preliminary results on graphs $\mathcal{G}$ with $n_2(\mathcal{G}) = 0$.

Lemma 14.2.3 *Let $\mathcal{G}$ be a vacuum graph such that $n_2(\mathcal{G}) = 0$.*

1. *If $\mathcal{G}$ has a face*[7] *of length 1 then $\omega(\mathcal{G}) \geq \frac{1}{2}$.*
2. *If $\mathcal{G}$ has a face of length 3, then $\omega(\mathcal{G}) \geq \frac{1}{2}$.*

Proof (i) $\mathcal{G}$ is either: a single-vertex graph with two tadpole lines i.e. the double-tadpole graph of Fig. 14.8, in which case we explicitly check that $\omega(\mathcal{G}) = \frac{1}{2}$; or it contains a non-trivial two-point graph with a single vertex (Fig. 14.9). The contraction of this two-point graph changes L to $L-2$, n_1 to $n_1 - 1$ and F to $F-1$. Hence this yields a new graph $\mathcal{G}'$ with degree $\omega(\mathcal{G}') = \omega(\mathcal{G}) - \frac{1}{2}$. The positivity of ω immediately implies that $\omega(\mathcal{G}) \geq \frac{1}{2}$.

(ii) The only way for $\mathcal{G}$ to have a face of length 3 without having also a face of length 1 is as pictured in Fig. 14.10. The jacket J_{ℓ_1} of this graph contains a ribbon with three twists and is therefore not orientable. This implies that at least one jacket, being not orientable, has a half-integer genus, and hence $\omega(\mathcal{G}) \geq \frac{1}{2}$.

Lemma 14.2.4 *Let $\mathcal{G}$ be a vacuum graph such that $n_2(\mathcal{G}) = 0$. If $\omega(\mathcal{G}) = 0$, then $\mathcal{G}$ has at least 6 faces of length 2.*

Proof Let F_p be the number of faces of length p in $\mathcal{G}$, and V the number of vertices. We have:

$$\sum_p F_p = F = 3 + \frac{3}{2}L - \frac{3}{2}V\,, \qquad \sum_p pF_p = 3L\,, \tag{14.25}$$

where the condition $\omega(\mathcal{G}) = 0$ was used in the first equation. From Lemma 14.2.3, we know that $F_1 = F_3 = 0$. Subtracting the second equation to four times the first therefore yields:

$$2F_2 + \sum_{p\geq 4}(4-p)F_p = 12 + 3L - 6V\,, \tag{14.26}$$

and, since moreover $L = 2V$ we conclude that

[7] We recall that the length of a face is defined as its number of colour-0 lines.

$$F_2 = 6 + \frac{1}{2}\sum_{p\geq 4}(p-4)F_p \geq 6. \tag{14.27}$$

This concludes the proof.

Lemma 14.2.5 *Let $\mathcal{G}$ be a vacuum LO graph. If $n_2(\mathcal{G}) \neq 0$, then $\mathcal{G}$ is of the form shown in Fig. 14.11.*

Proof One necessarily has a bubble b and a colour ℓ such that $\delta_\ell^{(b)} \geq 1$. Hence since $\mathcal{G}$ is LO, $\delta_\ell = \sum_{b|N_b>2} n_b \delta_\ell^{(b)} \geq 1$. A splitting of a jacket of colour ℓ in two connected components can only happen at a pillow vertex with colour ℓ, hence $\mathcal{G}$ is of the form shown in Fig. 14.11.

Proposition 14.2.6 *Any two-point LO graph contains a type I or type II elementary melon.*

Proof The proof is done by induction on $p = \lfloor \frac{n_1}{2} + n_2 \rfloor$.

$p = 1$: The only two possible graphs are shown on the left side of Figs. 14.6 and 14.7. The fact that there are no other possible graphs with $n_1 = 2$ and $n_2 = 0$ is easily deduced from the fact that all faces must have length 2 in this case (Lemma 14.2.4).

$p \geq 2$: If $n_2 \neq 0$, choose a pillow vertex in $\mathcal{G}$. If it does not directly provide an elementary melon of type II, then Lemma 14.2.5 ensures that the graph is of the form shown in Fig. 14.11, where the condition $\omega = 0$ imposes that $\tilde{\mathcal{G}}_1 \neq \emptyset$ is a two-point LO

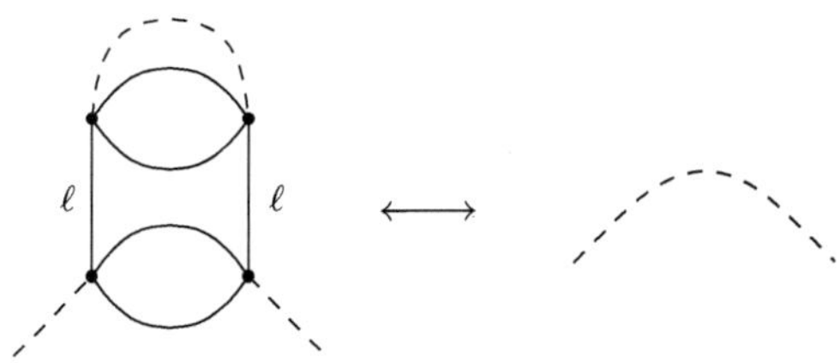

Figure 14.7 *Melonic move of type II*

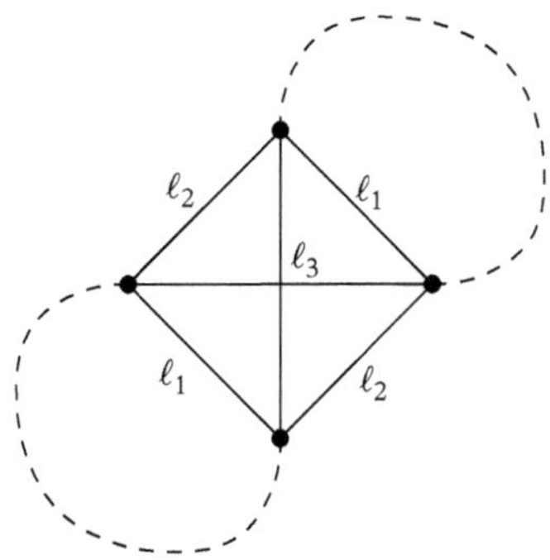

Figure 14.8 *Double tadpole graph (or infinity graph)*

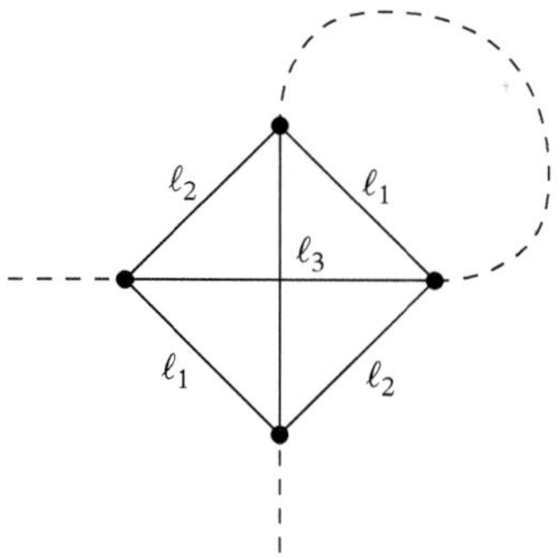

Figure 14.9 *A two-point graph with a face of length one*

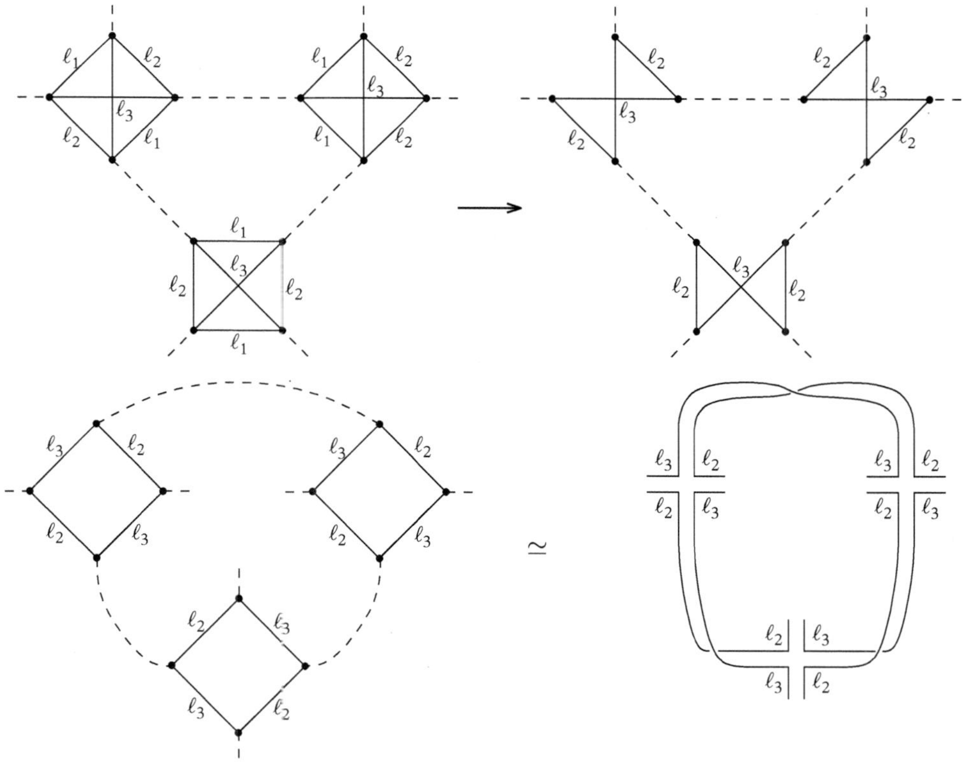

Figure 14.10 *Face of length three whose jacket yields a twisted ribbon graph*

graph with strictly lower p. By the induction hypothesis, it therefore itself contains an elementary melon of type I or II. If $n_2 = 0$, by Lemma 14.2.4 the graph is of the form shown in Fig. 14.12. If this does not already provide an elementary melon of type I, then we can perform the move shown in Fig. 14.13, which as is easily proved conserves the face structure and the degree. Now $n_2 \neq 0$ with p unchanged, and we can use again the previous argument to conclude the proof.

The main result of this subsection is thus:

Theorem 14.2.7 *The vacuum LO graphs of the quartic $O(N)^3$-invariant tensor model are the vacuum melonic graphs.*

Proof By Proposition 14.2.6, any LO graph can be reduced to one of the two graphs of Fig. 14.4 by successive contractions of melons. Hence, such a graph is melonic.

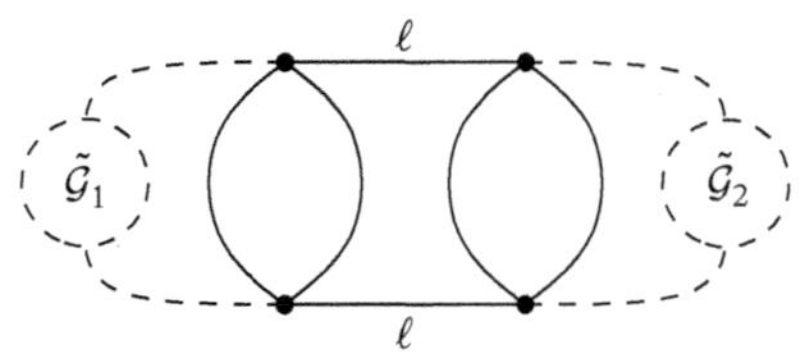

Figure 14.11 *Structure of a LO graph containing a pillow bubble*

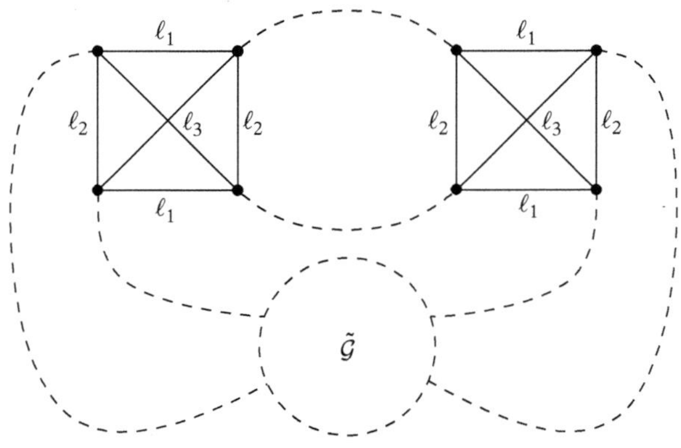

Figure 14.12 *Structure of a LO graph with $n_{4,2} = 0$*

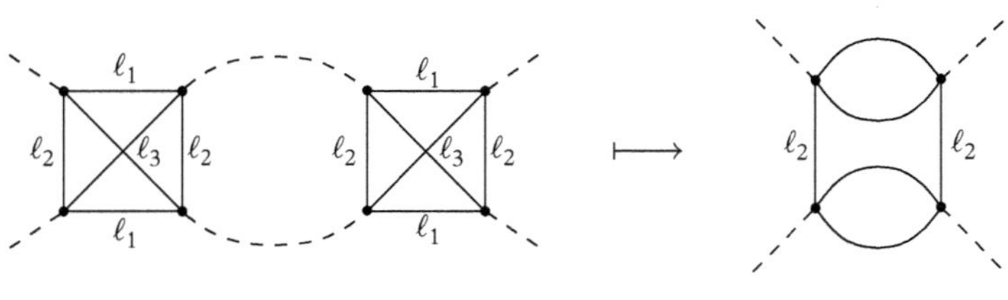

Figure 14.13 *A move leaving the degree invariant*

14.2.2 NLO

We now focus on the NLO graphs, which are characterized by $\omega = \frac{1}{2}$. From equation (14.17), we infer that graphs with degree $\omega = \frac{1}{2}$ must have a single non-trivial jacket of demigenus $k = 1$ and should also verify condition (14.20). Therefore Lemma 14.2.5 also holds for NLO vacuum graphs.

One can check that there are exactly three single-vertex graphs with $\omega = \frac{1}{2}$: there are the so-called 'infinity' graphs represented in Fig. 14.8 (there is one such graph for each value of ℓ_1). We readily obtain an infinite family of NLO graphs by insertion of non-trivial melonic two-point graphs, which as we have seen do not change the degree. In order to determine whether this family exhausts the set of NLO graphs, we follow Raasakka and Tanasa (2015) and introduce the notion of *core graph*.

Definition 14.2.8 *A* core graph *is a vacuum graph with no melonic two-point sub-graph.*

The question now is whether there exists more NLO core graphs than the three infinity graphs. Let us first prove the following lemma:

Lemma 14.2.9 *Let $\mathcal{G}$ be a NLO core graph. Then:*

1. $n_2(\mathcal{G}) = 0$; *and*
2. *if $\mathcal{G}$ is made of more than 1 vertex, then all its faces have lengths higher or equal to 3. Moreover, all the faces of length 3 must have the same colour as the non-planar jacket of $\mathcal{G}$.*

Proof (i) Assuming $n_2(\mathcal{G}) \neq 0$, condition (14.20) imposes that $\mathcal{G}$ has the structure shown in Fig. 14.11. Since $\mathcal{G}$ is a core graph, $\omega(\tilde{\mathcal{G}_1}) \geq \frac{1}{2}$ and $\omega(\tilde{\mathcal{G}_2}) \geq \frac{1}{2}$. But one can check (from the definition of ω) that $\omega(\mathcal{G}) = \omega(\mathcal{G}_1) + \omega(\mathcal{G}_2)$, which yields $\omega(\mathcal{G}) \geq 1$ in contradiction with the NLO character of $\mathcal{G}$.

(ii) Since there is no pillow vertex in $\mathcal{G}$, a face of length 1 has to be of the form shown in Fig. 14.9. Let us call $\tilde{\mathcal{G}}$ the non-trivial two-point subgraph which closes this figure. By hypothesis, $\omega(\tilde{\mathcal{G}}) \geq \frac{1}{2}$. One can also verify that $\omega(\mathcal{G}) = \omega(\tilde{\mathcal{G}}) + \frac{1}{2}$, which is inconsistent with $\mathcal{G}$ being NLO. Therefore $\mathcal{G}$ has no face of length 1. If $\mathcal{G}$ had a face of length 2, it would have the structure of Fig. 14.12. Performing the move of Fig. 14.13, which does not change the degree, would lead to a NLO graph with $n_2 \neq 0$. This graph would therefore have to contain a two-point melonic subgraph, and so would $\mathcal{G}$, but this cannot be since $\mathcal{G}$ is a core graph. Finally, just like before, the only way for $\mathcal{G}$ to have a face of length 3 without having also a face of length 1 is as pictured in Fig. 14.10. The existence of such a face of colour ℓ_1 imposes that the jacket of colour ℓ_1 is non-orientable. Since $\mathcal{G}$ has exactly one non-planar jacket, all the faces of length 3 must have colour ℓ_1.

The main result of this subsection is thus:

Theorem 14.2.10 *The NLO core graphs of the quartic model are the double-tadpole graphs of Fig. 14.8.*

Proof Let us assume that $\mathcal{G}$ is a NLO core graph with more than 1 vertex and look for a contradiction. By Lemma 14.2.9 (i), $\mathcal{G}$ cannot have any pillow vertex, and therefore its degree can be expressed as $\omega = 3 + \frac{3}{2}V - F$. We know that $\mathcal{G}$ has three jackets, two of which (say J_1 and J_2) are planar and the last one being of demigenus $k = 1$. We also know by Lemma 14.2.9 (ii) that the faces of length 3 have the same colour as the non-planar jacket, namely the colour 3. The total number of faces F can be split into the $f(J_1)$ faces of J_1 (which are of colour 2 and 3) and the F_1 faces of colour 1: $F = f(J_1) + F_1$. J_1 being planar, Euler's relation implies $f(J_1) = v(J_1) + 2 = V + 2$, hence:

$$\omega(\mathcal{G}) = 3 + \frac{3}{2}V - (V + 2 + F_1) = 1 + \frac{1}{2}V - F_1 . \tag{14.28}$$

The F_1 faces of colour 1 have lengths higher or equal to 4, and each line of $\mathcal{G}$ contains exactly one ribbon line of colour 1, therefore:

$$2V = L = \sum_{p \geq 4} p F_1^{(p)} \geq 4F_1 , \tag{14.29}$$

where $F_1^{(p)}$ is the number of faces of length p and colour 1. One concludes that,

$$\omega(\mathcal{G}) \geq 1 + \frac{1}{2} \times 2F_1 - F_1 = 1 , \tag{14.30}$$

which contradicts the fact that $\mathcal{G}$ is an NLO Feynman graph.

14.3 General quartic model: Critical behaviour

The LO connected and one particle irreducible Green functions are proportional to a product of Kronecker delta functions. Let us call $G_{LO}(g,\mu)$ (resp. $\Sigma_0(g,\mu)$) the proportionality factor of the connected (resp. one particle irreducible) Green function, in terms of the following parametrization of the coupling constants:

$$g := {\lambda_1}^2 , \qquad \mu := -\frac{\lambda_2}{{\lambda_1}^2} . \tag{14.31}$$

In this way, the variable g will allow to keep track of the total number of elementary melonic insertions, while μ will count the number of elementary melonic insertions of type II.

14.3.1 Explicit counting of melonic graphs

The melonic graphs of our model can be represented by unlabelled coloured trees. More precisely, with the weight we have chosen in the action, G_{LO} writes:

$$G_{LO}(g,\mu) = \sum_{p,q \in \mathbb{N}} C_{p,q}\, g^{p+q} \mu^q , \tag{14.32}$$

where $C_{p,q}$ is the number of melonic two-point graphs with p type I melons and q type II melons, up to local colour permutations of type II vertices.[8] Such melonic graphs can be recursively constructed by successive insertions of type I and type II elementary melons. They can therefore be represented by abstract trees with edge colour labels, which record the location and type of the successive melonic insertions. One can for instance adopt the convention of Fig. 14.14. Each equivalent class of melonic two-point graphs (up to colour relabelling at the type II vertices) is represented by a rooted coloured tree, an *admissible colouring* of the edges of a tree being as follows: the 4 lines outgoing (the notion of outgoing being defined with respect to the root) from a coordination 5 vertex are labelled with integers from 0 to 4, the 2 lines outgoing from a coordination 3 vertex have labels 0 and 1, and finally the unique edge outgoing from the root vertex has colour 0. In this manner, $C_{p,q}$ counts the number of *coloured* rooted trees with p vertices of coordination 5, q vertices of coordination 3, and $3p+q+2$ leaves (including the root vertex). Note also that a colouring is equivalent to a choice of local orientation around each vertex of the tree, hence $C_{p,q}$ equivalently counts the number of *binary-quaternary plane trees* with p quaternary and q binary vertices. See Fig. 14.15 for an example of a tree representation of a melonic two-point function.

By Cayley's theorem, the number of labelled trees with p vertices of valency 5, q vertices of valency 3, $3p+q+2$ leaves, and therefore a total number of $4p+2q+2$ vertices, is

$$\frac{(4p+2q)!}{(4!)^p\, 2^q}. \tag{14.33}$$

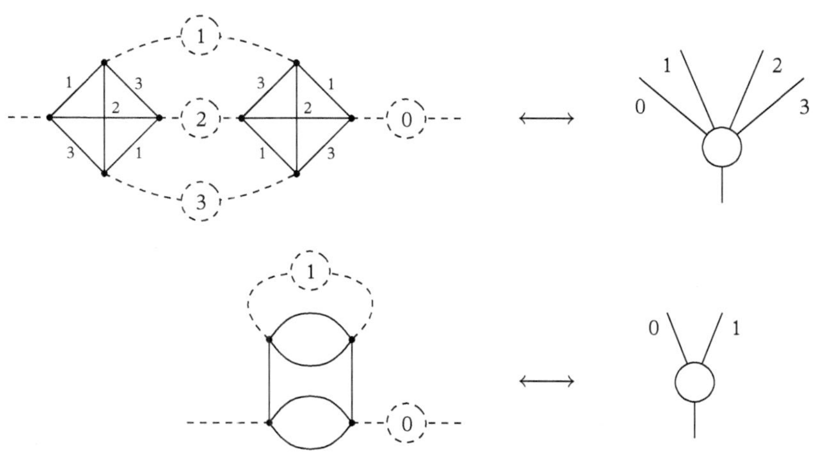

Figure 14.14 *Correspondence between elementary melons and plane tree vertices. Each labeled circular dashed region on the left-hand side corresponds to the open leg with the same label on the right-hand side*

[8] This is because and the reason why we have divided the corresponding interaction term by an extra factor 3 in the action.

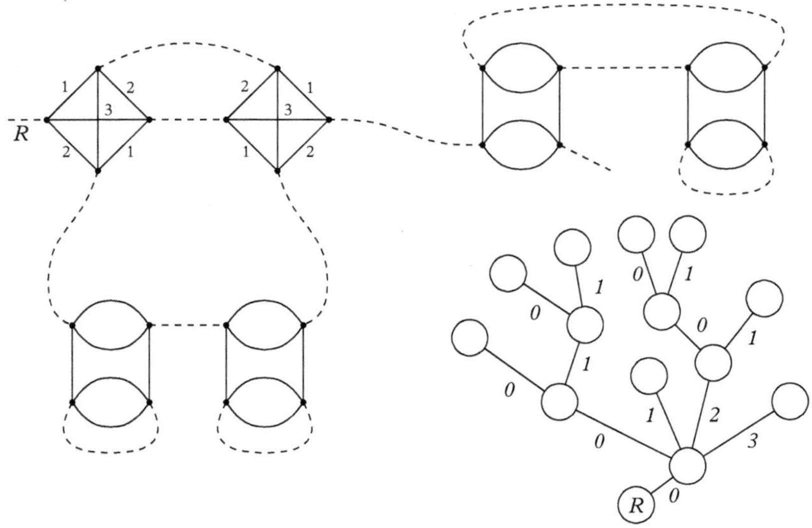

Figure 14.15 *A two-point melonic graph with root external leg R and its tree representation*

The outgoing edges of each coordination 5 vertex admit 4! distinct colourings or orientations, while the outgoing edges of coordination 3 vertices admit only 2 distinct colourings. This gives a total multiplicative factor of $(4!)^p 2^q$. Since our trees are furthermore unlabelled and rooted, one should divide by the number of possible permutations of the vertices except for the root: namely, one should divide by $p!$ (for the coordination 5 vertices), $q!$ (for the coordination 3 vertices), and $(3p+q+1)!$ (for all the coordination 1 vertice but the root). This yields:

$$C_{p,q} = \frac{[4p+2q]!}{p!\,q!\,(3p+q+1)!}\,. \tag{14.34}$$

Remember that tree structures can naturally be enumerated through QFT techniques, see for example (Abdesselam 2004; Bonzom 2011). In the present situation one can for instance define the partition function

$$Z(g,\mu,J) = \int d\mu_C \exp\left(\varphi J + g\varphi\psi^4 - \mu g\varphi\psi^2\right), \tag{14.35}$$

where the covariance C of the Gaussian measure μ_C is defined by:

$$C(\varphi,\varphi) = C(\psi,\psi) = 0, \qquad C(\varphi,\psi) = 1. \tag{14.36}$$

The inspection of the perturbative expansion of this auxiliary field theory shows that the melonic Green function G_{LO} evaluates as the connected expectation value:

$$G_{LO}(g,\mu) = \frac{\langle \psi_0 \rangle_c (g,\mu,J=1)}{Z(g,\mu,J=1)} \,. \tag{14.37}$$

One can easily compute $Z(g,\mu,J)$ and $\langle \psi_0 \rangle_c(g,\mu,J)$ perturbatively in μ and J, and hence reduce formula (14.34) to the computation of a product of two formal power series. Given the simplicity of the previous proof we will not give more details here, the interested reader is referred to Bonzom (2011) for a similar calculation.

In order to obtain a first crude understanding of the divergence structure of G_{LO}, one may resort to the following asymptotics of the coefficients $C_{p,q}$.

Proposition 14.3.1 *The coefficients $C_{p,q}$ have the following asymptotic behaviour:*

1. *For any $q_0 \in \mathbb{N}$:*

$$C_{p,q_0} \underset{p\to+\infty}{\sim} \frac{1}{3}\sqrt{\frac{2}{3\pi}\frac{1}{q_0!}} \left(\frac{16}{3}\right)^{q_0} p^{q_0-3/2} \left(\frac{4^4}{3^3}\right)^p \tag{14.38}$$

2. *For any $p_0 \in \mathbb{N}$:*

$$C_{p_0,q} \underset{q\to+\infty}{\sim} \frac{1}{\sqrt{\pi}} \frac{1}{p_0!} 16^{p_0} q^{p_0-3/2} 4^q \tag{14.39}$$

Proof These expressions are direct consequences of Stirling's formula.

As a consequence of Fubini's theorem, if the right-hand side of equation (14.32) is absolutely convergent then the partial sums over p and q respectively are absolutely convergent and therefore:

$$|g| < \frac{3^3}{4^4} \qquad \text{and} \qquad |\mu g| < \frac{1}{4} \,. \tag{14.40}$$

This already proves the existence of a critical regime for G_{LO}, but this is not enough to locate the singularities or compute the critical exponents. Indeed, G_{LO} can be rewritten as:

$$G_{LO}(g,\mu) = \sum_{n\in\mathbb{N}} \alpha_n(\mu) g^n \,, \tag{14.41}$$

with

$$\alpha_n(\mu) = \sum_{q=0}^{n} \frac{\mu^q}{3n-2q+1} \binom{n}{q} \binom{4n-2q}{n} , \tag{14.42}$$

and it is the asymptotic behaviour of $\alpha_n(\mu)$ which determines the critical behaviour of G_{LO}. Rather than directly evaluating this complicated sum, we will adopt an alternative strategy based on the analysis of the analytic properties of an algebraic equation for G_{LO}. Interestingly, not only this method does not require any evaluation of $\alpha_n(\mu)$ (or any prior knowledge of the coefficients $C_{p,q}$), it actually yields the asymptotic formula for $\alpha_n(\mu)$ we are missing as a bonus.

14.3.2 Diagrammatic equations, LO and NLO

We know from the standard relation between the connected and one particle irreducible Green functions that

$$G_{LO} = \frac{1}{1-\Sigma_0}. \tag{14.43}$$

From the structure of melonic graphs, we can furthermore infer the relation depicted in Fig. 14.16. Carefully taking combinatorial factors and signs into account, this leads to:

$$\Sigma_0 = {\lambda_1}^2 {G_{LO}}^3 - \lambda_2 G_{LO} = g {G_{LO}}^3 + g\mu G_{LO}. \tag{14.44}$$

We can therefore deduce the following equation for G_{LO} alone:

$$G_{LO} = 1 + g{G_{LO}}^2 \left({G_{LO}}^2 + \mu\right). \tag{14.45}$$

Let us now investigate the NLO behaviour. The graphs contributing to the connected NLO two-point function can be obtained from the NLO vacuum graph by cutting one of the graph's internal lines. The result of this cutting process thus follows, from a combinatorial point of view, from the characterization of the NLO vacuum graphs from the previous section. One has three distinct types of graphs, depending on the type of edge that is cut in a given graph $\mathcal{G}$:

1. one can cut an edge of a melon of type I;
2. one can cut an edge of a melon of type II; and
3. one can finally cut an edge associated to a tadpole line in the core graph of $\mathcal{G}$.

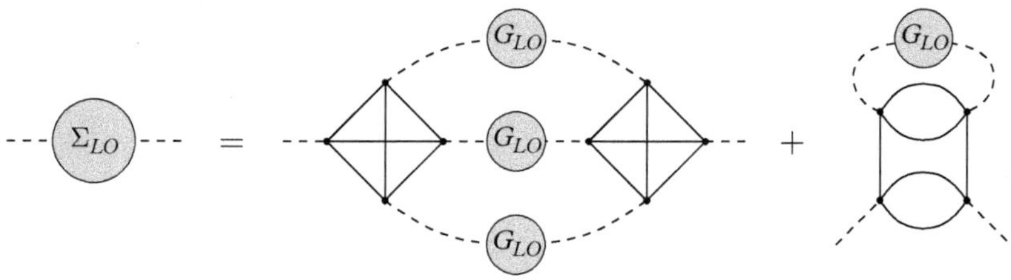

Figure 14.16 *Diagrammatic equation for the two-point function at LO*

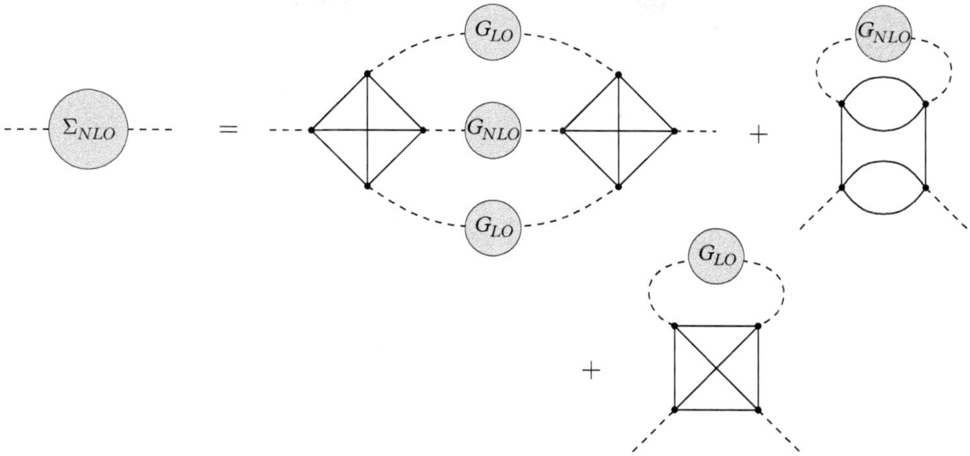

Figure 14.17 *Diagrammatic equation for the two-point function at NLO*

The diagrammatic equation obtained in this way is depicted in Fig. 14.17, and the analytic equation associated is written:

$$\Sigma_{NLO} = 3{\lambda_1}^2 {G_{LO}}^2 G_{NLO} - \lambda_2 G_{NLO} - \lambda_1 G_{LO} \tag{14.46}$$

$$= 3g {G_{LO}}^2 G_{NLO} + g\mu G_{NLO} - \sqrt{g} G_{LO}. \tag{14.47}$$

The combinatorial factor three in the first term above comes from the fact that one has three distinct choices for the NLO two-point insertion, choices corresponding to the three distinct colours of the coloured edges.

14.3.3 Singularity analysis

Let us assume for the moment that $\mu \geq 0$ and $g > 0$, so that all the series we consider have positive coefficients. By Pringsheim's theorem,[9] we know that G_{LO} is singular on the boundary of its domain of convergence. Using standard methods of analytic combinatorics, we are furthermore able to determine how G_{LO} behaves close to the critical curve.[10]

[9] Pringsheim's theorem stipulates that a power series $\sum_{n\in\mathbb{N}} a_n z^n$ with positive coefficients a_n and radius of convergence $R > 0$ has a singularity at $z = R$. See e.g. Flajolet and Sedgewick (2009).

[10] G_{LO} is a generating function of tree-like objects, which lead to square-root singularities on very general grounds Flajolet and Sedgewick (2009).

Proposition 14.3.2 *For any $\mu \geq 0$, define the critical value*

$$g_c(\mu) = \frac{G_c(\mu) - 1}{G_c(\mu)^2 \left(G_c(\mu)^2 + \mu\right)}, \tag{14.48}$$

where $G_c(\mu)$ is the unique real solution of the polynomial equation

$$-3x^3 + 4x^2 - \mu x + 2\mu = 0. \tag{14.49}$$

The (adherence of the) domain of convergence of the series defining G_{LO} is $\{(g,\mu) \in \mathbb{R}_+ \times \mathbb{R}, |g| \leq g_c(|\mu|)\}$. Moreover, for any $\mu \geq 0$, there exists a constant $K(\mu) > 0$ such that:

$$G_{LO}(g,\mu) \underset{g \to g_c(\mu)^-}{=} G_c(\mu) - K(\mu)\sqrt{1 - \frac{g}{g_c(\mu)}}\left(1 + \mathcal{O}(1 - \frac{g}{g_c(\mu)})\right). \tag{14.50}$$

Proof $G_{LO}(g,\mu)$ is a power series in g with positive coefficients, therefore by Pringsheim's theorem it has a singularity at some $g_c(\mu) > 0$. One moreover has the equation:

$$g = \frac{G_{LO}(g,\mu) - 1}{F(G_{LO}(g,\mu) - 1)} \equiv \Psi(G_{LO}(g,\mu) - 1), \tag{14.51}$$

where $F(u) := (1+u)^4 + \mu(1+u)^2$ and Ψ are both analytic around $G_{LO}(0,\mu) - 1 = 0$. Since the function $g \mapsto G_{LO}(g,\mu)$ is not analytic at $g_c(\mu)$, at $\tau = G_{LO}(g_c(\mu),\mu) - 1 > 0$ one must have $\Psi'(\tau) = 0$. Otherwise we could locally invert the previous equation to obtain an analytic dependence of G_{LO} in a neighbourhood of $g_c(\mu)$. This leads to the equation:

$$F(\tau) - \tau F'(\tau) = 0, \tag{14.52}$$

known to combinatorists as the characteristic equation of the generating function G_{LO} Flajolet and Sedgewick (2009). One can check that its unique real positive solution is $\tau = G_c(\mu) - 1$, where $G_c(\mu)$ is inferred from the value of τ given above. In particular, $g_c(\mu)$ is indeed defined by equation (14.48). Moreover, the second derivative does not cancel:[11]

$$\Psi''(\tau) = -\tau\frac{F''(\tau)}{[F(\tau)]^2} = -\tau\frac{12(1+\tau)^2 + 2\mu}{[F(\tau)]^2} < 0. \tag{14.53}$$

We therefore obtain by Taylor expansion

$$g_c(\mu) - g \approx -\frac{\Psi''(\tau)}{2}\left(G_{LO}(g,\mu) - G_c(\mu)\right)^2, \tag{14.54}$$

[11] We use the characteristic equation to obtain this formula.

which can be locally inverted (by use of the singular inversion theorem) to give formula (14.50) with

$$K(\mu) = \sqrt{-\frac{2g_c}{\Psi''(\tau)}} = \sqrt{\frac{2F(\tau)}{F''(\tau)}} = \sqrt{\frac{G_c(\mu)^2\left(G_c(\mu)^2+\mu\right)}{6G_c(\mu)^2+\mu}}. \tag{14.55}$$

This concludes the proof.

The singular behaviour of G_{LO} can be used to retrieve information about the asymptotics of the coefficients $C_{p,q}$ without explicitly enumerating the trees they count (as we have done in Section 14.3.1). This method is central in analytic combinatorics Flajolet and Sedgewick (2009), and here is an example of what one can infer.

Corollary 14.3.3 *For any $\mu \geq 0$, and with the same notations as in Proposition 14.3.2, the coefficients $\alpha_n(\mu)$ of the power series $G_{LO}(\cdot,\mu)$ behaves asymptotically as*

$$\alpha_n(\mu) \underset{n\to+\infty}{\sim} \frac{K(\mu)\, g_c(\mu)^{-n}}{2\sqrt{\pi}\, n^{3/2}}. \tag{14.56}$$

Proof The analytic function $F(z) = (1+z)^4 + \mu(1+z)^2$ is aperiodic.[12] Hence one can directly apply Theorem VI.6 of Flajolet and Sedgewick (2009). Let us however give the idea of the proof. The aperiodicity of F implies that of $G_{LO}(\cdot,\mu)$, and by Daffodil's lemma (see again Flajolet and Sedgewick (2009): p. 266), one can deduce that $G_{LO}(\cdot,\mu)$ has no other singularity than $g_c(\mu)$ on the circle $|g| = g_c(\mu)$. The application of Cauchy's formula:

$$\alpha_n(\mu) = \frac{1}{2i\pi}\int_\gamma \frac{G_{LO}(z,\mu)}{z^{n+1}}\mathrm{d}z \tag{14.57}$$

to a suitable contour γ around $g_c(\mu)$ (known as a Hankel contour) then shows that the asymptotics of the coefficients $\alpha_n(\mu)$ is dictated by the critical behaviour at $g_c(\mu)$. Therefore the known asymptotic expansion of $\sqrt{1-z}$ at $z=1$ directly yields the asymptotic estimate of $\alpha_n(\mu)$.

In particular, taking $\mu = 1$, we obtain an estimation of the number of binary–quaternary plane trees of size n (where n is the number of vertices which are neither leaves nor the root). Taking $\mu = 3$ yields, in turn, an estimation of the number of melonic two-point graphs with n elementary melons, which we denote by $\mathcal{M}_n$. A numerical application of the previous Corollary shows that:

[12] Let us define the support $\mathrm{Supp}(F) = \{n \in \mathbb{N} | F_n \neq 0\} = \{0,1,2,3,4\}$, where $F(z) = \sum_n F_n z^n$. F being aperiodic means that there exists no $r \in \mathbb{N}$ and no integer $d \geq 2$ such that $\mathrm{Supp}(F) \subset r + d\mathbb{N}$, which is clearly the case.

$$\mathcal{M}_n \underset{n \to +\infty}{\sim} \frac{\chi \beta^n}{n^{3/2}}, \tag{14.58}$$

with

$$\chi \approx 0.111 \quad \text{and} \quad \beta \approx 14.8. \tag{14.59}$$

Let us now briefly comment on the case $\mu < 0$. Given the form of our equation for G_{LO}, the negative sign of μ might in principle allow for multi-critical points, that is critical points at which $\frac{\mathrm{d}^2 g}{\mathrm{d}G_{LO}{}^2}$ also cancels out, therefore leading to different scaling behaviours[13]. Since our definition of G_{LO} is well-controlled in the region $g \leq g_c(|\mu|)$ only, we only have access to possible singularities on the boundary $(g = g_c(|\mu|))$. Exploring the phase space further would necessitate a careful study of possible analytic continuations, which is not the purpose of the present paper. We have checked, using the same method as in Proposition 14.3.2, that no new singularity is present. For instance, let us specialize to the case $\mu = -1$, which is easier to analyse. A factor $(G_{LO} - 1)$ can then be factored out from the relation between g and G_{LO}, which simplifies to:

$$g = \frac{1}{G_{LO}^2 (G_{LO} + 1)}. \tag{14.60}$$

At a critical point $(g = g_c, G_{LO} = G_c)$, $\frac{\mathrm{d}g}{\mathrm{d}G_{LO}}$ has to vanish, which leads to the characteristic equation:

$$3{G_c}^2 + 2G_c = 0. \tag{14.61}$$

$G_c = 0$ being excluded, we conclude that $G_c = -\frac{2}{3}$. But then $g_c = \frac{27}{4} > g_c(1)$, which would bring us outside the domain of convergence of G_{LO}. Hence there is no singular point when $\mu = -1$, and this conclusion actually holds for arbitrary $\mu < 0$.

This is summarized on the phase space representation of Fig. 14.18.

14.3.4 Critical exponents

We now use the critical behaviour of G_{LO} to infer that of the free energy and deduce the value of the susceptibility critical exponent.

[13] More precisely, one would get a behaviour in $\left(1 - \frac{g}{g_c}\right)^{1/p}$ whenever the first non-zero derivative of g is of order $p \geq 2$.

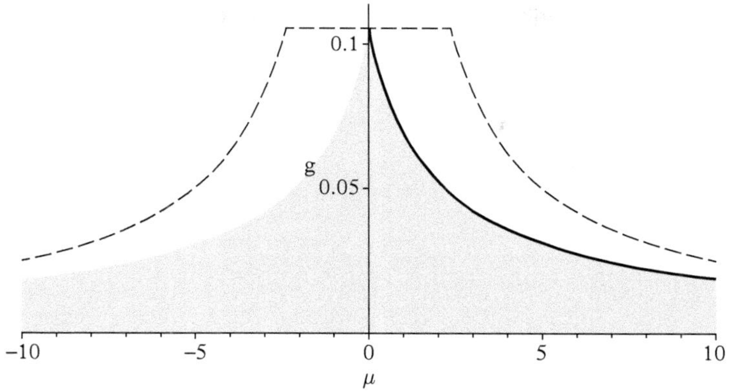

Figure 14.18 *Dashed line: crude bound on the domain of convergence of $G_{LO}(g,\mu)$; grey region: actual domain of convergence of $G_{LO}(g,\mu)$; black line: critical points*

LO

We parametrize the full connected two-point function as:

$$\frac{1}{Z_N}\int [\mathrm{d}T]\, T_{i_1i_2i_3}\, T_{j_1j_2j_3}\, e^{-S_N[T]} = \frac{C_N}{N^{3/2}} \prod_{k=1}^{3} \delta_{i_k j_k}, \tag{14.62}$$

in such a way that

$$C_N = G_{LO} + N^{-1/2} G_{NLO} + \dots \tag{14.63}$$

The free energy itself is defined as

$$F_N := \frac{1}{N^3} \ln Z_N = F_{LO} + N^{-1/2} F_{NLO} + \dots . \tag{14.64}$$

The relation between C_N and the LO free energy F_N is contained in the DSE:

$$0 = \frac{1}{Z_N} \sum_{i_1,i_2,i_3} \int [\mathrm{d}T] \frac{\delta}{\delta T_{i_1i_2i_3}} \left(T_{i_1i_2i_3}\, e^{-S_N[T]} \right) \tag{14.65}$$

$$= N^3 - N^3 C_N + \lambda_1 \partial_{\lambda_1} \ln Z_N + \lambda_2 \partial_{\lambda_2} \ln Z_N \tag{14.66}$$

which immediately yields

$$C_N(\lambda_1,\lambda_2) = 1 + (\lambda_1 \partial_{\lambda_1} + \lambda_2 \partial_{\lambda_2}) F_N(\lambda_1,\lambda_2). \tag{14.67}$$

Extracting the LO contributions, and resorting to the variables g and μ provides the looked for relation between G_{LO} and F_{LO}:

$$G_{LO}(g,\mu) = 1 + (2g\,\partial_g - \mu\,\partial_\mu)\,F_{LO}(g,\mu). \tag{14.68}$$

Close to the critical point $g_c(\mu)$, one can parametrize the most singular part of F_{LO} as

$$K_1(\mu)\left(1 - \frac{g}{g_c(\mu)}\right)^{2-\gamma_{LO}} \tag{14.69}$$

for some $K_1(\mu)$ independent of g. The critical exponent γ_{LO} is the LO susceptibility exponent and is, by equation (14.68), equal to

$$\gamma_{LO} = \frac{1}{2}. \tag{14.70}$$

This is the same critical exponent as for the $U(N)^3$-invariant and MO models. This indicates that all these models have the same universal properties in the critical regime and at LO.

NLO

In order to compute the susceptibility exponent γ_{NLO}, one may try to directly infer the critical behaviour at NLO order from the LO one. This can be achieved by means of equation (14.46), together with the standard QFT identity relating the connected two-point function G_{NLO} to the connected two-point function G_{LO}, and to the 1PI NLO two-point function Σ_{NLO}:

$$G_{NLO} = G_{LO}^2\,\Sigma_{NLO}. \tag{14.71}$$

Using these two equations, one gets

$$G_{NLO} = \frac{-\lambda_1 G_{LO}^3}{1 + \lambda_2 G_{LO}^2 - 3{\lambda_1}^2 G_{LO}^4} = \frac{-\sqrt{g}\,G_{LO}^3}{1 - g\mu G_{LO}^2 - 3g G_{LO}^4}. \tag{14.72}$$

Using the LO two-point function identity (14.45), one gets:

$$\frac{\partial G_{LO}}{\partial g} = \frac{G_{LO}^3(G_{LO}^2 + \mu)}{G_{LO} - 2g G_{LO}^2(2G_{LO}^2 + \mu)}. \tag{14.73}$$

Using again identity (14.45) to express the first term of the denominator, one has

$$\frac{\partial G_{LO}}{\partial g} = \frac{G_{LO}^3(G_{LO}^2 + \mu)}{1 - g\mu G_{LO}^2 - 3g G_{LO}^4}. \tag{14.74}$$

One can then re-express the NLO two-point function (14.72) as

$$G_{NLO} = \frac{-\sqrt{g}}{G_{LO}^2 + \mu} \frac{\partial G_{LO}}{\partial g}. \tag{14.75}$$

We can therefore use the same argument as at LO. First, the critical behaviour of G_{LO} implies that the most singular contribution of G_{NLO} is in

$$\left(1 - \frac{g}{g_c(\mu)}\right)^{-1/2}. \tag{14.76}$$

Second, as a consequence of relation (14.67), the most singular part of F_{NLO} behaves as

$$K_2(\mu) \left(1 - \frac{g}{g_c(\mu)}\right)^{1/2} \tag{14.77}$$

for some function $K_2(\mu)$ independent of g.

We thus find the same critical value of the coupling constant (i.e. the radius of convergence) for the NLO series (as series in the coupling constant g) as for the LO series. Nevertheless, one has a distinct value for the NLO susceptibility exponent:

$$\gamma_{NLO} = \frac{3}{2}. \tag{14.78}$$

This again coincides with the critical exponents found in the complex and MO models. Hence we expect these three types of theory to remain in the same universality class also at NLO.

14.4 Selected further reading

- S. Carrozza and A. Tansa. (2016). O(N) Random tensor models. *Lett. Math. Phys.*, 106:1531–1559.

  Original research paper which constitutes the backbone of the present chapter.

15

The Sachdev–Ye–Kitaev (SYK) holographic model

Quantum mechanical models involving q-body Hamiltonians were proposed in a nuclear physics setting in the 1970s (see the review article of Brody et al. (1981), and references therein). Similarly, Hamiltonians involving simultaneous interactions between many particles were introduced by French and Wong (1970). Many-particle eigenvalue densities generated by such q-body interactions in asymtotically large spectroscopic spaces were studied by Mon and French (1975).

In the 1990s, in the context of condensed matter physics, the paper Sachdev and Ye (1993) of Sachdev and Ye then led to an important amount of interest. In a series of talks in 2015, Kitaev introduced a simplified version of this model and showed it can be a particularly interesting toy model for AdS/CFT physics. The model, called ever since the Sachdev–Ye–Kitaev (SYK) model, has attracted a huge amount of interest for both condensed matter and high-energy physics—see, for example, (Parcollet and Georges 1999; Maldacena and Stanford 2016; Polchinski and Rosenhaus 2016; Gross and Rosenhaus 2017; Bonzom, Nador, and Tanasa 2016; Carrozza and Pozsgay 2018), or the review articles (Sarosi 2018); or Rosenhaus (2018).

In this chapter, we first review the SYK model, which is, as will be explained, a quantum mechanical model of N fermions. The model is quenched, this meaning that the coupling constant is a random tensor with Gaussian distribution. The SYK model is dominated in the large N limit again by melonic graphs, in the same way the tensor models presented in the previous three chapters are dominated by melonic graphs. We then give a purely graph-theoretical proof of the melonic dominance of the SYK model. This proof follows the one exposed in the research article (Bonzom, Nador, and Tanasa 2019). It is this property which led E. Witten (2019) to relate the SYK model to the coloured tensor model, see Chapter 16. In the rest of the chapter we deal with the so-called coloured SYK model, which is a particular case of the generalization of the SYK model given by D. Gross and V. Rosenhaus (2017). We first analyse in detail the LO and NLO vacuum, 2- and four-point graphs of this model. This follows the research article Bonzom, Lionni, and Tanasa (2017). We then exhibit a thorough asymptotic combinatorial analysis of the Feynman graphs at an arbitrary order in the large N

Combinatorial Physics: Combinatorics, quantum field theory, and quantum gravity models. Adrian Tanasa, Oxford University Press (2021). © Adrian Tanasa. DOI: 10.1093/oso/9780192895493.003.0015

expansion. This follows the research article Fusy, Lionni, and Tanasa (2018). We end this chapter by an analysis of the effect of non-Gaussian distribution for the coupling. This follows the research article Krajewski (2019) (see also the review article Laudonio, Pascalie, and Tanasa (2020)).

15.1 Definition of the SYK model: Its Feynman graphs

The SYK model has N Majorana fermions ψ_i $(i=1,\dots,N)$ coupled via a q-body random interaction (q being here an even integer)

$$S_{\text{SYK}} = \int d\tau \left(\frac{1}{2}\sum_{i=1}^{N} \psi_i \frac{d}{dt}\psi_i - \frac{i^{q/2}}{q!} \sum_{i_1,\dots,i_q=1}^{N} j_{i_1\dots i_q}\psi_{i_1}\dots\psi_{i_q} \right) \tag{15.1}$$

where $J_{i_1\cdots i_q}$ is the coupling constant. Furthermore, the model is quenched, which by definition means that the coupling J is a random tensor with a Gaussian distribution such that

$$\langle J_{i_1\cdots i_q}\rangle = 0 \qquad \text{and} \qquad \langle J_{i_1\cdots i_q} J_{j_1\cdots j_q}\rangle = 6J^2 N^{-(q-1)} \prod_{m=1}^{q} \delta_{i_m, j_m}. \tag{15.2}$$

The fields $\psi_i(t)$ satisfy fermionic anticommutation relations $\{\psi_i(t), \psi_j(t)\} = \delta_{i,j}$. This anticommutation property excludes graphs with tadpoles (also known as loops, in a graph-theoretical language); the model being $(0+1)$-dimensional, the Feynman amplitude of such a graph is zero.

It is worth emphasizing here that the SYK model has three remarkable properties:

1. Solvable at strong coupling: in the large N limit, one can sum all Feynman graphs and explicitly compute the correlation functions at strong coupling.
2. Maximally chaotic: quantum chaos is quantified by the so-called Lyapunov exponent. The Lyapunov exponent of a black hole in Einstein gravity and of the SYK model both saturate the maximal allowed bound.
3. Emergent conformal symmetry: the two-point function has an emergent conformal symmetry in the IR limit. This symmetry is spontaneously and explicitly broken by the mode saturating the chaos bound.

The SYK model is the only toy model known so far to enjoy all these properties (other known models have some of these properties but no other model has all of them). It is this fact which led to the huge amount of interest within the high-energy physics community, as previously mentioned.

In a Feynman graph of the SYK model, the interaction term is represented by a vertex with q incident fermionic lines. Each fermionic edge $m = 1,\dots,q$ carries an index $i_m = 1,\dots,N$ which is contracted at the vertex with a coupling constant $J_{i_1\cdots i_q}$. The free

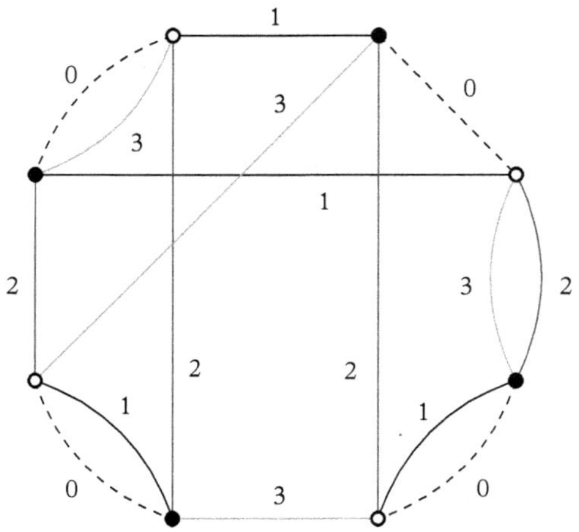

Figure 15.1 *An example of a Feynman graph of the $q = 4$ SYK model*

energy expands onto those connected, q-regular and tadpoleless (or loopless, in a graph theoretical language) graphs.

The so-called average over the disorder is done using standard QFT Wick contractions between pairs of Js, with covariance (**15.2**). An additional edge is thus adjacent to each vertex. We represent this additional edge as a dashed edge, and we call it a *disorder edge*. An example of such a Feynman graph of the SYK model is given in Fig. 15.1.

The previous description of the Feynman graphs ignores the indices of the random couplings. Indeed, a disorder edge propagates q field indices, where the field index of fermionic line incident on a vertex is identified with the index of a fermionic line at another vertex. We thus represent a disorder edge as an edge made of q strands, where each strand connects fermionic edges as follows:

$$\langle J_{i_1 \cdots i_q} J_{j_1 \cdots j_q} \rangle =$$ 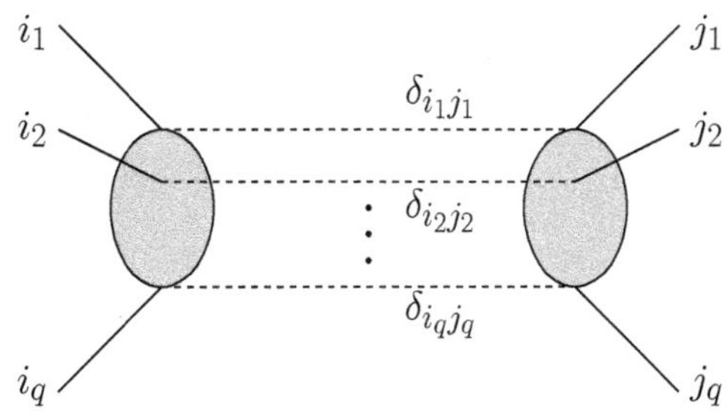 (15.3)

Here, the grey discs represent the Feynman vertices.

We denote by $\mathbb{G}$ the set of Feynman graphs of the SYK model. For $G \in \mathbb{G}$, we further denote $G_0 \subset G$ the q-regular graph obtained by removing the strands of the disorder lines, see Fig. 15.2. Note that the graph G_0 has to be connected. Moreover, each vertex

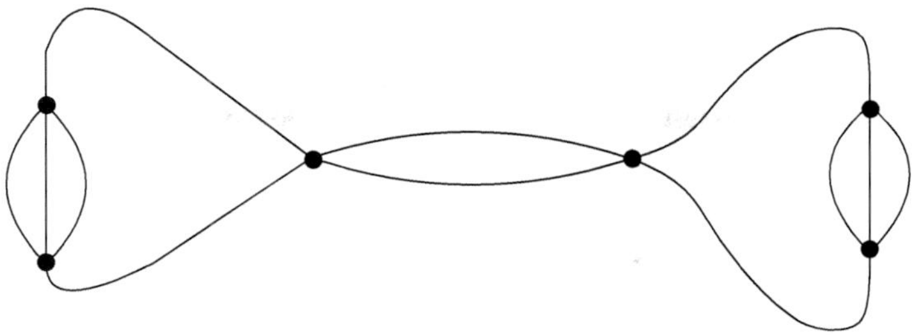

Figure 15.2 *The graph obtained after deleting of the disorder lines of the graph of Fig. 15.1*

of the graph G has exactly one adjacent disorder edge. This implies that the graphs G and G_0 have an even number of vertices.

Let us now give the following definition:

Definition 15.1.1 *A cycle made of alternating fermionic lines and strands of disorder lines is called a face. We denote $F(G)$ the number of faces of $G \in \mathbb{G}$.*

Let us consider graphs with two vertices and $G_{0,\text{min}} =$ 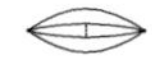 . There are $q!$ such graphs corresponding to permutations of the strands of the disorder line connecting these two vertices. However, among these $q!$ graphs there is only one which maximizes the number of vertices. This graph is

$G_{\text{min}} =$ 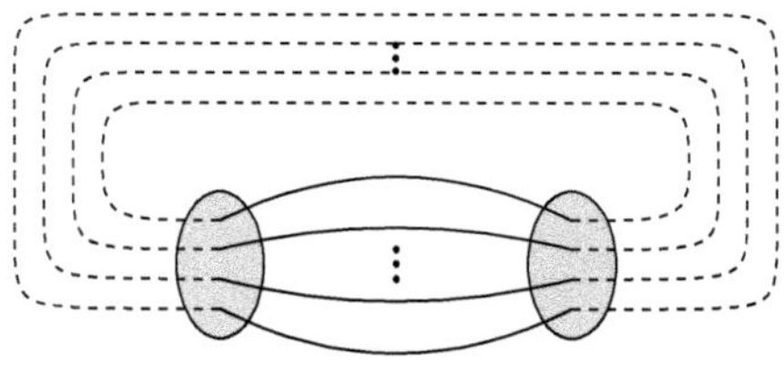 . (15.4)

When computing the Feynman amplitude of an SYK graph in the large N limit, there is a contribution of a factor N per face. The Feynman amplitude also receives a factor $N^{-(q-1)}$ for each disorder edge. The large N limit Feynman amplitude of an SYK graph is thus

$$N^{\delta(G)},$$

where

$$\delta(G) = F(G) - (q-1)V(G)/2, \tag{15.5}$$

where $V(G)$ is the number of vertices. We call the parameter $\delta(G)$ the **SYK degree** of the Feynman graph G. To find the dominant graphs in this large N limit, we thus need to find the graphs which maximize the number of faces at fixed number of vertices.

15.2 Diagrammatic proof of the large N melonic dominance

This section follows the original research article Bonzom, Nador, and Tanasa (2019). Let us first give the following definitions:

Definition 15.2.1 *We call* **dipole** *the following two-point graph:*

$D =$ 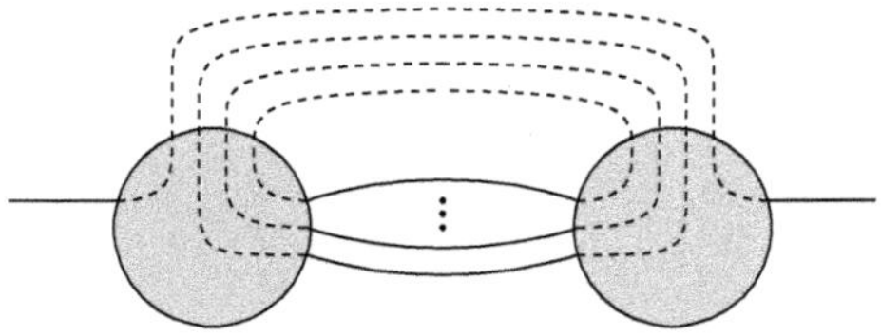 . (15.6)

Let us note that a dipole is made of two vertices connected by $(q-1)$ fermionic lines and a disorder line. A priori, there are $q!$ ways of connecting the strands of the disorder line. Among these $q!$ possibilities, we chose the one for D which creates the maximal number of faces.

Definition 15.2.2 *A* **melonic move** *is the insertion of a dipole on a fermionic line:*

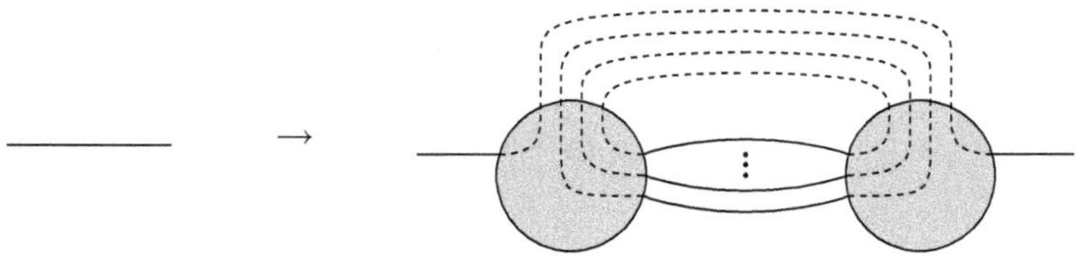 . (15.7)

Definition 15.2.3 *A* **melonic graph** *is a graph obtained from the graph $G_{\min}$ by iterated melonic moves, in any order.*

An example of such a melonic SYK graph is given in Fig. 15.3.

Some properties of melonic graphs

Proposition 15.2.4 *A melonic move adds two vertices and $q-1$ faces to a graph. The number of faces of melonic graphs is*

$$F(G) = q + (q-1)\frac{V(G)-2}{2}. \tag{15.8}$$

Thus, one has $\delta(G) = 1$ for melonic graphs.

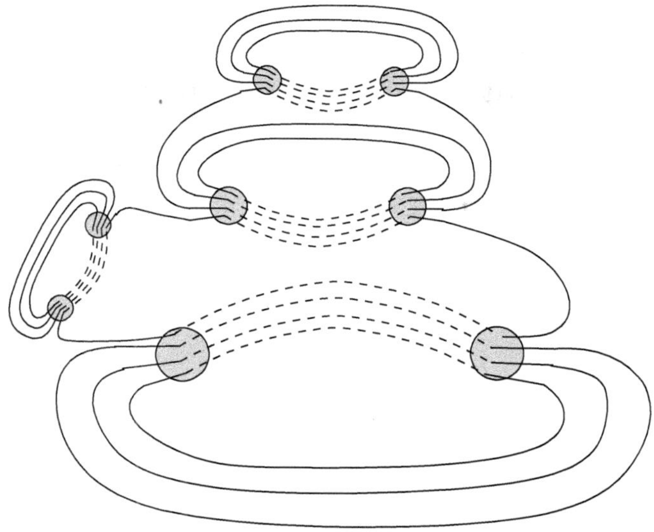

Figure 15.3 *An example of melonic graph*

Proof The first statement follows directly from the definition of the melonic move (see the previous Definition 15.2.2). The number of faces is then obtained by induction. Indeed, $F(G_{\text{min}}) = q$ at $V(G) = 2$ for G_{min}, the only melonic graph with two vertices. The induction is completed by using the first statement. The identity $\delta(G) = 1$ for melonic graphs follows from the expression of $\delta(G)$ in the Definition (15.5).

Note that, by definition, one can always find a dipole in a melonic graph. Let us also notice that there is always more than one such dipole.

Proposition 15.2.5 *A melonic graph with at least four vertices has at least two dipoles.*

Proof We proceed by induction on the number of vertices. There is a single melonic graph with four vertices,

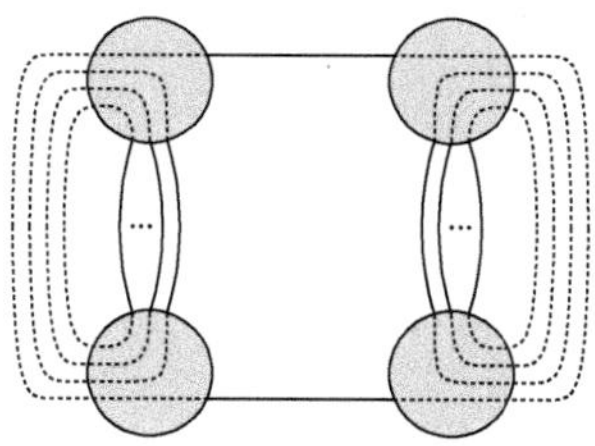 . (15.9)

One can directly check that this graph indeed has two dipoles.

Assume that the proposition holds for graphs with at most $V - 2 \geq 4$ vertices and let G be a melonic graph with V vertices. By construction, the graph G can be obtained by a melonic move on the fermionic edge e of the graph G', a melonic graph with $V-2$ vertices. From the induction hypothesis, the graph G' has at least two dipoles. If e is not an edge connecting the two vertices of a dipole, then the melonic move $G' \to G$ increases the number of dipoles. If e connects two vertices of a dipole in the graph G', then the total number of dipoles is unchanged between G' and G. This comes from the fact that cutting the edge e destroys one dipole, but the melonic move itself adds one. This concludes the proof.

Melonic graphs satisfy a gluing rule which generalizes the melonic move. Let $G_1, G_2 \in \mathbb{G}$ be two melonic graphs and e_1 in G_1, e_2 in G_2 two fermionic lines. If one cuts open e_1 in G_1 and e_2 in G_2, then there are two ways to glue the half-edges of e_1 with those of e_2. To avoid this ambiguity, we use orientations.

Definition 15.2.6 *If (G,e) is a graph G with an oriented fermionic line e, denote $G^{(e)}$ the two-point graph obtained by cutting e into two half-edges with their induced orientations. For two such graphs (G_1,e_1) and (G_2,e_2), denote $G_1^{(e_1)} \star G_2^{(e_2)}$ the unique connected graph obtained by gluing $G_1^{(e_1)}$ with $G_2^{(e_2)}$ in the only way which respects the orientations of the half-edges,*

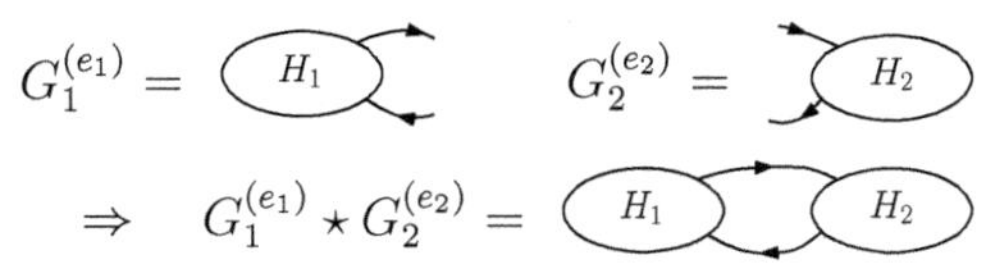

(15.10)

Proposition 15.2.7 *Let $G_1, G_2 \in \mathbb{G}$ be two melonic graphs and e_1 in G_1, e_2 in G_2 two oriented fermionic lines. Then $G_1^{(e_1)} \star G_2^{(e_2)}$ is a melonic graph.*

Proof The result is proved by induction on the number of vertices of the graph G_1. If G_1 is melonic graph and has two vertices, then $G_1 = G_{\text{min}}$ and the insertion of $G_1^{(e_1)}$ is the melonic move on e_2 (for any orientations of e_1 and e_2).

Assume the proposition holds for graphs G_1' with $V-2$ vertices and consider a new melonic graph G_1 with V vertices. It is obtained from a melonic move performed on a fermionic edge e_1' of the melonic graph G_1'. One then needs to find the edge e_1 in G_1', form $G_1'^{(e_1)} \star G_2^{(e_2)}$, which is melonic from the induction hypothesis, and then perform the melonic move on e_1' to get $G_1^{(e_1)} \star G_2^{(e_2)}$, which will thus be a melonic graph also. This is summarized in the following commutative diagram:

$$
\begin{array}{ccc}
G_1' & \xrightarrow[]{\text{Cutting } e_1 \text{ and inserting } G_2^{(e_2)}} & G_1'^{(e_1)} \star G_2^{(e_2)} \\
\text{Melonic move on } e_1' \downarrow & & \downarrow \text{Melonic move on } e_1' \\
G_1 & \xrightarrow[\text{Cutting } e_1 \text{ and inserting } G_2^{(e_2)}]{} & G_1^{(e_1)} \star G_2^{(e_2)}
\end{array}
\qquad (15.11)
$$

We thus want to use the path from G'_1 to $G_1^{(e_1)} \star G_2^{(e_2)}$ which goes right and then down. When e_1 and e'_1 are distinct in G'_1, this is straightforward:

$$G_1^{'(e_1)} \star G_2^{(e_2)} = $$ 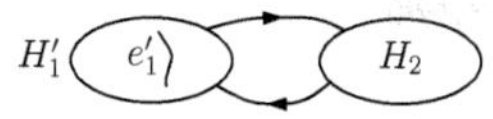 . (15.12)

By the induction hypothesis, this graph is melonic. By definition of the melonic insertion the graph remains meloinic after the melonic insertion on e'_1.

However, if e_1 is incident to or part of the dipole which is inserted from G'_1 to G_1, it means it does not exist in G'_1, as it is created by the melonic move. We distinguish two cases.

- e_1 is a fermionic line connecting the two vertices of the dipole. Then from Proposition 15.2.5 we know that G_1 has at least one other dipole. Therefore, one can redefine G'_1 has the melonic graph obtained from G_1 by removing the latter. Then, the fermionic line e_1 can be identified without issues in G'_1 and the previous reasoning applies.
- e_1 is incident to the dipole, i.e. $(G_1, e_1) =$ 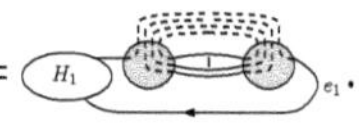. Then $G^{(e_1)} \star G_2^{(e_2)}$ has the form

  $$G^{(e_1)} \star G_2^{(e_2)} = $$ 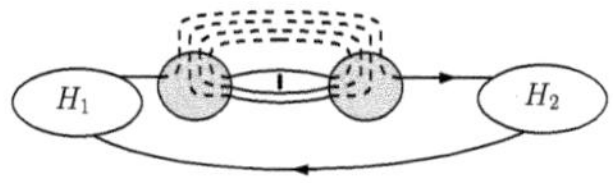 (15.13)

  The graph G'_1 is $G'_1 =$ 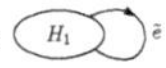 and is melonic. From the induction hypothesis $G_1^{'(\tilde{e})} \star G_2^{(e_2)} =$ 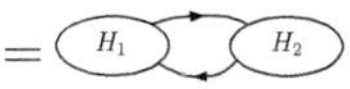 is melonic. Then so is $G^{(e_1)} \star G_2^{(e_2)}$, since it is obtained by a melonic move on $G_1^{'(\tilde{e})} \star G_2^{(e_2)}$.

2-cuts

Recall that, following the definition (15.5) of an SYK degree, we need to identify the graphs which maximize the number of faces at fixed number of vertices. Let us denote the maximal number of faces on V vertices by

$$F_{\max}(V) = \max_{\{G \in \mathbb{G}, V(G) = V\}} F(G), \tag{15.14}$$

and the set of graphs maximizing $F(G)$ at fixed V by

$$\mathbb{G}_{\max}(V) = \{G \in \mathbb{G} \text{ s.t. } V(G) = V \quad \text{and} \quad F(G) = F_{\max}(V)\}. \tag{15.15}$$

Let us now give the following definition:

Definition 15.2.8 *A* **2-cut** *is a pair of edges in a graph whose removal (or equivalently cutting) disconnects the graph.*

Let us now prove that, if there exist two edges in the same face which do not form a 2-cut, the graph is not dominant at large N:

Proposition 15.2.9 *Let $G \in \mathbb{G}$ and e_1, e_2 two fermionic lines in G which belong to the same face. If $\{e_1, e_2\}$ is not a 2-cut in G, then $G \notin \mathbb{G}_{\max}(V(G))$.*

Proof There are two cases to distinguish: whether $\{e_1, e_2\}$ is a 2-cut or not in G_0.

1. $\{e_1, e_2\}$ is not a 2-cut in G_0.
   We draw G as

   $G =$ 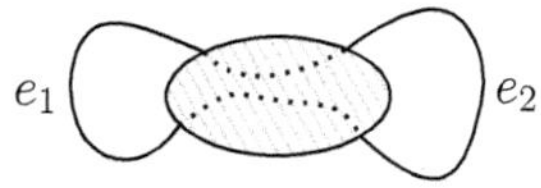

   (15.16)

   where the dotted line represents the paths alternating fermionic lines and strands of disorder lines which constitute the face of e_1 and e_2.

   Now consider G' obtained by cutting e_1 and e_2 and regluing the half-lines in the unique way which creates one additional face,

   $G' =$ 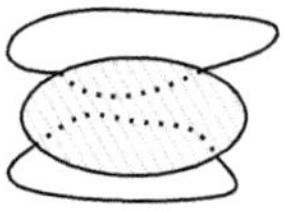 . (15.17)

   G'_0 is connected, since $\{e_1, e_2\}$ is not a 2-cut in G_0, and hence $G' \in \mathbb{G}$. No other faces of G are affected. Therefore, $F(G') = F(G) + 1$ and thus $G \notin \mathbb{G}_{\max}(V(G))$.
2. $\{e_1, e_2\}$ is a 2-cut in G_0.
   An example of this situation is when G_0 is melonic, but G is not because the disorder lines are added in a way which does not respect melonicity.

   In this case, G looks like

   $G =$ 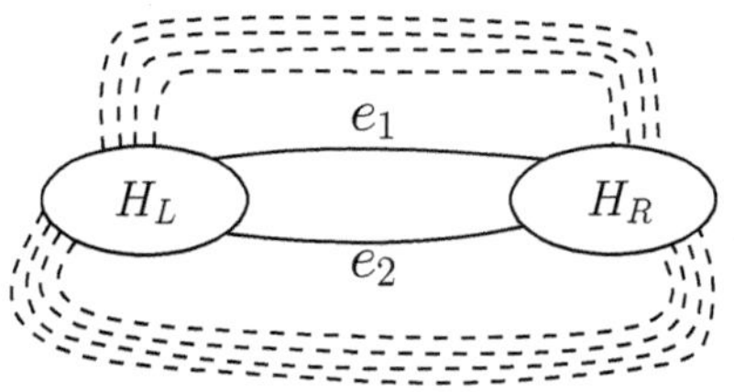

   , (15.18)

   i.e. H_L and H_R are both connected, and the only lines between them are e_1, e_2, and some disorder lines. Consider G' obtained by cutting e_1 and e_2 and regluing the half-lines as follows

   $G' =$ 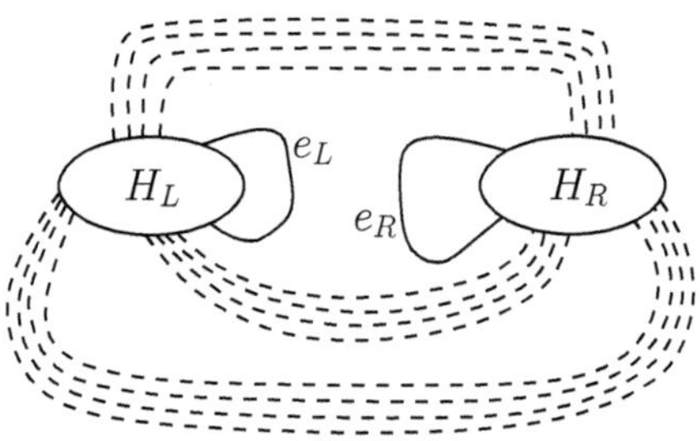

   . (15.19)

Notice that $G' \notin \mathbb{G}$, since G'_0 consists of two connected components G'_{0L} and G'_{0R}.

Consider a disorder line e_0 between them. It joins two vertices v_L in G'_{0L} and v_R in G'_{0R}. We perform the contraction of the disorder line e_0 as follows:

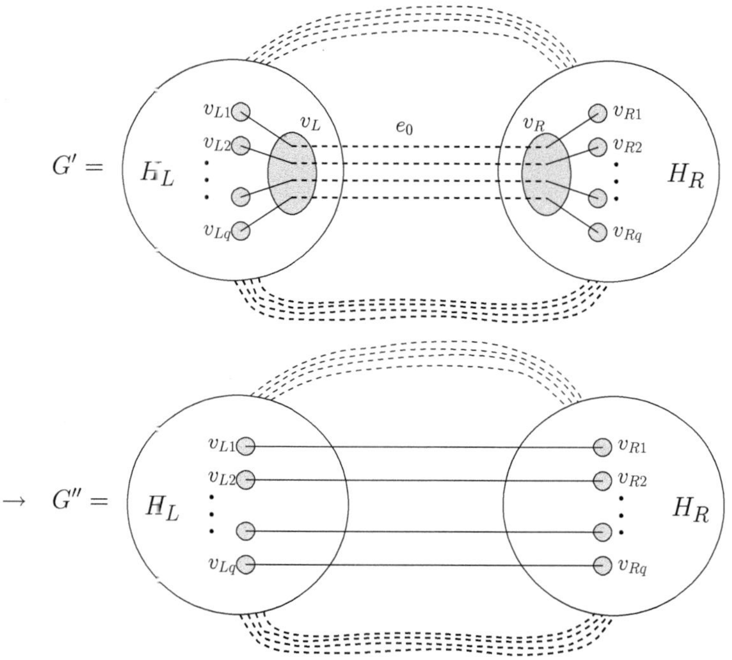

. (15.20)

It removes v_L, v_R and e_0 and joins the pending fermionic lines which were connected by the strands of e_0. The key point is that $G'' \in \mathbb{G}$ now, since the contraction of e_0, connects the two disjoint components of G'_0 by q fermionic lines.

Let us now analyse the variations of the number of faces from G to G''. First, from G to G': in G the lines e_1, e_2 belong to the same face, while e_L and e_R may or may not belong to the same face in G', hence

$$F(G) \leq F(G'). \tag{15.21}$$

Then the contraction of e_0 does not change the number of faces. Indeed, the faces of G' which do not go along e_0 are not affected. As for those which go along e_0, they follow paths,

$$v_{Li} \to v_L \to v_R \to v_{Ri} \tag{15.22}$$

for $i = 1, \ldots, q$ (some of those q paths may belong to common faces). In G'', they become paths going directly from v_{Li} to v_{Ri}. There is thus a one-to-one correspondence between the faces of G' and those of G''. Therefore, $F(G) \leq F(G'')$.

To conclude the proof, notice that G'' has two vertices less than G. Therefore, we can perform a melonic insertion on any fermionic line of G'' to get a graph $\tilde{G} \in \mathbb{G}$ with $V(G) = V(\tilde{G})$ and

$$F(\tilde{G}) = F(G'') + q - 1 \tag{15.23}$$

as in Proposition 15.2.4. For $q > 1$ it comes that $F(G) < F(\tilde{G})$ and thus $G \notin \mathbb{G}_{\max}(V(G))$.

Let us now prove that

$$\bigcup_{V \text{ even}} \mathbb{G}_{\max}(V) = \{G \in \mathbb{G} \text{ s.t. } \delta(G) = 1\} = \{\text{melonic graphs}\}. \tag{15.24}$$

This, is equivalent to the main theorem of this subsection, which states:

Theorem 15.2.10 *The weight of $G \in \mathbb{G}$ is bounded by:*

$$\delta(G) \leq 1 \tag{15.25}$$

Moreover, the graphs such that $\delta(G) = 1$ are the melonic graphs.

Proof We proceed by induction. The graph $G_{\min}$ is melonic by definition. It has $F(G_{\min}) = q$ and $V(G_{\min}) = 2$ hence satisfies $\delta(G_{\min}) = 1$. Since it is the only graph on two vertices, the theorem indeed holds on two vertices.

Let $V \geq 4$ even. We assume the theorem is true up to $V - 2$ vertices and consider $G \in \mathbb{G}$ with $V(G) = V$ vertices.

We need to investigate pairs $\{e_1, e_2\}$ with e_1, e_2 two fermionic lines belonging in a common face. Notice that such a pair exists. If it was not the case, then all faces would be of length 2 (i.e. one fermionic line and one disorder line), which implies $G = G_{\min}$, which is impossible since G has $V \geq 4$ vertices.

Let $\{e_1, e_2\}$ be a pair of fermionic edges belonging to the same face. Proposition 15.2.9 implies that this pair of fermionic edges is a 2-cut in G. The graph therefore takes the form

$G =$ 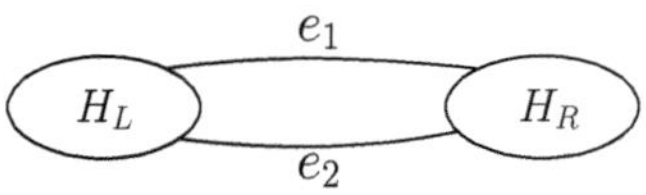

(15.26)

where H_L, H_R are connected, two-point graphs (in the sense that e_1 and e_2 are hanging out). We cut e_1 and e_2 and glue the resulting half-lines to close H_L and H_R into G_L, G_R, and use 'reverse' orientations as follows:

$G_L =$ 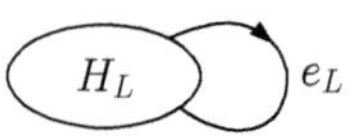

$G_R =$ 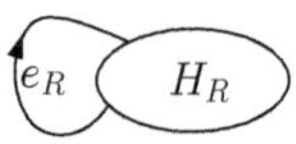

. (15.27)

We thus have

$$G = G_L^{(e_L)} \star G_R^{(e_R)}. \tag{15.28}$$

Since the edges e_1 and e_2 belong to the same face, we have

$$F(G) = F(G_L) + F(G_R) - 1, \tag{15.29}$$

and $F(G)$ is maximal iff $F(G_L)$ and $F(G_R)$ are. From the induction hypothesis, this requires G_L and G_R to be melonic. Then $G = G_L^{(e_L)} \star G_R^{(e_R)}$ is melonic too according to Proposition 15.2.7.

Let us end this subsection with the following result:

Corollary 15.2.11 *A graph $G \in \mathbb{G}$ is melonic iff all pairs $\{e_1, e_2\}$ of fermionic lines which belong in a common face are 2-cuts.*

The large N melonic dominance of the SYK model being now proved, in the rest of this review we represent SYK graphs with disorder edges as regular edges, without the stranded structure explained in this section.

15.3 The coloured SYK model

15.3.1 Definition of the model, real, and complex versions

The SYK generalization we study in this section contains q flavours of fermions. Moreover, each fermion of a given flavour appears exactly once in the interaction and the Lagrangian couples q fermions together. The action writes:

$$S = \int d\tau \left(\frac{1}{2} \sum_{f=1}^{q} \sum_{i=1}^{N} \psi_i^f \frac{d}{dt} \psi_i^f - \frac{i^{q/2}}{q!} \sum_{i_1, \ldots, i_q = 1}^{N} j_{i_1 \ldots i_q} \psi_{i_1}^1 \ldots \psi_{i_q}^q, \right). \tag{15.30}$$

Note that we use superscripts to denote the flavour. Moreover, in order to simplify the notations, the model has $q \cdot N$ fermions - we have N fermions of a given flavour.

The SYK generalization introduced previously is a particular case of the Gross–Rosenhaus generalization Gross and Rosenhaus (2017). Indeed, Gross and Rosenhaus took a number f of flavours, with N_a fermions of flavour a, each appearing q_a times in the interaction, such that $N = \sum_{a=1}^{f} N_a$ and $q = \sum_{a=1}^{f} q_a$. In our case, the number f of flavours is equal to q, $q_a = 1$ and $N_a = N$ (recall that we have now a total of $q \cdot N$ fermions). The action (15.30) is close in spirit to the action of the coloured tensor model Gurau (2011a), hence the name coloured SYK model.

Thus, the Feynman graphs obtained through perturbative expansion of the action (15.30) are edge-coloured graphs where the colours are the flavours. At each vertex, each of the q fermionic fields which interact has one of the q flavours, and each flavour is present exactly once.

An example of such a Feynman graph is given on the right of Fig. 15.4 (while on the left side we have a melonic graph of the SYK model). The disorder edge, represented,

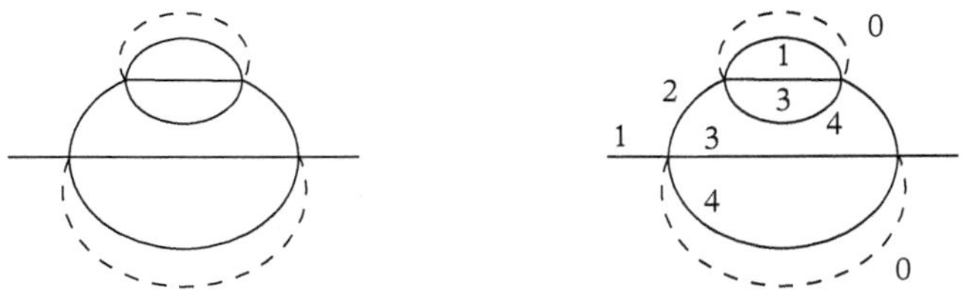

Figure 15.4 *Melonic graphs of the SYK and coloured SYK models*

as in the previous section, as a dashed edge, can be considered to have the fictitious flavour 0.

There is also a complex version of the model (15.30), a version initially mentioned in Gurau (2017a). This latter version can be easily obtained by using complex fields and by considering the interacting term in (15.30) as well as its complex conjugate:

$$\int d\tau \left(\frac{1}{2}\sum_{f=1}^{q}\sum_{i=1}^{N} \bar{\psi}_i^f \frac{d}{dt}\psi_i^f - \frac{i^{q/2}}{q!}\sum_{i_1,\ldots,i_q=1}^{N} j_{i_1\ldots i_q}\psi^1_{i_1}\ldots\psi^q_{i_q} - \frac{(-i)^{q/2}}{q!}\sum_{i_1,\ldots,i_q=1}^{N} \bar{j}_{i_1\ldots i_q}\bar{\psi}^1_{i_1}\ldots\bar{\psi}^q_{i_q}, \right). \tag{15.31}$$

The Feynman graphs obtained through perturbative expansion of the complex action have the same structure as the one previously explained for the real model (15.30). However, in the complex case, one has two types of vertices, which we can refer to as white and black, as it is done in the tensor model literature (see, for example, the book Gurau (2017a), and references within). Each edge connects a white to a black vertex. The Feynman graphs of (15.31) are thus the subset of the Feynman graphs of (15.30) which are bipartite. This is a feature which simplifies the diagrammatic analysis of the complex model.

15.3.2 Diagrammatics of the real and complex model

For each colour $i \in \{1,\ldots,q\}$, a Feynman graph has cycles (i.e. closed paths) which alternate the colours 0 and i. We call them **faces of colours** $0i$. This terminology is an extension of matrix models where those cycles are faces of ribbon graphs.

We denote by $F_{0i}(G)$ the number of faces of colours $0i$ for $i = 1,\ldots,q$ of a graph G,

$$F_0(G) = \sum_{i=1}^{q} F_{0i}(G), \tag{15.32}$$

the total number of faces which have the colour 0. We further denote by $E_0(G)$ the number of edges of colour 0 of the graph G (which is, as in the SYK case of the previous section, half the number of vertices of the graph G). In the large N limit, the Feynman amplitude of a coloured SYK graphs is given by

$$N^{\chi_0(G)}$$

where the **coloured SYK degree** is

$$\chi_0(G) = F_0(G) - (q-1)E_0(G). \tag{15.33}$$

Using coloured tensor model results (see, for example Bonzom, Gurau, and Rivasseau (2012)), one can prove:

$$\chi_0(G) = F_0(G) - (q-1)E_0(G) \leq \begin{cases} 1 & \text{if } G \text{ is a vacuum graph,} \\ 0 & \text{if } G \text{ is a two-point graph.} \end{cases} \tag{15.34}$$

The case of four-point graphs will be discussed later.

In the language of Bonzom, Gurau, and Rivasseau (2012), the graphs of the coloured SYK models have a single bubble, i.e. a single connected component after removing the edges of colour 0, since this bubble is the underlying, connected fermionic graph at fixed couplings.

LO, NLO of a vacuum, and two-point graphs

Notice that all two-point graphs are obtained by cutting an edge e of colour $i \in \{1, \dots, q\}$ in a vacuum graph G. Since there is a single face, with colours $0i$, which goes through e in G, cutting it decreases the exponent of N by one exactly.

To study $\chi_0(G)$, we perform in G the contraction of the edges of colour 0 to get the graph $G_{/0}$,

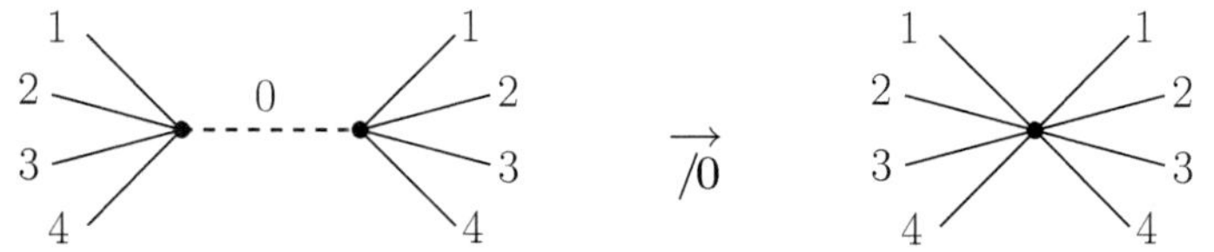

(15.35)

This means that two vertices of G connected by an edge of colour 0 become a single vertex in $G_{/0}$. The map $G \mapsto G_{/0}$ is not one-to-one because of this. Nevertheless, in a complex case, where G is bipartite, it can be made one-to-one by orienting the edges from, say, ψ^i to $\bar{\psi}^i$, i.e. from white to black vertices. Then the edges of $G_{/0}$ are oriented, and this is sufficient to reconstruct G. In the real case, G is not always bipartite and there are typically several graphs G for the same $G_{/0}$.

The main property of $G_{/0}$ is that all q colours are incident exactly twice on each vertex. Therefore, the edges of colour i form a disjoint set of cycles (we recall that a cycle is a closed path which visits its vertices only once). Let $\ell_i(G_{/0})$ be the number of cycles of edges of colour i. From the construction of $G_{/0}$, its cycles of colour i are the faces of colours $0i$ of G,

$$F_{0i}(G) = \ell_i(G_{/0}). \tag{15.36}$$

Let us introduce $L(G_{/0})$ the cyclomatic number of $G_{/0}$, i.e. its number of independent cycles, or first Betti number. As is well-known, it is the number of edges of $G_{/0}$ minus its number of vertices plus one. The number of edges of $G_{/0}$ is the number of edges of G with colours in $\{1,\dots,q\}$, thus $qE_0(G)$. The number of vertices of $G_{/0}$ simply is $E_0(G)$, so that

$$L(G_{/0}) = (q-1)E_0(G) + 1. \tag{15.37}$$

This shows that

$$\chi_0(G) = \sum_{i=1}^{d} \ell_i(G_{/0}) - L(G_{/0}) + 1-, \tag{15.38}$$

which has a simple graphical interpretation: it is minus the number of multicoloured cycles. Indeed, a cycle can be single-coloured or multicoloured. The former are counted by $\sum_{i=1}^{q} \ell_i(G_{/0})$, while $L(G_{/0})$ counts the total number of cycles. Therefore, their difference leaves precisely the number of cycles $\ell_m(G_{/0})$ which are multi-coloured, up to a sign,

$$\chi_0(G) = -\ell_m(G_{/0}) + 1. \tag{15.39}$$

The classification of graphs G with respect to $\chi_0(G)$ is therefore obtained from $\ell_m(G_{/0})$.

The LO large N limit graphs are graphs which satisfy $\ell_m(G_{/0}) = 0$, i.e. $G_{/0}$ has no multicoloured cycles. It means that it is made of single-coloured cycles which are glued without forming additional cycles. The corresponding graphs G are easily seen to be melonic. Indeed, one starts from $G_{/0}$ being a simple single-coloured cycle of colour $i \in \{1,\dots,q\}$ with loops of all other colours on its vertices. Then, each vertex of $G_{/0}$ is replaced with a pair of vertices and each loop becomes an edge between them. The colour 0 from the average over disorder is added between the vertices of each pair too. One gets a melonic cycle as follows:

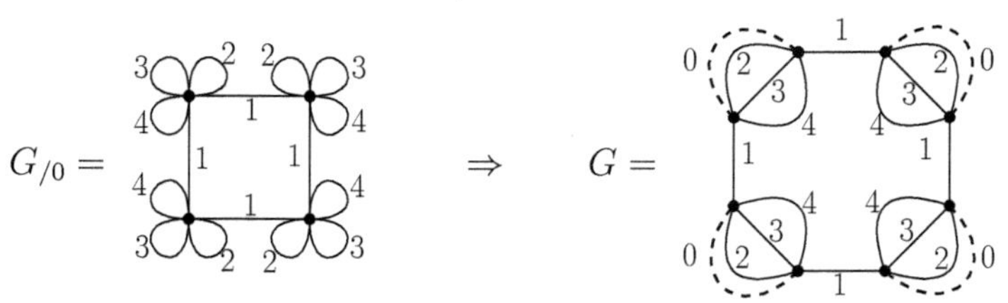

(15.40)

More general $G_{/0}$ are obtained by cutting a loop, say of colour 2, and replacing it with a cycle and loops attached to its vertices. This corresponds to cutting an edge of colour

2 in G and gluing another melonic cycle. This recursive process generates all the graphs corresponding to the large N limit.

The large N two-point function is simply obtained by cutting an edge of colour $i \in \{1, \ldots, q\}$. From this recursive process, one finds the following description of the large N, fully dressed propagator

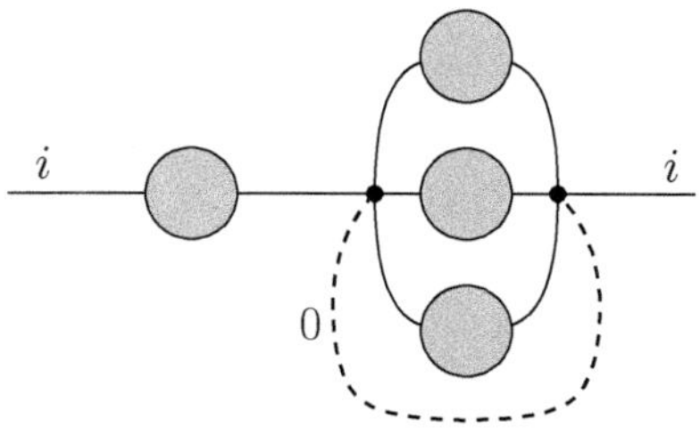

(15.41)

where each grey blob reproduces the same structure.

One can check that replacing an edge in G with any LO two-point function of the form (15.41) does not change $\chi_0(G)$. Therefore, *all solid edges in the remaining of the article are large N, fully dressed propagators.*

Two-point functions in the representation as $G_{/0}$ are simply obtained by contracting all edges of colour 0 of two-point graphs G. Therefore, solid edges in $G_{/0}$ will also represent fully dressed propagators from now on.

At the NLO, one finds graphs such that $\ell_m(G_{/0}) = 1$, i.e. $G_{/0}$ has a single multicoloured cycle. Compared to the large N limit, this means that one obtains $G_{/0}$ by gluing single-coloured cycles (with loops attached to their vertices) so as to form a single multicoloured cycle.

Considering that solid edges are fully dressed two-point functions, the NLO graphs $G_{/0}$ are completely characterized by the length n of the multicoloured cycle with colours $i_1, i_2, \ldots, i_n$. For instance, at length $n = 6$:

$G_{/0}^{\mathrm{NLO}} =$ 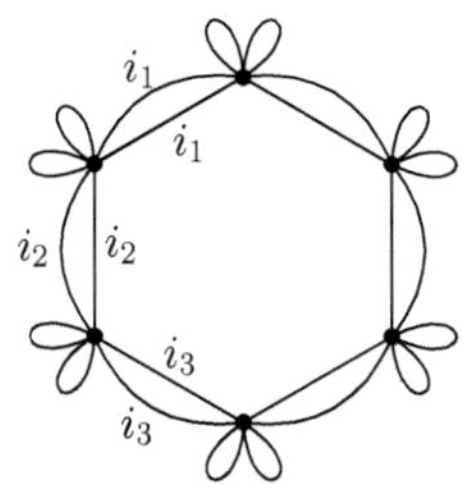 .

(15.42)

To find the corresponding graphs G, one splits each vertex of $G_{/0}$ into two vertices connected by an edge of colour 0 and so that each colour is incident exactly once on each vertex. There are several ways to connect the edges of colour i_j and i_{j+1} to a pair of vertex. Overall, this leads to the two following families of graphs

$G^{\rm NLO} =$ 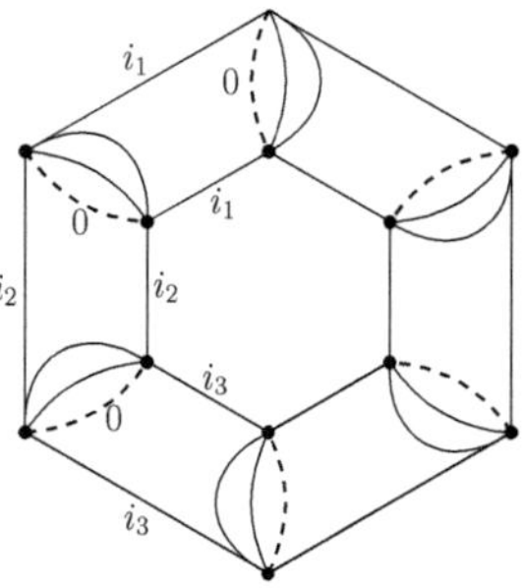

and $\tilde{G}^{\rm NLO} =$ 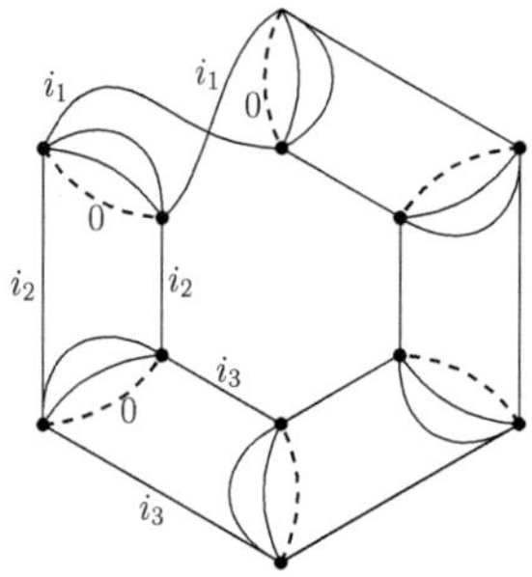

. (15.43)

Notice that $G^{\rm NLO}$ is bipartite (recall that the melonic two-point functions are) while $\tilde{G}^{\rm NLO}$ is not. The graph $\tilde{G}^{\rm NLO}$ is obtained from $G^{\rm NLO}$ by crossing two edges, say with colours i_1. Adding more crossings is always equivalent to $G^{\rm NLO}$ (for an even number of crossings) or $\tilde{G}^{\rm NLO}$ (for an odd number of crossings).

To remember that the previous graphs can have arbitrary lengths, and also to offer a convenient representation of NLO two-point functions (to come), we introduce *chains* which are four-point graphs,

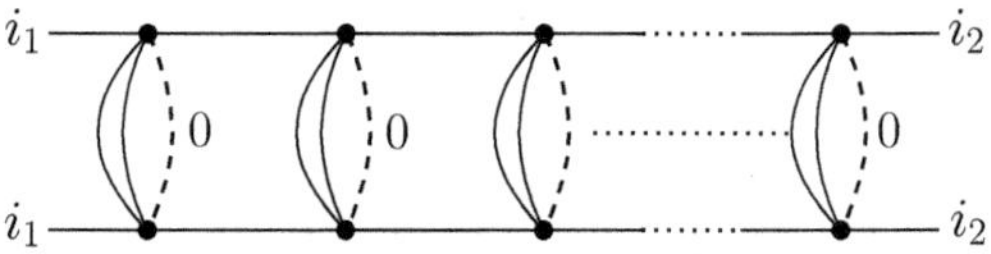

. (15.44)

A combinatorial detail of importance is that a chain can have down to two vertices only and has at least two vertices unless stated otherwise. We will represent arbitrary choices of chains as boxes,

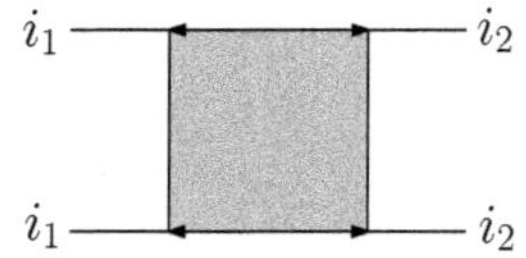

. (15.45)

where the arrows indicate the direction of the chain (the box by itself being symmetric).

This enables to represent the two families of NLO vacuum graphs as

$G^{\rm NLO} =$ 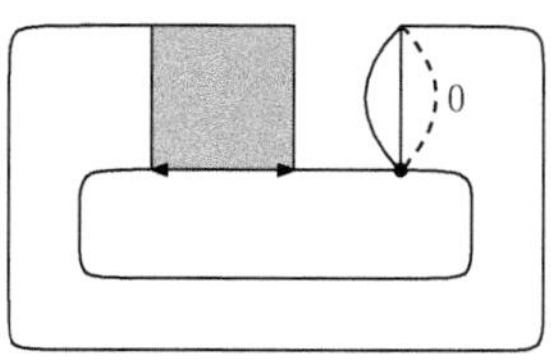

and $\tilde{G}^{\rm NLO} =$ 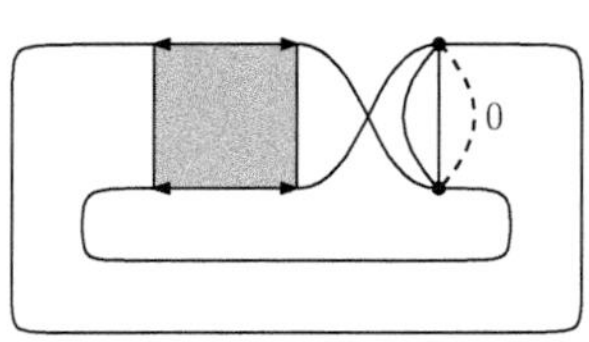

. (15.46)

To get the two-point NLO graphs from vacuum graphs, it is sufficient to cut an edge of a given colour $i \in \{1, \ldots, q\}$ in a NLO vacuum graph. However, we have to remember that we have used dressed propagators in (15.43). For instance, G^{NLO} really is

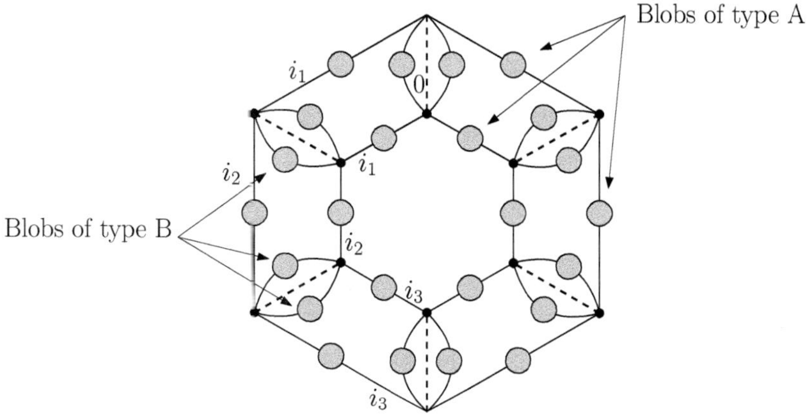

, (15.47)

where the grey blobs represent arbitrary LO, two-point functions. One might (not necessarily but typically) cut an edge which is contained in a grey blob of (15.47). There are two cases to distinguish depending on where an edge is cut in (15.47), because there are two types of blobs in (15.47).

- Blobs of type A are inserted on the $2n$ edges of colours $i_1, \ldots, i_n$ which are characterized as follows: such an edge connects two vertices which are not incident to the same edge of colour 0.
- Blobs of type B are the others: they are inserted on the edges whose end points are incident to the same edge of colour 0.

If the cut edge is chosen within a blob of type A, then there are two types of NLO two-point graphs:

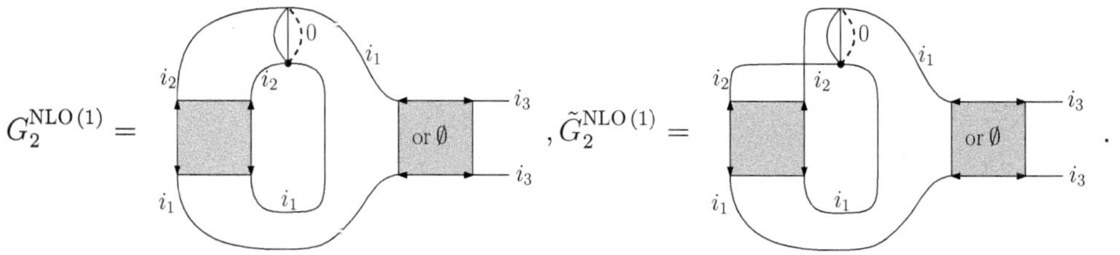

(15.48)

Only $G_2^{\mathrm{NLO}(1)}$ is bipartite (and would thus contribute in a complex model).

If the cut edge is within a blob of type B, then the NLO two-point contributions are

$$G_2^{\mathrm{NLO}\,(2)} =$$

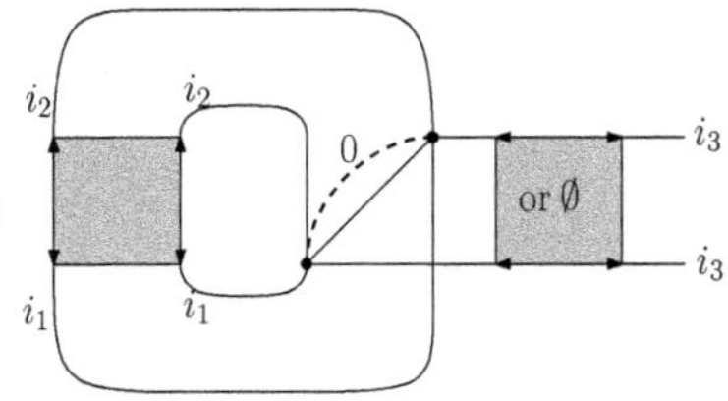

$$,\ \tilde{G}_2^{\mathrm{NLO}\,(2)} =$$

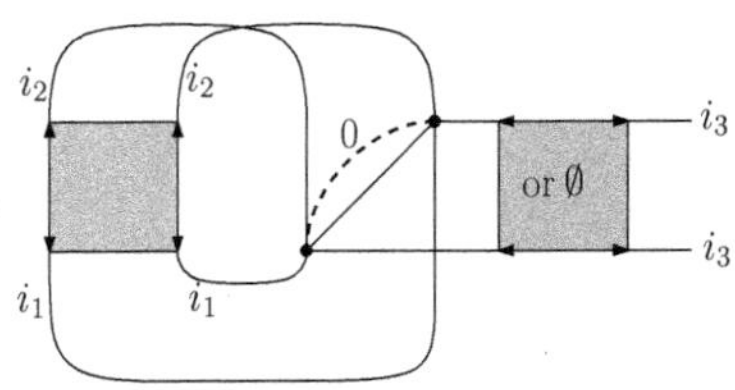

$$. \tag{15.49}$$

Again, only $G_2^{\mathrm{NLO}(2)}$ is bipartite.

Let us give a specific example of a Feynman diagram which can be obtained as a particular case of $G_2^{\mathrm{NLO}(2)}$ previously illustrated:

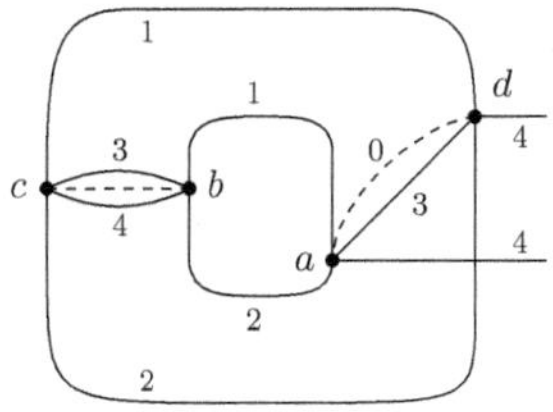

$$=$$

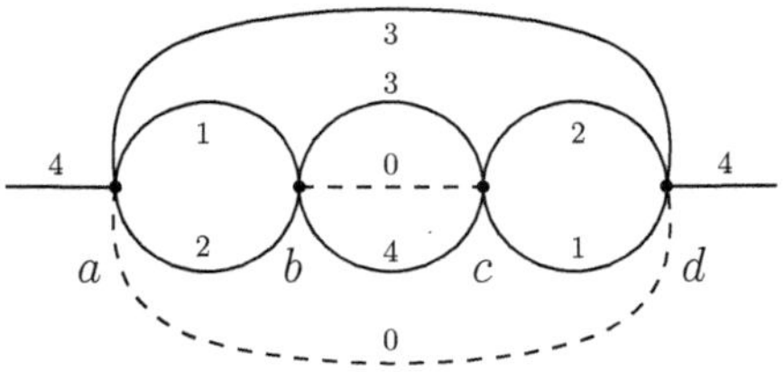

$$. \tag{15.50}$$

Thus, on the left-hand side of (15.50) one has the Feynman diagram obtained if the chain on the left-hand side of $G_2^{\mathrm{NLO}(2)}$ (i.e. the chain attached to internal edges) has only two vertices, while the chain on the right-hand side of $G_2^{\mathrm{NLO}(2)}$ (i. e. the chain attached to external edges) is empty. On the right-hand side of (15.50) we redraw the Feynman diagram thus obtained.

The method we have used to identify LO and NLO contributions to the free energy and two-point function can in principle be applied at any order. However, the number of diagrams grows importantly, and the description becomes tedious. Here, we therefore only give the diagrams which contribute to the NNLO of the partition function.

Graphs contributing to the NNLO are such that the corresponding $G_{/0}$ have exactly two independent multicoloured cycles,

$$\ell_m(G_{/0}^{\mathrm{NNLO}}) = 2. \tag{15.51}$$

A reasoning similar to the NLO case of Section 15.3.2 leads to families of graphs such as the following ones:

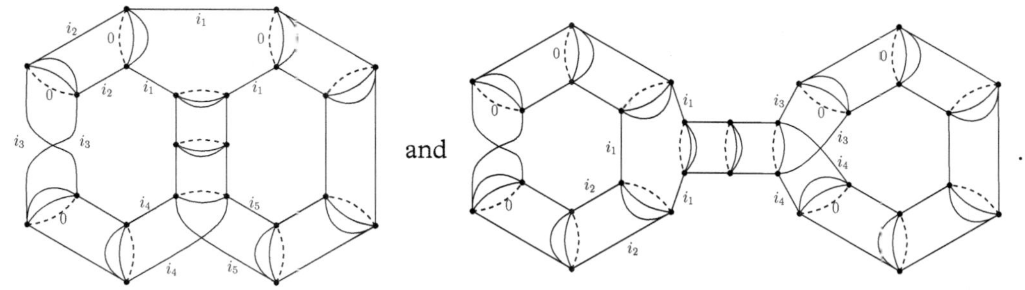

(15.52)

A collection of diagrams is pictured below. To obtain all such graphs, one has to consider one crossing or no crossing in every loop in every possible way.

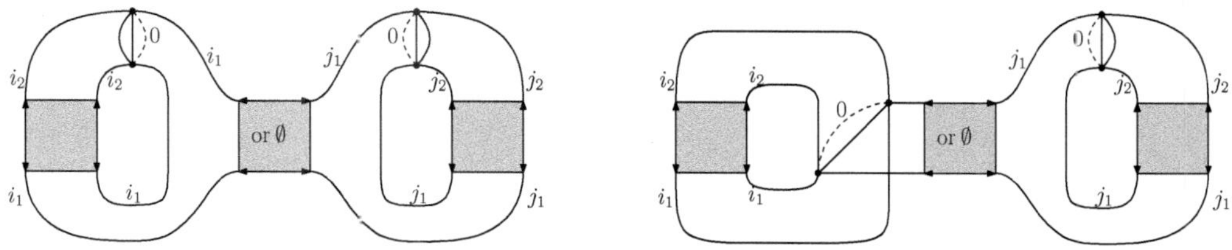

(15.53)

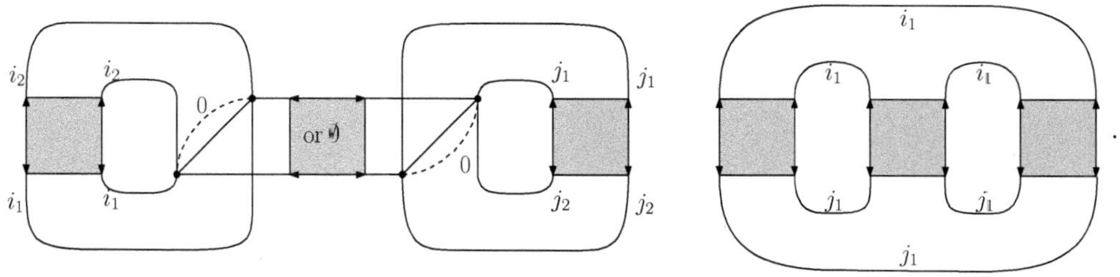

(15.54)

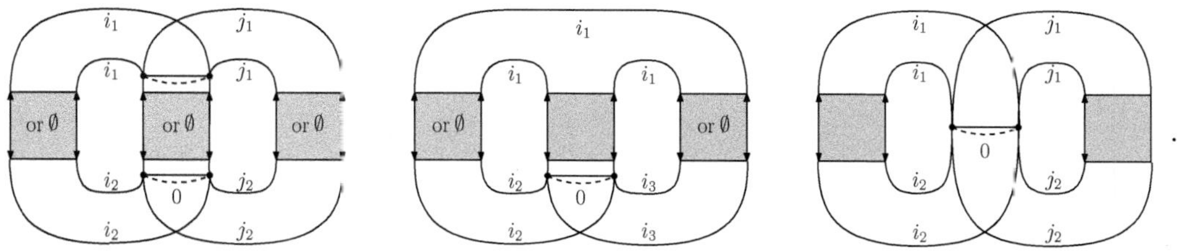

(15.55)

LO and NLO of four-point functions

One can check that the external edges of four-point graphs come in pairs where two edges of a pair share the same colour. This gives two sets of two-point functions, depending on whether all external edges have the same colour or not,

$$\langle\psi_i\psi_i\psi_i\psi_i\rangle \quad \text{for } i \in \{1,\dots,q\}, \text{ and } \quad \langle\psi_i\psi_i\psi_j\psi_j\rangle \quad \text{for } i \neq j, \tag{15.56}$$

where ψ_i, ψ_j are fermions of colours i and j. Here we have dropped the time dependence since we are only concerned with the diagrammatics.

There are no major diagrammatic differences between the two types of four-point functions. We will thus treat both simultaneously.

Four-point graphs can be obtained by cutting two edges in a vacuum graph. They can be two edges with the same colour or two different colours in $\{1,\dots,q\}$. If G is a vacuum graph, we denote $G_{e,e'}$ the four-point graph obtained by cutting e and e'. Obviously, if G_4 is a four-point graph, there is a (possibly non-unique) way to glue the external lines two-by-two, creating two edges e, e', and to thus get a vacuum graph G such that $G_4 = G_{e,e'}$.

Faces of G and $G_{e,e'}$ are the same except for those which go along e and e'. When e and e' have distinct colours, two different faces go along them in G, and are thus broken in $G_{e,e'}$. When e and e' have the same colour, there can be one or two faces along them. Therefore, the weight received by $G_{e,e'}$ reads

$$w_N(G_{e,e'}) = N^{\chi_0(G_{e,e'})}, \qquad \text{with} \qquad \chi_0(G_{e,e'}) = \chi_0(G) - \eta(G_{e,e'}) \leq 1 - \eta(G_{e,e'}), \tag{15.57}$$

where $\eta(G_{e,e'}) \in \{1,2\}$ is the number of faces broken by cutting e and e' in G.

The classification thus seems a little intricate because of the two possible values for $\eta(G_{e,e'})$. We however claim that it is sufficient to only consider the graphs G with edges e, e' such that

$$\eta(G_{e,e'}) = 2. \tag{15.58}$$

This is always the case when e and e' have different colours. Let us thus focus on the case where e and e' have the same colour $i \in \{1,\dots,q\}$. Let G_4 be a four-point graph with four external legs of colour i. We claim that there is always one way to connect the external legs pairwise into two edges e and e' with two different faces along them. Denoting G in this vacuum graph, we thus interpret G_4 as the graph $G_{e,e'}$ with $\eta(G_{e,e'}) = 2$.

With the same notations, we have thus found that

$$\chi_0(G_{e,e'}) = \chi_0(G) - 2. \tag{15.59}$$

The strategy is thus for both types of four-point functions:

- use the classification of vacuum graphs which we have established: LO, NLO graphs, etc.; and
- cut two edges of them such that $\eta(G_{e,e'}) = 2$.

In the large N limit, cutting two edges in melonic graphs (such that $G_{e,e,'}$ remains connected, as well as $G_{e,e'}$ minus its edges of colour 0) leads precisely to the chains introduced in (15.44) (one might add two-point insertions on the external legs).

At the NLO, one finds:

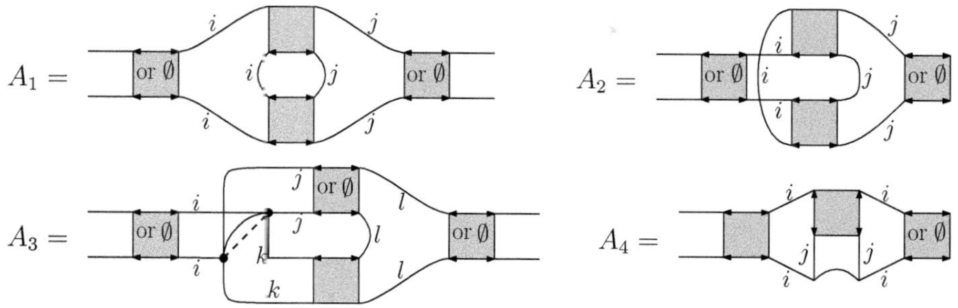

(15.60)

by cutting an edge in $G_2^{\mathrm{NLO}(1)}$ in (15.48),

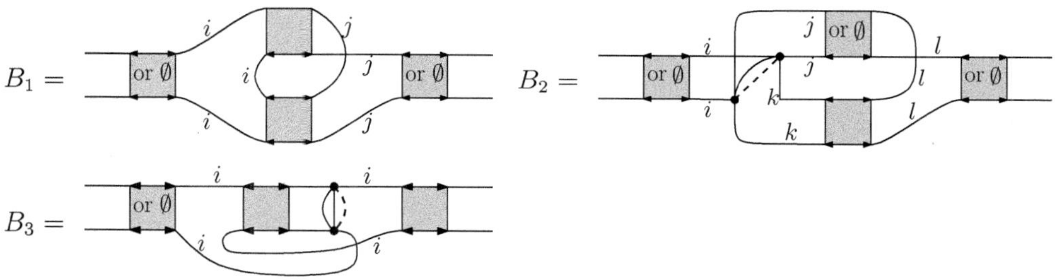

(15.61)

by cutting an edge in $\tilde{G}_2^{\mathrm{NLO}(1)}$ in (15.48),

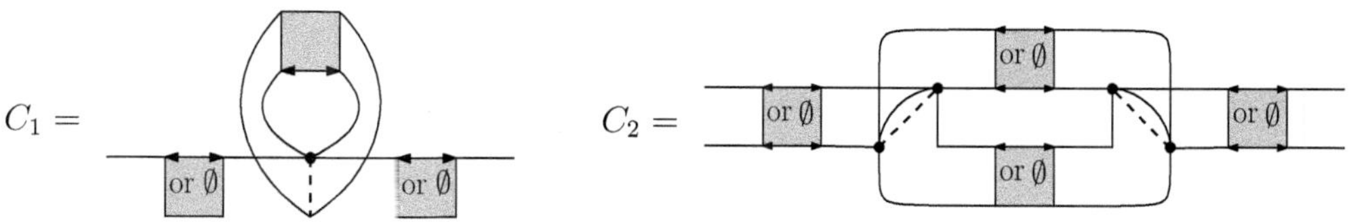

(15.62)

by cutting an edge in $G_2^{\mathrm{NLO}(2)}$ in (15.49),

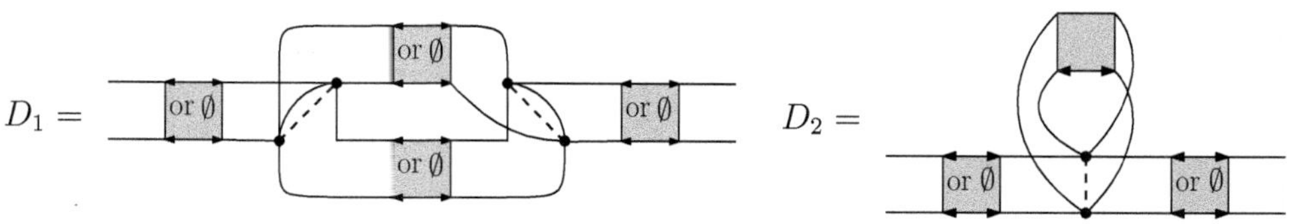

(15.63)

by cutting an edge in $\tilde{G}_2^{\mathrm{NLO}(2)}$ in (15.49), and finally the two following families

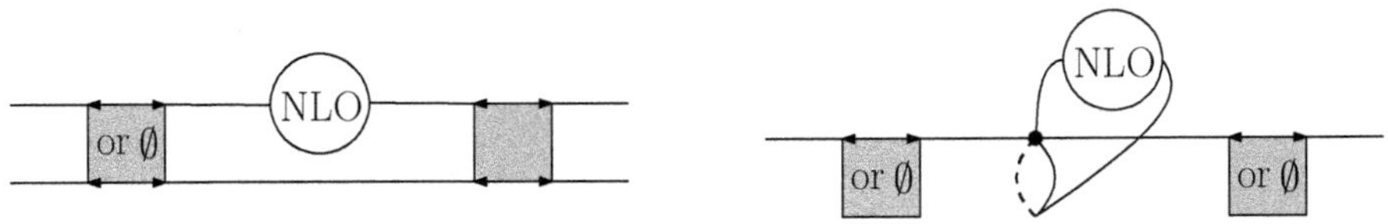

(15.64)

obtained by performing a two-point insertion of the NLO two-point function into the LO four-point chains.

15.3.3 More on the coloured SYK Feynman graphs

In this subsection we present an asymptotic combinatorial analysis of the Feynman graphs of the coloured SYK model. This subsection follows the original research article Fusy, Lionni, and Tanasa (2018). Let us start with the following definition:

Definition 15.3.1 *A connected regular $(q+1)$-edge-coloured graph has edges carrying colours in $\{0,\cdots,q\}$, so that each of the colours reaches every vertex precisely once. Throughout the text, we will simply refer to such a graph as a coloured graph.*

A coloured graph is said to be rooted if one of its colour-0 edges is distinguished and oriented. It is said to be bipartite if its vertices are coloured in black and white so that every edge links a black and a white vertex. For a rooted bipartite graph, we take the convention that the vertex at the origin of the root-edge is black. We denote by $\tilde{\mathbb{G}}^q$ the family of connected rooted $(q+1)$-edge-coloured graphs, and $\mathbb{G}^q$ the subfamily of bipartite graphs from $\tilde{\mathbb{G}}^q$.

As already mentioned in the previous sections, a connected coloured graph is an SYK graph if it remains connected when all the colour-0 edges are deleted. We denote by $\tilde{\mathbb{G}}^q_{\mathrm{SYK}}$ the family of rooted $(q+1)$-edge-coloured SYK graphs, and $\mathbb{G}^q_{\mathrm{SYK}}$ the subset of bipartite SYK graphs.

In Fig. 15.5 we give an example of a generic coloured graph and of a bipartite SYK graph.

Denoting the number of vertices of a coloured graph G by $V(G)$ we define its *order*,

$$\delta_0(G) = 1 + \frac{q-1}{2} V(G) - F_0(G). \tag{15.65}$$

We let $\tilde{g}_{n,\delta}$ (resp. $g_{n,\delta}$) be the number of rooted (resp. rooted bipartite) coloured graphs of fixed order δ with $2n$ vertices, and $\tilde{c}_{n,\delta}$ (resp. $c_{n,\delta}$) be the number of rooted (resp. rooted bipartite) SYK graphs of fixed order δ with $2n$ vertices.

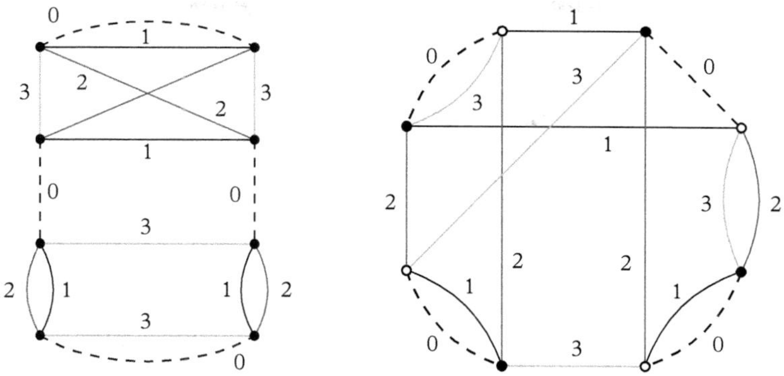

Figure 15.5 *A generic four-coloured graph and a bipartite SYK graph*

We also denote by m_δ the number of rooted[1] trivalent maps with $2\delta-2$ vertices, which is given by (see A062980 in OEIS) the recurrence

$$m_1 = 1,\ m_2 = 5,\ m_\delta = (6\delta - 8)m_{\delta-1} + \sum_{k=1}^{\delta-1} m_k \cdot m_{\delta-k} \quad \text{for } \delta \geq 3,$$

and define the constant κ_δ as $\kappa_0 = \sqrt{q/2\pi(q-1)^3}$ and

$$\kappa_\delta = \frac{1}{\Gamma\left(\frac{3\delta-1}{2}\right)} \frac{2}{q(q-1)} \left(\frac{q-1}{2q^3}\right)^{\frac{3\delta-1}{2}} \left(\frac{q^4}{4}\right)^{\delta} m_\delta, \quad \text{for } \delta \geq 1.$$

Let us now state the main result of this subsection:

Theorem 15.3.2 *For $\delta \geq 0$ and $q \geq 3$, the numbers $g_{n,\delta}$, $c_{n,\delta}$, $\frac{1}{2^\delta}\tilde{g}_{n,\delta}$, and $\frac{1}{2^\delta}\tilde{c}_{n,\delta}$ behave asymptotically as*

$$\kappa_\delta \cdot n^{3(\delta-1)/2} \cdot \gamma^n, \tag{15.66}$$

where $\gamma = \frac{(q+1)^{q+1}}{q^q}$, and κ_δ has been defined.

Proof See the original article Fusy, Lionni, and Tanasa (2018).

As a consequence, for $q \geq 3$, if we let $G_{n,\delta}$ be a random rooted $(q+1)$-edge-coloured graph of order δ and with $2n$ vertices, then for δ fixed and $n \to \infty$:

$$\mathbb{P}(G_{n,\delta} \text{ is SYK}) \to 1, \qquad \text{and} \qquad \mathbb{P}(G_{n,\delta} \text{ is bipartite} \mid G_{n,\delta} \text{ is SYK}) \to 2^{-\delta}. \tag{15.67}$$

[1] A map is called rooted if it has a marked edge that is given a direction.

15.3.4 Non-Gaussian disorder average in the complex model

In this subsection we consider non-Gaussian disorder for the SYK coupling. We consider the complex version of the SYK model containing q flavours with the non-Gaussian disorder, whose action is given by (15.31). Following Krajewski et. al. (2019), one can derive the effective action for this model and show that the effect of this non-Gaussian averaging is a modification of the variance of the Gaussian distribution of couplings at leading order in N.

In order to obtain these results, we first need to average the partition function over the non-Gaussian disorder. The most convenient way to perform this is through the use of replicas. We thus add an extra replica index $r = 1,\dots,n$ to the fermions. One has:

$$\langle \log Z(\jmath)\rangle_\jmath = \lim_{n\to 0}\frac{\langle Z^n(\jmath)\rangle_\jmath - 1}{n}, \tag{15.68}$$

with

$$Z^n(\jmath) = \int \prod_{1\le r\le n} [d\psi_r][d\overline{\psi}_r]\exp\sum_r S_\jmath(\psi_r,\overline{\psi}_r), \tag{15.69}$$

where $S_\jmath(\psi,\overline{\psi})$ is given by (15.31). The angle brackets stand for the averaging over $\jmath$, which is performed with a non-Gaussian weight of the type

$$\langle Z^n(\jmath)\rangle_\jmath = \frac{\int d\jmath d\bar{\jmath}\, Z^n(\jmath)\exp\left[-\left[\frac{N^{q-1}}{\sigma^2}\jmath\bar{\jmath} + V_N(\jmath,\bar{\jmath})\right]\right]}{\int d\jmath d\bar{\jmath}\exp\left[-\left[\frac{N^{q-1}}{\sigma^2}\jmath\bar{\jmath} + V_N(\jmath,\bar{\jmath})\right]\right]}. \tag{15.70}$$

We further impose that the potential V_N is invariant under independent unitary transformations:

$$\jmath_{i_1,\dots,i_q} \to \sum_{j_1,\dots,j_q} U^1_{i_1 j_1}\cdots U^q_{i_q j_q}\jmath_{j_1,\dots,j_q}, \qquad \bar{\jmath}_{i_1,\dots,i_q} \to \sum_{j_1,\dots,j_q} \overline{U}^1_{i_1 j_1}\cdots \overline{U}^q_{i_q j_q}\bar{\jmath}_{j_1,\dots,j_q}. \tag{15.71}$$

Assuming that the potential V_N is a polynomial (or an analytic function) in the couplings $\jmath$ and $\bar{\jmath}$, this invariance imposes that the potential can be expanded over not-necessarily connected graphs. These graphs are made up by black and white vertices of valence q, whose edges connect only black to white vertices (bipartite graphs) and are labelled by a colour $a = 1,\dots,q$ in such a way that, at each vertex, the q incident edges carry distinct colours (we thus have edge-coloured graphs). Each of these graphs can be canonically associated by a particular contraction of the tensors $\jmath$ and $\bar{\jmath}$. The contraction of their indices means that each white vertex carries a tensor $\jmath$, each black vertex a tensor $\bar{\jmath}$, and that the indices have to be contracted by identifying two indices on both sides of an edge, the place of the index in the tensor being defined by the colour of the edge denoted by $c(e)$. We will refer to the graph G, using the shorthand $\langle \jmath,\bar{\jmath}\rangle_G$, which is given by

$$\langle j,\bar{j}\rangle_G = \sum_{1\leq i_{v,a},\ldots,i_{\bar{v},a}\leq N} \prod_{\substack{\text{white}\\ \text{vertices } v}} j_{i_{v,1},\ldots,i_{v,q}} \prod_{\substack{\text{black}\\ \text{vertices } \bar{v}}} \bar{j}_{i_{\bar{v},1},\ldots,i_{\bar{v},q}} \prod_{\substack{\text{edges}\\ e=(v,\bar{v})}} \delta_{i_{v,c(e)},i_{\bar{v},c(e)}}. \tag{15.72}$$

The most general form of the potential V_N is then expanded over these graphs as

$$V_N(j,\bar{j}) = \sum_{\text{graph } G} \lambda_G \frac{N^{q-k(G)}}{\text{Sym}(G)} \langle j,\bar{j}\rangle_G. \tag{15.73}$$

In this expression, λ_G is a real number, $k(G)$ is the number of connected components of G and $\text{Sym}(G)$ its symmetry factor. The Gaussian term corresponds to a dipole graph (a white vertex and a black vertex, connected by q lines) and reads

$$\frac{N^{q-1}}{\sigma^2} j\bar{j} = \frac{N^{q-1}}{\sigma^2} \sum_{1\leq i_1,\ldots,i_q\leq N} j_{i_1,\ldots,i_q} \bar{j}_{i_1,\ldots,i_q}. \tag{15.74}$$

Introducing the pair of complex conjugate tensors K and $\overline{K}$ defined by

$$K_{i_1,\ldots,i_q} = \mathrm{i}^{\frac{q}{2}} \sum_r \int dt \psi^1_{i_1,r} \cdots \psi^q_{i_q;r} \qquad \overline{K}_{i_1,\ldots,i_q} = \mathrm{i}^{\frac{q}{2}} \sum_r \int dt \overline{\psi}^1_{i_1,r} \cdots \overline{\psi}^q_{i_q;r}, \tag{15.75}$$

the averaged partition function reads

$$\langle Z^n(j)\rangle_G = \frac{\int [d\psi][d\overline{\psi}] \exp\left[-\int dt \sum_{a,i_a} \overline{\psi}^a_{i_a} \partial_t \psi^a_{i_a}\right] \int dj d\bar{j} \exp\left[-\left[\frac{N^{q-1}}{\sigma^2} j\bar{j} + V_N(j,\bar{j}) + j\overline{K} + \bar{j}K\right]\right]}{\int dj d\bar{j} \exp\left[-\left[\frac{N^{q-1}}{\sigma^2} j\bar{j} + V_N(j,\bar{j})\right]\right]}. \tag{15.76}$$

In order to study the large N limit of the average (15.76), we introduce the background fields $L = -\frac{\sigma^2}{N^{q-1}} K$ and $\overline{L} = -\frac{\sigma^2}{N^{q-1}} \overline{K}$. Let us shift the variables j and $\bar{j}$ by the background fields L and $\overline{L}$. The numerator in the integral (15.76) reads

$$\exp\left[-\frac{\sigma^2}{N^{q-1}} K\overline{K}\right] \int dj d\bar{j} \exp - \left[\frac{N^{q-1}}{\sigma^2} j\bar{j} + V_N\left(j - \frac{\sigma^2}{N^{q-1}} K, \bar{j} - \frac{\sigma^2}{N^{q-1}} \overline{K}\right)\right] \tag{15.77}$$

and the effective potential in the shifted variables is

$$V_N(s,L,\overline{L}) = -\log \int dj d\bar{j} \exp - \left[\frac{N^{q-1}}{s} j\bar{j} + V_N\left(j + L, \bar{j} + \overline{L}\right)\right] + N^q \log \frac{\pi s}{N^{q-1}}. \tag{15.78}$$

In this framework, s is a parameter that interpolates between the integral we have to compute, at $s=\sigma^2$ (up to a trivial multiplicative constant) and the potential we started with at $s=0$ (no integration and $j=\overline{j}=0$). The inclusion of the constant ensures that the effective potential remains zero when we start with a vanishing potential. This comes to

$$\int dj d\overline{j} \exp\left[-\left[\frac{N^{q-1}}{\sigma^2} j\overline{j} + V_N\left(j - \frac{\sigma^2}{N^{q-1}}K, \overline{j} - \frac{\sigma^2}{N^{q-1}}\overline{K}\right)\right]\right]$$
$$= \left(\frac{N^{q-1}}{\pi s}\right)^{N^q} \exp\left[-V_N\left(s=\sigma^2, L=-\frac{\sigma^2}{N^{q-1}}K, \overline{L}=-\frac{\sigma^2}{N^{q-1}}\overline{K}\right)\right]. \quad (15.79)$$

Using standard QFT manipulations (see for example, the book Jean Zinn-Justin (2020)), one can show that the effective potential $V_N(s,L,\overline{L})$ in equation (15.78) obeys the following differential equation:

$$\frac{\partial V}{\partial s} = \frac{1}{N^{q-1}} \sum_{1\le i_1,\dots,i_q\le N} \left(\frac{\partial^2 V}{\partial L_{i_1,\dots,i_q}\partial \overline{L}_{i_1,\dots,i_q}} - \frac{\partial V}{\partial L_{i_1,\dots,i_q}}\frac{\partial V}{\partial \overline{L}_{i_1,\dots,i_q}}\right). \quad (15.80)$$

One can represent this equation in a graphical way as shown in Fig. 15.6. The first term on the right-hand side corresponds to an edge closing a loop in the graph and the second term in the right-hand side corresponds to a bridge (also known as 1PR) edge or q cut.

Since the effective potential is also invariant under the unitary transformations defined in equation (15.71), it may also be expanded over graphs as in (15.72),

$$V_N(s,L,\overline{L}) = \sum_{\text{graph } G} \lambda_G(s) \frac{N^{q-k(q)}}{\text{Sym}(G)} \langle L, \overline{L}\rangle_G, \quad (15.81)$$

with s dependent couplings $\lambda_G(s)$. Inserting this graphical expansion in the differential equation (15.80), we obtain a system of differential equations for the couplings,

$$\frac{d\lambda_G}{ds} = \sum_{G'/(\overline{v}v)=G} N^{k(G)-k(G')+e(v,\overline{v})-q+1}\lambda_{G'} - \sum_{(G'\cup G'')/(\overline{v}v)=G} \lambda_{G'}\lambda_{G''}. \quad (15.82)$$

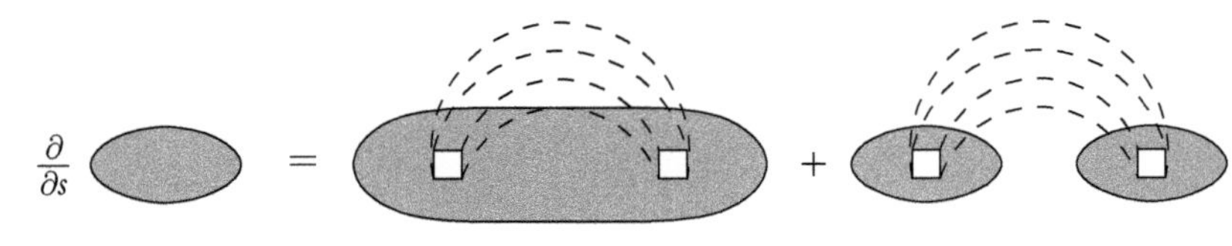

Figure 15.6 *Graphical representation of equation (15.80) for* $q=4$

A derivation of the potential V_N with respect to $L_{i_1,\dots,i_q}$ (resp. $\overline{L}_{i_1,\dots,i_q}$) removes a white vertex (resp. a black vertex). Then, the summation over the indices in $i_1,\dots,i_q$ in (15.80) reconnects the edges, respecting the colours.

In the first term on the right-hand side of (15.80), given a graph G in the expansion of the left-hand side, we have to sum over all graphs G' and pairs of a white vertex v and a black vertex $\overline{v}$ in G' such that the graph $G'/(\overline{v}v)$ obtained after reconnecting the edges (discarding the connected components made of single lines) is equal to G—see Figs. 15.7 and 15.8.

The number $e(v,\overline{v})$ is the number of edges directly connecting v and $\overline{v}$ in G. After summation over the indices, each of these lines yields a power of N, which gives the factor of $N^{e(v,\overline{v})}$.

The operation of removing two vertices and reconnecting the edges can at most increase the number of connected components (including the graphs made of single closed lines) by $q-1$, so that we always have $k(G)-k(G')+e(v,\overline{v})-q+1\leq 0$. We obtain the equality if and only if G' is a melonic graph. Therefore, in the large N limit, only melonic graphs survive in the first term on the right-hand side of (15.82) (this is further proof of the melonic dominance in the SYK model already shown in Theorem 15.2.10).

In the second term, we sum over graphs G' and white vertices $v\in G'$ and graphs G'' and black vertices $\overline{v}\in G''$, with the condition that the graph obtained after removing the

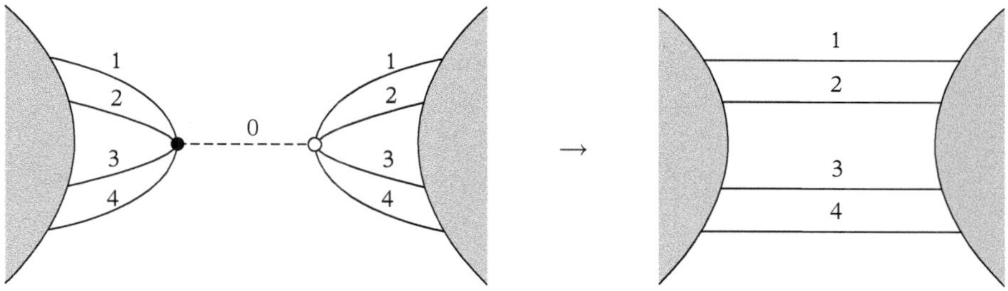

Figure 15.7 *Removal of a white and a black vertex and re-connection of the edges*

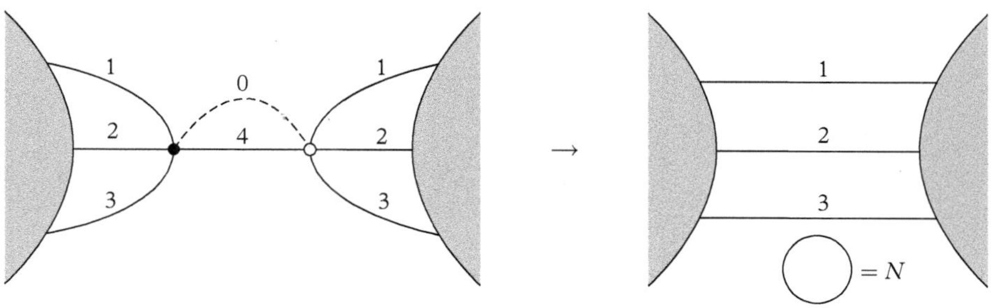

Figure 15.8 *Removal of a white and a black vertex and re-connection of the edges creating a loop*

vertices and reconnecting the lines $(G' \cup G'')/(\overline{v}v)$ is equal to G. In that case, the number of connected components necessarily diminishes by 1, so that all powers of N cancel.

The crucial point in the system (15.82) is that only negative (or null) powers of N appear. It can be written as

$$\frac{d\lambda_G}{ds} = \beta_0(\{\lambda_G\}) + \frac{1}{N}\beta_1(\{\lambda_G\}) + \dots. \tag{15.83}$$

As a consequence, if $\lambda_G(s=0)$ is bounded, then $\lambda_G(s)$ is also bounded for all s (i.e. it does not contain positive powers of N).

Let us now substitute $L = -\frac{\sigma^2}{N^{q-1}}K$ and $\overline{L} = -\frac{\sigma^2}{N^{q-1}}\overline{K}$ in the expansion of the effective potential (15.72),

$$\begin{aligned} &V_N\left(s=\sigma^2, L = -\frac{\sigma^2}{N^{q-1}}K, \overline{L} = -\frac{\sigma^2}{N^{q-1}}\overline{K}\right) \\ &= \sum_{\text{graph } G} \lambda_G(\sigma^2)\frac{(-\sigma^2)^{v(G)}N^{q-k(q)-(q-1)v(G)}}{\text{Sym}(G)}\langle K, \overline{K}\rangle_G. \end{aligned} \tag{15.84}$$

Here, $v(G)$ is the number of vertices of G. The exponent of N can be rewritten as $(q-1)(1-v(G)) + 1 - k(G)$. It has a maximal value for $v(G) = 2$ and $k(G) = 1$, which corresponds to the dipole graph. This is a re-expression the Gaussian universality property of random tensors.

Taking into account the non-Gaussian quenched disorder, we now derive the effective action for the bilocal invariants

$$\widetilde{G}^a_{r,r'}(t,t') = \frac{1}{N}\sum_i \psi^a_{i,r}(t_1)\overline{\psi}^a_{i,r'}(t'). \tag{15.85}$$

Note that these invariants carry one flavour label a and two replica indices r, r'.

To this end, let us come back to the partition function (15.76). We then express the result of the average over j and $\overline{j}$ as a sum over graphs G using the expansion of the effective potential (15.84) and replacing the tensors K and $\overline{K}$ in terms of the fermions ψ and $\overline{\psi}$ (see equation (15.75)).

Then, each graph G involves the combination

$$\begin{aligned} \langle K, \overline{K}\rangle_G = &\sum_{1\leq i_{v,a},\dots,i_{\overline{v},a}\leq N} \prod_{\substack{\text{white} \\ \text{vertices } v}} \sum_{r_v} \int dt_v \psi^1_{i_v,1,r_v}(t_v)\cdots\psi^q_{i_v,q,r_v}(t_v) \\ &\prod_{\substack{\text{black} \\ \text{vertices } \overline{v}}} \sum_{r_{\overline{v}}} \int dt_{\overline{v}} \overline{\psi}^1_{i_{\overline{v}},1,r_{\overline{v}}}\cdots\overline{\psi}^q_{i_{\overline{v}},q,\overline{v}}(t_{\overline{v}}) \prod_{\substack{\text{edges} \\ e=(v,\overline{v})}} \delta_{i_{v,c(e)},i_{\overline{v},c(e)}}. \end{aligned} \tag{15.86}$$

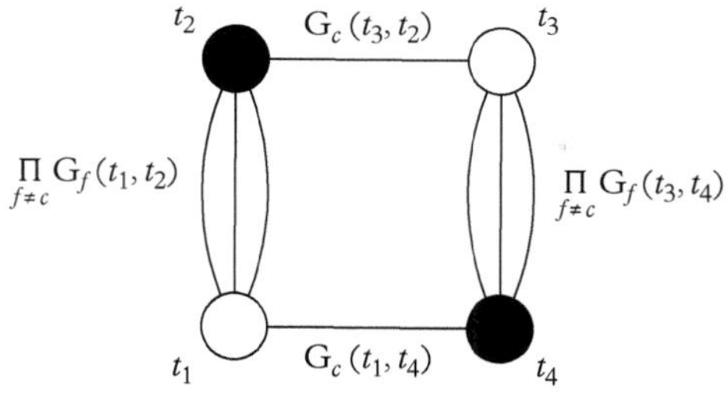

Figure 15.9 *Graphical representation of the term $\langle G\rangle_G$ for the quartic melonic graph for $q = 4$*

After introducing the Lagrange multiplier $\widetilde{\Sigma}$ to enforce the constraint (15.85) and assuming a replica symmetric saddle point, the effective action of our model can be written:

$$\frac{S_{eff}[G,\Sigma]}{N} = -\sum_{f=1}^{q} \log\det\left(\delta(t_1-t_2)\partial_t - \widetilde{\Sigma}_f(t_1,t_2)\right) + \int dt \sum_{f=1}^{4} \widetilde{\Sigma}_f(t)\widetilde{G}_f(t) \tag{15.87}$$

$$-\sum_{G} N^{-(v(G)-2)(q/2-1)+1-k(G)} \mu_G(\sigma^2, \{\lambda_{G'}\})\langle\widetilde{G}\rangle_G. \tag{15.88}$$

The term $\langle\widetilde{G}\rangle_G$ associated to a graph G is constructed as follows:

- to each vertex associate a real variable t_v;
- to an edge of colour c joining v to v' associate $\widetilde{G}_c(t_v, t_{v'})$; and
- multiply all edge contributions and integrate over vertex variables.

We then add up these contributions, with a weight λ_G and a power of N given by

$$N^{q-k(G)} \times (N^{-(q-1)})^{v(G)} \times N^{e(G)} = N \times N^{-(v(G)-2)(q/2-1)+1-k(G)}, \tag{15.89}$$

with $e(G)$ the number of edges of G, obeying $2e(G) = qv(G)$.

At leading order in N, only the Gaussian terms survives (i.e. the graph G with $v(G) = 2$ and $k(G) = 1$), except for the matrix model case ($q = 2$). In this case, all terms corresponding to connected graphs survive. Let us emphasize that the variance of the Gaussian distribution of coupling is thus modified as a consequence of the non-Gaussian averaging of our model. Remarkably, for $q > 2$, this is the only modification at leading order in N.

The actual value of the covariance (which we denote by σ') induced by non-Gaussian disorder is most easily computed using a DSE. In our context, the latter arises from

$$\sum_{i_1\ldots i_q} \int dj d\bar{j} \frac{\partial}{\partial \bar{j}_{i_1\ldots i_q}} \left\{ j_{i_1\ldots i_q} \exp -\left[\frac{N^{q-1}}{\sigma^2} j\bar{j} + V_N(j,\bar{j})\right]\right\} = 0. \tag{15.90}$$

At large N, it leads to the algebraic equation

$$1 = \frac{\sigma'^2}{\sigma^2} + \sum_{\text{melonic graph } G} \frac{\lambda_G}{\text{Sym}(G)} (\sigma')^{v(G)}. \tag{15.91}$$

Finally, it is interesting to note that this effective action, despite being non-local, is invariant under reparametrization (in the IR) at all orders in $1/N$:

$$G(t,t') \to \left(\frac{d\phi}{dt}(t)\right)^{\Delta} \left(\frac{d\phi}{dt'}(t')\right)^{\Delta} G(\phi(t),\phi(t')). \tag{15.92}$$

Indeed, changing the vertex variables as $t_v \to \phi(t_v)$, the Jacobians exactly cancel with the rescaling of G since $\Delta = 1/q$ and all vertices are q-valent.

15.4 Selected further reading

- J. Maldacena and D. Stanford. (2016). Comments on the Sachdev–Ye–Kitaev model, p. J. Maldacena and D. Stanford. (2016). *Phys. Rev. D*, 94. arxiv:1604.07818.
  Based on the original KITP talks of Kitaev, this is a very good physics article introducing and analysing the SYK model.

16

SYK-like tensor models

In 2016, E. Witten proposed a reformulation of the SYK model (see Chapter 15) using real fermionic coloured tensor fields without quenched disorder Witten (2019). This comes from the fact that, as we have seen from the previous chapters, both the SYK model and the coloured tensor models have the same class of dominant graphs in the large N limit, the melonic graphs.

Gurau (2017c) complemented Witten's results with some further results of random tensor theory (albeit using a complex version of Witten (2019)). He gave a classification of the Feynman graphs of the model at all orders in the $1/N$ expansion for the free energy and the two-point function, based on the Gurau–Schaeffer classification of coloured graphs Gurau and Schaeffer (2016), which is the same type of classification as the one used in this book for the MO tensor model in Section 13.3 or the one used in Subsection 15.3.3 for the coloured SYK model. Note that these classifications are somewhat formal, in the sense that they do not give the actual Feynman graphs which contribute at a given order of the $1/N$ expansion explicitly (as it is done for example in Subsection 15.3.2 for the LO and NLO of the coloured SYK model).

Following the terminology introduced since then in the mathematical physics literature, we will refer to the model of (Witten 2019; Gurau 2017) as the Gurau–Witten model.

In Klebanov and Tarnopolsky (2017), I. Klebanov and G. Tarnopolsky related the SYK model to the tensor model with an $O(N)^3$-symmetry of Chapter 14. This model, whose purely combinatorial part was originally introduced in the article Carrozza and Tanasa (2016), will be referred in this chapter as the $O(N)^3$-invariant model (although it is only one, with quartic interactions, of all possible $O(N)^3$-invariant models). Also, in Klebanov and Tarnopolsky (2017), the SYK model was related to the MO tensor model presented in Chapter 13.

For the sake of completeness, let us also mention SYK-like quantum mechanics with $Sp(N)$ symmetry has been investigated in Carrozza and Tanasa (2016).

In this chapter we analyse in detail the diagrammatics of various such SYK-like tensor models: the Gurau–Witten model (in the following section), and the MO and $O(N)^3$-invariant tensor model, in the rest of the chapter.

Combinatorial Physics: Combinatorics, quantum field theory, and quantum gravity models. Adrian Tanasa, Oxford University Press (2021). © Adrian Tanasa. DOI: 10.1093/oso/9780192895493.003.0016

16.1 The Gurau–Witten model and its diagrammatics

We focus in this subsection on the Gurau–Witten (Gurau 2017c; Witten 2019) model. The action of this model is written:

$$S = \int dt \left(\frac{\imath}{2} \sum_{f=1}^{4} \psi^f \frac{d}{dt} \psi^f + \frac{\lambda}{N^{\frac{D(D-1)}{4}}} \psi^1 \psi^2 \psi^3 \cdots \psi^q \right). \tag{16.1}$$

Note that the $q = D+1$ fields used ψ^f, $f = 1, \ldots, q$ are now rank $q-1$ tensor fields. The notation $\psi^1\psi^2 \cdots \psi^q$ in the interaction has a specofic pattern of index contraction. The tensor ψ^i has $q-1$ indices $(n_{ii-1}, \ldots, n_{i1}, n_{iq}, \ldots, n_{ii+1})$ where each n_{ij}, $i \neq j$, has range $n_{ij} = 1, \ldots, N$. The contraction in the interaction is defined by setting $n_{ij} = n_{ji}$, which identifies one index of ψ^i with one index of ψ^j.

The Feynman graphs obtained through perturbative expansion are stranded graphs where each strand represents the propagation of an index n_{ij}, alternating stranded edges of colours i and j. However, it is important to emphasize here that since no twists among the strands are allowed, one can easily represent the Feynman tensor graphs as standard Feynman graphs with additional colours on the edges. Those colours are just the labels f of the tensor fields ψ^f. A vertex has degree q and is incident to exactly one edge of each colour in $\{1, \ldots, q\}$. An example of such a Feynman diagram is given in Fig. 16.1 for $q = 4$. *The graphs are thus exactly those of the coloured SYK model described in Section 15.3.2 at fixed couplings* j. However, the scaling with N is different.

The partition function perturbatively expands as a $1/N$ expansion. In the large N limit, the Feynman amplitude of a graph is written as

$$w_N(G) = N^{\chi(G)}$$

where

$$\chi(G) = F(G) - \frac{(q-1)(q-2)}{4} V(G). \tag{16.2}$$

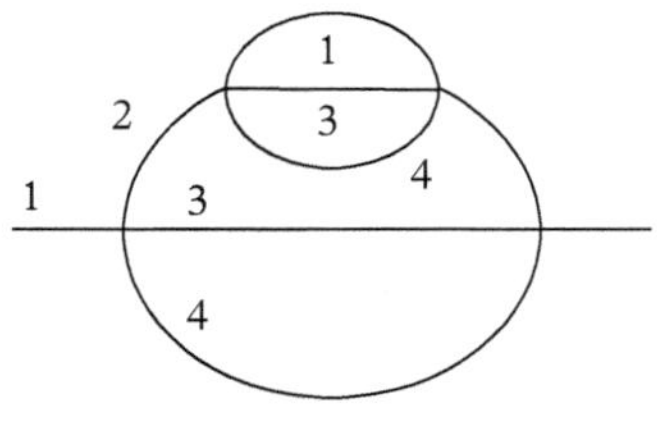

Figure 16.1 *A melonic graph of the Gurau–Witten model*

We have denoteded the number of vertices of G by $V(G)$, which is also twice its number of edges of any colour i. Let us also call a **face of colours** ij, each cycle alternating the colours $i < j$. We denote their numbers by $F_{ij}(G)$. The total number of faces is

$$F(G) = \sum_{\substack{i,j=1 \\ i<j}}^{q} F_{ij}(G). \tag{16.3}$$

Instead of $\chi(G)$ one can introduce the reduced degree:

$$\delta(G) = q - 1 - \chi(G). \tag{16.4}$$

One has

$$\delta(G) = q - 1 + \frac{(q-1)(q-2)}{4} V(G) - F(G), \tag{16.5}$$

for a connected graph.

One can prove that the reduced degree for such a coloured graph is positive (see Gurau (2012)) and vanishes for melonic graphs only. This implies that χ is bounded,

$$\chi(G) = F(G) - \frac{(q-1)(q-2)}{4} V(G) \leq \begin{cases} q-1 & \text{if } G \text{ is a vacuum graph,} \\ 0 & \text{if } G \text{ is a two-point graph.} \end{cases} \tag{16.6}$$

The case of four-point graphs will now be analysed.

16.1.1 Two-point functions: LO, NLO, and so on

As already mentioned, a four-point graph is obtained by cutting a line from a vacuum graph. If G is the vacuum graph and e the edge being cut, we denote G_e the two-point graph. The other way around, there is a unique way to glue the external lines of a two-point graph together to form a vacuum graph, so that all two-point graphs can be thought of as a graph G_e. We say that G is the closure of G_e.

The powers of N in G and G_e differ by $q - 1$,

$$\chi(G_e) = \chi(G) - (q-1), \tag{16.7}$$

since $q - 1$ faces are broken by cutting e. For a two-point graph G_e, we will refer to its reduced degree as the reduced degree $\delta(G)$ of its closure G.

Dominant graphs in the large N expansion are again melonic graphs (see Fig. 16.1). The classification of two-point coloured graphs with respect to the degree has been performed in Gurau and Schaeffer (2016) for bipartite graphs (e.g. which are graph obtained through perturbative expansion of a complex tensor model) and we thus only need to incorporate non-bipartite graphs.

The main result of this subsection is:

Theorem 16.1.1 *NLO, two-point graphs of the Gurau–Witten model are bipartite.*

Proof See Appendix H.

This implies that the NLO, two-point graphs of the Gurau–Witten model are given by the Gurau–Schaeffer classification included in Gurau and Schaeffer (2016). These graphs have reduced degree $\delta(G) = q - 3$ and exponent $\chi(G) = 2$, and the corresponding G_e are in Gurau and Schaeffer (2016),

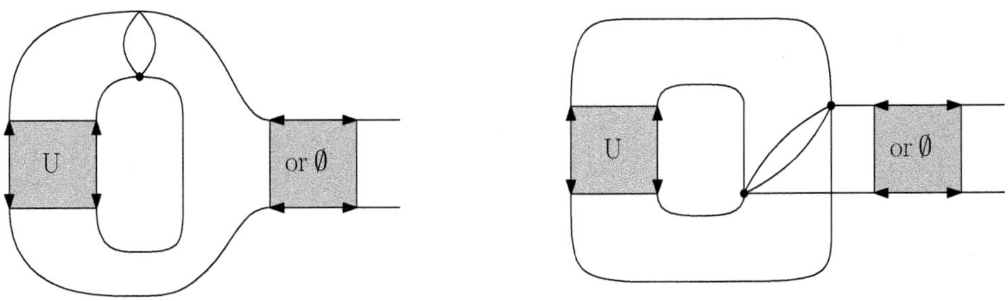

(16.8)

Here, the box with a U inside is an *unbroken* chain, corresponding to any four-point graph of the form

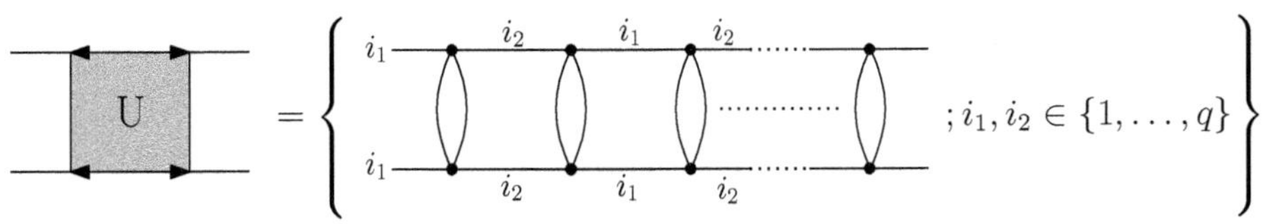

(16.9)

where the same two colours i_1, i_2 alternate along the chain (its right-hand side edges are of colour i_1 or i_2 depending on the parity of the number of pairs of vertices) and it is allowed to have a single pair of vertices. A box without a U inside is any chain, i.e. any sequence of colours can be found along the chain.

Non-bipartite graphs first appear at the NNLO, with $\chi(G) = 1$ ($\delta(G) = q - 2$). The only known examples are the following:

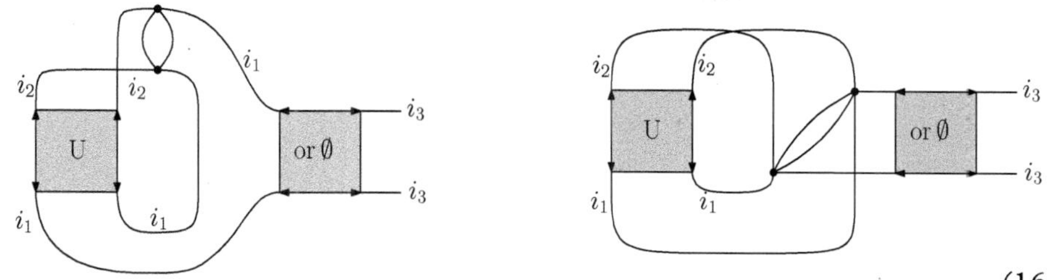

(16.10)

The graphs in (16.8) and (16.10) look very much like the NLO two-point graphs (15.48) and (15.49) of the coloured SYK model (with unbroken chains being just SYK chains). However, in the SYK case, the bipartite and non-bipartite graphs have the same exponent of N, which is not the case here.

16.1.2 Four-point function: LO, NLO, and so on

For generic q, the four external legs of a four-point graph have either all the same colour, or two legs have one colour while the other two have another colour. Cutting two edges e and e' in a vacuum graph G gives a four-point graph $G_{e,e'}$. The other way around, there is always a way to glue the external legs of a four-point graph two-by-two, so as to obtain a vacuum graph. Therefore, all four-point graphs can be realized (non-uniquely) as a graph type $G_{e,e'}$.

We denote by $\eta(G_{e,e'})$ the number of faces which are broken by cutting e and e', so that the graph $G_{e,e'}$ contributes at order

$$\chi(G_{e,e'}) = \chi(G) - \eta(G_{e,e'}). \tag{16.11}$$

In the SYK model, we found $\eta(G_{e,e'}) \in \{1,2\}$ since it only probes the faces with colours $0i$. Here however, η can have more values.

Number of broken faces with two external colours. We first restrict attention to the four-point function with two colours in its external legs. Cutting the first e of colour $i \in \{1,\dots,q\}$ in G opens up $q-1$ faces (those of colours (ik)). Cutting the e' of colour $j \neq i$ then breaks the $q-2$ faces of colour (jk) with $k \neq i$ and possibly another face of colour (ij). Therefore,

$$\eta(G_{e,e'}) = 2q - 3 + \epsilon_{e,e'} \qquad \text{with } \epsilon_{e,e'} = 0,1, \tag{16.12}$$

and $\chi(G_{e,e'}) = \chi(G) - 2q + 3 - \epsilon_{e,e'} = -q + 2 - \delta(G) - \epsilon_{e,e'}$.

Number of broken faces with one external colour. When all external legs have the same colour, there can be more than a single face running along both e and e', which can make η less than in the case of two external colours. We can write $\eta(G_{e,e'}) = 2(q-1) - \phi_{e,e'}$ where $\phi_{e,e'}$ is the number of faces shared by e and e'. In the SYK model, we could always restrict our attention to cases where $\eta(G_{e,e'}) = 2$ and disregard the cases where it is 1. Here, in a similar way, we can restrict attention to the cases where it is maximal in the following sense.

Given a four-point graph G_4 with the same colour on all external lines, we have three ways to glue them two-by-two and obtain a vacuum graph so that G_4 can be obtained from different vacuum graphs G with edges e, e' cut. So, we want a prescription to identify a graph G such that $G_4 = G_{e,e'}$ in a canonical way. The prescription is to choose one of the three ways to connect the external lines which maximizes the total number of faces created, i.e. maximizes η. This gives a vacuum graph G with distinguished edges e, e' and only then we say that the initial four-point graph is $G_4 = G_{e,e'}$.

Combinatorial considerations[1] show that

$$\eta(G_{e,e'}) \geq \frac{4}{3}(q-1), \tag{16.13}$$

in general, and,

$$\eta(G_{e,e'}) \geq \frac{3}{2}(q-1), \tag{16.14}$$

for bipartite graphs. This is equivalent to saying that the minimal number of faces shared by e and e' satisfies $\phi_{e,e'} \leq 2(q-1)/3$ in general and $\phi_{e,e'} \leq (q-1)/2$ for bipartite graphs. The reason is that when those inequalities are not satisfied, then it implies that the same four-point graph can in fact be obtained from another G_{e_1,e_2} with $\eta(G_{e_1,e_2}) > \eta(G_{e,e'})$.

Broken and unbroken chains. At the LO, the cases of one and two colours on the external legs are similar. This is due to a property of melonic graphs (those satisfying $\delta(G) = 0$). In a vacuum melonic graph, two edges e and e' may belong to 1 or no common face. This is obviously true for any (not necessarily melonic) graph if e and e' have different colours. If they have the same colour i, one might add edges of colour 0 on the canonical pairs and contract them as we did in (15.35) for the SYK model. One obtains a graph with no multicoloured cycles if G is melonic. The only pairs of vertices which may belong to more than one common face are on the same monocoloured cycle and opening them would, therefore separate the graph into two connected components, which is excluded.

This implies that equation (16.12) is true both when there are two colours on the external legs and at LO when there is a single colour on the external legs. Setting $\delta(G) = 0$, we get two types of contributions depending on the value of $\epsilon_{e,e'} = 0, 1$. They were described in (Gurau 2017c; Gurau and Schaeffer 2016) as unbroken (U) and broken (B) chains,

[1] The proof consists in counting the number of broken faces going from one external line, say the leg a, to the others labelled b, c, d. There are, respectively, $\epsilon_b, \epsilon_c, \epsilon_d$ broken faces going from the leg a to the legs b, c, d. The total number of broken faces going along the external line a is $\epsilon_b + \epsilon_c + \epsilon_d = q - 1$. The three ways to reconnect the external lines two by two can create $f_b = 2\epsilon_b + \epsilon_c + \epsilon_d$ faces, or $f_c = \epsilon_b + 2\epsilon_c + \epsilon_d$, or $f_d = \epsilon_b + \epsilon_c + 2\epsilon_d$ faces. Then $\eta = \max(f_b, f_c, f_d)$ with the constraint $f_b + f_c + f_d = 4(\epsilon_b + \epsilon_c + \epsilon_d) = 4(q-1)$.

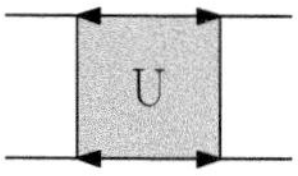

$\sim N^{2-q}$ 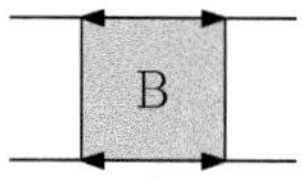

$\sim N^{1-q}$. (16.15)

Unbroken chains were defined in (16.9). A chain is broken if it is not unbroken, or equivalently no faces propagate from one end of the chain to the other.

To prove that there are no other contributions with the same exponents of N, $\chi(G_{e,e'}) \geq 1-q$, we simply apply the bounds previously established on the number of broken faces. Assume that G is not melonic. It has exponent $\chi(G) \leq 2$ or even $\chi(G) \leq 1$ for non-bipartite graphs (see section 16.1.1). We distinguish the following cases:

- $\chi(G) \leq 1$ and one external colour; then the bound (16.13) leads to

$$\chi(G_{e,e'}) \leq 1-q+\frac{1}{3}(4-q). \tag{16.16}$$

  This shows that for $q > 4$ all contributions coming from cutting two edges of the same colour in a vacuum graph G with $\chi(G) \leq 1$ can be disregarded at orders up to $\chi(G_{e,e'}) \geq 1-q$ (either they contribute to $\chi(G_{e,e'}) < 1-q$ or do not maximize $\eta_{e,e'}$ and are thus obtained by cutting edges in different graphs).

  However, at $q = 4$, contributions coming from cutting two edges of the same colour in a vacuum graph G with $\chi(G) = 1$ can be expected at order $\chi(G_{e,e'}) = 1-q = -3$. We will in fact exhibit such a contribution in (16.22).

- $\chi(G) = 2$ and one external colour: then the graphs are bipartite and the bound (16.14) leads to

$$\chi(G_{e,e'}) \leq 1-q+\frac{1}{2}(5-q). \tag{16.17}$$

  The conclusion is similar to the previous case. For $q > 4$, all contributions coming from cutting two edges of the same colour in a vacuum graph G with $\chi(G) = 2$ can be disregarded at orders up to $\chi(G_{e,e'}) \geq 1-q$ (either they contribute to $\chi(G_{e,e'}) < 1-q$ or do not maximize $\eta_{e,e'}$ and are thus obtained by cutting edges in different graphs).

  However, at $q = 4$, contributions coming from cutting two edges of the same colour in a vacuum graph G with $\chi(G) = 2$ can be expected at order $\chi(G_{e,e'}) = 1-q = -3$. We will give examples of four-point graphs with $\chi(G_{e,e'}) = 5-2q$ which is the same as $1-q$ at $q = 4$.

- $\chi(G) \leq 2$ and two external colours: then the bound (16.12) leads to

$$\chi(G_{e,e'}) \leq 5-2q = 1-q+(4-q). \tag{16.18}$$

  The conclusion is similar to the previous two cases. For $q > 4$, all contributions coming from cutting two edges of different colours in a vacuum graph G with

$\chi(G) \leq 2$ can be disregarded at orders up to $\chi(G_{e,e'}) \geq 1-q$ (they contribute to $\chi(G_{e,e'}) < 1-q$).

However, at $q=4$, contributions coming from cutting two edges of different colours in a vacuum graph G with $\chi(G)=2$ can be expected at order $\chi(G_{e,e'}) = 1-q=-3$. We are going to identify all of them in the next paragraph (since $5-2q=1-q$ at $q=4$).

Beyond chains, two external colours. To go further in the $1/N$ expansion, we restrict attention to the case with two colours on the external legs which is much easier to handle thanks to (16.12). If G is a vacuum NLO graph, $\chi(G)=2$ and thus $\chi(G_{e,e'}) = 5-2q-\epsilon_{e,e'}$. In particular, one can obtain contributions which scale like $\chi(G_{e,e'}) = 5-2q$ by cutting edges such that $\epsilon_{e,e'}=0$. With the same reasoning, it is clear that vacuum graphs of higher orders, i.e. $\chi(G) \leq 1$, contribute only to four-point graphs of higher order, $\chi(G_{e,e'}) \leq 4-2q$.

Let us detail the graphs with $\chi(G_{e,e'}) = 5-2q$: they are obtained by cutting two edges e,e' of different colours i and j in a vacuum NLO graph, with one face of colour (ij) going along e,e'. One can equivalently cut an edge along a broken face in (16.8) and find

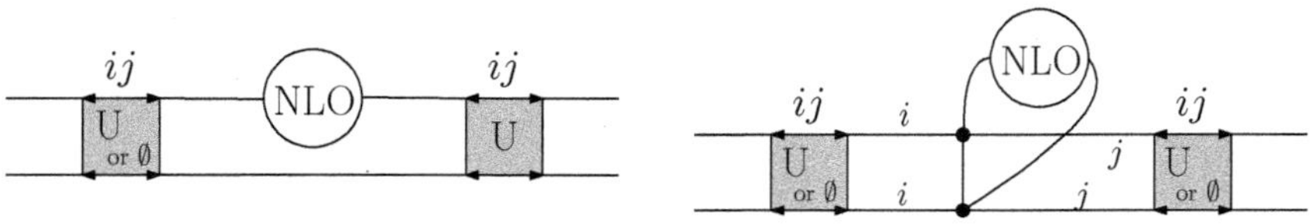

(16.19)

as well as

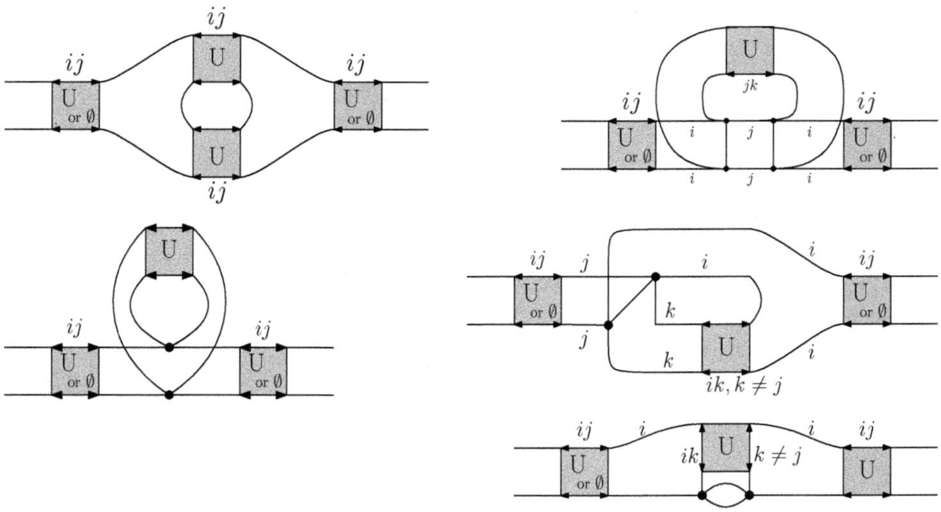

(16.20)

where i,j are the two external colours (it does not matter on which side i is).

Note that for $q = 4$, those contributions have the same exponent $\chi(G_{e,e'}) = -3$ as the broken chains since $1 - q = 5 - 2q = -3$. They are higher-order contributions for $q > 4$.

A single external colour. The case with the same colour on all four external legs is more difficult as we have not been able to identify the contributions following directly (16.15) in the $1/N$ expansion. We have found a family of graphs with $\chi(G_{e,e'}) = 6 - 2q$, for any $q > 4$ (so, it first appears at $q = 6$ in the Gurau–Witten model),

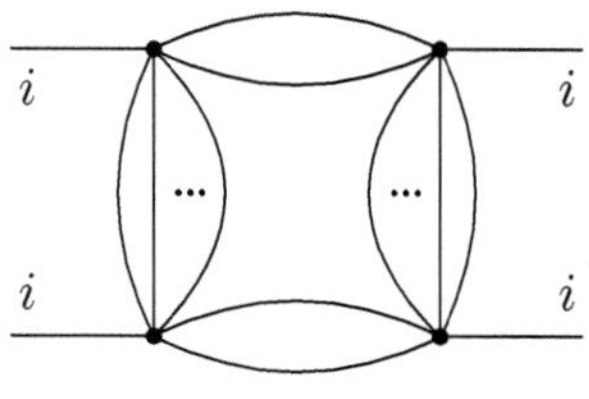

(16.21)

for which there are two faces running along both e and e'. The graphs of (16.19) and (16.20), amended to have the same colours on the external lines, contribute at $\chi(G_{e,e'}) = 5 - 2q$. Notice that at $q = 4$, they actually contribute with the same exponent of N as the broken chains (16.15), since then $1 - q = 5 - 2q = -3$.

Moreover, still at $q = 4$, there are NNLO vacuum graphs which, after cutting two edges, also contribute to the four-point function at the same order $\chi(G_{e,e'}) = -3$. An example is the following:

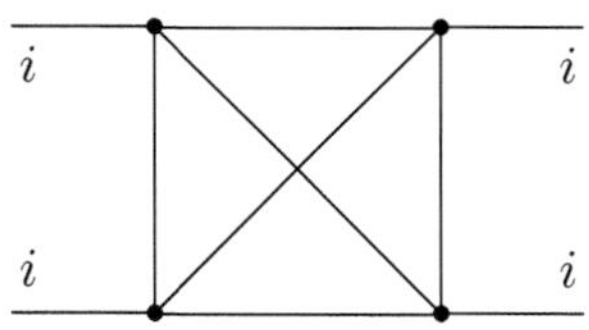

(16.22)

However, we have not proved that there are no graphs with $\chi(G_{e,e'})$ between $1 - q$ and $6 - 2q$ for $q > 4$ and that there are no other graphs contributing at $\chi(G_{e,e'}) = -3$ for $q = 4$. We expect these issues to be particularly difficult, given the already long proof that (16.8) are the NLO two-point graphs in Gurau and Schaeffer (2016).

Four external colours. There is an exceptional four-point function at $q = 4$, i.e. four colours, with four distinct colours on the external lines, one per leg. A graph contributing to this exceptional four-point function can be turned into a vacuum graph by adding a vertex and attaching the four external lines to that vertex. Therefore, all graphs with four external colours are obtained by removing a vertex from a vacuum graph. This breaks exactly one face of each colour type (ij), for $i < j \in \{1, 2, 3, 4\}$. The $1/N$ expansion of this four-point function thus follows that of the free energy.

At NLO, one finds the following diagrams:

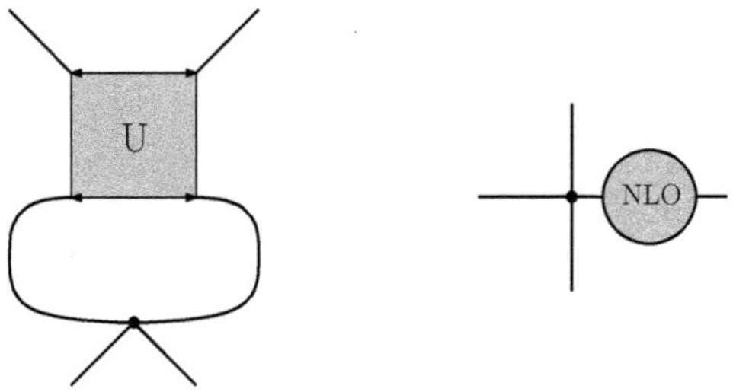

(16.23)

In Appendix I we give a list of the diagrammatic results of this section, as well as the analogous results of coloured SYK model (see Subsection 15.3.2).

Let us conclude this section by mentioning that the diagrammatic analysis is much simpler in the coloured SYK model than in the Gurau–Witten model, because of the method we developed in Section 15.3.2. Furthermore, let us note that the analysis of the four-point function in the Gurau–Witten model depends on the number of distinct colours on the external lines.

It is known that the (coloured) SYK model and the Gurau–Witten models are different per se, but the huge interest in the Gurau–Witten model comes from the fact that its LO coincide with the LO of the SYK model (as mentioned earlier in this chapter). However, we show here that at the NLO (for the free energy or the two-point function), the two models start to differ, although the diagrams remain quite similar.

In fact, we have found that diagrams which contribute to the same order in the SYK model can contribute to different orders in the Gurau–Witten model. For example, in the case of the four-point function for the Gurau–Witten model, we have a distinction between broken and unbroken chains, phenomenon which does not exist in the case of the coloured SYK model. The Gurau–Witten thus sort of lifts a degeneracy of Feynman graphs of the SYK (at least at low orders). This is obviously due to the fact that both models have different exponents of N, taking into account different faces. This phenomenon is likely to be more and more important when one studies further and further orders (NNLO, NNNLO, and so on) in the diagrammatics of the large N expansions of the two models.

16.2 The $O(N)^3$-invariant SYK-like tensor model

This model consist of a single real fermionic tensor ψ_{ijk} of size N with $O(N)$ symmetry on each of its indices:

$$\psi_{abc} \to \psi'_{a'b'c'} = O1^{a}_{a'} O2^{b}_{b'} O3^{c}_{c'} \psi_{abc}, \qquad O_i \in O(N). \tag{16.24}$$

With this symmetry, we can build two different invariants of order 4 in ψ_{ijk} depending on how indices of the tensor are contracted. We can either have a **tetrahedral** interaction,

$$I_1 = \psi_{abc}\psi_{ade}\psi_{fbe}\psi_{fdc} \tag{16.25}$$

or a 'pillow', also known as **melonic** interaction,

$$I_2 = \psi_{abc}\psi_{dbc}\psi_{aef}\psi_{def}. \tag{16.26}$$

As we will mention later, here we only study the tetrahedral interaction.

The interaction terms (16.25) and (16.26) can be represented diagrammatically by edge-coloured graphs with four vertices of degree 3 given in Fig. 16.2. This diagrammatic correspondence is such that each vertex represents a tensor and each incident edge represents an index. Each edge carries a colour corresponding to its position on the tensor. An edge of colour i connecting two vertices denotes a summation on the i-th index between two tensors. This makes sure that the interactions are $O(N)^3$-invariant. More generally, $O(N)^d$- and $U(N)^d$-invariant polynomials are called bubbles in the literature.

Notice that I_1 is fully symmetric under permutations of the colours, while there are in fact three different versions of I_2 (where the two dipoles can be connected by edges of colour 1, or 2, or 3).

As mentioned previously in this chapter, as in Klebanov and Tarnopolsky (2017), we only study the tetrahedral interaction. The SYK-like $(0+1)$-dimensional action is written:

$$S_{CTKT} = \int dt \left(\frac{\imath}{2}\psi_{abc}\partial_t\psi_{abc} + \frac{\lambda}{4}\psi_{abc}\psi_{ade}\psi_{fbe}\psi_{fdc} \right). \tag{16.27}$$

Therefore, a Feynman graph of the model is represented as bubbles connected by propagators, which are represented as dashed edges, also referred to as 0 coloured edge. Therefore, all the vertices of the Feynman graph are of valency 4 and have exactly one half-edge of each colour $i \in \{0,1,2,3\}$.

The **jacket** J_i of a graph G, for $i \in \{1,2,3\}$, is the graph obtained by deleting all edges of colour i. If j,k denotes the complementary colours, $\{i,j,k\} = \{1,2,3\}$, then J_i is a 3-coloured graph whose vertices are those of G and have degree 3, and whose edges have colours $0,j,k$. An example of a Feynman graph and its three jackets is given in

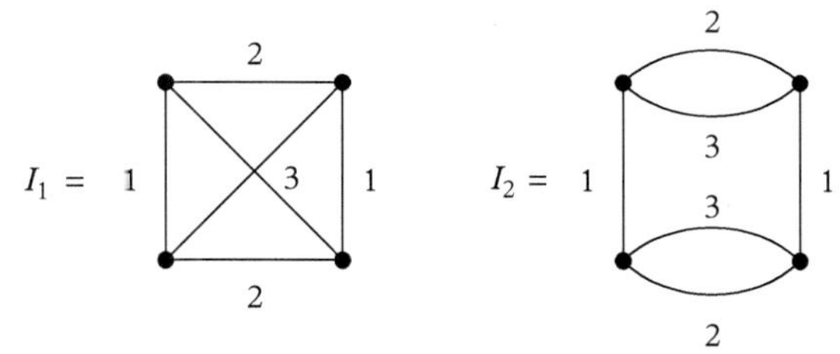

Figure 16.2 *On the left, the tetrahedral bubble. On the right, a melonic bubble with 4 vertices*

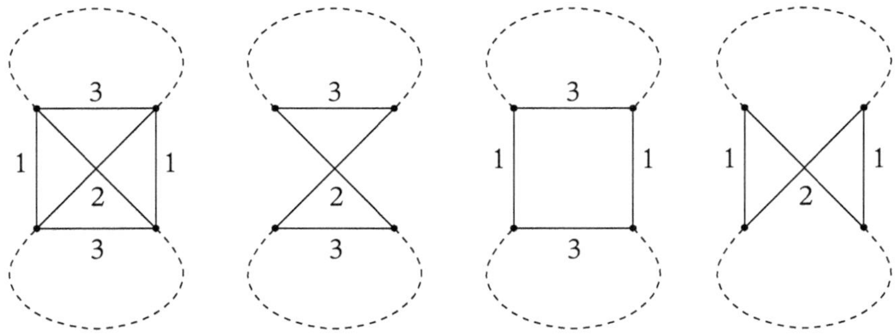

Figure 16.3 *An example of a Feynman graph of the model and its three jackets*

Fig. 16.3. Due to the structure of the tetrahedral bubble, all the jackets of a connected graph are connected. Therefore, a jacket represents a connected surface, whose genus is given by the Euler characteristic formula,

$$\chi_i = 2 - 2g_i = V - E(J_i) + F(J_i), \tag{16.28}$$

where g_i can be a half-integer. Here, V is the number of vertices of G, $E(J_i)$ the number of edges of J_i and $F(J_i)$ the number of faces of J_i. For example, the jackets of Fig. 16.3 have $V = 4$, $E = 6$.

Since J_i has vertices of degree 3, $E(J_i) = 3V/2$. Moreover, for graphs encoding surfaces, the faces correspond to the bicoloured cycles, with colour pairs $\{0,j\}$, $\{0,k\}$, and $\{j,k\}$. Those bicoloured cycles can be read either on J_i or directly on G. Notice that the bicoloured cycles with colours $\{j,k\}$ lie within the bubbles and there is exactly one for each bubble. Denoting F_i, F_j, F_k, the number of bicoloured cycles with colours $\{0,i\}$, $\{0,j\}$, $\{0,k\}$, we have

$$F(J_i) = F_j + F_k + V/4. \tag{16.29}$$

We then get

$$2 - 2g_i = F_j + F_k - n, \tag{16.30}$$

where $n = V/4$ is the number of bubbles.

As mentioned previously, the bicoloured cycles of a jacket represent the faces of the corresponding discrete surface. Those bicoloured cycles are also objects of graph G and we will still call them faces for G. In particular, we call a **face of colour** i of G a bicoloured cycle with colours $\{0,i\}$. A face has even length, with an equal number of edges of both its colours. The length of a face is defined by this number.

The degree of the graph which organizes the perturbative expansion of the model is given by the sum of the genera of the jackets,

$$\omega = g_1 + g_2 + g_3, \tag{16.31}$$

from which we know that $\omega \geq 0$. Using (16.30) for all three colours, one finds

$$\omega = 3 + \frac{3}{2}n - (F_1 + F_2 + F_3). \tag{16.32}$$

16.3 The MO SYK-like tensor model

The MO tensor model has been already been introduced as analysed in Chapter 13. In this section we recall some its key properties and show it can used to define a SYK-like tensor model.

As already mentioned in Section 13.1, the model has a complex Fermionic tensor field ψ_{ijk}. However, the symmetry associated to this tensor is $U(N) \times O(N) \times U(N)$. This stems from the interaction term of this model which is

$$I = \psi_{abc}\bar{\psi}_{ade}\psi_{fbe}\bar{\psi}_{fdc}. \tag{16.33}$$

The interaction term is represented as a four valent vertex, where a field ψ is an incident half-edge decorated with the sign $+$ and a field $\bar{\psi}$ is an incident half-edge with the sign $-$. To represent the indices, it is customary (see again Section 13.1) to blow up the half-edges into three strands, one for each index, and then connect the strands according to the contraction pattern

$I =$ 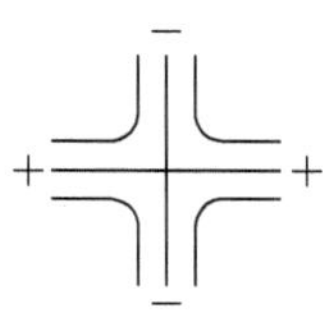 (16.34)

As already explained in Section 13.1, the propagator of this model is a 3-stranded edge which propagates each index of ϕ, connecting two half-edges of different signs. As in Chapter 13, then arriving onto a sign $+$ at a vertex, we call the **left** strand the one which goes to the left, the **right** strand the one which goes to the right, and the **straight** strand the one going straight. Denote by F the number of closed strands, $F = F_L + F_S + F_R$. The degree of the MO model for a connected graph with n vertices is

$$\omega_{MO} = 3 + \frac{3}{2}n - F. \tag{16.35}$$

Note that this formula is similar to the one obtained in the case of the model of the previous section, except for the replacing of the faces of colour $1, 2, 3$, with closed strands of type L, R, S.

Recall from Chapter 13 that the MO model also has a notion of jackets. The jacket J_i is the 2-stranded graph, which are proven to be ribbon graphs, (see again Chapter 13, obtained by deleting the type $i \in \{S, L, R\}$ strand. This leads to a similar formula as in the $O(N)^3$-invariant case, in terms of the genus of the jackets:

$$\omega_{MO} = g_L + g_S + g_R. \tag{16.36}$$

16.4 Relating MO graphs to $O(N)^3$-invariant graphs

In this section we first recall the construction from Chapter 14 allowing to associate an $O(N)^3$-invariant graph to any MO graph. The reciprocal proposition is not true—counter-examples are easily found and the class of $O(N)^3$-invariant graphs is strictly larger than the class of MO graphs (see again Chapter 14). Moreover, we give here a sufficient condition under which the reciprocal holds. It can be formulated topologically, as a jacket being orientable, or combinatorially, as the genus being an integer.

Theorem 16.4.1 *There is an explicit bijection between $O(N)^3$-invariant graphs with a marked, orientable jacket and MO graphs supplemented with a colour in $\{1,2,3\}$. This bijection maps bubbles to vertices and faces to closed strands.*

Proof We first build a map from MO graphs to a subset of $O(N)^3$-invariant graphs. To do so, we identify the interaction of the MO model with a bubble of the $O(N)^3$-invariant model in the following way:

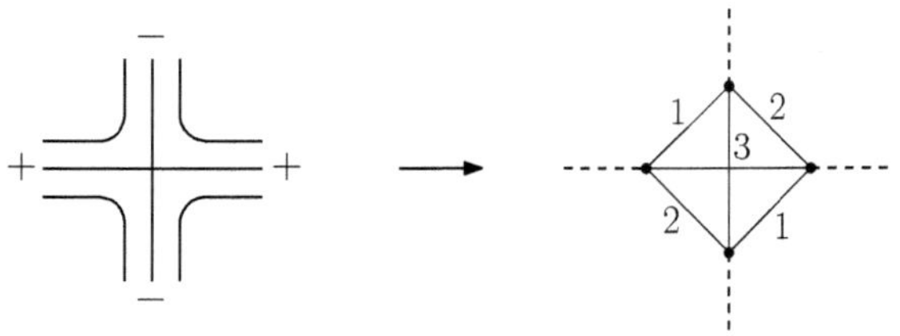

(16.37)

which induces the following correspondence.

MO interaction	$O(N)^3$-invariant bubble
Left strands	Edges of colour 1
Right strands	Edges of colour 2
Straight strands	Edges of colour 3
Edges	Edges of colour 0

Propagators then extends this correspondence to the whole graph, so that left strands become faces of colour 1, right strands faces of colour 2, and straight strands faces of colour 3. If G is an MO graph, then denote $\tilde{G}$ the corresponding $O(N)^3$-invariant graph. Moreover, if $i \in \{1,2,3\}$ is a colour, define $\tilde{G}_i$ by exchanging the colours i and 3 in $\tilde{G}$.

We now show that the jacket J_i in $\tilde{G}_i$ is orientable. By definition, this is the same as J_3 being orientable in $\tilde{G}$. Due to the correspondence we have just described, this is in turn equivalent to the jacket J_S being orientable in G. The latter statement is now proved, as J_S in G is a 2-stranded graph obtained by removing all straight strands. This means that it is represented as a ribbon graph with 4-valent ribbon vertices and ribbon edges which do not twist the ribbons,

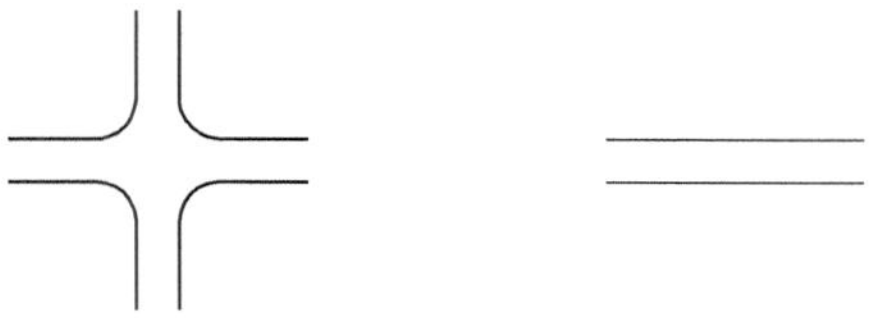
(16.38)

These local rules generate orientable surfaces. This means that the ribbon graphs J_i in $\tilde{G}_i$ is orientable.

Next we invert the map starting from $\tilde{G}_i$ an $O(N)^3$-invariant graph with orientable, marked jacket J_i for $i \in \{1,2,3\}$. We exchange the colour i with 3 and keep $i \in \{1,2,3\}$ as additional data for the MO graph G that we are going to get. We thus consider $\tilde{G}$ an $O(N)^3$-invariant graph whose jacket J_3 is orientable. The difficulty in inverting (16.37) is to find how the signs $+/-$ can appear.

This is due to the orientability of the ribbon graph J_3. The latter is an edge-coloured graph, with colours $\{0,1,2\}$. Recall here that this orientability is equivalent to the edge-coloured graph being bipartite. The jacket J_3 is thus bipartite. We colour its vertices, say, black and white. Since the vertices of G and J_3 are the same, one obtains a colouring of the vertices of G (G is not bipartite however). One can then apply the following mapping:

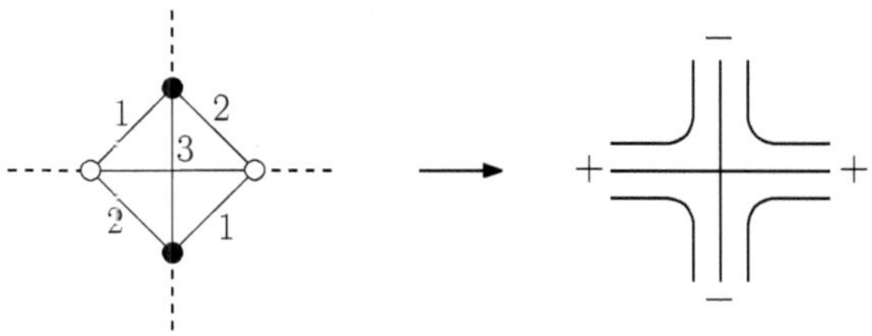

(16.39)

In other words, white/black vertices inherited from orientability become the $+/-$ signs we were looking for. Edges of colour 0 connect white to black vertices in G. This implies that $+$ can only connect to $-$ and the other way around. This way, $\tilde{G}$ is mapped to an MO graph G, with the colour i of the marked jacket needing to be stored in addition with G.

16.5 Diagrammatic techniques for $O(N)^3$-invariant graphs

In this section, we develop explicit diagrammatic techniques which we use to study the graphs of degrees 1 and $3/2$ of the $O(N)^3$-invariant tensor model.

16.5.1 Two-edge-cuts

Consider G a 2-particle-reducible (2PR) graph (see Chapter 3), i.e. a graph with a 2-edge-cut: a pair of edges $\{e,e'\}$ of colour 0 whose removal disconnects G,

$G =$ 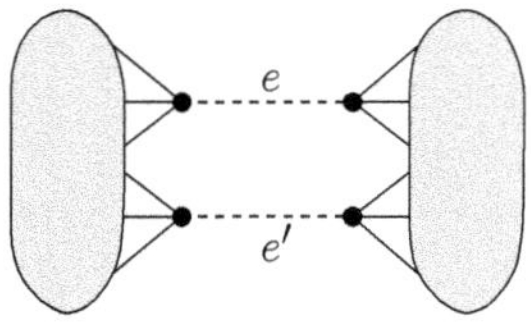 (16.40)

There is a natural flip operation which turns G into a pair of graphs,

$G_L =$ 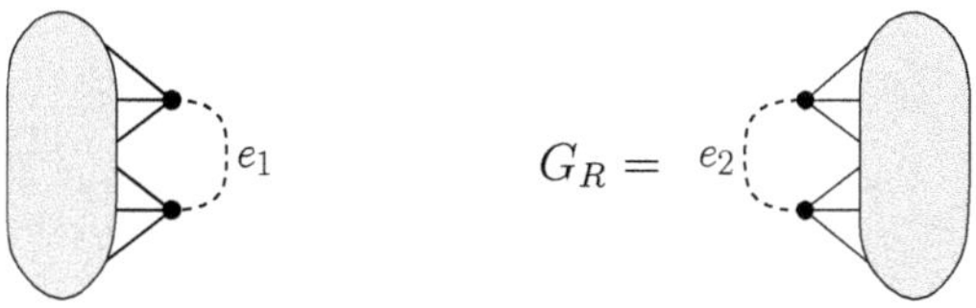 $G_R =$ (16.41)

by cutting e, e' into half-edges and gluing them as previously illustrated.

The following proposition is well-known.

Proposition 16.5.1 *With the previous notations;*

$$\omega(G) = \omega(G_L) + \omega(G_R). \tag{16.42}$$

This is the additivity of the degree for 2PR graphs.

Proof Recall that each edge of colour 0 contributes to exactly one face of each colour 1, 2, 3. Since G is 2PR, it is the same face of colour i which goes along e and e' for all $i = 1,2,3$. However, there are different faces along e_1 and e_2 since they live in different connected components. One has:

$$F(G) = F(G_L) + F(G_R) - 3. \tag{16.43}$$

The result then follows from the formula (16.32) for the degree.

Therefore, if one is interested in finding all graphs at a fixed value ω of the degree, it is possible to distinguish the cases of 2PR and 2PI graphs. The 2PR graphs of degree ω are given by:

- a 2PI graph of degree ω with insertions of two-point graphs of vanishing degree ($\omega(G_R) = 0$). Any such two-point graph is obtained by recursive insertions of a fundamental graph called the *melonic insertion*,

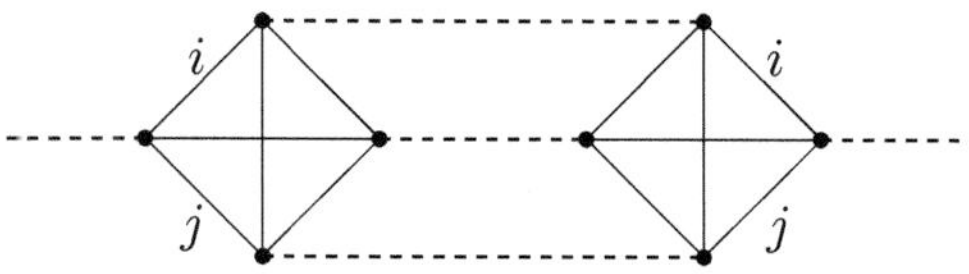

(16.44)

  to be inserted on any edge of colour 0.
- G_L and G_R both have degrees less than ω.

16.5.2 Dipole removals

Dipoles were introduced in Fusy and Tanasa (2014) to facilitate the analysis of graphs at fixed degree in the MO model. Here we adapt the notion to the $O(N)^3$-invariant model. Just as in Fusy and Tanasa (2014), dipoles are defined as the minimal subgraphs which have an internal face of length two. There are no dipoles with single bubbles. With two bubbles, one has three types of dipoles:

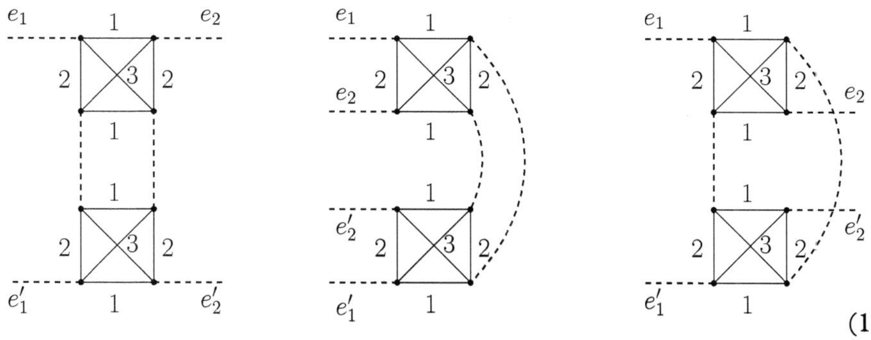

(16.45)

We label them with the colour of the their internal face of length 2: the left one has colour 1, the middle one has colour 2, and the right one has colour 3.

If a 2PI graph contains a dipole, it can be removed while preserving connectedness:[2]

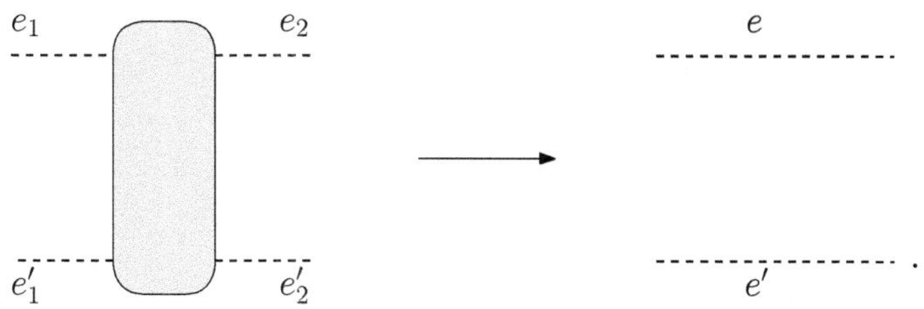

(16.46)

[2] If $\{e_1, e_2\}$ (or $\{e'_1, e'_2\}$) forms a 2-edge cut, then the dipole removal disconnects the graph.

Proposition 16.5.2 *There are four types of dipole removals, which are described in the following proof. Three of them decrease the degree, by 4, 2, or 1, and the other one does not change it.*

Proof Let us calculate the variation of the degree. Without loss of generality, we consider the removal of a dipole of colour 1. The new graph G' has two bubbles less. It also loses a face of colour 1. As for the faces of colour 2 and 3, it depends on their paths in G. Up to symmetry, there are the following four possibilities:

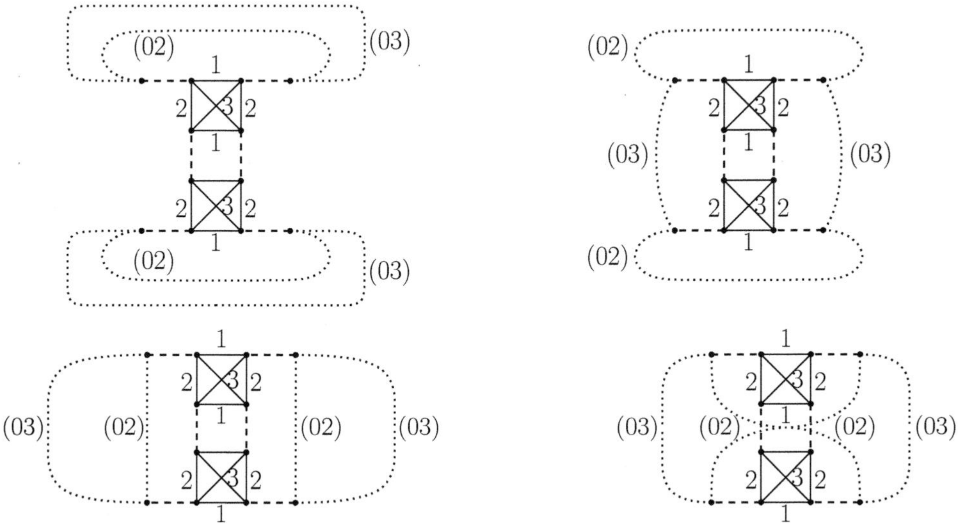

(16.47)

Here, the coloured lines do not represent edges but instead the paths of the faces, i.e. bicoloured cycles with alternating colours 0 and i, which go along the bubbles of the dipole. Denote by $\epsilon_2, \epsilon_3 \in \{1,2\}$ the number of faces of colour 2 and 3 in G which go along the dipole, and by ϵ_2', ϵ_3' the number of those faces left in G'.

- In the first case (top left), $\epsilon_2 = \epsilon_3 = 1$. Then, each of those faces gets split into two[3] in G', i.e. $\epsilon_2' = \epsilon_3' = 2$. One has:

$$\begin{aligned}\omega(G') &= 3 + \frac{3}{2}(n(G) - 2) - (F_1(G) - 1 + F_2(G) + 1 + F_3(G) + 1) \\ &= \omega(G) - 4.\end{aligned} \tag{16.48}$$

  In particular, since the degree is positive, it means that $\omega(G) \geq 4$.
- In the second case (top right), $\epsilon_2 = 1$ and $\epsilon_3 = 2$. The face of colour 2 is split into two in G' while those of colour 3 are merged, i.e. $\epsilon_2' = 2$ and $\epsilon_3 = 1$. Therefore, the number of faces of colour 2 and 2 is globally unchanged and

[3] Notice that our drawing becomes disconnected after the removal, but this is misrepresentation due to the fact that our drawing does not contain other edges which would leave G' connected.

$$\begin{aligned}\omega(G') &= 3+\frac{3}{2}(n(G)-2)-(F_1(G)-1+F_2(G)+1+F_3(G)-1). \\ &= \omega(G)-2.\end{aligned} \tag{16.49}$$

In particular, since the degree is positive, it means that $\omega(G)\geq 2$.

- In the third case (bottom left), $\epsilon_2=\epsilon_3=2$, and for both colours the two faces are merged into one in G', leaving $\epsilon_2'=\epsilon_3'=1$ and

$$\begin{aligned}\omega(G') &= 3+\frac{3}{2}(n(G)-2)-(F_1(G)-1+F_2(G)-1+F_3(G)-1) \\ &= \omega(G),\end{aligned} \tag{16.50}$$

i.e. the degree is unchanged by the dipole removal.

- In the fourth case (bottom right), $\epsilon_2=1$ and $\epsilon_3=2$. However, in contrast with the previous cases, the face of colour 2 is not split into two, i.e. $F_2(G)=F_2(G')$. The two faces of colour 3 are merged into one, so, $\epsilon_2'=\epsilon_3'=1$. This leads to

$$\begin{aligned}\omega(G') &= 3+\frac{3}{2}(n(G)-2)-(F_1(G)-1+F_2(G)+F_3(G)-1) \\ &= \omega(G)-1.\end{aligned} \tag{16.51}$$

This concludes the proof.

16.5.3 Dipole insertions

One can consider the reverse operation: the dipole insertion on the edges e and e'. By inspecting the previous proof, one finds the following lemma which provides a necessary condition for a dipole insertion not to increase the degree by more than 1.

Lemma 16.5.3 *A dipole insertion on $\{e,e'\}$ which increases the degree by 1 or preserves it, requires that there are two colours $\{i,j\}\subset\{1,2,3\}$, such that the face of colour i along e also goes along e', and similarly for j.*

Proof A variation of the degree of at most 1 corresponds to the third and fourth cases of the previous proof. After the dipole removals in those cases, the edges e and e' have two colours (2 and 3 in our pictures) for which the same face goes along e and e'.

We recall that melonic insertions do not change the degree of a graph. It is thus common, in order to analyse the degree, to remove at first all melonic two-point functions, and then add the full melonic two-point function on all edges of colour 0 at the end of the analysis. For instance, here we will aim at identifying all graphs of degree 1, so that we can try and identify all melon-free graphs for degree 1 and then add all melonic decorations on the edges of colour 0. However, since we also consider dipole removals and insertions, the question arises as to whether new dipole insertions which change the degree by at most 1 would be allowed after melonic insertions. The following lemma answers this question.

Lemma 16.5.4 *Let G' be obtained from G by a melonic insertion on the edge e. Then, there exists a possible dipole insertion preserving the degree on the pair $\{e', e''\}$ in G iff it exists in G', for $e', e'' \neq e$.*

This means that melonic insertions do not 'offer' new dipole insertions which preserve the degree or increase it by one.

Proof Let G be a graph and G' obtained by a melonic insertion on e,

$G =$ 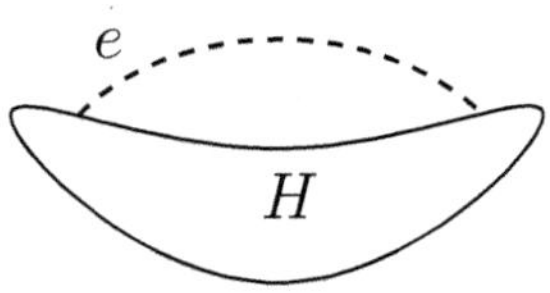 $G' =$ 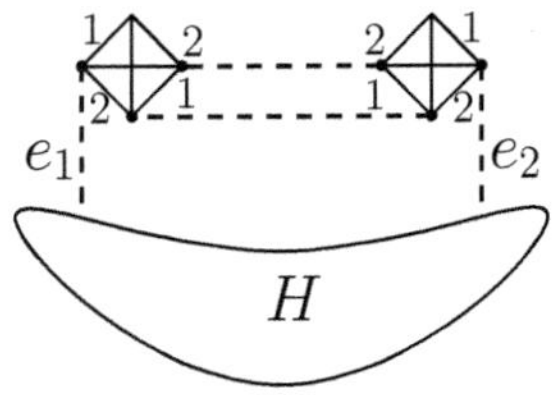 (16.52)

To investigate the insertion of dipoles which increase the degree by one or preserve it, we use Lemma 16.5.3 and consider the pairs of edges of colour 0 $\{e', e''\}$ such that there are two colours for which the faces going along e' and e'' are the same. Also, denote $\tilde{G}$ the graph obtained from G by performing the dipole insertion on $\{e', e''\}$.

If e' and e'' lie in H in G, that is also the case in G' so that the same dipole insertion can be made before or after the melonic insertion.

If $e' = e$ (or $e'' = e$), then the corresponding dipole insertion gives rise to two possible insertions in G', one with e_1 and e'', the other with e_2 and e''. In fact, the resulting graphs only differ by the position of the melonic insertion in $\tilde{G}$, so the melonic insertion does not provide new dipole insertions.

All the faces going along e_1 also go along e_2, so that it seems that a new dipole insertion is possible. However, because e_1 and e_2 share three faces instead of two, the dipole insertion is in fact a melonic insertion.

Finally, it is easy to check that for each of the three edges of colour 0 connecting the bubbles of the melonic insertions, there are no other edges which would share at least two faces with it. Therefore, there is no possible dipole insertion which preserves the degree or change it by one.

16.5.4 Chains of dipoles

A chain of dipoles is a sequence of dipoles of arbitrary length. Such a chain can be of fixed colour,

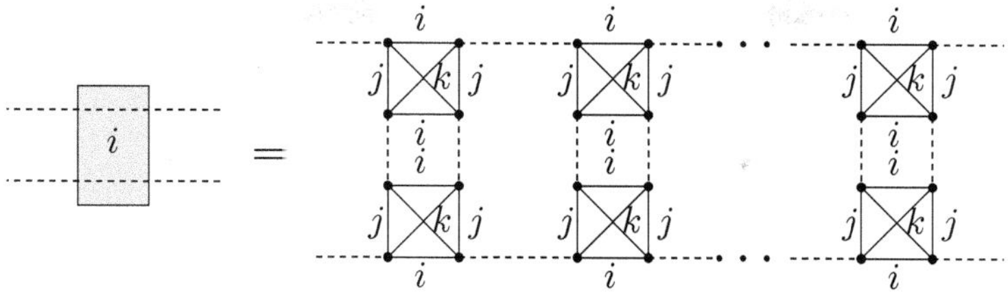

(16.53)

or it can have multiple colours, in which case we use the same box diagram without any colour label.

Lemma 16.5.5 1. *If there is a pair $\{e, e'\}$ in G where a dipole of colour i can be inserted without changing the degree, then a chain of colour i can be inserted without changing the degree.*

2. *If, in addition, a dipole of a different colour can also be inserted without changing the degree, then any chain can be inserted without changing the degree.*

Proof Without loss of generality, let $i = 1$. Inserting a dipole without changing the degree means that we are in the third case of Proposition 16.5.2, see (16.47). There, the paths of the faces of colour 2 and 3 had been drawn, but not those of colour 1. There are two possible paths for the faces of colour i going along e and e': they are either different, or the same faces.

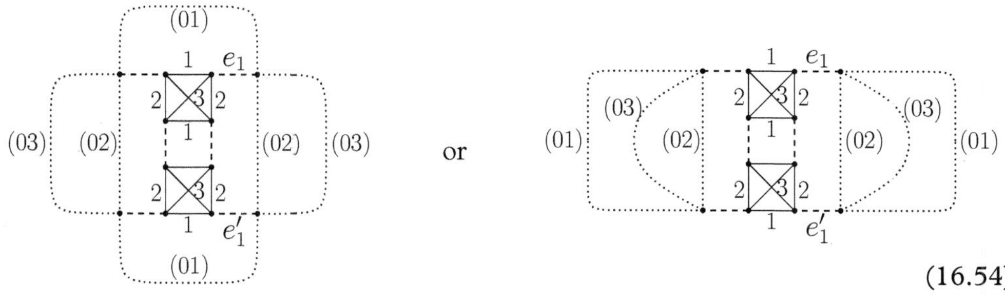

(16.54)

In the first case, the edges e_1 and e'_1 share their faces of colour 2 and 3, meaning that indeed a new dipole of colour 1 can be inserted without changing the degree. However, a dipole of another colour would change the degree because the faces of colour 1 going along e_1 and e'_1 are different but would be merged after a dipole insertion of colour 2 or 3.

In the second case, e_1 and e'_1 share their 3 faces so that any dipole insertions can be performed without changing the degree.

Chains of dipoles are related to melonic graphs in the following way. If one opens up a melonic subgraph by cutting an edge of colour 0, this creates a chain of dipoles. The following lemma is useful to glue two graphs along a 2-edge cut as in Section 16.5.1.

Lemma 16.5.6 *If G has the form*

$G =$ 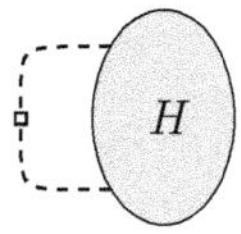 (16.55)

where the square represents an arbitrary melonic two-point function, then cutting an arbitrary edge of colour 0 within this two-point function results in a two-point graph of the form

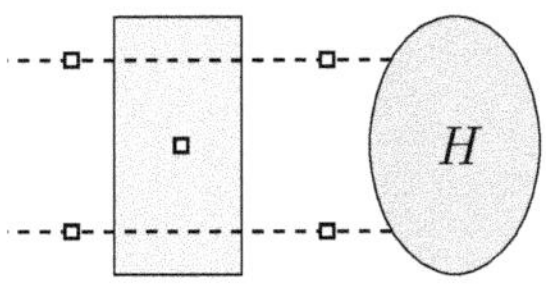 , (16.56)

where the square inside the chain means that all edges of colour 0 can have melonic two-point function insertions.

Proof Let e be the edge of colour 0 which is cut. By definition of melonic two-point functions, the most generic situation (up to colour relabelling of each bubble) is like

$G =$ 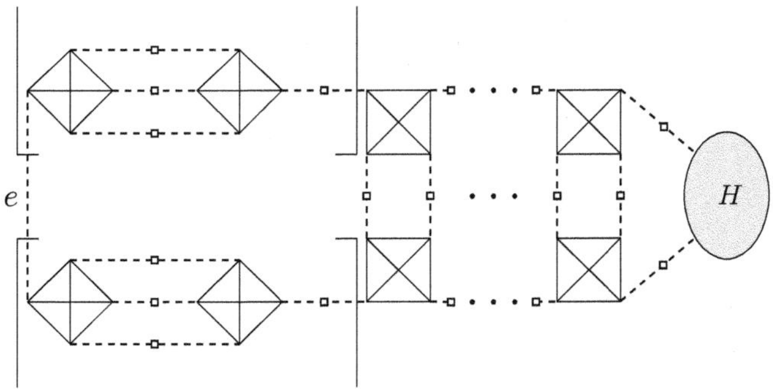 (16.57)

where the parts between brackets may be empty in which case e is extended and the three dots indicate potential dipole repetitions. Cutting the edge e leads to the expected result.

16.5.5 Face length

In this section we gather some simple results on the faces of $O(N)^3$-invariant graphs.

Lemma 16.5.7 *If G has degree ω, n bubbles and jackets J_i of genus g_i, then*

$$F_i = 1 - \omega + 2g_i + \frac{n}{2}. \tag{16.58}$$

The degree is usually given in the literature in terms of the number of faces. Here we have an inverse, simple relation.

Proof Recall that jackets are defined such that there are as many jackets as colours. As it turns out, the linear system made of the three equations (16.30) can be inverted to get an expression for the number of faces of a given colour in terms of the Euler characteristics of the jackets,

$$F_i = \frac{1}{2}(\chi_j + \chi_k - \chi_i + n). \tag{16.59}$$

Using the definition of the degree $\omega = \sum g_i$, leads to identity (16.58), which concludes the proof.

A standard method in the analysis of coloured graphs is to analyse the length of faces.

- A face of length one is equivalent to a tadpole, i.e. the subgraph

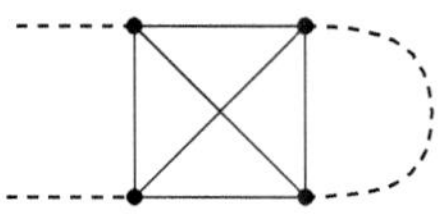

(16.60)

for any colour assignment. Then the graph is 2PR and after performing the flip as in (16.40), (16.41), we have $G_R =$  which has degree $1/2$ and

$$\omega(G) = \omega(G_L) + 1/2. \tag{16.61}$$

There are obviously no tadpoles in 2PI graphs.

- A face of length 2 is part of dipole (see the previous definition of dipoles)

We can thus focus on 2PI, dipole-free graphs, which in particular only have faces of length greater than or equal to three.

Lemma 16.5.8 *If G is 2PI and dipole-free, then*

$$\sum_{l\geq 5}(l-4)\,F_{i,l} = F_{i,3} + 4(\omega - 2g_i - 1) \tag{16.62}$$

where $F_{i,l}$ is the number of faces of colour i and length l.

Proof By noticing that each edge of colour 0 contributes to exactly one face of each colour, and that there are $2n$ such edges, one obtains

$$\sum_{l\leq 1} l\,F_{i,l} = 2n. \tag{16.63}$$

Here $F_{i,l}$ is the number of faces of colour i and length l, and n is the number of bubbles of the graph. Combining equations (16.58) and (16.63), the number of bubbles can be eliminated,

$$\sum_{l\leq 1}(l-4)\,F_{i,l} = 4(\omega - 2g_i - 1). \tag{16.64}$$

Restricting this equation to 2PI, dipole-free graphs implies that $l \geq 3$. The term $l = 3$ in the sum is in fact the only term with a negative sign in the left-hand side and we extract it to move it to the right-hand side.

If a jacket is orientable, one has the following property:

Lemma 16.5.9 *If G has its jacket J_i orientable, then its faces of colour i are of an even length, for $i \in \{1,2,3\}$.*

Proof Without loss of generality, assume $i = 3$. As already noticed in the proof of Theorem 16.4.1, when J_3 is orientable, there is a colouring of the vertices of G such that J_3 is bipartite. G itself is not bipartite since the edges of colour 3 connect white vertices to white vertices and black vertices to black vertices,

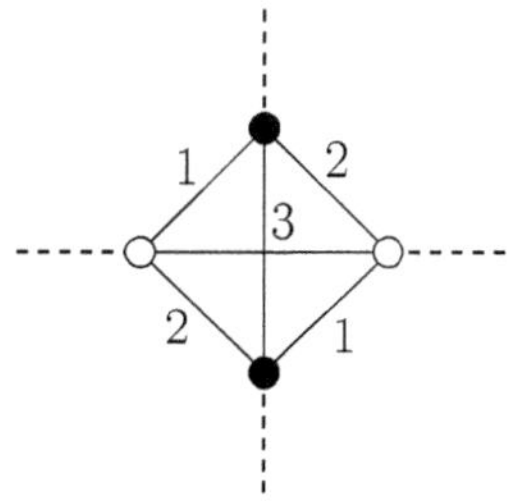

(16.65)

The edges of colour 0 (here dashed) then connect white to black and black to white vertices. We conclude that the faces of colour 3 have even lengths.

With the different operations already described, we will list in the following sections all the graphs of degree 1 and 3/2. For each degree, we first give the graphs that can be obtained through composition of smaller degree graphs. Then we give all the 2PI graphs with no melons nor tadpoles, and finally we study the possible dipole insertions. Our proof relies on degree 0 and degree 1/2 $O(N)^3$-invariant graphs.

16.5.6 The strategy

To identify all graphs at a given degree ω, we propose the following strategy.

1. Consider all graphs up to melonic two-point functions on edges of colour 0, since they leave the degree invariant.
2. Put aside the 2PR graphs of degree ω. They are indeed obtained by composing two graphs G_L, G_R of smaller degree and such that $\omega(G_L)+\omega(G_R) = \omega$. This leaves the 2PI graphs to be determined.

3. Find all 2PI, dipole-free graphs of degree ω.
4. Consider all (chains of) dipole insertions which preserve the degrees on the graphs found previously of degree ω, and (chains of) dipole insertions on graphs of smaller degrees which bring it up to ω. Then consider dipole insertions on the newly found graphs, and so on, iteratively.
5. Finally, an induction on the number of bubbles allows to conclude. This last step is necessary because of the iterative nature of the previous step.

Notice that the steps 2 and 4 are 'automatic': they rely on simple operations (composition forming a 2-edge cut, and dipole insertions) to be performed on graphs already known and can be in practice automated. Only step 3 requires an independent analysis, which has to be performed 'by hand'.

Recall from Chapter 14, that the degree 0 and respectively $1/2$ graphs of the $O(N)^3$-invariant model are the melonic graphs and tadpoles. Melonic graphs are obtained by recursive melonic insertions, i.e. insertions of the two-point graph (16.44), starting from the 2-bubble graph

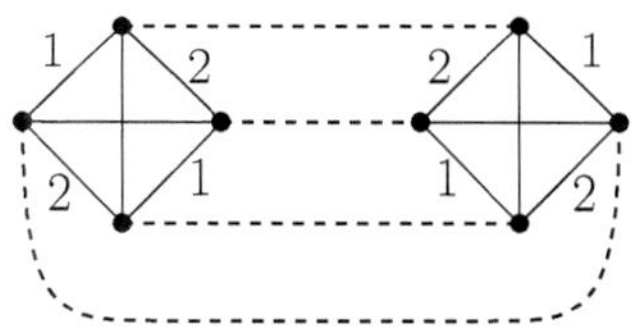

(16.66)

which is invariant under colour permutations.

The graphs of degree $1/2$ are obtained by recursive insertions of the same two-point graph (16.44) starting with the double-tadpole

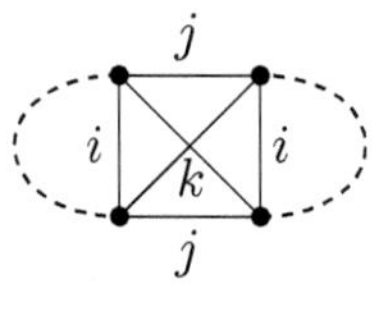

(16.67)

Notice that there are three different double-tadpoles, depending on the colour i of the two faces of length one.

Before ending this section, let us remark that a clear distinction needs to be made between fermionic and bosonic models. In the case of fermionic models, the tadpole graph of (16.67) vanishes. Since all degree $1/2$ graph are obtained from this particular graph (see again Sylvain Carrozza and Adrian Tanasa (2016) for details), this implies that, for the fermionic models, there are no degree $1/2$ graphs.

16.6 Degree 1 graphs of the $O(N)^3$-invariant SYK-like tensor model

In this section we focus on the $\omega = 1$ case and apply the strategy proposed in Subsection 16.5.6. We thus find all the Feynman graphs of degree 1 of the $O(N)^3$-invariant SYK-like tensor model. Notice that this also gives all the graphs of the MO model.

As already emphasized, the only non-automatic step of our strategy is Step 3 whose goal is to obtain the 2PI, dipole-free graphs of degree one. This is what we start with, in Subsection 16.6.1. We then go directly to Step 5 since all the other steps are automatic, in Subsection 16.6.2.

16.6.1 2PI, dipole-free graph of degree one

Theorem 16.6.1 *There is a unique 2PI, dipole-free, $O(N)^3$-invariant graph of degree one. This graph is:*

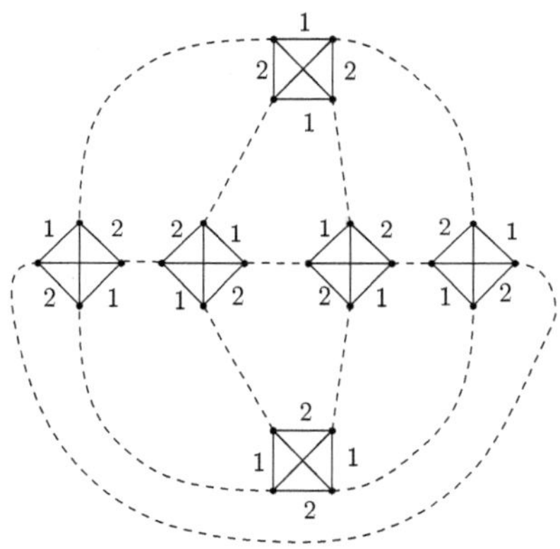

$$F_1 = 4 \qquad F_2 = 4 \qquad F_3 = 3 \qquad \omega = 3 + \tfrac{3}{2}6 - 11 = 1. \tag{16.68}$$

Proof Let G be a 2PI graph of degree 1 with no dipoles and with n bubbles. Considering (16.31), the genera of its three jackets are either $(g_1, g_2, g_3) = (0, \frac{1}{2}, \frac{1}{2})$ or $(g_1, g_2, g_3) = (0, 0, 1)$ up to colour permutations. In both cases, G has a planar jacket which we assume, without loss of generality, to be J_3, hence $g_3 = 0$.

Since J_3 is planar, it is bipartite and thus admits a canonical embedding obtained by using the counter-clockwise orientation for the colours (012) around white vertices, and clockwise around black vertices. This in turn provides a canonical embedding for G itself, obtained by adding the edges of colour 3 at the corners between the colours 1 and 2. We use this representation in the remaining of the proof.

Moreover, among the two other jackets, at least one, say J_1, has non-zero genus, i.e. $g_1 > 0$. Using equation (16.62) with $\omega = 1$, we find:

$$F_{1,3} = 8g_1 + \sum_{l \geq 5} (l - 4) F_{1,l} > 0. \tag{16.69}$$

Let us thus focus on a face of colour 1 in G,

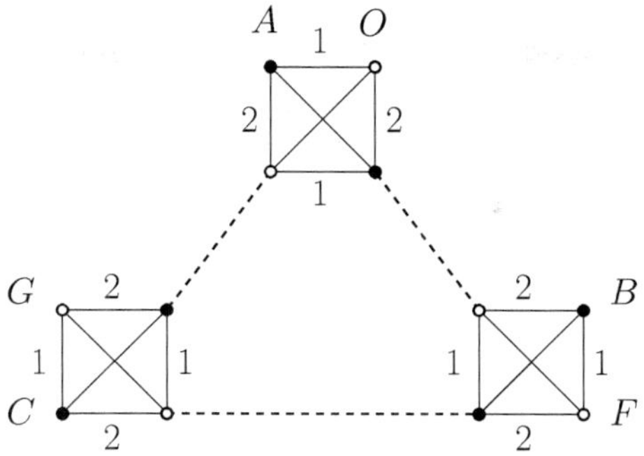

(16.70)

where the letters label vertices.

To complete the graph, we need to investigate the faces of colour 3. Lemma 16.5.9 applies to J_3: the faces of G of colour 3 are of even lengths. Since G is 2PI, it has no tadpoles, so, no faces of length 1, and since it is dipole-free, it also has no faces of length 2. The faces of colour 3 thus have length at least 4. Equation (16.62) with $g_3 = 0$ and $\omega = 1$ becomes

$$\sum_{l \geq 5} (l-4)\, F_{3,l} = 0, \tag{16.71}$$

meaning that all the faces of colour 3 of G have to be of length exactly 4. We will thus complete G so that all its faces of colour 3 have length 4.

Consider the face of colour 3 going through the vertex O and recall that edges of colour 0 only connect black to white vertices. If O is connected to A, a tadpole would be created which is forbidden. If O is connected B, a dipole would be formed, which is also forbidden, and similarly for C. The vertex O must therefore be connected to a vertex of another bubble:

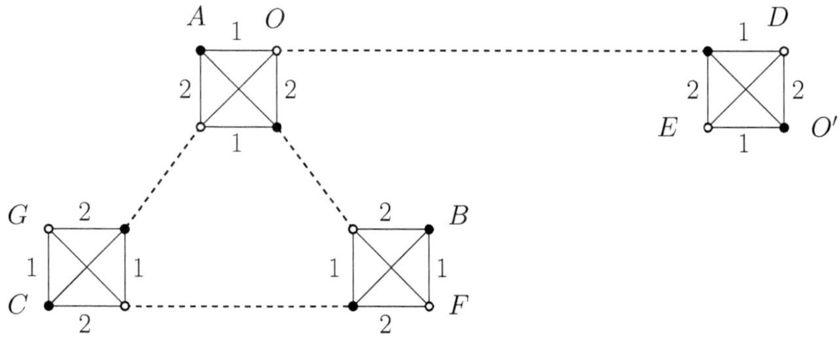

(16.72)

The vertex O' cannot be connected to D or E because this would create tadpoles. It cannot connect to F or G because the face would have length greater than 4. Therefore, O' must be connected to another bubble:

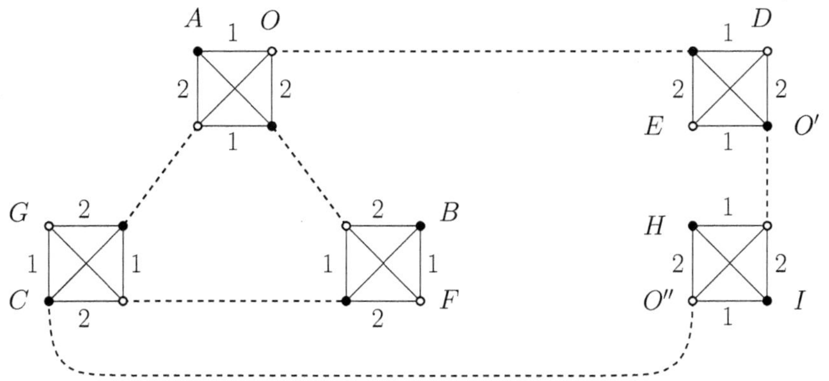

(16.73)

where the connection of O'' to C is forced to get a face of length 4.

Let us now consider the face of colour 3 going through A and F. First, assume that face connects to a new bubble and let us prove that this cannot be true. Planarity of J_3 requires the new bubble to lie in the same region as follows:

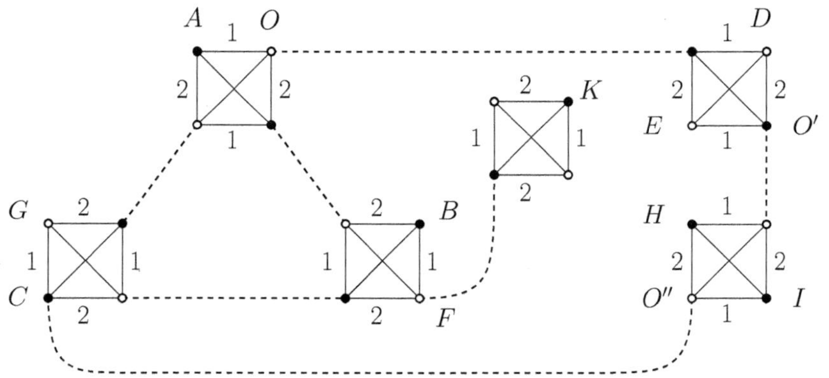

(16.74)

Again, K cannot be connected to the white vertices of its bubbles as that would create tadpoles. If it is connected to E, this then forces an edge of colour 0 from D to A so that the face be of length 4, but this creates a dipole. Any other connection would be non-planar for J_3, except if K is connected to another new bubble. However, it is then easy to check that this new bubble would have to be connected to A to close the face at length 4 and that would be non-planar.

The only possibility is thus that F is connected to H instead. Then, the vertex I cannot, however, be connected to the vertices D or G without creating a dipole. This requires I to be connected to a new bubble,

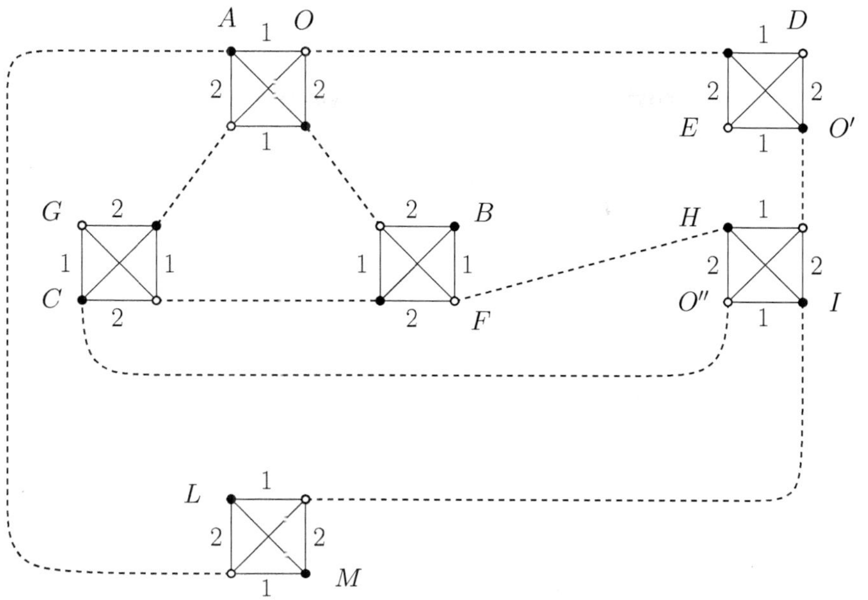

(16.75)

which is in turn connected to A to close the face of colour 3 at length 4.

Finally, consider the face of colour 3 going through G and B. We explained previously that F could not be connected to a new bubble. The same reasoning applies to B, as can be checked directly, so that it has to be connected to E. Then we must create a path with two edges of colour 0 and one edge of colour 3 between D and G. It is straightforward to check that if D connects to a new bubble then one has to break planarity of J_3 to close the face. Eventually, the only possibility is to connect D to M and L to G. This is the graph given in the theorem.

16.6.2 The graphs of degree 1

The main result of this section is the following:

Theorem 16.6.2 *The graphs of degree 1 are the graph given in* (16.68) *and the following graphs:*

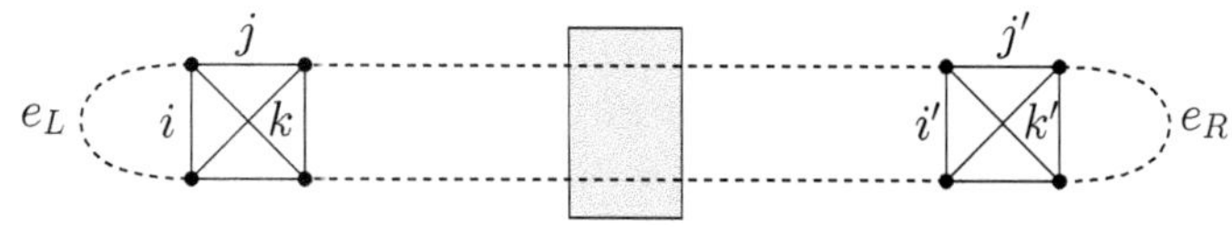

(16.76)

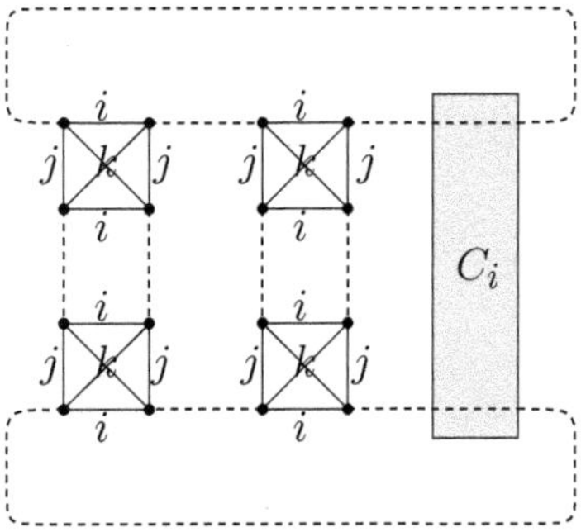

(16.77)

In this Theorem, dashed edges represent **any melonic two-point function,** *including in* (16.68), *and the boxes represent possibly empty chains of dipoles.*

We have drawn explicitly two dipoles of colour i in (16.77) so that when the chain C_i is empty the graph has four bubbles and is indeed of degree 1 (if it had only two bubbles it would be melonic).

Proof We use an induction on the number of bubbles.

Case G 2PR. Then we perform the flip as in (16.41) which turns G into the pair $\{G_L, G_R\}$ such that $\omega(G) = \omega(G_L) + \omega(G_R)$.

- If G_L (or G_R) has vanishing degree, the other graph has degree 1 and fewer bubbles so we can apply the induction hypothesis. The graph G_L is melonic and we find that G is according to the theorem upon re-inserting G_L.
- If $\omega(G_L) = \omega(G_R) = 1/2$, then they are described by (16.67): double tadpoles with arbitrary melonic insertions on the edges of colour 0. The graph G is then a composition of G_L and G_R. From Lemma 16.5.6 it is easy to see that this gives rise to the family (16.76).

All the other cases correspond to G being 2PI.

Case G 2PI and dipole-free. Then G is exactly the graph (16.68).

Case G 2PI with a dipole. Then the dipole can be eliminated as in (16.46) leading to G' connected, with two bubbles less than G. As seen in Proposition 16.5.2, dipole removals decrease or preserve the degree. Since G has degree 1, only the third and fourth cases of the proof of Theorem 16.5.2 may appear. They are the cases where the Lemma 16.5.3 and Lemma 16.5.4 apply.

- If the dipole insertion decreases the degree by 1, then G' is melonic. It is thus obtained by melonic insertions on the 2-bubble graph (16.66). From Lemma 16.5.4, we can study the dipole insertions directly on (16.66). Then, for any pair of edges of colour 0 in (16.66), it is easy to check that a dipole insertion is in fact a melonic insertion, which in particular preserves the degree.
- Therefore, the dipole removal from G must preserves the degree: $\omega(G') = 1$. From the induction hypothesis, G' has the form given in the theorem. The last step of the proof is thus to verify that performing a dipole insertion which

preserves the degree on any of those graphs does not create a graph which is not already in this family. Using Lemma 16.5.4, we can consider the three types of graphs of the theorem without any melonic two-point graph.

– G' has the form (16.76). First, consider the cases with only two bubbles. There are two of them, starting with

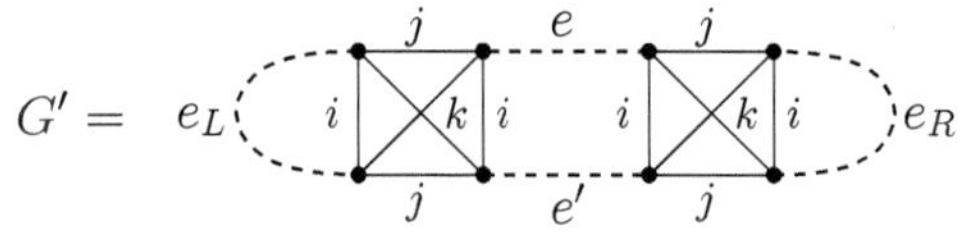

(16.78)

where the same faces of colour j and k go along e_L and e_R. Then a dipole insertion which preserves the degree can be performed on these two edges and this leads to the graph (16.77) with four bubbles, i.e.

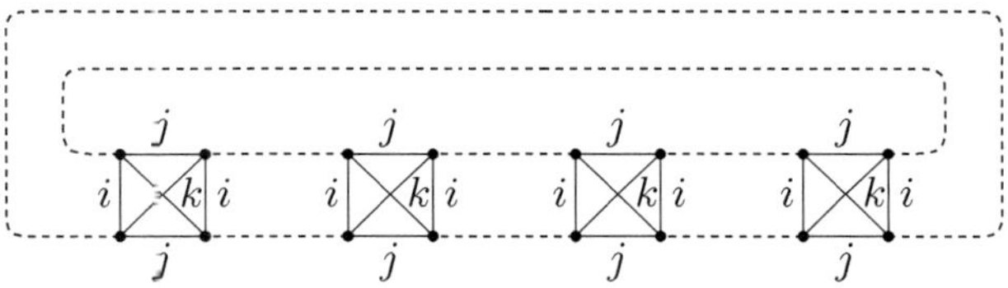

(16.79)

The other case is

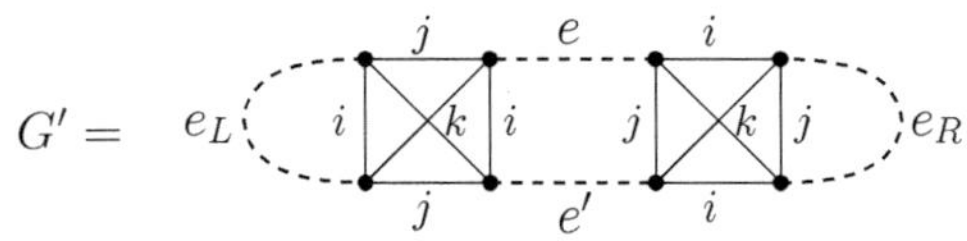

(16.80)

where no dipole insertion preserving the degree can be done on e_L, e_R this time. However, both in (16.78) and (16.80), the edges e and e' have their faces of each colour in common and a dipole of any colour can be inserted without changing the degree. This leads to a graph such as (16.76), with 4 bubbles.

Consider now a graph (16.76) with a non-empty chain. The dipole in position l has colour j_l,

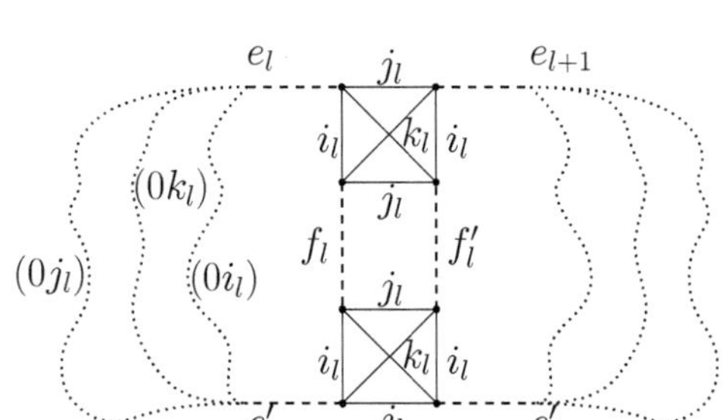

(16.81)

The dotted lines represent the bicoloured paths which close the faces of each colour going along $e_l, e_{l+1}, e'_l, e'_{l+1}$. There can be dipole insertions which preserve the degree on the pair $\{e_l, e'_l\}$ and on the pair $\{e_{l+1}, e'_{l+1}\}$, both resulting in lengthening the chain of dipoles. These dipole insertions can be of any colour. Notice that the dotted paths on the left are disjoint from those on the right. This implies that no other dipole insertions preserving the degree can be done.

- The graph G' has the form (16.77). First, it can be the graph with four bubbles pictured in (16.79). Only one dipole preserving the degree can be inserted, of colour i,

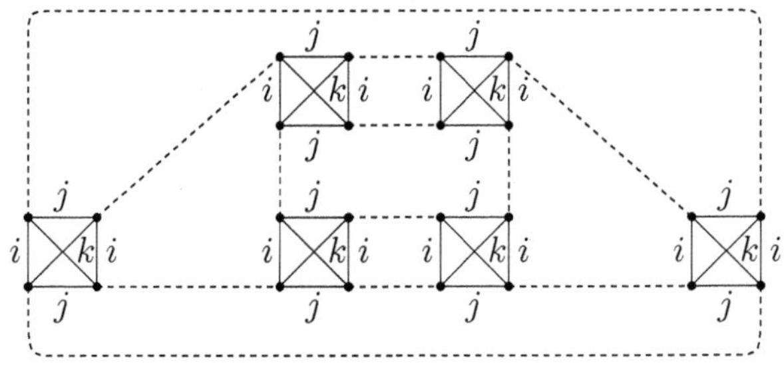

(16.82)

One can then add dipoles of colour i to create a chain of dipoles. This corresponds to (16.77).

No other dipole preserving the degree can be inserted. Indeed, a dipole of this chain is type

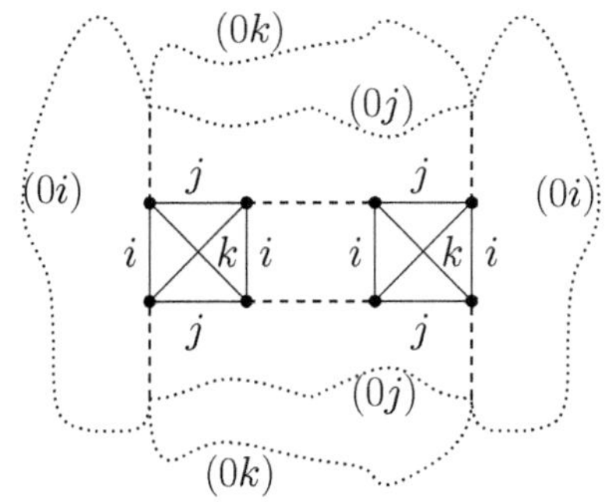

(16.83)

where the dotted lines represent the bicoloured paths closing the faces. Those paths are disjoint and therefore, only dipoles of colour i extending the chain are allowed.

- The graph G' can be the graph (16.68). It can be checked directly that no two edges have more than one face in common, meaning that no dipole insertion preserving the degree exists.

This ends the induction and thus concludes the proof.

16.7 Degree 3/2 graphs of the $O(N)^3$-invariant SYK-like tensor model

Our strategy can be applied to graphs of higher degrees, provided one can identify the 2PI, dipole-free graphs (Step 3 of the strategy in Section 16.5.6). We do so in the following case of degree $\omega = 3/2$. The other steps of the strategy are briefly discussed at the end of the section and are left as an exercise for the interested reader.

Theorem 16.7.1 *There is a unique $O(N)^3$-invariant, dipole-free, 2PI graph of degree 3/2 which is*

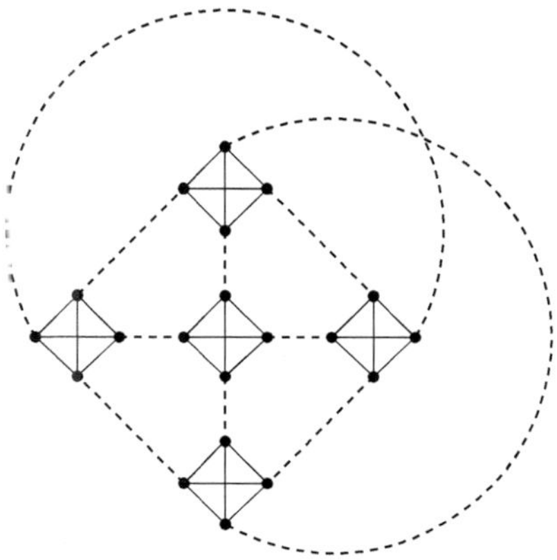
(16.84)

Its three jackets have genus 1/2 (topological projective planes). For each colour, there are exactly two faces of length 3 and one of length 4.

Proof Let G be a graph of degree 3/2. A priori, its jackets can have genera:

$$(g_1, g_2, g_3) = (3/2, 0, 0), \text{ or } (g_1, g_2, g_3) = (1, 1/2, 0), \text{ or } (g_1, g_2, g_3) = (1/2, 1/2, 1/2), \tag{16.85}$$

up to colour permutations. The first two cases have at least one planar jacket, while the third case has not. We first show that the jackets cannot be planar.

Assume one jacket is planar, say J_3. From Lemma 16.5.9, the faces of colour 3 have even lengths, $F_{3,2l+1} = 0$. This in turn simplifies (16.62) to

$$\sum_{l \geq 2} (l-2) F_{3,2l} = 1, \tag{16.86}$$

further implying that there is one face of length 6, all others being of length 4.

It can be checked that the face of length 6 has to have six distinct bubbles. For instance, we investigate the cases where it has only five bubbles. It could be as follows:

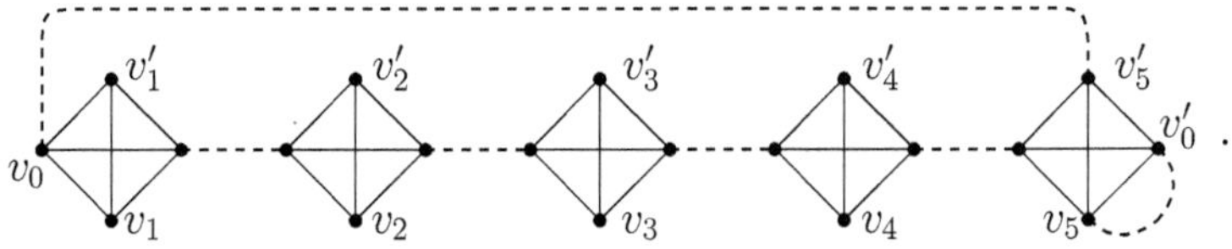

(16.87)

where v_0 is connected v_5', but then it forces a face of length 1 on the colour 1 or 2 and G would be 2PR.

If v_0 is connected to v_4', then the planarity of J_3 forces v_5 to be connected to a one-point function, and v_1', v_2', v_3' to a three-point function,

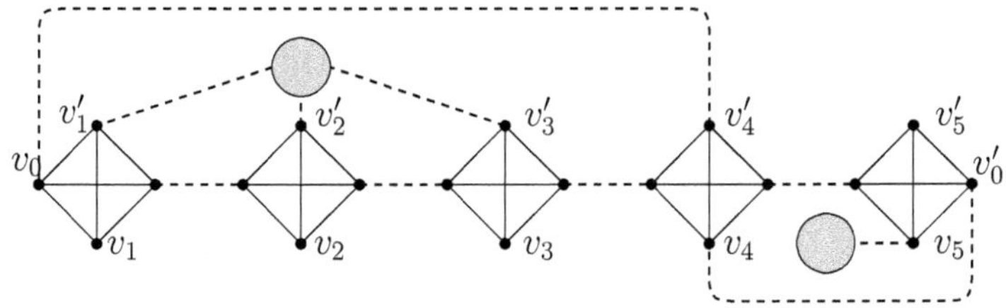

(16.88)

but they do not exist, since the number of vertices is always even (there are no odd-point functions).

If v_0 is connected to v_3', then planarity of J_3 forces the edges of colour 0 connected to v_4 and v_5 to form a 2-cut, making G 2PR, or to connect v_4 to v_5 by a single edge of colour 0 which then forms a dipole and this is forbidden as well,

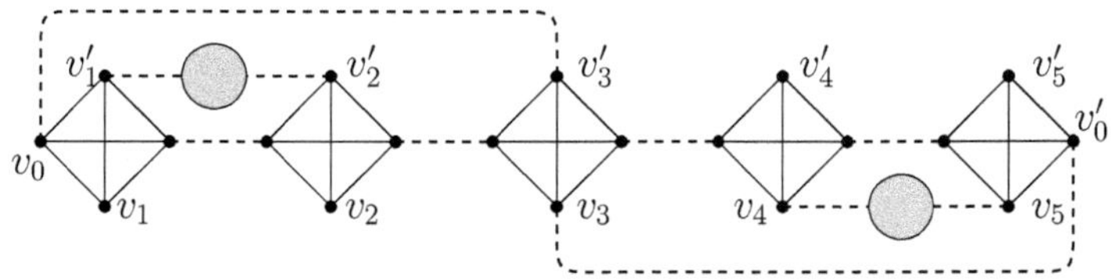

(16.89)

(and similarly, between v_1' and v_2').

Let us thus denote $b_1, \ldots, b_6$ the six distinct bubbles forming a face of colour 3 and length 6. The other edges of colour 3 of those bubbles must belong to faces of length 4. The same arguments as those just used to prove that the bubbles are distinct can be used to prove that the vertex v_1 of b_1 cannot be connected to a vertex of $b_2, \ldots, b_6$. It, therefore connects to another bubble b_7, and by symmetry v_1' too, to b_8. A face of length 4 is then obtained by 'crossing' one of the bubbles $b_2, \ldots, b_6$. Say it crosses b_3,

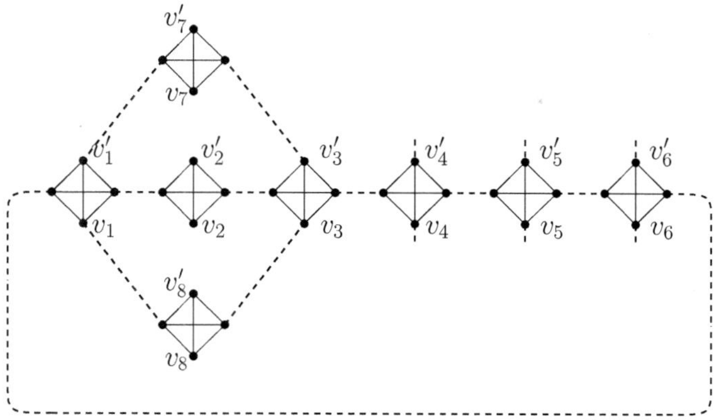

(16.90)

then planarity of J_3 forces a two-point function between v_2' and v_7, and G being 2PI forces this two-point function to be trivial, i.e. a single edge of colour 0. Same thing between v_2 and v_8'. The face of colour 3 going along those edges must be of length 4 too. To do that, it is necessary to cross one of the bubbles b_4, b_5 or b_6. In the case it crosses b_4

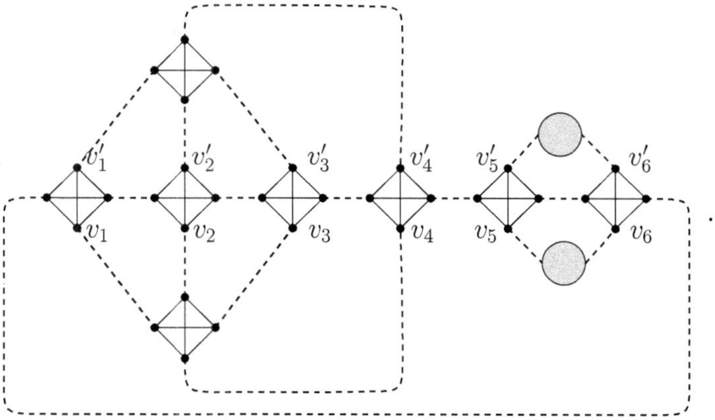

. (16.91)

Planarity of J_3 again forces a two-point function between v_5 and v_6, making G 2PR, or directly an edge of colour 0 creating a dipole. If the face of colour 3 going through v_2 and v_2' crosses b_6, this is similar. If it crosses b_5 instead, planarity of J_3 would force one-point function connected to v_4 and v_4', which is impossible.

All other cases are treated similarly. The conclusion is that G 2PI, dipole-free, of degree $3/2$, cannot have a planar jacket.

All jackets therefore have genus $1/2$. We can thus embed any jacket into the projective plane without crossings. Adding the edges of the missing colour to the jacket, an embedding of G is obtained. As a convention, we will consider that J_3 is embedded without crossings, and the projective plane is represented as a disc with opposite points identified.

Next thing is to study the face lengths. From Lemma 16.5.8, one finds

$$F_{i,3} = 2 + \sum_{l \geq 5} (l-4) F_{i,l} \geq 2 \tag{16.92}$$

meaning that there are at least two faces of length 3 for each colour.

The faces of length 3. Consider a face of length 3 and colour 3 and try to represent it using the embedding. It might be like:

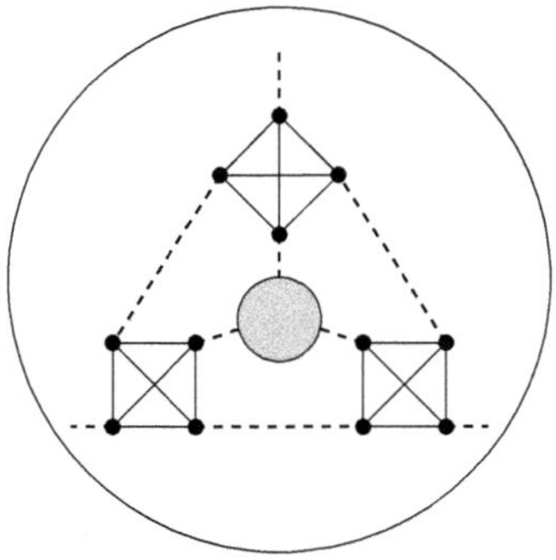

(16.93)

where the interior regions must then have a three-point function which is impossible. Therefore, a face of length 3 and colour 3 must follow the (unique up to homotropy) non-contractible cycle of the projective plane like:

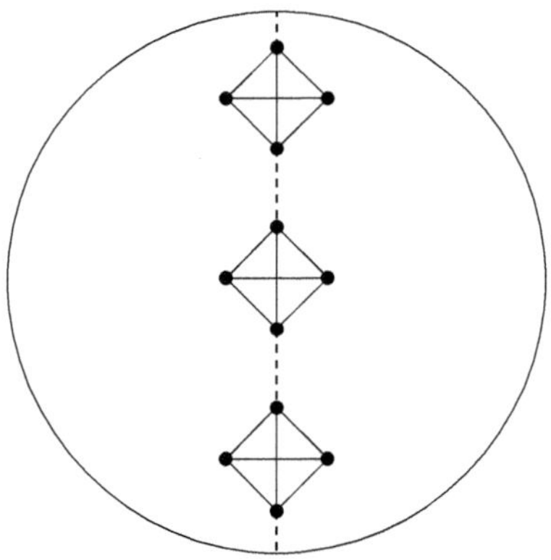

(16.94)

We can then study the location of a second face of length 3 and colour 3. It must obviously also wrap around the non-contractible cycle and must therefore 'cross' the previous face. This gives

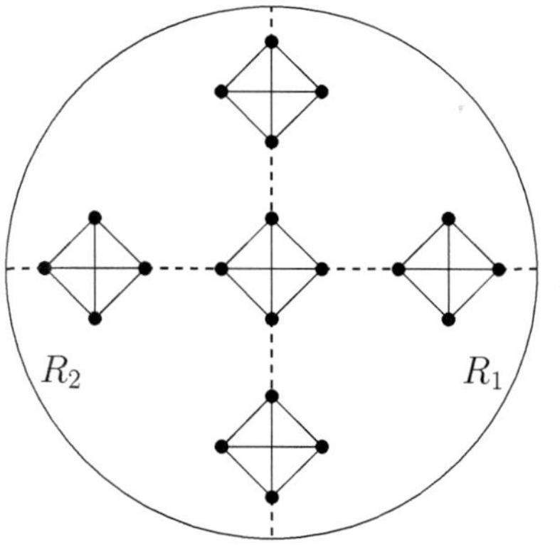

(16.95)

In J_3, the exterior of the bubbles consists in two regions R_1 and R_2. A third face of colour 3 and length 3 would have to go through both regions and thus be like

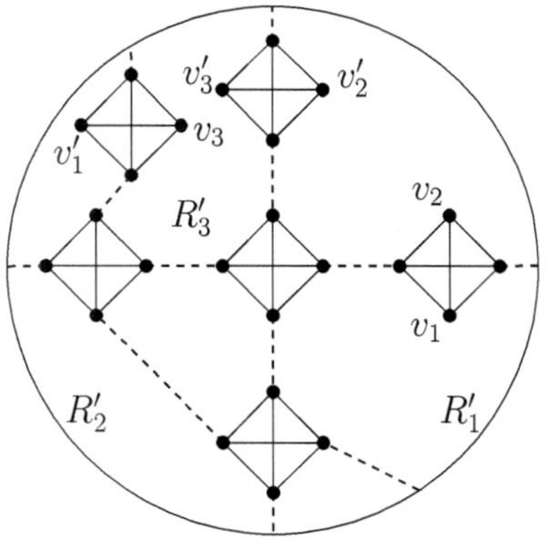

(16.96)

The only unpaired vertices in the region R'_1 are v_1 and v'_1. Only a two-point function can thus connect them and it in fact has to be just an edge of colour 0 for G to be 2PI. For the same reason, v_2 and v'_2 must be connected by an edge of colour 0, as well as v_3 and v'_3. All vertices are then adjacent to edges of colour 0. However, the counting of faces reveals that this graph is of degree 1 (its jacket J_2 is planar). We conclude that there are only two faces of colour 3 and length 3. The other faces of colour 3 are thus all of length 4.

The faces of length 4. Consider one, denoted f, going through one of the bubbles already drawn in (16.95). It has to go through the two regions R_1 and R_2 and must therefore go through at least two of the bubbles already drawn.

Say it goes through v_1 and v_2', with new notations depicted, with an additional bubble between them. Then the vertex v of the new bubble lies in a region with no other unpaired vertex, so it has to connect to a one-point function which is zero,

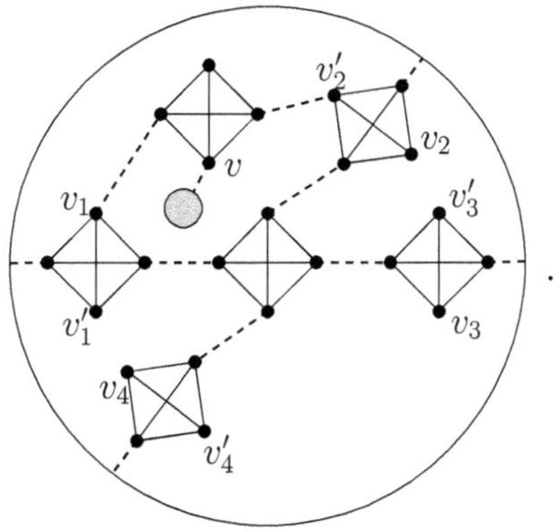

(16.97)

Similarly, if there are two new bubbles between v_1 and v_2', then their vertices must form a non-trivial two-point function or a dipole, which are both forbidden. It goes similarly, if one tries to add a bubble between v_1 and v_3 with a path of colours 0 and 3 between them,

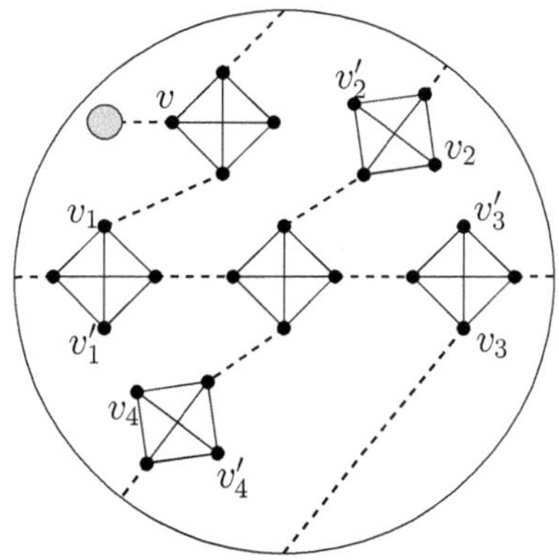

(16.98)

with a one-point function, and if two bubbles are inserted, they must create a non-trivial two-point function or a dipole.

Finally, let us try and add a bubble connecting v_1 to v_4' via a path of colours 0 and 3. The embedding of J_3 in the projective plane enforces a trivial two-point function between v and v_2' and between v' and v_3', and a four-point function in the last region (it may be two two-point functions).

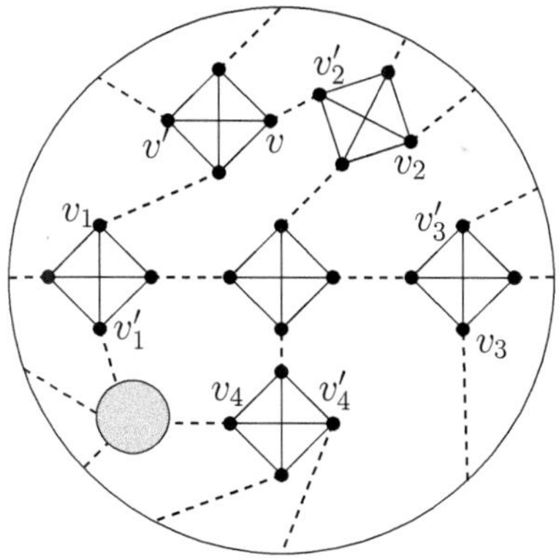

(16.99)

However, there is no way to close the two open faces of colour 3 so that they are of length 4. Indeed, this four-point function has to contain a path of colours 0 and 3 between v_1' and v_4 with exactly one bubble, which then enforces a one-point function. If two bubbles connecting v_1 to v_4' via a path with colours 0 and 3, then it is direct to see that this enforces three-point functions in the regions R and R' in the following

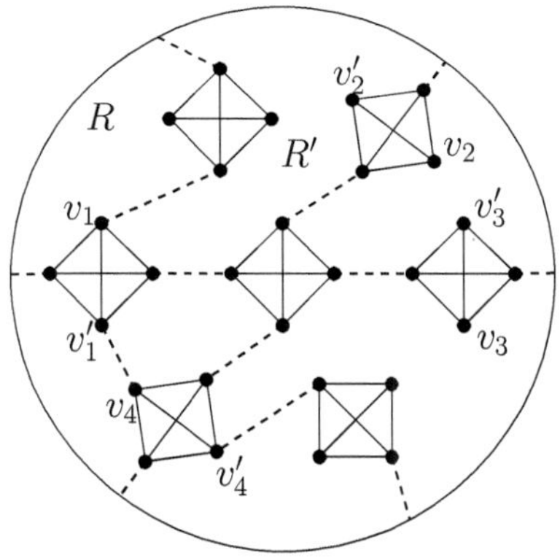

(16.100)

and this is thus impossible.

This leaves no other option than to connect the unpaired vertices of (16.95) without adding other bubbles and while maintaining the embedding of J_3 in the projective plane. If there is an edge of colour 0 between v_1 and v_3, this creates a dipole. If v_1 is connected to v_4' instead, then one might connect v_4 to v_2 but this creates a dipole. Instead, v_4 can be connected to v_3' but then v_3 needs a one-point function to avoid crossings, which is impossible,

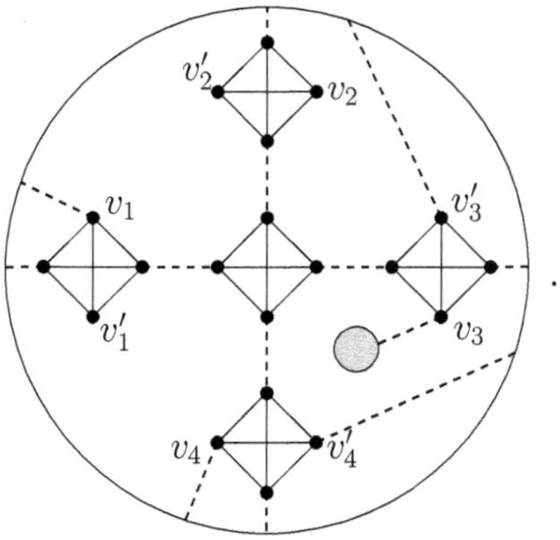

(16.101)

Therefore, v_1 can only be connected to v_2'. Then v_2 cannot be connected to v_4 without forming a dipole and cannot be connected to v_1' because the face of colour 3 would have length 2 only. Thus, v_2 can only be connected to v_3', and then v_3 to v_4' and finally v_4 to v_1'

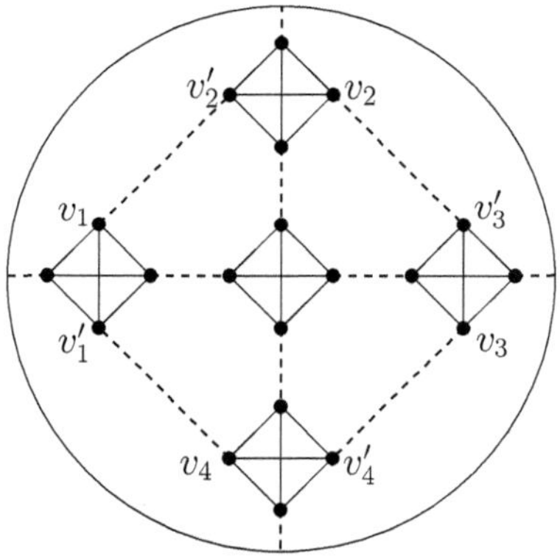

(16.102)

which is the graph of the Theorem. This concludes the proof.

Let us end this section by adding a few comments on the other steps of the strategy to find all graphs of degree 3/2.

- As for 2PR graphs without melonic insertions (Step 2), it is enough to consider all compositions of G_L and G_R with $\omega(G_L) = 1$ and $\omega(G_R) = 1/2$. Those graphs have been detailed throughout this article. We leave the compositions to the interested reader.
- As for the dipole insertions, as in the case of degree 1, only those which preserve the degree and those which increase the degree by one have to be considered.

Appendix A
Examples of tree weights

A.1 Symmetric weights—complete partition

Symmetric weights correspond to the symmetric partition Π^s of V into $|V|$ singletons. Let us check directly that this is indeed the case, namely that $w_s(G,T)$, as defined in (4.7) is equal to $w^{\Pi^s}(G,T)$ defined by formula (4.21):

Lemma A.1.1 *The symmetric weights $w_s(G,T)$ are the partition weights for the partition Π^s of V into $|V|$ singletons:*

$$w_s(G,T) = \sum_{\sigma|T(\sigma)=T} 1/|E(G)|! = \sum_{\tau} \prod_{i=0}^{|V|-2} \frac{1}{k_i} = w^{\Pi^s}(G,T), \tag{A.1}$$

where the sum over τ is performed over all the orderings of T and k_i is the number of trans-block edges for the partition Π_i^s in the partition sequence corresponding to T_τ, starting from the all-singletons partition Π^s.

Proof Notice that, in the symmetrical case, every sector is admissible, hence there is no restriction on the sum over τ. We work by induction on the number of vertices of G in (A.1). Let us start with an initial general graph $G = G_0$, and suppose it has a certain set L_0 of tadpole edges. Consider the graph $G_0' = G_0 - L_0$ with all tadpoles of G deleted. Since the weights $w_s(G,T)$ cannot depend on the position of the tadpoles edges in the Hepp sector σ, we have $w_s(G_0,T) = w_s(G_0',T)$. In G_0', which has no tadpoles, all edges are trans-block at first step (since $\Pi_0 = \Pi^s$ is made of singletons). Therefore the factor k_0 in (4.21) is $k_0 = |E(G_0')|$. We can write

$$\begin{aligned} w_s(G,T) &= \sum_{\ell_1=\sigma(1)\in T} \frac{1}{k_0} \sum_{\sigma_1 \mid T(\sigma_1)=T-\ell_1} \frac{1}{(k_0-1)!} \\ &= \sum_{\ell_1=\sigma(1)\in T} \frac{1}{k_0} w_s(G_1,T_1) \\ &= \sum_{\ell_1=\sigma(1)\in T} \frac{1}{k_0} w^{\Pi^s}(G_1,T_1) = w^{\Pi^s}(G,T). \end{aligned} \tag{A.2}$$

where ℓ_1 is the first edge of T_τ; G_1 is obtained by contracting the edge ℓ_1 in G_0', T_1 is obtained by contracting the first edge ℓ_1 in T_τ and the sum over σ_1 runs over the Hepp sector of G_1. Since G_1 has one vertex less than G_0 we used the induction hypothesis in the last line of (A.2) to conclude.

Consider the particular example of the graph of Fig. 2. The symmetric weights are:

$$\begin{aligned} w_s(G,T_{125}) &= w_s(G,T_{126}) = w_s(G,T_{156}) = w_s(G,T_{256}) = 1/15, \\ w_s(G,T_{135}) &= w_s(G,T_{136}) = w_s(G,T_{235}) = w_s(G,T_{236}) = w_s(G,T_{145}) \\ &= w_s(G,T_{146}) = w_s(G,T_{245}) = w_s(G,T_{246}) = 11/120. \end{aligned} \tag{A.3}$$

These weights were computed in Rivasseau and Wang (2014) (note the different labeling we use here with respect to the one of Rivasseau and Wang (2014)).

A.2 One singleton partition—rooted graph

The next case we deal with is the one when the partition is made of a certain number p of singletons plus a single block with all other remaining vertices. As already mentioned, when $p = 1$, the partition corresponds to work on a rooted graph. When the number p of singletons is at least two, the weights correspond to multi-rooted weights (see the example in Subsection A.3.

Consider the graph of Figure A.1. Let us find the tree weights in the case of root at v_1. This correspond to the partition

$$\Pi_1 = [\{v_1\}, \{v_2, v_3\}]. \tag{A.4}$$

There are six admissible ordered trees, namely T_{12}, T_{13}, T_{14}, T_{21}, T_{23}, and finally T_{24}. The weights are

$$w^{\Pi_1}(G,T_{12}) = w^{\Pi_1}(G,T_{21}) = \int_0^1 du_1 du_2 u_1.(u_2)^2 = 1/6, \tag{A.5}$$

since $i(l_1) = 0, j(l_1) = 1; i(l_2) = 0, j(l_2) = 2; i(l_3) = 1, j(l_3) = 2; i(l_4) = 1, j(l_4) = 2$, where, from now on, we have simplify the notations for the contact indices (since an edge is identified with a pair of (non-necessarily distinct) vertices of the graph).

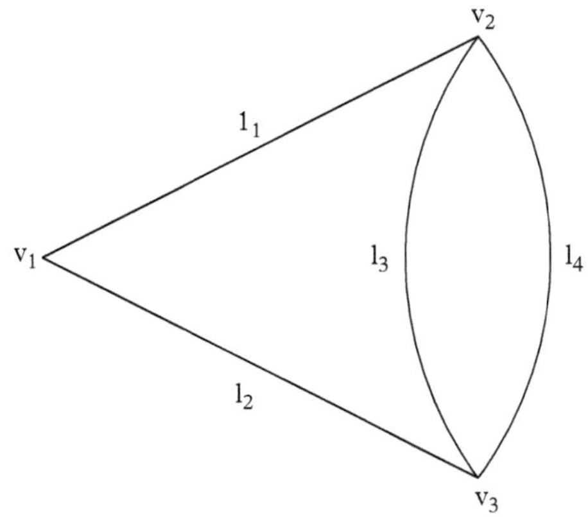

Figure A.1 *An example of a three vertex graph*

Similarly, one has

$$w^{\Pi_1}(G,T_{13}) = w^{\Pi_1}(G,T_{14}) = w^{\Pi_1}(G,T_{23}) = w^{\Pi_1}(G,T_{24}) = \int_0^1 du_1 du_2 (u_1 u_2).(u_2) = 1/6, \tag{A.6}$$

since $i(l_1)=0, j(l_1)=1; i(l_2)=0, j(l_2)=2; i(l_3)=1, j(l_3)=2; i(l_4)=1, j(l_4)=2$.

We can check that $\sum_{T\subset G} w^{\Pi_1}(G,T)=1$.

Let us now count the factors in the case of the root at v_2. This corresponds, in the formalism of this paper, to consider the partition:

$$\Pi_2 = [\{v_2\},\{v_1,v_3\}]. \tag{A.7}$$

As previously presented, there are seven admissible ordered trees, namely $T_{12}, T_{13}, T_{14}, T_{31}, T_{32}, T_{41}$, and T_{42}. The associated weights compute to:

$$w^{\Pi_2}(G,T_{12}) = \int_0^1 du_1 du_2 [1.1.(u_1 u_2).(u_1 u_2)] = 1/9 \tag{A.8}$$

since $i(l_1)=0, j(l_1)=1; i(l_2)=1, j(l_2)=2; i(l_3)=0, j(l_3)=2; i(l_4)=0, j(l_4)=2$, and

$$w^{\Pi_2}(G,T_{13}) = w^{\Pi_2}(G,T_{14}) = \int_0^1 du_1 du_2 [1.u_2.u_1.(u_1 u_2)] = 1/9 \tag{A.9}$$

since $i(l_1)=0, j(l_1)=1; i(l_2)=1, j(l_2)=2; i(l_3)=0, j(l_3)=2; i(l_4)=0, j(l_4)=2$. and

$$w^{\Pi_2}(G,T_{31}) = w^{\Pi_2}(G,T_{41}) = \int_0^1 du_1 du_2 u_1.u_2.1.u_1 = 1/6 \tag{A.10}$$

since $i(l_1,T_{31})=0$, $j(l_1,T_{31})=2$; $i(l_2,T_{31})=1$, $j(l_2,T_{31})=2$; $i(l_3,T_{31})=0$, $j(l_3,T_{31})=1$; $i(l_4,T_{31})=0, j(l_4,T_{31})=1$. and

$$w^{\Pi_2}(G,T_{32}) = w^{\Pi_2}(G,T_{42}) = \int_0^1 du_1 du_2 u_1 u_2.1.1.u_1 = 1/6 \tag{A.11}$$

since $i(l_1,T_{32})=0$, $j(l_1,T_{32})=2$; $i(l_2,T_{32})=1$, $j(l_2,T_{32})=2$; $i(l_3,T_{32})=0$, $j(l_3,T_{32})=1$; $i(l_4,T_{32})=0, j(l_4,T_{32})=1$. As expected, we have again: $\sum_{T\subset G} w^{\Pi_2}(G,T)=1$.

A.3 Two singleton partition—multi-rooted graph

We consider the graph of Fig. A.2 with four vertices and six edges, for the partition

$$\Pi = [\{v_1\};\{v_2\};\{v_3,v_4\}]. \tag{A.12}$$

This corresponds to considering a graph with *two roots*, the first at the vertex v_1 and the second at the vertex v_2. This graph has twelve trees:

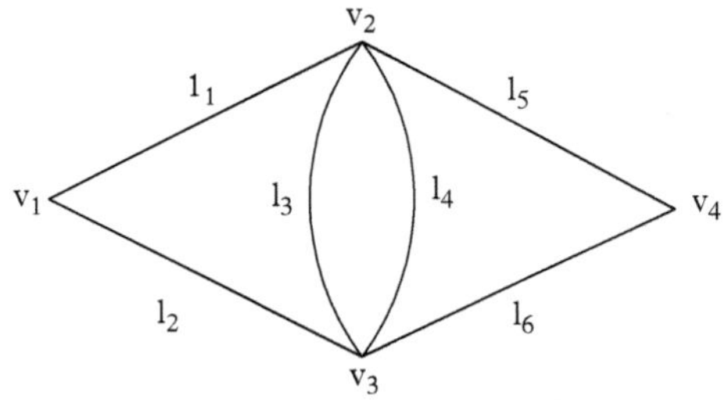

Figure A.2 *An example of a four vertex graph*

$$\begin{aligned}&T_{125}, T_{135}, T_{145}, T_{235}, T_{245},\\ &T_{236}, T_{246}, T_{256},\\ &T_{126}, T_{136}, T_{146}, T_{156}.\end{aligned} \tag{A.13}$$

The first five trees in this list (the trees in the first line of (A.13)) can be each endowed with six admissible orders (each order is for these trees admissible). The next three trees in the list (the trees in the second line of (A.13)) can be endowed with only four admissible orders, while the last four trees (the trees in the last line of (A.13)) can be endowed with three admissible orders. This makes up for a total of fifty-four ordered trans-block trees to consider for this graph.

Let us explicitly consider the first of these ordered trees, namely the (l_1, l_2, l_5) ordered tree. For the three tree lines, the contact indices are: $i(l_1) = 0, j(l_1) = 1$, $i(l_2) = 0, j(l_2) = 2$ and $i(l_5) = 0, j(l_5) = 3$. For the remaining three loop lines, the contact indices are: $i(l_3) = 0, j(l_3) = 2$, $i(l_4) = 0, j(l_4) = 2$ and finally $i(l_6) = 0, j(l_6) = 1$. This leads to the contribution:

$$\int_0^1 du_1 du_2 du_3 u_1^4 u_2^3 u_3 = \frac{1}{40}. \tag{A.14}$$

The other five admissible orders for this tree lead to the weights $1/80$, $1/50$, $1/100$, $1/100$ and finally, again $1/100$. Thus, for the total of six admissible orders that one can endow this tree with, one obtains a total weight of $7/80$.

After a tedious but straightforward computation, we obtain all forty-eight admissible order contributions and find the complete list of all tree weights for this partition:

$$\begin{aligned}&w^{\Pi}(G, T_{135}) = w^{\Pi}(G, T_{145}) = 47/400,\\ &w^{\Pi}(G, T_{235}) = w^{\Pi}(G, T_{245}) = 11/100,\\ &w^{\Pi}(G, T_{236}) = w^{\Pi}(G, T_{246}) = 2/25,\\ &w^{\Pi}(G, T_{136}) = w^{\Pi}(G, T_{146}) = 3/40,\\ &w^{\Pi}(G, T_{256}) = 1/20, \quad w^{\Pi}(G, T_{126}) = 11/200,\\ &w^{\Pi}(G, T_{125}) = 7/80, \quad w^{\Pi}(G, T_{156}) = 17/400.\end{aligned} \tag{A.15}$$

Note that these tree weights are different from the symmetric weights of (A.3). Finally, one can check that $\sum_{T \in G} w^{\Pi}(G, T) = 1$, as expected.

Appendix B
Renormalization of the Grosse–Wulkenhaar model, one-loop examples

We now consider in detail the behaviour of a planar, regular, and irregular, four-point function.

Following Tanasa and Kreimer (2012), we first analyze the example of the 1-loop graph of Fig. B.1. The Feynman integral to analyze is written:

$$\int dx_1 \dots dx_4 dy_1 \dots dy_4 \phi(x_1)\dots\phi(x_4)\delta(x_1 - x_2 + y_2 - y_1)\delta(y_3 - y_4 + x_3 - x_4)$$
$$C(y_1, y_3)C(y_2, y_4)e^{\imath(x_1 \wedge x_2 + y_2 \wedge y_1)}e^{\imath(y_3 \wedge y_4 + x_3 \wedge x_4)}, \tag{B.1}$$

where we have used the second form of (9.6) to express the Moyal oscillating phase of the two vertices. Note that, as in the commutative case, previous the integral has to be considered in the short-distance (high-momenta) regime for the internal edges. As previously described, we introduce the set of short and long variables associated with the graph:

$$u_1 = y_3 - y_1,\ v_1 = y_3 + y_1,\ u_2 = y_2 - y_4,\ v_2 = y_4 + y_2. \tag{B.2}$$

Conversely, this can be written:

$$y_1 = \frac{1}{2}(v_1 - u_1),\ y_2 = \frac{1}{2}(v_2 + u_2),\ y_3 = \frac{1}{2}(u_1 + v_1),\ y_4 = \frac{1}{2}(v_2 - u_2). \tag{B.3}$$

We perform the change of variables,

$$(y_1, \dots, y_4) \to (u_1, v_1, u_2, v_2)$$

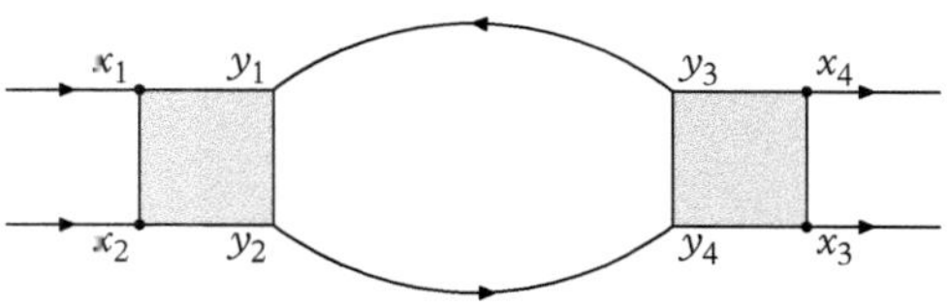

Figure B.1 *A planar regular four-point graph in the Moyal ϕ^4 model*

and integrate over v_1 (using the first of the δ-functions in (B.1)). After some algebra, the Feynman integral becomes

$$\int dx_1 \dots dx_4 du_1 du_2 dv_2 \phi(x_1)\dots\phi(x_4)\delta(x_1 - x_2 + x_3 - x_4 + u_1 + u_2)$$
$$C(u_1, 2(x_1 - x_2) + u_1 + u_2 + v_2)C(u_2, v_2)e^{\imath(x_1\wedge x_2 + x_3\wedge x_4)}$$
$$e^{\imath\left(\frac{1}{2}(x_2 - x_1)\wedge u_2 - \frac{1}{4}u_1\wedge u_2 - \frac{1}{4}u_1\wedge v_2 + \frac{1}{2}u_2\wedge v_2\right)}, \tag{B.4}$$

where we have dropped the nonessential constant coming from the change of variable (B.2).

The integral rewrites as:

$$\int dx_1 \dots dx_4 du_1 du_2 dv_2 \phi(x_1)\dots\phi(x_4)\delta(x_1 - x_2 + x_3 - x_4 + t(u_1 + u_2))$$
$$C(u_1, 2t(x_1 - x_2) + u_1 + u_2 + v_2)C(u_2, v_2)e^{\imath(x_1\wedge x_2 + x_3\wedge x_4)}$$
$$e^{\imath\left(\frac{1}{2}t(x_2 - x_1)\wedge u_2 - \frac{1}{4}u_1\wedge u_2 - \frac{1}{4}u_1\wedge v_2 + \frac{1}{2}u_2\wedge v_2\right)}|_{t=1}. \tag{B.5}$$

This formula is designed so that at $t = 0$ all dependence on the external variables x factorizes out of the u, v integral giving the desired vertex form. One then has to perform a Taylor expansion with respect to the t variable:

$$f(1) = f(0) + \int_0^1 dt f'(t).$$

The first term, $f(0)$ is of the desired parallelogram-like form; furthermore, the remainder term is proven to be irrelevant (see Gurau et al. (2006) for details).

This is thus a generalization of the commutative situation, where one has replaced the notion of locality of the vertex with the more involved notion of a non-local but parallelogram-shaped vertex of a Moyal field theory. The same type of arguments also hold for the model (9.41).

Let us now give some insights on the planar irregular case in order to better understand why these graphs are finite and thus *are not* primitively divergent (and hence do not have to be taken into consideration when defining the coproduct). As before, we work in position space and we consider the example of a 1-loop graph, namely the one in Fig. B.2. With the conventions detailed previously, the Feynman integral to analyze is written:

$$\int dx_1 \dots dx_4 dy_1 \dots dy_4 \phi(x_1)\dots\phi(x_4)\delta(x_1 - x_2 + y_2 - y_1)\delta(x_4 - y_4 + y_3 - x_3)$$
$$C(y_1, y_3)C(y_2, y_4)e^{\imath(x_1\wedge x_2 + y_2\wedge y_1)}e^{\imath(x_4\wedge y_4 + y_3\wedge x_3)}. \tag{B.6}$$

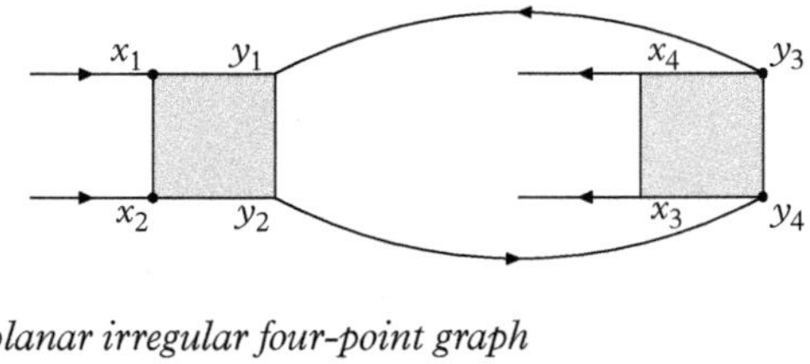

Figure B.2 *A one-loop planar irregular four-point graph*

We perform the same change of variables (B.2) as in the planar regular case; as previously given, we integrate over v_1 (using the first of the $\delta-$functions in (B.6)). We then focus on the variables x_4 and v_2 which are the ones leading to an improved behaviour with respect to the planar regular case (B.5). Thus, the factor involving these variables in the oscillation in the final form of the integral is:

$$e^{\imath(-v_2 \wedge x_4 + \frac{1}{2} v_2 \wedge (x_3 + x_1 - x_2))}.$$

This factor is different from the one found in (B.5) and when integrated against x_4 and v_2 (taking into account the v_2 contribution of the propagators (9.11)) leads to an improvement in the UV behaviour of the integral (see again Gurau et al. (2006) for details). Thus, the Feynman integral (B.5), logarithmically divergent in the planar regular case, becomes convergent, as announced.

Appendix C
The B_+ operator in Moyal QFT, two-loop examples

C.1 One-loop analysis

One has:

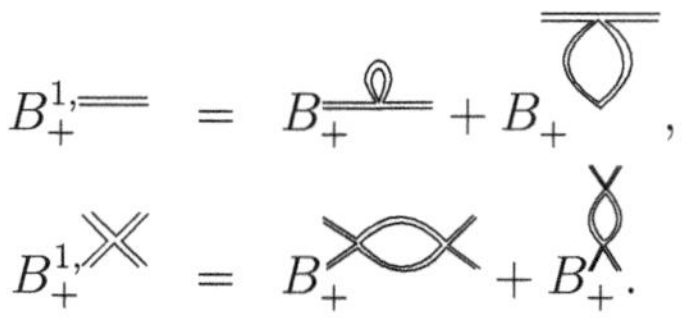

(C.1)

Applying this map on $1_{\mathcal{H}}$ leads to:

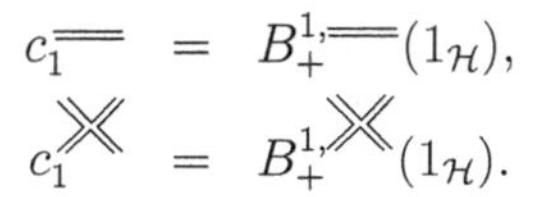

(C.2)

Applying (9.61) and (C.1) leads directly to:

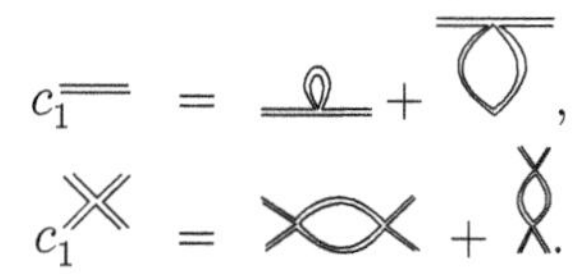

(C.3)

C.2 Two-loop analysis

We first work out the easier two-loop two-point function and then proceed with the four-point one. Using the definition (9.63), one gets:

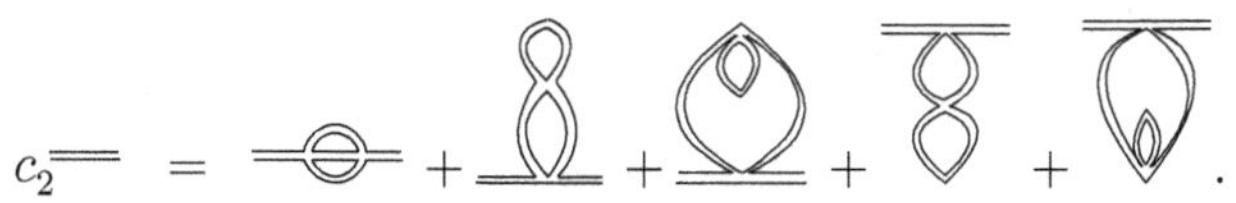

(C.4)

Applying the coproduct Δ on each of these ribbon graphs, one has (for the non-trivial part):

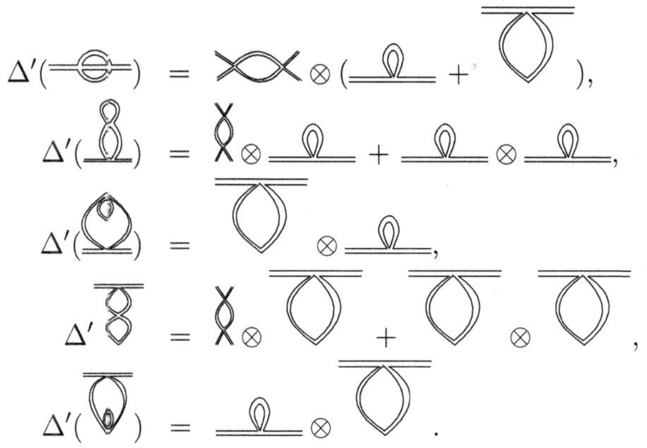

(C.5)

Putting all this together leads to:

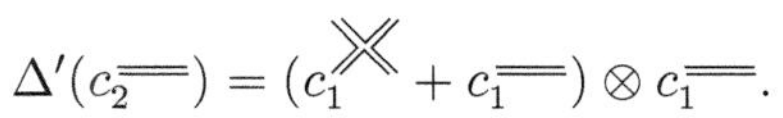

(C.6)

Let us now focus on the more involved case of the four-point function. Using again the definition (9.63), one gets:

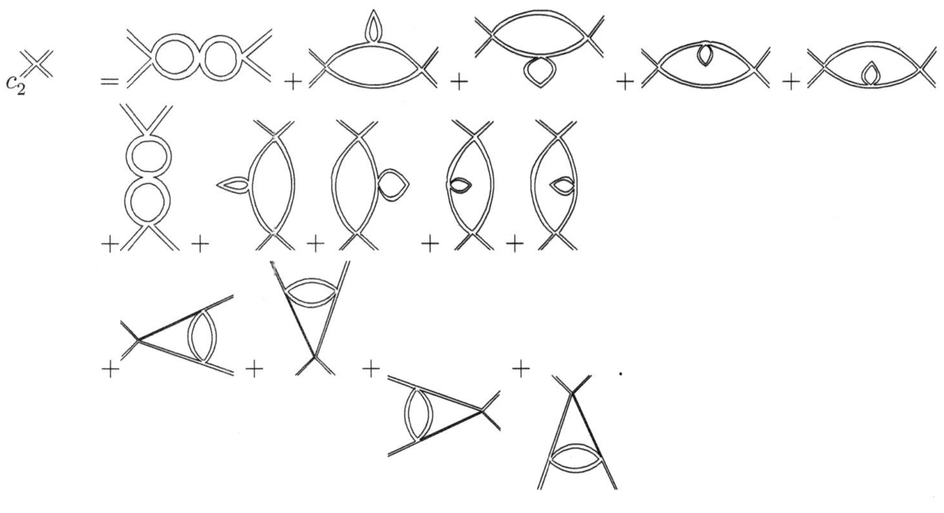

(C.7)

Applying the coproduct Δ on each of these ribbon graphs, one has (for the non-trivial part):

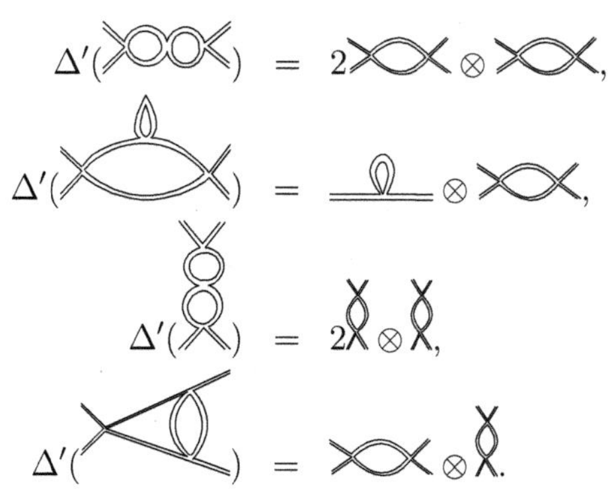

(C.8)

The remaining ten ribbon graphs on the right-hand side of (C.7) are treated analogously, finally leading to

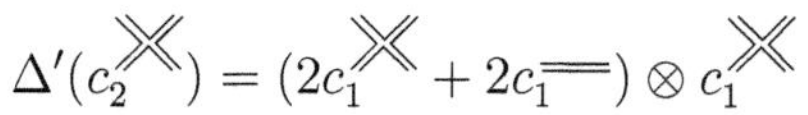

(C.9)

(where we have used (C.3)). The results (C.6) and (C.9) are thus illustrations of Theorem 9.8.6; these equations further give the expressions of the polynomials $P_1^{=}$ and $P_1^{\times}$.

Let us now show that each such two-loop graph lies in the image of the Hochschild one-cocycles $B_+^{1,=}$. Using the definition (9.59) one has:

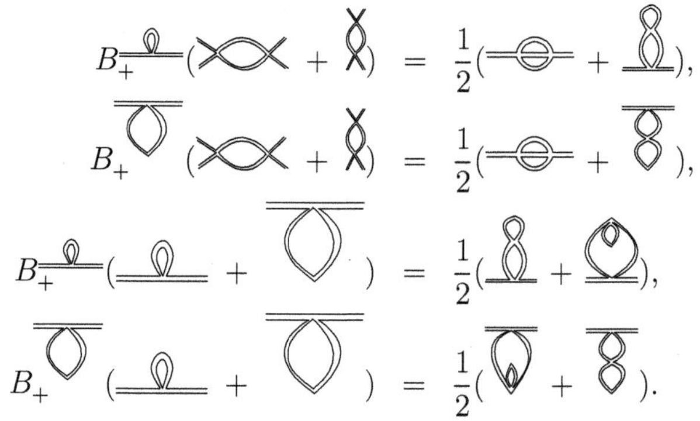

(C.10)

The $\frac{1}{2}$ coefficients previously provided come from the computation of the permutation of external edges, number of bijections and number of maximal forests for each of the resulting ribbon graphs, as explained in subsection 9.8.2. When adding up all this, one does not obtain $c_2^{=}$ (as given by (C.4)). This comes from the fact that we have not yet included the planar irregular sector, which gives birth through some insertions to planar regular graphs (as explained in Subsection 9.8.2). One has:

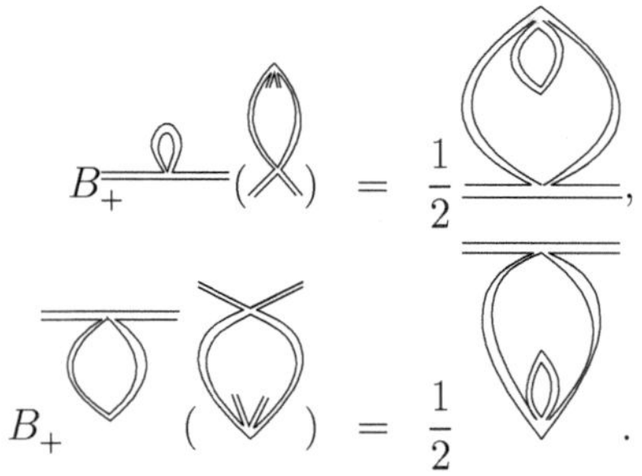

(C.11)

The two new graphs here belong to

(C.12)

(which corresponds to the planar irregular sector). Note that the rest of the planar irregular graphs belonging to (C.12) do not lead to planar regular graphs when one acts on them with the operator $B_+^{1,=}$. Furthermore, one can analogously define

(C.13)

Acting on this graph with $B_+^{1,=}$ does not lead to a planar regular graph. Thus, adding up (C.10) and (C.11), one obtains $c_2^{=}$, as expected. We have thus proved that

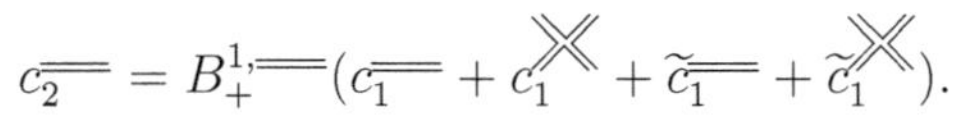

(C.14)

Let us now explicitly show the necessity of adding the irregular sector also when writing down Theorem 9.8.4 at this two-loop level. In order to do this, we first consider the four planar regular graphs:

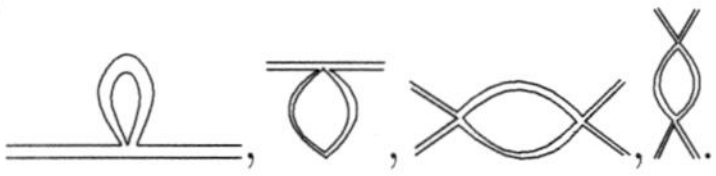

(C.15)

Using (C.10), one can write down the left-hand side and the right-hand side of Theorem 9.8.4. Let us right down the contribution of the graphs of (C.15) on the left-hand side of Theorem 9.8.4. One has the following relations:

1.

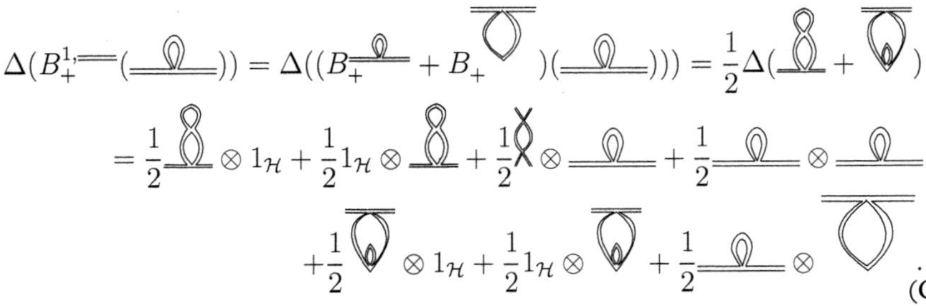

(C.16)

2.

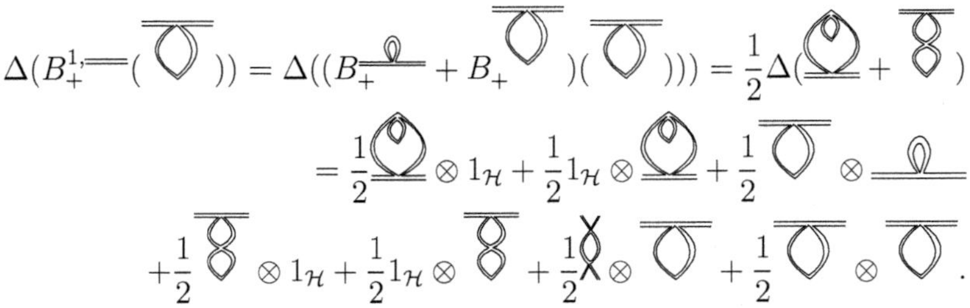

(C.17)

3.

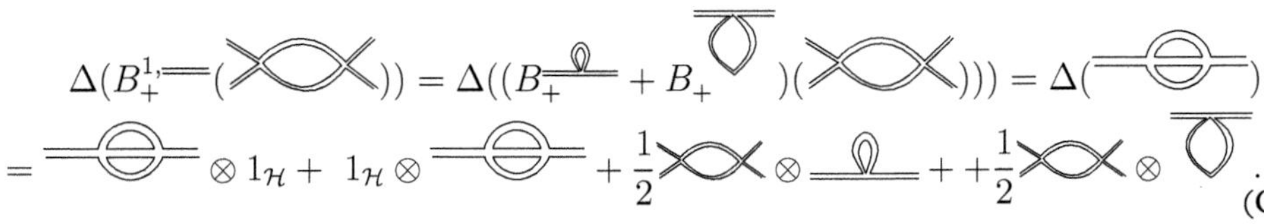

(C.18)

4.

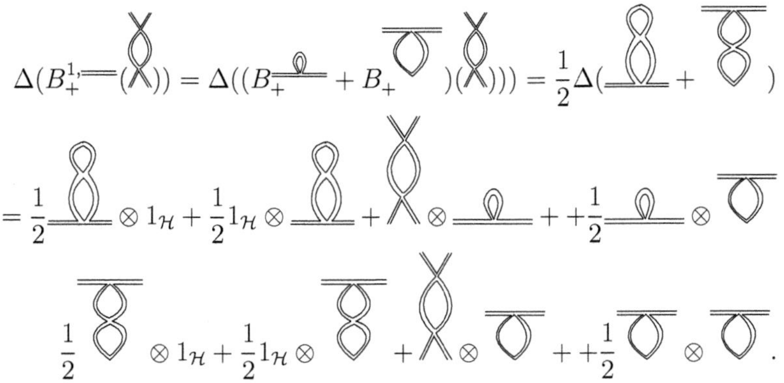

(C.19)

Adding together the contributions of equations (C.16) to (C.19) one has, for the left-hand side of Theorem 9.8.4, the following results (corresponding to the planar regular graphs listed in equation (C.15)):

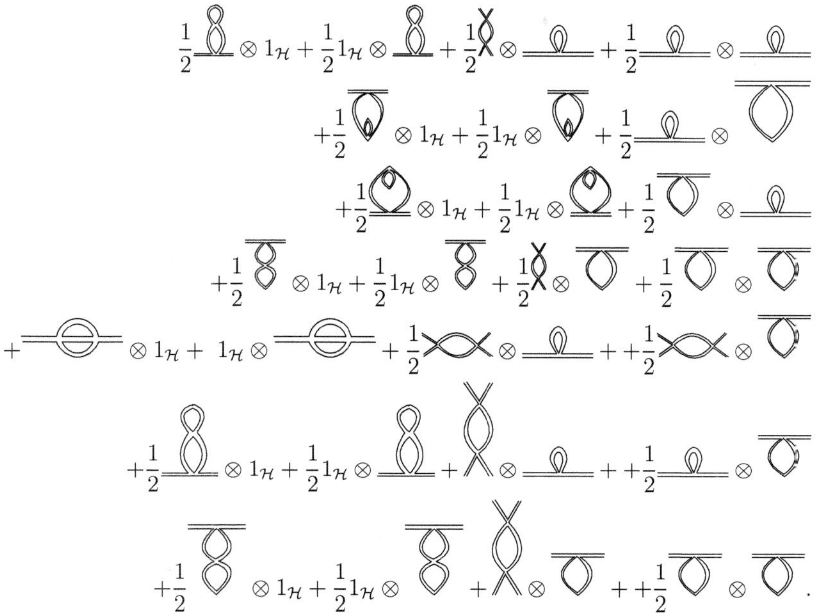

(C.20)

The right-hand side of Theorem 9.8.4, corresponding to the total contribution of the planar regular sector listed in (C.15), is worked out analogously, leading to

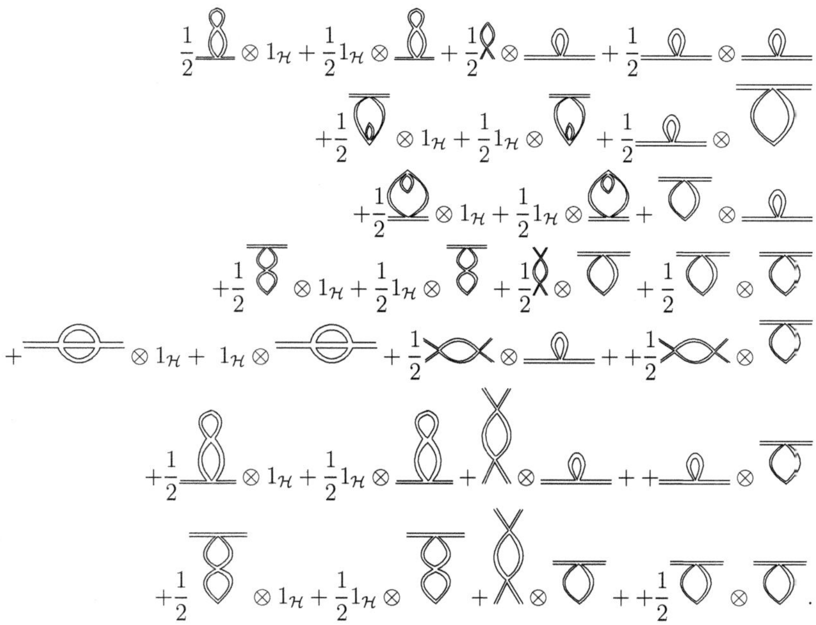

(C.21)

Comparing the previous equations (C.20) and (C.21), one is left on the right-hand side with

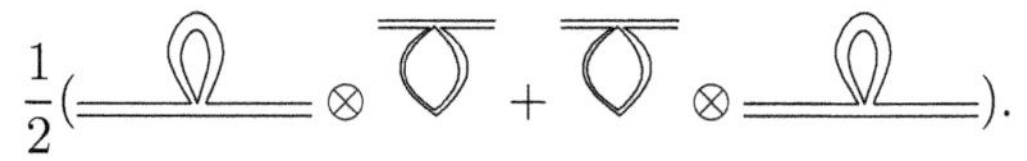

(C.22)

As previously mentioned, the planar irregular sector $\tilde{c}_1^{\times}$ saves the day. Indeed, when computing the left-hand side contribution associated with this new sector, one has (using (C.11))

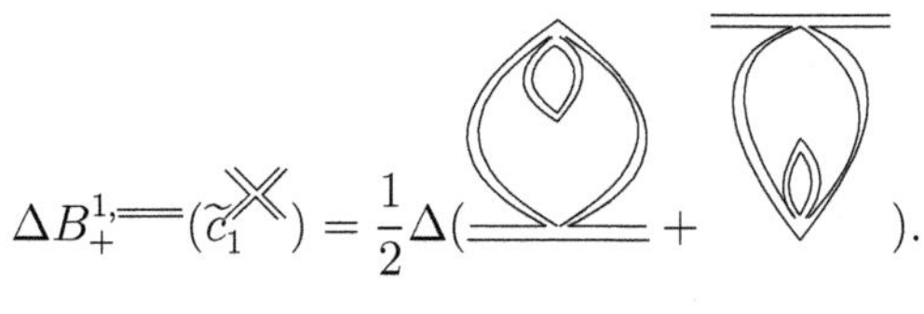

(C.23)

Equation (C.23) then leads to

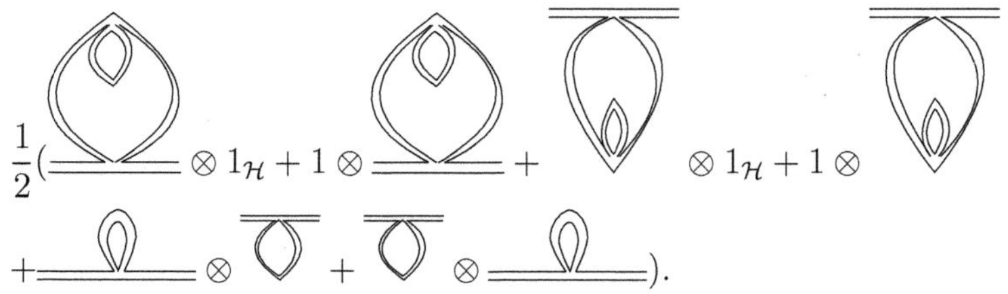

(C.24)

Let us now explicitly calculate the contribution of the new planar irregular sector $\tilde{c}_1^{\times}$ to the RHS of Theorem 9.8.4. Again using (C.11) and discarding the planar irregular graphs from the final list, one gets

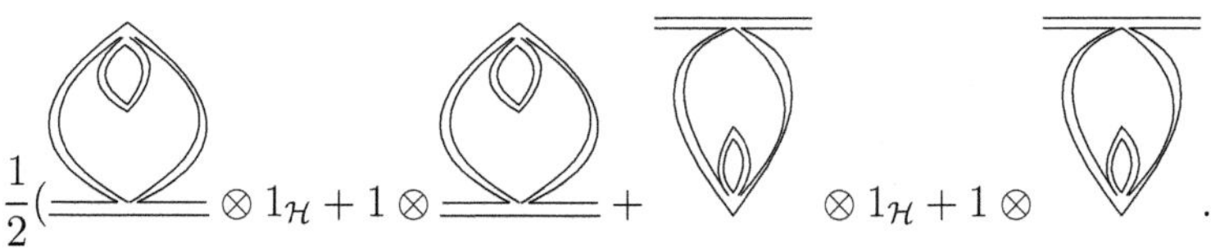

(C.25)

This cancels out the left-hand side contribution of (C.24). One can thus see that the planar irregular sector has finally led to a total contribution

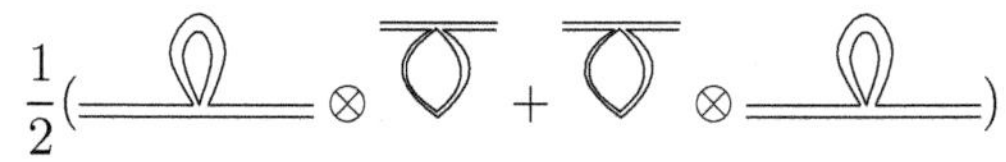

(C.26)

on the left-hand side of Theorem 9.8.4. This cancels out the rest (C.22) of the planar regular sector. Let us remark that taking into consideration the planar irregular tadpole (C.13) does not change the situation, since the insertion of this graph leads directly to non-planar graphs which are to be discarded. We have thus completely checked out Theorem 9.8.4 at the two-loop level, as announced.

Appendix D
Explicit examples of GFT tensor Feynman integral computations

In this appendix, we first investigate the behavior of a non-colourable but MO tensor graph, then that of a colourable (and MO) tensor graph and finally, that of a non-colourable and non-MO tensor graph.

D.1 A non-colourable, MO tensor graph integral

Let us now calculate the Feynman integral of the GFT graph of Fig. D.1. We denote by h_1 and by h_2 the two group elements associated to the internal edges 1 and 2, respectively.

One has

$$\int dh_1 dh_2 \delta(g_1 h_1 h_2^{-1} g_5)\delta(g_4 h_1 g_{4'}^{-1})\delta(g_2 h_2 g_{2'}^{-1})\delta(g_{1'}^{-1} h_1^{-1} h_2 g_{5'}^{-1}). \tag{D.1}$$

Performing the integral on h_2 using the third δ function in (D.1) and performing the integral on h_1 using the second δ function in (D.1) leads to the following result:

$$\delta(g_1 g_{4'} g_4^{-1} g_{2'}^{-1} g_2 g_5)\delta(g_{1'}^{-1} g_4 g_{4'}^{-1} g_2^{-1} g_{2'} g_{5'}^{-1}). \tag{D.2}$$

As expected the Feynman integral (D.1) is not divergent (this could have been directly stated from the fact there is no internal bubble of the GFT graph).

Nevertheless, an interesting phenomenon of some kind of 'ultraviolet/infrared' mixing on the group manifold takes place. Thus, for

$$g_1 = g_5^{-1} g_2^{-1} g_{2'} g_4 g_{4'}^{-1} g_1^{-1} \text{ or } g_{1'} = g_4 g_{4'}^{-1} g_2^{-1} g_{2'} g_{5'}^{-1} \tag{D.3}$$

the Feynman integral (D.2) becomes divergent. This comes from the fact that one has a non-trivial dependence of the integral on the external group momenta.

D.2 A colourable, multi-orientable tensor graph integral

However, this phenomenon is not specific to MO non-colourable tensor graphs. In the case of the graph in Fig. D.2, which is colourable (and MO) the Feynman integral writes:

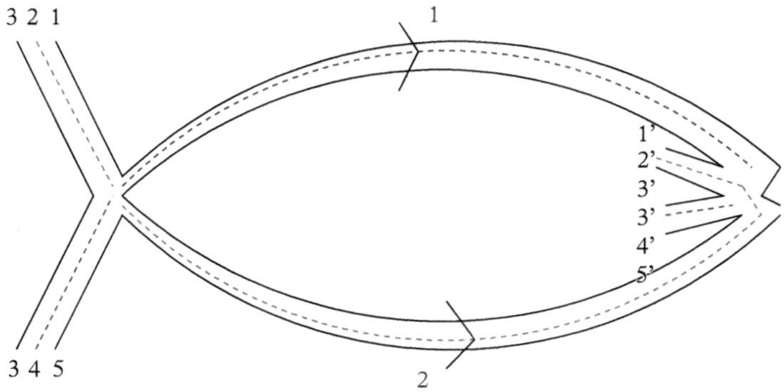

Figure D.1 *An example of a non-colourable, MO four-point GFT graph. The indices in blue label the two internal edges. The rest of the indices refer to the group elements of the external edges*

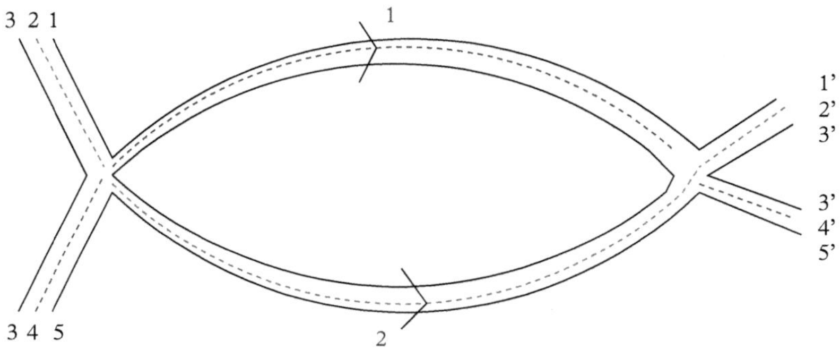

Figure D.2 *An example of a colourable, MO four-point GFT graph. The indices in blue label the two internal edges. The rest of the indices refer to the group elements of the external edges*

$$\int dh_1 dh_2 \delta(g_1 h_1 g_{1'}^{-1}) \delta(g_4 h_1 g_{4'}^{-1}) \delta(h_1^{-1} h_2) \delta(g_2 h_2 g_{2'}^{-1}) \delta(g_5 h_2 g_{5'}^{-1}). \tag{D.4}$$

As previously calculated, we integrate on h_2 using the third δ function in (D.4) and we then integrate on h_1 using the first of the δ functions in (D.4). The result can be written:

$$\delta(g_4 g_1^{-1} g_{1'} g_{4'}^{-1}) \delta(g_2 g_1^{-1} g_{1'} g_{2'}^{-1}) \delta(g_5 g_1^{-1} g_{1'} g_{5'}^{-1}). \tag{D.5}$$

The group 'ultraviolet/infrared' mixing already described is still present, with several independent directions in the group:

$$g_{4'} = g_4 g_1^{-1} g_{1'} \text{ or } g_{2'} = g_2 g_1^{-1} g_{1'} \text{ or } g_{5'} = g_5 g_1^{-1} g_{1'}, \tag{D.6}$$

turning the product (D.5) divergent.

D.3 A non-colourable, non-multi-orientable tensor graph integral

Let us analyze the non-colourable, non-MO GFT graph in Fig. D.3. The Feynman integral of this graph is written:

$$\int dh_1 dh_2 \delta(g_1 h_1 g_{1'}^{-1}) \delta(g_4 h_1 h_2^{-1} g_2) \delta(g_{1'} h_1^{-1} h_2 g_{4'}) \delta(g_2 h_2 h_1^{-1} g_4) \delta(g_5 h_2 g_{3'}^{-1}) \tag{D.7}$$

First, integrating on h_1 using the first of the δ functions in (D.7) and then on h_2 using the last of the δ function in (D.7) leads to:

$$\delta(g_{6'} g_{1'}^{-1} g_1 g_5^{-1} g_{3'} g_{4'}) \delta(g_4 g_1^{-1} g_{1'} g_{3'}^{-1} g_5 g_2) \delta(g_2 g_5^{-1} g_{3'} g_{1'}^{-1} g_1 g_4). \tag{D.8}$$

This product of δ-functions on external group elements can become infinite for any of the following independent group directions:

$$g_{6'} = g_{4'}^{-1} g_{3'}^{-1} g_5 g_1^{-1} g_{1'} \text{ or } g_4 = g_2^{-1} g_5^{-1} g_{3'} g_{1'}^{-1} g_1 \text{ or } g_2 = g_4^{-1} g_1^{-1} g_{1'} g_{3'}^{-1} g_5. \tag{D.9}$$

The same type of phenomenon takes place when computing the Feynman integrals of the tadpoles of Figs. 10.2 and 10.3, which are non-colourable but MO and non-colourable, non-multi-orientable, respectively.

Let us end this appendix with the following remark. As announced in Chapter 8, the presence of the 'middle' strand of these GFT edges (which makes the difference with respect to the combinatorial maps or ribbon maps of non-commutative QFT) is necessary for defining the bubbles. Moreover, this third strand is required for the computation of Feynman integrals (which are related to the new concept of bubbles), as we have seen in detail in this appendix.

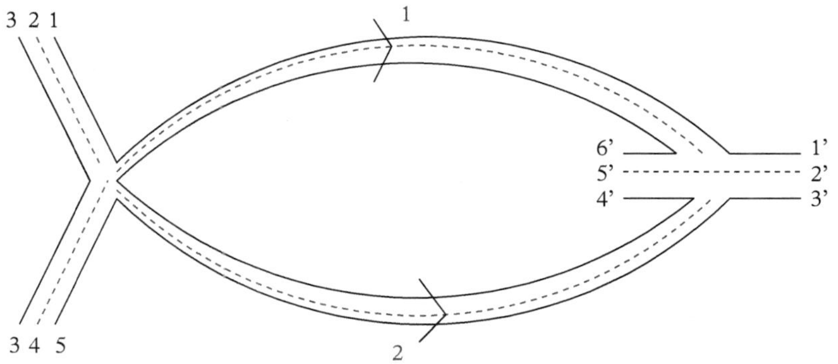

Figure D.3 *An example of a non-colourable, non-MO four-point GFT graph. The indices in blue label the two internal edges. The rest of the indices refer to the group elements of the external edges*

Appendix E
Coherent states of *SU*(2)

We denote by $R^{(j)m}{}_k(g)$ the matrix element of the group element g in the representation of spin (or highest weight) j, computed between the states $\langle j,m|$ and $|j,k\rangle$. We have

$$\mathbf{1}_j = \mathrm{d}_j \sum_{mm'} |j,m\rangle\langle j,m'| \int_{\mathrm{SU}(2)} dg\, R^{(j)m}{}_j(g)\overline{R}^{(j)}{}_j{}^{m'}(g) = \mathrm{d}_j \int_{\mathrm{SU}(2)} dg\, |j,g\rangle\langle j,g|, \qquad \text{(E.1)}$$

where we have introduced the notation:

$$|j,g\rangle \equiv g|j,j\rangle = \sum_m |j,m\rangle R^{(j)m}{}_j(g). \qquad \text{(E.2)}$$

The states $|j,g\rangle$ are the coherent states Perelomov (1986), and the last expression in (E.1) is a decomposition of the identity in terms of these coherent states.

Appendix F
Proof of the double-scaling limit of the $U(N)^D$-invariant tensor model

In this appendix we establish the equation (12.78) (we will actually derive (12.77) on our way).

At the end of the day, we do not know how to derive these equalities solely via the DSE. We shall utilize, in this section, a different analysis, (utilized first in Dartois, Gurau, and Rivasseau (2013) for the two-point function), to probe these sub-leading terms.

In this proof, we shall restrict ourselves to the quartic model described earlier and use a universality argument, presented in the next section, to extend the proof to the generic model. Our main reason for doing so is that the universality argument is succinct yet powerful. As one will see, the following analysis is quite involved and tailored for the quartic model only. To prove these results for generic models directly is an arduous task deserving its own paper.

The quartic model is defined by the action and the generating function:

$$S_{(\mathrm{quart})}(\mathbb{T},\overline{\mathbb{T}}) = \tilde{\rho}_{\mathcal{B}_2}(\mathbb{T},\overline{\mathbb{T}}) - \sum_{c=1}^{D} \frac{z}{2} \tilde{\rho}_{\mathcal{B}_{4,\{c\}}}(\mathbb{T},\overline{\mathbb{T}}) ,$$

$$Z_{(\mathrm{quart})}(\mathbb{J},\overline{\mathbb{J}}) = \int \left(\prod_{\vec{c}} N^{D-1} \frac{\mathrm{d}\mathbb{T}_{\vec{a}} \mathrm{d}\overline{\mathbb{T}}_{\vec{a}}}{2\pi i} \right) e^{-N^{D-1} S_{(\mathrm{quart})}(\mathbb{T},\overline{\mathbb{T}}) + \tilde{\rho}_{\mathcal{B}_2}(\mathbb{T},\overline{\mathbb{J}}) + \tilde{\rho}_{\mathcal{B}_2}(\mathbb{J},\overline{\mathbb{T}})} .$$

The $2p$-point cumulants are sums over connected $(D+1)$-coloured graphs with $2p$ edges of colour 0 adjacent to $2p$ vertices of degree 1 called the external vertices. Moreover, the connected components of the sub-graphs with colours $1,\dots,D$ are bubbles $\mathcal{B}_{4,\{c\}}$.

We recall that the amplitude of such a Feynman graph $\mathcal{G}$ is easily evaluated: each subgraph $\mathcal{B}_{4,\{c\}}$ brings a N^{D-1} scaling factor and a trace-invariant operator. Each edge of colour 0 brings a $\frac{1}{N^{D-1}} \prod_{c=1}^{D} \delta_{a_c b_c}$ factor. It follows that the indices are identified along the faces of colour $0c$ of $\mathcal{G}$. The indices corresponding to the internal faces are summed and bring a factor N each. The indices corresponding to the external faces of $\mathcal{G}$ reconstitute the trace-invariant operator associated to $\partial\mathcal{G}$, $\delta^{\partial\mathcal{G}}_{\vec{a}^v,\vec{b}^v}$. Denoting $E^0(\mathcal{G})$ the number of edges of colour 0 of $\mathcal{G}$ (including the external edges), $B(\mathcal{G})$ the number of subgraphs of colours $\{1,\dots,D\}$ of $\mathcal{G}$, and $F^{0c}_{\mathrm{int}}(\mathcal{G})$ the number of internal faces of colour $0c$ of $\mathcal{G}$, the $2p$-point cumulant is written:

$$W^{(2p)}_{\vec{a}^1\dots\vec{a}^p,\vec{b}^1\dots\vec{b}^p}\left(N,z,\{t_i\}\right) = \sum_{n\geq 0} \frac{1}{n!} z^n \sum_{\substack{\mathcal{G},p(\partial\mathcal{G})=p \\ B(\mathcal{G})=n}} \frac{N^{B(\mathcal{G})(D-1)} N^{\sum_{c=1}^{D} F^{0c}_{\mathrm{int}}(\mathcal{G})}}{N^{(D-1)E^0(\mathcal{G})}} \delta^{\partial\mathcal{G}}_{\vec{a}^v,\vec{b}^v} , \tag{F.1}$$

where the sum runs over graphs $\mathcal{G}$ with labelled sub-graphs $\mathcal{B}_{4,\{c\}}$. The contribution of an invariant $\mathcal{B}$ to the $2p$-point cumulant is obtained by restricting to graphs whose boundary is $\mathcal{B}$, that is $\partial\mathcal{G} = \mathcal{B}$:

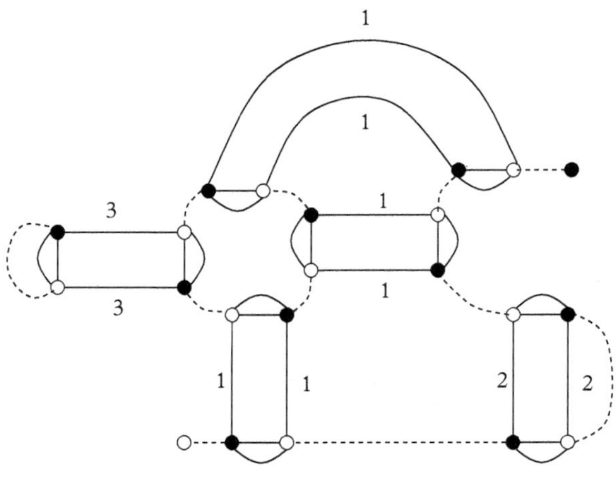

Figure F.1 *A Feynman graph with two external legs*

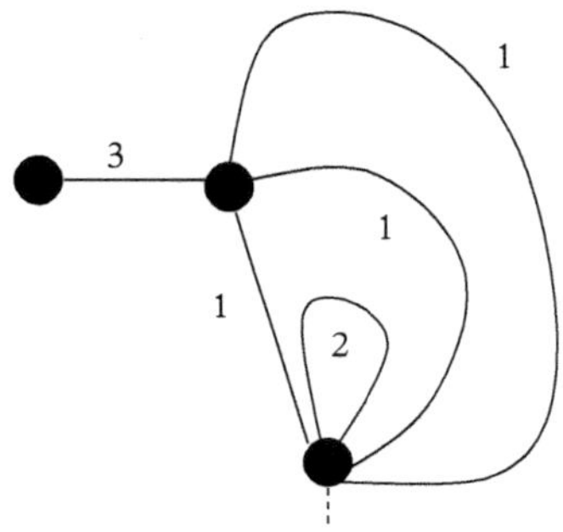

Figure F.2 *The associated edge-coloured map with a (dashed) cilium*

$$W\left(\mathcal{B};N,z\right)=\sum_{n\geq 0}\frac{1}{n!}z^n\sum_{\substack{\mathcal{G},\partial\mathcal{G}=\mathcal{B}\\ B(\mathcal{G})=n;}}\frac{N^{B(\mathcal{G})(D-1)}N^{\sum_{c=1}^{D}F^{0c}_{\text{int}}(\mathcal{G})}}{N^{(D-1)E^0(\mathcal{G})}},\tag{F.2}$$

The graphs and amplitudes of this model can be recast in terms of an *intermediate field representation* (see Chapter 3). Although somewhat lengthy to introduce, this representation clarifies greatly the $1/N$ expansion. The intermediate field representation can be obtained by introducing intermediate fields, integrating out $\mathbb{T},\overline{\mathbb{T}}$, and deriving the new Feynman rules of the theory. Here we do not need this full machinery, but we will take advantage of the fact that the graphs of this intermediate field representation are in a one-to-one correspondence with the Feynman graphs of the tensor model.

The intermediate field representation

We will call *effective graphs* the graphs of the intermediate field representation. The mapping to the Feynman graphs of our tensor model is illustrated in Fig. F.2. The effective graphs are obtained from the regular edge-coloured graphs by contracting all the edges of colour 0 and all the edges

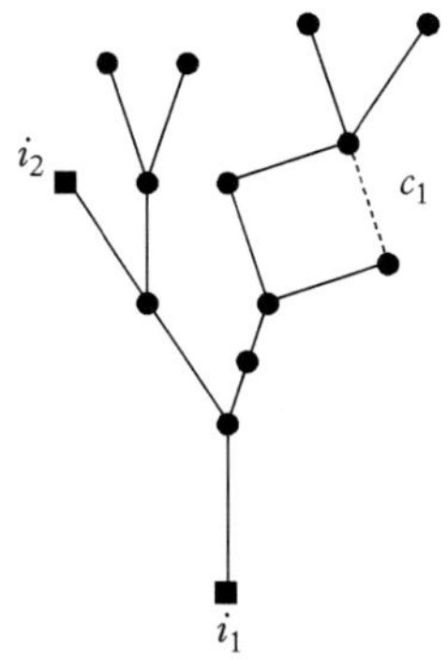

Figure F.3 *An effective graph with two marked vertices* $\{i_1, i_2\}$. *The e-edges have a colour, as stressed out by the dashed edge with colour* c_1

of colour $c' \neq c$ in each $\mathcal{B}_{4,\{c\}}$ to constitute *effective vertices* (e-vertices for short), while associating an *effective edge* of colour c (e-edges of colour c for short) to the couple of edges of colour c in each $\mathcal{B}_{4,\{c\}}$.

The external edges of colour 0 will then be partitioned into pairs associated to some of the e-vertices. We decorate those e-vertices by a mark, or a *cilium*, to signal such a couple. An e-vertex can have at most one cilium. This mapping is obviously bijective. A typical example of a contribution to the four-point cumulant is presented in Fig. F.3.

We use boxes to represent the marked vertices when there is no ambiguity as for the cilium positions around the vertices. Note indeed that, in this intermediate field representation, the order of the e-edges adjacent to an e-vertex is specified. It means that the effective graphs are in fact combinatorial maps (i.e. graphs with ascribed order of the edges at a vertex) with edges coloured $\{1, \dots, D\}$. We call a *corner* the piece of an e-vertex comprised between two consecutive e-edges. Note that a cilium is incident to two corners (or to a unique corner, if the graph has one ciliated vertex and no edges). We will denote the maps thus obtained by $\mathcal{M}$.

Every $\mathcal{M}$ has D canonical sub-maps $\mathcal{M}_c$ obtained by deleting all the edges of colour $c' \neq c$ in $\mathcal{M}$. All the vertices of $\mathcal{M}$ belong to $\mathcal{M}_c$. The sub-maps $\mathcal{M}_c$ have a well-defined notion of *faces*. They fall in two categories: the *internal faces* of $\mathcal{M}_c$ are the circuits obtained by going along the e-edges (of colour c) and along the corners of the e-vertices of $\mathcal{M}_c$, while the *external faces* of $\mathcal{M}_c$ are the open paths obtained by going along the corners and the e-edges (of colour c) of $\mathcal{M}_c$ from one cilium to another. By convention, all the faces are oriented clockwise. We define the *faces of colour* c of $\mathcal{M}$ as the faces of $\mathcal{M}_c$. Note that some of the faces can be reduced to a single corner on an isolated vertex.

All the elements present in the formula (F.2) are faithfully represented within effective graphs:

- each subgraph $\mathcal{B}_{4,\{c\}}$ of $\mathcal{G}$ corresponds to an e-edge of colour c of $\mathcal{M}$;
- each edge of colour 0 of $\mathcal{G}$ corresponds to a corner of $\mathcal{M}$; and
- each (internal or external) face of colour $0c$ of $\mathcal{G}$ corresponds to an (internal or external) face of colour c of $\mathcal{M}$.

The boundary graph $\partial\mathcal{G}$ can be reconstructed from $\mathcal{M}$. To do so, one draws a black and a white vertex for every cilium of $\mathcal{M}$ and for each external face of colour c of $\mathcal{M}$ going from a source cilium to a target cilium, one connects the white vertex corresponding to the source cilium with the black vertex of the target cilium by an edge of colour c. We denote $\partial\mathcal{M}$ the boundary graph of the effective map $\mathcal{M}$.

Let us denote $E^c(\mathcal{M})$, $V(\mathcal{M})$ and $F^c_{\rm int}(\mathcal{M})$ the numbers of e-edges of colour c, e-vertices and *internal* faces of colour c of $\mathcal{M}$. Furthermore, let us define $E(\mathcal{M}) = \sum_{c=1}^{D} E^c(\mathcal{M})$ and $\sum_{c=1}^{D} F^c_{\rm int}(\mathcal{M}) = F_{\rm int}(\mathcal{M})$. The numbers of corners of $\mathcal{M}$ is $p(\mathcal{B}) + 2E(\mathcal{M})$. The number of cilia of $\mathcal{M}$ is half the number of vertices of $\mathcal{B} = \partial\mathcal{M}$. The equation (F.2) becomes:

$$
\begin{aligned}
W\Big(\mathcal{B}; N, z\Big) &= \sum_{v\geq 0} \frac{1}{v!} \sum_{\substack{\mathcal{M},\partial\mathcal{M}=\mathcal{B}\\ V(\mathcal{M})=v}} z^{E(\mathcal{M})} \frac{N^{E(\mathcal{M})(D-1)} N^{F_{\rm int}(\mathcal{M})}}{N^{(D-1)[p(\mathcal{B})+2E(\mathcal{M})]}} \\
&= N^{-(D-1)p(\mathcal{B})} \sum_{v\geq 0} \frac{1}{v!} \sum_{\substack{\mathcal{M},\partial\mathcal{M}=\mathcal{B}\\ V(\mathcal{M})=v}} z^{E(\mathcal{M})} N^{-E(\mathcal{M})(D-1)+F_{\rm int}(\mathcal{M})},
\end{aligned}
\tag{F.3}
$$

where the sum runs over edge-coloured maps with labelled vertices such that $\partial\mathcal{M} = \mathcal{B}$.

Lemma F.0.1 *We have the bound:*

$$
-E(\mathcal{M})(D-1) + F_{\rm int}(\mathcal{M}) \leq D - (D-1)p(\mathcal{B}) - \rho(\mathcal{B}) - (D-2)\Big[E(\mathcal{M}) - V(\mathcal{M}) + 1\Big].
\tag{F.4}
$$

Proof The external faces are open paths. They can be represented as cycles by adding *external strands* to our drawings. For each external face of colour c, we connect its source and target cilia by an external strand of colour c (which can for instance be represented as a dashed edge). By convention we orient the external strands form the target cilium to the source cilium. The external faces now become cycles, obtained by going between the cilia along the corners and e-edges of the graph and closing the path into a cycle by following the external strands.

Note that the external strands encode the boundary $\partial\mathcal{M}$ of $\mathcal{M}$. For each cilium we draw a black and a white vertex and for each strand of colour c we connect the white vertex of its target cilium with the black vertex of its source cilium. Henceforth, we use this as the definition of the boundary graph of a map.

We denote by $F^c_{\rm ext}(\mathcal{M})$ the number of external faces of colour c of the map $\mathcal{M}$, and $\sum_{c=1}^{D} F^c_{\rm ext}(\mathcal{M}) = F_{\rm ext}(\mathcal{M})$. Initially, $\mathcal{M}$ has exactly $Dp(\mathcal{B})$ external faces, hence a total of:

$$
Dp(\mathcal{B}) + F_{\rm int}(\mathcal{M}),
$$

faces either internal or external. We are interested in finding a bound on this total number of faces.

An e-edge belongs to either one or two faces (internal or external). By deleting an e-edge and merging the corners of the two e-vertices to which it is hooked, the total number of faces of the map can not increase by more than 1. Remember that, while in the initial map every external face contained exactly one external strand, by deleting an edge we can create external faces containing several external strands. However, as the deletion does not affect the connectivity of the external strands, the latter still encode the boundary of the initial map $\partial\mathcal{M}$.

We choose a spanning tree in $\mathcal{M}$ and iteratively delete the e-edges in its complement. We denote the map obtained at the end of this procedure (which is a tree decorated with external strands) by $\mathcal{M}^{(0)}$ and we have:

$$Dp(\mathcal{B}) + F_{\text{int}}(\mathcal{M}) \leq F_{\text{ext}}(\mathcal{M}^{(0)}) + F_{\text{int}}(\mathcal{M}^{(0)}) + \Big[E(\mathcal{M}) - V(\mathcal{M}) + 1\Big],$$

and $\partial\mathcal{M} = \partial\mathcal{M}^{(0)}$.

Starting from $\mathcal{M}^{(0)}$ we build the maps $\mathcal{M}^{(s)}$ obtained by eliminating one by one the e-vertices of $\mathcal{M}$ of coordination one. Choose a univalent e-vertex (hence hooked to a unique e-edge, say of colour c) in $\mathcal{M}^{(s)}$ having no cilium. The map $\mathcal{M}^{(s+1)}$ is obtained by deleting this e-vertex and the e-edge it is adjacent to. The boundary graph is unchanged by this procedure, $\rho(\mathcal{M}^{(s)}) = \rho(\mathcal{M}^{(s+1)})$, and $D-1$ internal faces are deleted (all the faces of colour $c' \neq c$ contained in the e-vertex), hence:

$$F_{\text{ext}}(\mathcal{M}^{(s)}) + F_{\text{int}}(\mathcal{M}^{(s)}) + \rho(\partial\mathcal{M}^{(s)}) = F_{\text{ext}}(\mathcal{M}^{(s+1)}) + F_{\text{int}}(\mathcal{M}^{(s+1)}) + \rho(\partial\mathcal{M}^{(s+1)}) + (D-1). \tag{F.5}$$

If the univalent e-vertex (hooked to a unique e-edge, say of colour c) on $\mathcal{M}^{(s)}$ is ciliated then there are D incoming and D outgoing external strands at this cilium. Let us denote the cilium by i. We build first the map $\tilde{\mathcal{M}}^{(s)}$ by deleting i and all the external strands which *start and end* at i, and reconnecting the remaining external strands incident at i respecting the colours. The map $\mathcal{M}^{(s+1)}$ is then obtained from $\tilde{\mathcal{M}}^{(s)}$ by deleting the resulting univalent e-vertex and the e-edge to which it is hooked.

Going from $\mathcal{M}^{(s)}$ to $\tilde{\mathcal{M}}^{(s)}$ changes the boundary graph: $\partial\mathcal{M}^{(s)} \neq \partial\tilde{\mathcal{M}}^{(s)}$, while going from $\tilde{\mathcal{M}}^{(s)}$ to $\mathcal{M}^{(s+1)}$ preserves it. There are several cases:

- The black and white vertices associated to the cilium i in $\partial\mathcal{M}^{(s)}$ belong to two different connected components of $\partial\mathcal{M}^{(s)}$. Then the number of connected components of the boundary graph decreases by 1, $\rho(\mathcal{M}^{(s)}) = \rho(\tilde{\mathcal{M}}^{(s)}) + 1$. At the same time, a new face is created for every colour $c' \neq c$ (this new face is contained in the e-vertex of interest). For the colour c (of the e-edge hooked to the e-vertex), at most one face can be deleted, thus:

$$F_{\text{ext}}(\mathcal{M}^{(s)}) + F_{\text{int}}(\mathcal{M}^{(s)}) + \rho(\partial\mathcal{M}^{(s)}) \leq F_{\text{ext}}(\tilde{\mathcal{M}}^{(s)}) + F_{\text{int}}(\tilde{\mathcal{M}}^{(s)}) + \rho(\tilde{\partial}\mathcal{M}^{(s)}) + 1 - (D-1) + 1. \tag{F.6}$$

- The black and white vertex associated to the cilium i in $\partial\mathcal{M}^{(s)}$ belong to the same connected component of $\partial\mathcal{M}^{(s)}$, but not all of the external strands starting at i end at i. Then the number of connected components of the boundary graph can only increase, $\rho(\mathcal{M}^{(s)}) \leq \rho(\tilde{\mathcal{M}}^{(s)})$. The faces of colour $c' \neq c$ containing the external strands starting and ending at i become internal. The other faces of colour $c' \neq c$ remain external. If the external strand of colour c starting at i ends also at i, the face containing it survives. If not, the number of faces of colour i can at most decrease by 1. Thus:

$$F_{\text{ext}}(\mathcal{M}^{(s)}) + F_{\text{int}}(\mathcal{M}^{(s)}) + \rho(\partial\mathcal{M}^{(s)}) \leq F_{\text{ext}}(\tilde{\mathcal{M}}^{(s)}) + F_{\text{int}}(\tilde{\mathcal{M}}^{(s)}) + \rho(\tilde{\partial}\mathcal{M}^{(s)}) + 1. \tag{F.7}$$

- all the external strands starting at i end at i. Then the number of connected components of the boundary graph decreases by 1, $\rho(\mathcal{M}^{(s)}) = \rho(\tilde{\mathcal{M}}^{(s)}) + 1$, but none of the faces is affected, hence:

$$F_{\text{ext}}(\mathcal{M}^{(s)}) + F_{\text{int}}(\mathcal{M}^{(s)}) + \rho(\partial\mathcal{M}^{(s)}) \leq F_{\text{ext}}(\tilde{\mathcal{M}}^{(s)}) + F_{\text{int}}(\tilde{\mathcal{M}}^{(s)}) + \rho(\tilde{\partial}\mathcal{M}^{(s)}) + 1\,. \tag{F.8}$$

When going from $\tilde{\mathcal{M}}^{(s)}$ to $\mathcal{M}^{(s+1)}$, $D-1$ faces are deleted. Taking into account that $D \geq 3$,

$$F_{\text{ext}}(\mathcal{M}^{(s)}) + F_{\text{int}}(\mathcal{M}^{(s)}) + \rho(\partial\mathcal{M}^{(s)}) \leq F_{\text{ext}}(\mathcal{M}^{(s+1)}) + F_{\text{int}}(\mathcal{M}^{(s+1)}) + \rho(\partial\mathcal{M}^{(s+1)}) + D. \tag{F.9}$$

Eliminating all the e-vertices we obtain the map $\mathcal{M}^{(s_f)}$ with $s_f = V(\mathcal{M}) - 1$, having only one e-vertex and exactly D faces. The final e-vertex can be ciliated or not, hence we obtain the bound

$$F_{\text{ext}}(\mathcal{M}^{(0)}) + F_{\text{int}}(\mathcal{M}^{(0)}) \leq D + \rho(\partial\mathcal{M}^{(s_f)}) - \rho(\partial\mathcal{M}^{(0)}) + (D-1)(V(\mathcal{M})-1) + \begin{cases} p(\mathcal{B}) - 1 & \text{if ciliated} \\ p(\mathcal{B}) & \text{if not} \end{cases},$$

hence, taking into according that $\partial\mathcal{M}^{(0)} = \partial\mathcal{M} = \mathcal{B}$ in both cases

$$\begin{aligned} &F_{\text{ext}}(\mathcal{M}^{(0)}) + F_{\text{int}}(\mathcal{M}^{(0)}) \leq D + (D-1)(V(\mathcal{M})-1) + p(\mathcal{B}) - \rho(\mathcal{B}) \Rightarrow \\ &F_{\text{int}}(\mathcal{M}) \leq D + (D-1)(V(\mathcal{M})-1) - (D-1)p(\mathcal{B}) - \rho(\partial\mathcal{B}) + \Big[E(\mathcal{M}) - V(\mathcal{M}) + 1\Big]. \end{aligned}$$

This lemma proves in particular in the sense of perturbation theory the scaling behaviour in the equation (12.29):

$$K\Big(\mathcal{B}; N, z\Big) = \sum_{v \geq 0} \sum_{\substack{\mathcal{M}, \partial\mathcal{M} = \mathcal{B} \\ V(\mathcal{M}) = v}} z^{E(\mathcal{M})} \frac{N^{-(D-1)p(\mathcal{B}) - E(\mathcal{M})(D-1) + F_{\text{int}}(\mathcal{M})}}{N^{D-2(D-1)p(\mathcal{B}) - \rho(\mathcal{B})}},$$

$$\left| \frac{N^{-(D-1)p(\mathcal{B}) - E(\mathcal{M})(D-1) + F_{\text{int}}(\mathcal{M})}}{N^{D-2(D-1)p(\mathcal{B}) - \rho(\mathcal{B})}} \right| \leq N^{-(D-2)\left[E(\mathcal{M}) - V(\mathcal{M}) + 1\right]}, \tag{F.10}$$

where the sum runs over maps $\mathcal{M}$ with v *unlabeled* vertices (cancelling the $1/v!$ factor), and $D \geq 3$ and $E(\mathcal{M}) - V(\mathcal{M}) + 1 \geq 0$ for a connected map.

LO of the four-point cumulants in the $1/N$ *expansion*

The relevance of the intermediate field representation is now transparent. Indeed, equation (F.10) teaches us that (in the sense of perturbation theory):

- the functions $K(\mathcal{B}; N, z)$ are finite for all N and admit a large N limit.
- the LO of $K(\mathcal{B}; N, z)$ is given by trees $\mathcal{M}$ such that $\partial\mathcal{M} = \mathcal{B}$ (in particular $\mathcal{M}$ must have $p(\mathcal{B})$ cilia).
- the NLO order is given by trees decorated by a loop edge and is suppressed by $N^{-(D-2)}$.
- the first q orders in the $1/N$ series are given (at most) by trees decorated with up to q loop edges.

From now on we concentrate on the four-point contributions $K\Big(\mathcal{B}_{4,\mathcal{C}}; N, z\Big)$. They are represented by maps with two cilia, $\{i_1, i_2\}$. At LO only trees with two cilia contribute. If all the edges in the tree connecting the two ciliated vertices have the same colour c, the boundary graph of $\mathcal{M}$ is $\mathcal{B}_{4,\{c\}}$. If not, the boundary graph of $\mathcal{M}$ is $\mathcal{B}_{4,\emptyset}$. Thus, the last statement of equation (12.77):

$$K\Big(\mathcal{B}_{4,\mathcal{C}}; N, z\Big)\Big|_{|\mathcal{C}| \geq 2} = \mathcal{O}\left(\frac{1}{N^{D-2}}\right), \tag{F.11}$$

is proven.

The other two statements are obtained as follows. Recall that $T(z) = \frac{1-\sqrt{1-4Dz}}{2Dz}$ is the physical solution of the equation $1 - T(z) + DzT(z)^2 = 0$, and counts rooted plane trees with a weight Dz per edge (i.e. tress with a weight z per edge and an arbitrary colour $c \in \{1, \ldots, D\}$).

The graph $\mathcal{B}_{4,\{c\}}$ is obtained from trees such that the path between i_1 and i_2 is formed by edges of the same colour. The simplest example is the tree with only two vertices separated by an edge of colour c. Any other tree, contributing at LO is obtained by inserting a (possibly empty) tree with coloured edges at any one of the four corners of the vertices i_1 or i_2 and inserting d additional intermediary vertices on the path between i_1 and i_2, each equipped with two corners on which arbitrary trees are inserted, thus:

$$K\left(\mathcal{B}_{4,\{c\}}; z\right) = zT(z)^4 \sum_{d=0}^{\infty} [zT(z)^2]^d = \frac{zT(z)^4}{1 - zT(z)^2} \,. \tag{F.12}$$

For $K(\mathcal{B}_{4,\emptyset}; z)$ the simplest tree has two edges of different colours hooked to i_1 and i_2 joined at an intermediary bi valent vertex. Denoting d the number of additional vertices inserted on the path between i_1 and i_2 and taking into account that only paths in which not all edges have the same colour contribute we have:

$$K(\mathcal{B}_{4,\emptyset}; z) = T(z)^4 \left[Dz \sum_{d=0}^{\infty} [DzT(z)^2]^d - Dz \sum_{d=0}^{\infty} [zT(z)^2]^d \right] = \frac{D(D-1)z^2 T(z)^6}{\left(1 - DzT(z)^2\right)\left(1 - zT(z)^2\right)}, \tag{F.13}$$

which reproduces equation (12.77) taking into account that $1 - DzT(z)^2 = 2 - T = T\sqrt{1 - 4Dz}$.

Reduced maps

The explicit resummation we performed for the LO in the previous subsection can be extended to all orders in the $1/N$ series and ultimately leads to the double scaling limit of tensor models.

It emerges that we can partition the maps $\mathcal{M}$, with $\partial\mathcal{M} = \mathcal{B}$ into classes, each possessing a canonical representative $\overline{\mathcal{M}}$, which we call the *pruned, reduced map* (or simply, reduced map). There are infinitely many maps $\mathcal{M}$ in the original sum, which are related through pruning and reduction to the same reduced map $\overline{\mathcal{M}}$. Moreover, the amplitude for the entire class can be resummed and thus assigned to this representative. This process is illustrated in Fig. F.4.

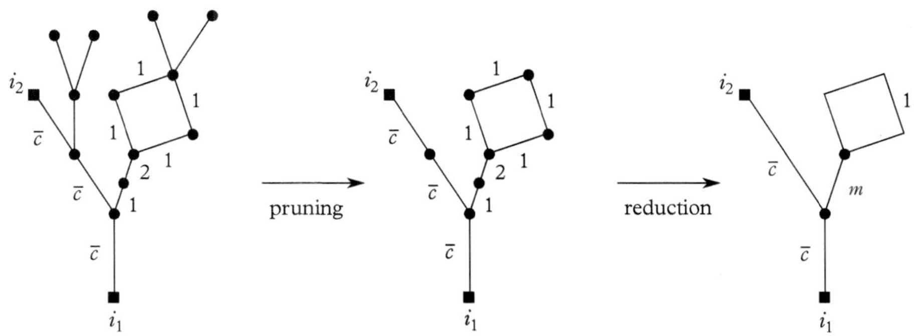

Figure F.4 *The process of pruning and reduction*

Let us start with a map $\mathcal{M}$.

- *Pruning* is the iterative removal of non-ciliated e-vertices of degree one. This procedure effectively removes tree-like sub-structures from the map $\mathcal{M}$. From the point of view of the original $(D+1)$-coloured graph, pruning is equivalent to the iterative removal of elementary melons.
- *Reduction* is the removal of all non-ciliated e-vertices of degree two, which from the point of view of the original $(D+1)$-coloured graphs, is equivalent to the contraction of certain chains of $(D-1)$–dipoles. This procedure effectively replaces chains of bivalent vertices by new edges, which we call *bars*. Those bars come in two types: *i) Type-c bars* represent chains of e-vertices of degree two where the connecting e-edges all have the same colour c; and *ii) Type-m bars* (m for multicoloured) represent chains of e-vertices of degree two, where the connecting e-edges have at least two different colours.

A type-m bar is a sequence of type-c bars connected by vertices of degree two and at least one change of colour. Any-vertex of the reduced map, except possibly the ciliated ones, therefore, has degree of at least three.

It is easy to show Dartois, Gurau, and Rivasseau (2013) that all the maps $\mathcal{M}$ associated to the reduced map $\overline{\mathcal{M}}$ possess the same scaling in N. The scaling exponent of a map in equation (F.10):

$$-E(\mathcal{M})(D-1)+F_{\text{int}}(\mathcal{M}),$$

is clearly invariant under the deletion of e-vertices of degree one with no cilia and of the e-edges adjacent to them (as exactly $(D-1)$ internal faces are formed only by this e-vertex). Also, exactly $(D-1)$ internal faces are formed by an e-vertex of degree two with no cilium and adjacent to two e-edges of the same colour.

Type-c bars bring the same scaling as regular e-edges of colour c, i.e. $N^{-(D-1)}$. However, packing up chains of such bars into type-m bars changes the scaling with an extra N^{-1}. Thus, a type-m bar comes with N^D. The faces and boundary of the reduced map $\overline{\mathcal{M}}$ are defined as before but taking into account that $\overline{\mathcal{M}}_c$ is obtained by deleting not only all the bars of colours different from c, but also *all the multicoloured bars*. We denote $E^m(\overline{\mathcal{M}})$ the number of multicoloured bars, and $E^u(\overline{\mathcal{M}})=\sum_{c=1}^{D}E^c(\overline{\mathcal{M}})$ the total number of type-c bars of $\overline{\mathcal{M}}$.

In addition to its scaling with N, a reduced map has a z-dependent amplitude. Following the process of pruning and reduction, it is found that this amplitude is evaluated via local weights assigned in the following way:

- corners are dressed with the LO full two-point function $T(z)$, reflecting the summation of arbitrary tree-like structures,
- type-c bars represent chains of bubbles $\mathcal{B}_{4,\{c\}}$ with the same colour, hence get the weight

$$z\sum_{k\geq 0}[zT(z)^2]^k=\frac{z}{1-zT(z)^2},$$

- type-m bars represent chains of bubbles $\mathcal{B}_{4,\{c\}}$ with at least one change of colours, hence the weight

$$Dz\sum_{k\geq 0}[DzT(z)^2]^k-Dz\sum_{k\geq 0}[zT(z)^2]^k=\frac{D(D-1)\,z^2T(z)^2}{(1-DzT(z)^2)\,(1-zT(z)^2)}.$$

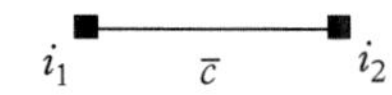

Figure F.5 *The leading reduced map for $K(\mathcal{B}_{4,\{c\}};N,z)$*

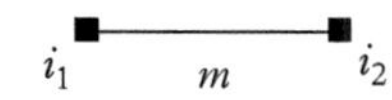

Figure F.6 *The leading reduced map for $K(\mathcal{B}_{4,\emptyset};N,z)$*

The perturbative expansion of $K(\mathcal{B};N,z)$ can be reorganized in terms of reduced maps $\overline{\mathcal{M}}$ with unlabeled vertices. The leading $1/N$ terms for $K(\mathcal{B}_{4,\{c\}};N,z)$ and $K(\mathcal{B}_{4,\emptyset};N,z)$ we computed in the previous sections are exactly the contributions of the reduced maps in Figs. F.5 and F.6.

The double scaling limit

Each type-m bar comes with a factor $1/(1-DzT(z)^2) = 1/(T(z)\sqrt{1-4Dz})$ which diverges as $z \to z_c = 1/4D$. Therefore, the amplitude of a reduced map has a singular part of the form $(1-4Dz)^{-\frac{E^m(\overline{\mathcal{M}})}{2}}$ close to criticality. We henceforth look for the most singular reduced maps at each fixed order in $1/N$, by maximizing the number of multicoloured bars at that order.

Consider a reduced map $\overline{\mathcal{M}}$ with boundary $\mathcal{B} = \partial\overline{\mathcal{M}}$. No face goes along a multicoloured edge, neither internal nor external. We delete all the multicoloured bars. The reduced map $\overline{\mathcal{M}}$ splits into several connected components. We denote $\overline{\mathcal{M}}(\nu)$, $\nu = 1,\dots,r$ the connected components which contain ciliated vertices, and $\overline{\mathcal{M}}(\mu)$, $\mu = 1,\dots,q$ those which do not contain any ciliated vertex. Remark that these connected components *are not* reduced maps, as they can contain vertices of degree two. However, they are edge-coloured maps.

As no face goes along the multicoloured bars, the boundary graph $\mathcal{B} = \partial\overline{\mathcal{M}}$ also splits into several connected components $\mathcal{B}(\nu) = \partial\overline{\mathcal{M}}(\nu)$ and $\mathcal{B}$ is the disjoint union of $\mathcal{B}(\nu)$. The type-c bars and internal faces are partitioned between the $\overline{\mathcal{M}}(\nu)$s and $\overline{\mathcal{M}}(\mu)$s, hence:

$$-DE^m(\overline{\mathcal{M}}) - E^u(\overline{\mathcal{M}})(D-1) + F_{\text{int}}(\overline{\mathcal{M}}) = -DE^m(\overline{\mathcal{M}}) + \sum_{\nu=1}^{r}\left[-E^u\left(\overline{\mathcal{M}}(\nu)\right)(D-1) + F_{\text{int}}\left(\overline{\mathcal{M}}(\nu)\right)\right]$$
$$+\sum_{\mu=1}^{q}\left[-E^u\left(\overline{\mathcal{M}}(\mu)\right)(D-1) + F_{\text{int}}\left(\overline{\mathcal{M}}(\mu)\right)\right]. \tag{F.14}$$

The components $\overline{\mathcal{M}}(\mu)$, $\mu = 1,\dots,q$, not containing any cilium are treated as follows. Either:

- $\overline{\mathcal{M}}(\mu)$ *is a tree*, hence $-E^u\left(\overline{\mathcal{M}}(\mu)\right)(D-1) + F_{\text{int}}\left(\overline{\mathcal{M}}(\mu)\right) = D$. There are two cases.
  - Either $\overline{\mathcal{M}}(\mu)$ has a unique e-vertex, incident to at least three multicoloured bars,
  - or $\overline{\mathcal{M}}(\mu)$ has more than one e-vertex. Then, $\overline{\mathcal{M}}(\mu)$ is incident to at least four multicoloured bars. The reduced map $\overline{\mathcal{M}}$ has the same scaling in N and the same singular behaviour as the map where $\overline{\mathcal{M}}(\mu)$ has been contracted to a unique e-vertex.

  Moreover, when $\overline{\mathcal{M}}(\mu)$ is an e-vertex of degree at least four, one can always build a reduced map with the same scaling in N, and strictly more type-m bars, by splitting the e-vertex into a binary tree whose edges are type-m bars, as in the Fig. F.7. Indeed, as no faces go all along a type-m bar, an e-vertex incident to only type-m bars exactly closes D faces. This

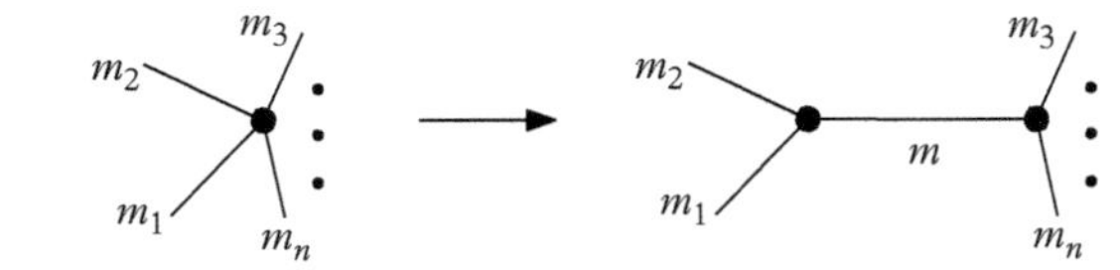

Figure F.7 *From the left to the right drawing, the scaling with N is preserved but a type-m bar is created. One can proceed until the initial e-vertex has become a binary tree*

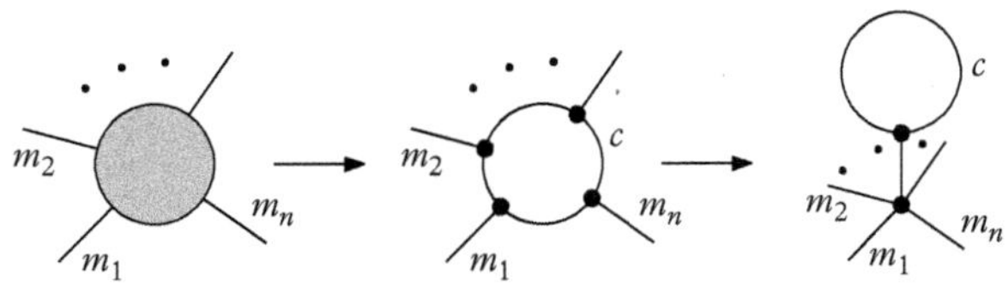

Figure F.8 *From the left to the middle drawing, one changes the gray blob representing $\overline{\mathcal{M}}(\mu)$ with a unicoloured loop, which cannot decrease the exponent of N. From the middle to the right, the scaling with N is preserved but a type-m bar is created*

way, adding both a vertex and a type-m bar does not affect the scaling, while bringing an additional power of $1/\sqrt{z_c - z}$. Therefore, the non-ciliated components $\overline{\mathcal{M}}(\mu)$ which are trees have to be e-vertices of degree three.

- Or $\overline{\mathcal{M}}(\mu)$ *is not a tree* and it is incident to at least one multicoloured bar. In this case, from Lemma F.0.1:

$$-E^u\left(\overline{\mathcal{M}}(\mu)\right)(D-1)+F_{\text{int}}\left(\overline{\mathcal{M}}(\mu)\right) \leq D-(D-2)\left[E^u\left(\overline{\mathcal{M}}(\mu)\right)-V\left(\overline{\mathcal{M}}(\mu)\right)+1\right] \leq D-(D-2),$$

and this bound is saturated by the unicoloured loop. It follows that $\overline{\mathcal{M}}$ scales at most like the same map where $\overline{\mathcal{M}}(\mu)$ has been replaced with a unicoloured loop.

Moreover, if $\overline{\mathcal{M}}(\mu)$ is a unicoloured loop incident to more than one type-m bar, then one can always build a reduced map with the same scaling in N but with strictly more type-m bars (hence more singular). Ones detaches the loop and attaches the bars to a common vertex which is then connected to the loop through a new type-m bar (see Fig. F.8). If $\overline{\mathcal{M}}(\mu)$ has n incident bars, the scaling is N^{2-nD} in both cases, but one gets a new singular factor in the second case.

Following Dartois, Gurau, and Rivasseau (2013), we call a type-c loop hooked to a single type-m bar a *cherry*, represented like 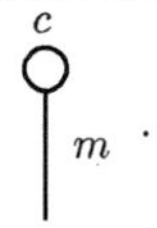

.

We now analyse the connected components $\overline{\mathcal{M}}(\nu), \nu = 1, \ldots, r$, which contain the cilia. If several type-m bars are incident to the same component $\overline{\mathcal{M}}(\nu)$, one can build a reduced map with the same scaling in N but more type-m bars by a process similar to the one applied on the components $\overline{\mathcal{M}}(\mu)$

which are not trees. One indeed detaches all the type-m bars incident to $\overline{\mathcal{M}}(\nu)$ but one, say m_1, and reconnects them on a new e-vertex created along m_1. This splits m_1 into two type-m bars, enhancing the singular behaviour at criticality. It has to be mentioned that upon detaching all but one of the type-m bars hooked to $\overline{\mathcal{M}}(\nu)$, the degree of some e-vertices in $\overline{\mathcal{M}}(\nu)$can drop down to two, hence the resulting drawing is not a reduced map. This however is not a problem, as the reduced maps in which those e-vertices of degree two are appropriately exchanged for bars do exist and scale as advertised, hence strictly dominate $\overline{\mathcal{M}}$.

Thus, for any boundary graph $\mathcal{B}$, the leading singular behaviour when $z \to z_c$ is captured by the reduced maps whose non-ciliated unicoloured components are either vertices of degree three or cherries, and each ciliated unicoloured component is incident to exactly one type-m bar. We denote $\mathbf{V}^3$ and $\mathbf{V}^{\text{cherry}}$ the number of e-vertices of degree 3 and the number of cherries. We need the following combinatorial relations:

- $\mathbf{V}^{\text{cherry}} + \mathbf{V}^3 = q$;
- as the type-m bars must connect all the unicoloured components in a connected way, we also get $E^m(\overline{\mathcal{M}}) = q + r - 1 + l = \mathbf{V}^{\text{cherry}} + \mathbf{V}^3 + r - 1 + l$, for some non-negative integer l; and
- as all the cherries are hooked to one type-m bar, all the vertices of degree three to three bars, and all the (ciliated) components $\overline{\mathcal{M}}(\nu)$ to one bar, we also have $2E^m(\overline{\mathcal{M}}) = \mathbf{V}^{\text{cherry}} + 3\mathbf{V}^3 + r$.

  There is however, one exception to this relation, namely when $r = 1$ and $\mathbf{V}^{\text{cherry}} = \mathbf{V}^3 = 0$, there exists a reduced map with $E^m(\overline{\mathcal{M}}) = 0$. It does not diverge at criticality (goes to a constant), and one needs to check its scaling with N separately.

From the previous three relations, we extract $\mathbf{V}^3$ and $E^m(\overline{\mathcal{M}})$ as a function of $\mathbf{V}^{\text{cherry}}, l, r$:

$$\begin{aligned} \mathbf{V}^3 &= \mathbf{V}^{\text{cherry}} + r - 2 + 2l \\ E^m(\overline{\mathcal{M}}) &= 2\mathbf{V}^{\text{cherry}} + 3l + 2r - 3. \end{aligned} \tag{F.15}$$

Moreover, we can rewrite the exponent of N due to non-ciliated unicoloured components in (F.14) as

$$\sum_{\mu=1}^{q} \Big[- E^u\Big(\overline{\mathcal{M}}(\mu)\Big)(D-1) + F_{\text{int}}\Big(\overline{\mathcal{M}}(\mu)\Big)\Big] = D\mathbf{V}^3 + 2\mathbf{V}^{\text{cherry}}, \tag{F.16}$$

Let us now address the double scaling limit of cumulants in the quartic model. The most singular contributions are selected by maximizing $2\mathbf{V}^{\text{cherry}} + 3l$ while keeping $(D-2)\mathbf{V}^{\text{cherry}} + Dl$ fixed (which is a linear programme similar to the one used for the double-scaling limit in Gurau and Schaeffer (2016)). For $D < 6$ the dominant singular behaviour is obtained by setting l to zero and introducing the new coupling $x = N^{D-2}(\frac{1}{4D} - z)$ to be held fixed, leading to the generic double scaling behaviour

$$N^{(D-2)\frac{2r-3}{2} - (D-2)\sum_{\nu=1}^{r}\left[E^u\left(\overline{\mathcal{M}}(\nu)\right) - V\left(\overline{\mathcal{M}}(\nu)\right) + 1\right]} f(\mathcal{B}; N, x)m \tag{F.17}$$

where $f(\mathcal{B}; N, x)$ as a function of N is bounded by a constant.

We now apply this formula to the two– and four–point cumulants.

The two-point cumulant. The two-point function is represented by reduced maps with a unique cilium, hence $r=1$. Furthermore, the unique connected component containing the ciliated vertex can be chosen to have no loops (i.e. it is formed only by the ciliated vertex). Separating the contribution of the reduced map with only the ciliated vertex (which, as we have already mentioned, must be evaluated separately), we get a double scaling ansatz:

$$K(\mathcal{B}_2;N,x) = f^{(0)}(\mathcal{B}_2;N,x) + N^{-\frac{D-2}{2}} f^{(-\frac{D-2}{2})}(\mathcal{B}_2;N,x)\,, \tag{F.18}$$

in agreement with our ansatz (12.74).

The four-point cumulant. For $\mathcal{B}_{4,\emptyset}$, the leading double scaling contributions come from the reduced maps in the Fig. F.9 having $r=2$, while for $\mathcal{B}_{4,\{c\}}$ the leading double scaling contributions come from the reduced maps in the Fig. F.10 having $r=1$ (and one must remember to treat the contribution of the map with only one edge of colour c separately). The maps contributing to $\mathcal{B}_{4,\mathcal{C}}$ with $|\mathcal{C}|\geq 2$ have $r=1$, but $\partial\overline{\mathcal{M}}(\nu)=\mathcal{B}_{4,\mathcal{C}}$ imposes that $\mathcal{M}(\nu)$ possesses loop edges. Thus:

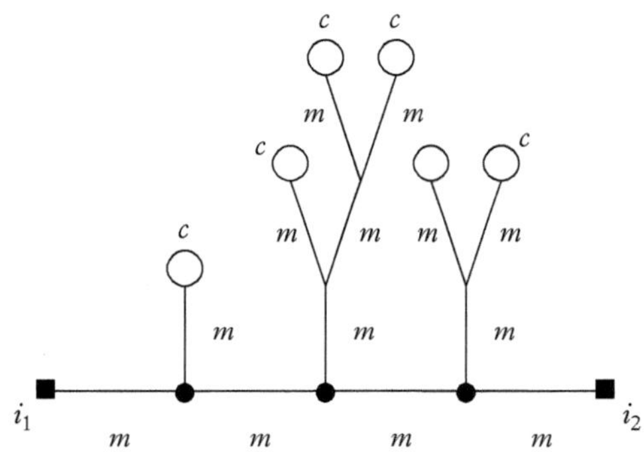

Figure F.9 *Leading contribution to* $K(\mathcal{B}_{4,\emptyset};N,x)$

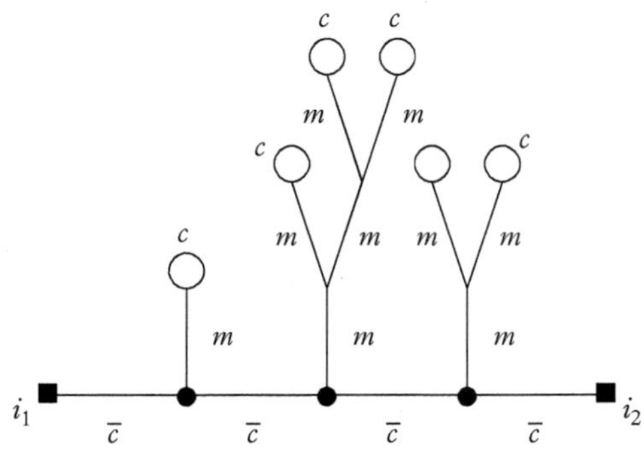

Figure F.10 *Leading contribution to* $K(\mathcal{B}_{4,\{c\}};N,x)$

$$
\begin{aligned}
K\left(\mathcal{B}_{4,\emptyset};N,x\right) &= N^{\frac{D-2}{2}} f^{(\frac{D-2}{2})}(\mathcal{B}_{4,\emptyset};N,x), \\
K\left(\mathcal{B}_{4,\{c\}};N,x\right) &= f^{(0)}(\mathcal{B}_{4,\{c\}};N,x) + N^{-\frac{D-2}{2}} f^{(-\frac{D-2}{2})}(\mathcal{B}_{4,\{c\}};N,x), \\
K\left(\mathcal{B}_{4,\mathcal{C}};N,x\right) &= \mathcal{O}\left(\frac{1}{N^{D-2}}\right),
\end{aligned}
\tag{F.19}
$$

which is precisely (12.78).

Appendix G
Proof of Theorem 15.3.2

A first important tool for proving Theorem 15.3.2 is a bijection (introduced in Lionni (2017)) between coloured graphs and so-called *constellations* (certain partially embedded graphs) such that the order of a coloured graph corresponds to the excess of the associated constellation. We recall this bijection in Section G.1 and also explain how it can be adapted to the non-orientable setting, giving the simple relation $\tilde{g}_{n,\delta} = 2^{\delta}\ g_{n,\delta}$ (which is also reflected previously by the fact that $g_{n,\delta}$ and $\tilde{g}_{n,\delta}/2^{\delta}$ have the same asymptotic estimate). Then, in Subsection G.2.1, we use the classical method of kernel extraction to obtain an explicit expression for the generating function $\mathcal{G}_{\delta}(z)$ of bipartite coloured graphs of fixed order δ. Singularity analysis of the obtained expression then gives us the asymptotic estimate of $g_{n,\delta}$. In Section G.3, we then give sufficient conditions (in terms of the kernel decomposition) for a coloured graph of order δ to be an SYK graph and deduce from it that asymptotically almost all coloured graph of order δ are SYK graphs. This implies that $c_{n,\delta}$ has the same asymptotic estimate as $g_{n,\delta}$; similar arguments in the non-orientable case ensure that $\tilde{c}_{n,\delta}$ has the same asymptotic estimate as $\tilde{g}_{n,\delta}$.

G.1 Bijection with constellations

We recall here a bijection introduced in Lionni (2017) from coloured graphs of order δ to so-called *constellations*[1] of excess δ. Thanks to this bijection, computing the generating function of coloured graphs of fixed-order amounts to computing the generating function of constellations of fixed excess (which can classically be done using kernel extraction, as we will show in Section G.2.1). We also explain in Section G.1.2 how the bijection can be adapted to the non-bipartite case.

G.1.1 Bijection in the bipartite case

Given a bipartite coloured graph $G \in \mathbb{G}^q$, we first orient all the edges from black to white. We then contract all the colour-0 edges, so that the pairs of black and white vertices they link collapse into $2q$-valent vertices which have one outgoing and one ingoing edge of each colour $i \in [\![1, q]\!]$.

(G.1)

[1] In Lionni (2017), constellations were called *stacked maps*, as a central quantity in that work was the sum of faces over a certain set of submaps. Here we use the term *constellations*, as it is a common name in the combinatorics literature for (the duality) of these objects.

The vertex resulting from the contraction of the distinguished colour-0 edge is itself distinguished. The obtained Eulerian graph, $G_{/0}$, is such that the subgraph obtained by keeping only the colour-i edges is a collection of directed cycles. For each such colour-i cycle containing p vertices, we add a colour-i vertex, and p colour-i edges between that vertex and the p vertices of the cycle, and then we delete the original colour-i edges, as illustrated.

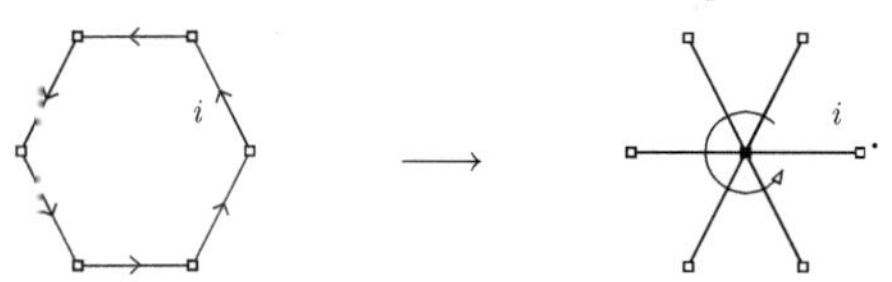

(G.2)

The cyclic ordering of the p edges around the cycle translates into a cyclic counterclockwise ordering of the p edges around the colour-i vertex, and each of these (deleted) edges corresponds to a corner of the colour-i vertex.

Doing this operation at every colour-i cycle, we obtain a connected diagram $S = \Psi(G)$ having

- non-embedded white vertices of valency q, with one incident edge of each colour $i \in [\![1,q]\!]$;
- embedded colour-i vertices;
- colour-i edges, which connect a white vertex to a colour-i vertex; and
- one distinguished white vertex (resulting from the contraction of the root-edge).

We denote by $\mathbb{S}^q$ the set of such diagrams, which we call (rooted) q-constellations.[2] An example is shown in Fig. G.1. We recall that the excess of a connected graph G is defined as $L(G) = E(G) - V(G) + 1$; it corresponds to its number of independent cycles. We let $\mathbb{S}^q_{n,\delta}$ be the set of constellations in $\mathbb{S}^q$ with n white vertices and excess δ, and let $\mathbb{G}^q_{n,\delta}$ be the set of rooted bipartite $(q+1)$-edge-coloured graphs with $2n$ vertices and order δ.

Theorem G.1.1 *Lionni (2017). The map Ψ described previously gives a bijection between $\mathbb{G}^q_{n,\delta}$ and $\mathbb{S}^q_{n,\delta}$, for every $q \geq 2, n \geq 1$ and $\delta \geq 0$.*

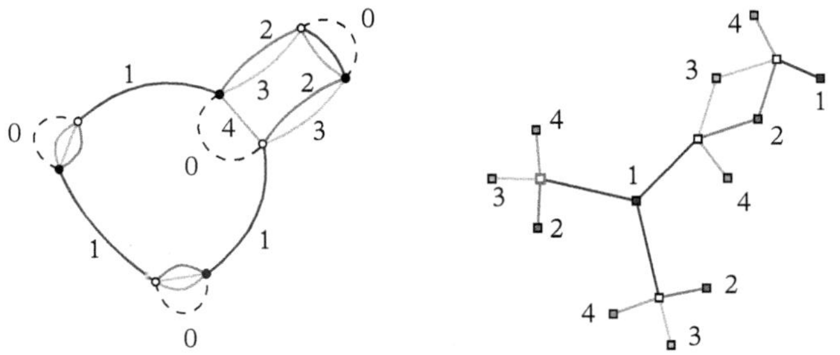

Figure G.1 *A bipartite SYK 5-coloured graph and the corresponding 4-constellation*

[2] We stress that in the usual definition of constellations, the white vertices are embedded, and the cyclic ordering of the edges is given by the ordering of their colours. However, we will not need here to view constellations as equipped with this canonical embedding.

Proof The map Ψ is clearly invertible, and hence gives a bijection from $\mathbb{G}^q$ to $\mathbb{S}^q$. Regarding the parameter correspondence, for $S = \Psi(G)$, the colour-0 edges of G correspond to the white vertices of S, and these edges form a perfect match in G, hence if S has n white vertices then G has $2n$ vertices. Finally, note that $\delta_0(G) = 1 + (q-1)n - F_0(G)$, and $L(S) = E - n - m + 1$, with E the number of edges and m the total number of coloured vertices in S. Since each colour-$0i$ cycle of G is mapped to a colour-i vertex of S, we have $F_0(G) = m$. Since each edge of S has exactly one white extremity and since white vertices have degree q, we have $E = qn$. Hence, $L(S) = \delta_0(G)$.

Trees and melonic graphs. The coloured graphs of vanishing order are melonic graphs. Recall that these graphs are obtained by recursively inserting pairs of vertices linked by q edges, as shown in Fig. G.2, starting from the only coloured graph with two vertices (left of Fig. G.3). Melonic graphs can equivalently be defined as the coloured graphs in $\tilde{\mathbb{G}}^q$ which satisfy the following identity:

$$q + (q-1)\big(\frac{1}{2}V(G) - R_0(G)\big) - F_0(G) = 0, \tag{G.3}$$

or equivalently,

$$\delta_0(G) = (q-1)\big(R_0(G) - 1\big). \tag{G.4}$$

The coloured graphs of order $\delta_0 = 0$, i.e. those which Ψ maps to trees, are melonic graphs which in addition have a single 0-residue (they are SYK graphs) Bonzom, Lionni, and Tanasa (2017). Indeed, in the recursive construction of Fig. G.2 for an SYK melonic graph, the edge on which a pair of vertices is inserted must not be of colour 0, and it is easily seen that a white vertex in a q-constellation incident to $q-1$ leaves whose colours are not $i \in [\![1,q]\!]$ corresponds to an insertion as in Fig. G.2 on an edge of colour $i \neq 0$ in the coloured graph.

Proposition G.1.2 *The bijection Ψ maps the melonic SYK graphs in $\mathbb{G}^q$ to the trees in $\mathbb{S}^q$.*

An example of a rooted melonic SYK graph is shown on the right of Fig. G.3 for $q = 3$.

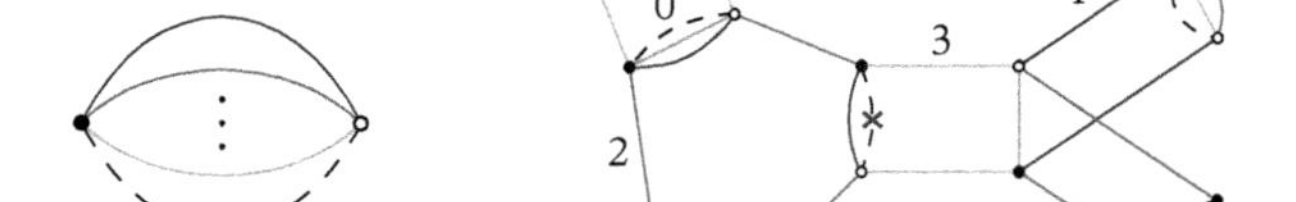

Figure G.2 *Insertion of a pair of vertices linked by q edges*

Figure G.3 *Melonic graphs*

Note that this is also true for $q=2$, although in that case a melonic SYK 3-coloured graph is dual to a unicellular planar bipartite map. Another remark is that the 0-residue $G_{\hat{0}}$ is also a melonic graph, and in general, deleting the edges of a given colour in a $(q+1)$-coloured melonic graph, one is left with a collection of melonic graphs with q colours.

G.1.2 The non-bipartite case

Consider a (not-necessarily bipartite) $(q+1)$-edge-coloured graph $G \in \tilde{\mathbb{G}}^q$. It has an even number of vertices, since colour-0 edges form a perfect matching. We assign an orientation to each non-root colour-0 edge. If G has $2n$ vertices, there are 2^{n-1} ways of doing so. A vertex is called an in-vertex (resp. out-vertex) if it is the origin (resp. end) of its incident colour-0 edge. We then orient canonically the remaining half-edges (those on colour-i edges for $i \in [\![1,q]\!]$); those at out-vertices are oriented outward and those at in-vertices are oriented inward.

Contracting the colour-0 edges into white vertices as in (G.1), we obtain a Eulerian graph such that for each $i \in [\![1,q]\!]$ every vertex has exactly one ingoing half-edge and one outgoing half-edge of colour i. We choose an arbitrary orientation for each cycle of colour $i \in [\![1,q]\!]$. For each white vertex and for each colour i, the orientation of its incident half-edges either coincides with the orientation of the colour-i cycle it belongs to, or they are opposite. We perform a star subdivision as in (G.2), with the difference that now, each newly added colour-i edge carries a $\pm$ sign, $+$ if the orientations of the colour-i half-edges agree with that of the colour-i cycle, $-$ otherwise, as illustrated.

(G.5)

We call *signed coloured graph* a rooted coloured graph together with a choice of orientation of each non-root colour-0 edge, and a choice of orientation of each colour-0i cycle for $i \in [\![1,q]\!]$. We call *signed constellation* a constellation (with a distinguished white vertex) together with a choice of $\pm$ sign for every edge. The previous transformation defines a bijection $\widehat{\Psi}$ between the set $\widehat{\mathbb{G}}^q_{n,\delta}$ of signed coloured graphs of order δ with $2n$ vertices and the set $\widehat{\mathbb{S}}^q_{n,\delta}$ of signed constellations with n white vertices and excess δ. Let $\widetilde{\mathbb{G}}^q_{n,\delta}$ be the set of rooted coloured graphs of order δ with $2n$ vertices. Since a coloured graph $G \in \widetilde{\mathbb{G}}^q_{n,\delta}$ has $n-1$ non-root colour-0 edges and satisfies $F_0(G) = 1+(q-1)n-\delta$, we have $\widehat{\mathbb{G}}^q_{n,\delta} \simeq 2^{qn-\delta}\widetilde{\mathbb{G}}^q_{n,\delta}$. Furthermore, since a constellation in $\mathbb{S}^q_{n,\delta}$ has qn edges, we have $\widehat{\mathbb{S}}^q_{n,\delta} \simeq 2^{qn} \times \mathbb{S}^q_{n,\delta}$. Hence we obtain

$$2^{qn-\delta} \times \widetilde{\mathbb{G}}^q_{n,\delta} \simeq_{\widehat{\Psi}} 2^{qn} \times \mathbb{S}^q_{n,\delta} \simeq_{\Psi} 2^{qn} \times \mathbb{G}^q_{n,\delta}. \tag{G.6}$$

As a consequence, the generating function $\widetilde{\mathcal{G}}_\delta(z)$ of non-necessarily bipartite rooted $(q+1)$-edge-coloured graphs of order δ, with z dual to the half number of vertices, satisfies:

$$\widetilde{\mathcal{G}}_\delta(z) = 2^\delta \mathcal{G}_\delta(z). \tag{G.7}$$

We can thus focus on the bipartite case when dealing with the enumeration of coloured graphs of fixed order.

G.2 Enumeration of coloured graphs of fixed order

In this section we compute the generating function $\mathcal{G}_\delta(z)$ of bipartite $(q+1)$-coloured graphs of fixed order δ. By Theorem G.1.1 this is the generating function of constellations of excess δ, with z dual to the number of white vertices. We classically rely on the method of kernel extraction to obtain an explicit expression of $\mathcal{G}_\delta(z)$. Then singularity analysis on this expression will allow us (in Subsection G.2.2) to obtain the asymptotic estimate of $g_{n,\delta} = [z^n]\mathcal{G}_\delta(z)$, stated in Theorem 15.3.2.

G.2.1 Exact enumeration

For a constellation S, the *core* C of S is obtained by iteratively deleting the non-root leaves (and incident edges) until all non-root vertices have degree at least 2. This procedure is shown in Fig. G.4 for the example of Fig. G.1. The core diagrams satisfy the following properties:

- white vertices are non-embedded while i-coloured vertices (for $i \in [\![1,q]\!]$) are embedded;
- white vertices have valency at most q, with incident edges of different colours;
- all non-root vertices (white or coloured) have valency at least 2; and
- each edge carries a colour $i \in [\![1,q]\!]$, and connects a white vertex to a colour-i vertex.

We now focus on the maximal sequences of non-root valency-two vertices:

Definition G.2.1 *A chain vertex of a core diagram is a non-root vertex of valency two. A core chain is a path whose internal vertices are chain vertices, but whose extremities are not chain vertices. The type of a core chain is given by the colours of its two extremities (coloured or white), and by the colour of their incident half-edges in the chain.*

Replacing all the core chains by edges whose two half-edges retain the colours of the extremal edges on each side of the chain, we obtain the *kernel* K of the constellation S. Note that K is a diagram that has a distinguished white vertex (the root vertex) and satisfies the following conditions:

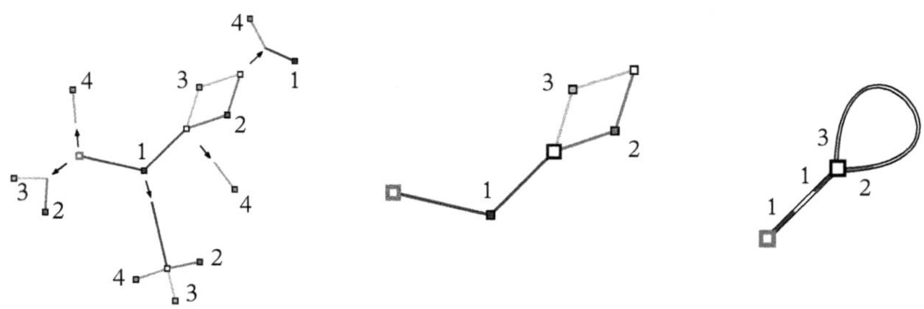

Figure G.4 *Cutting out tree contributions in the example of Fig. G.1 (left) leads to its core diagram (centre). The corresponding kernel is shown on the right of the figure*

- white vertices are non-embedded while i-coloured vertices (for $i \in [\![1,q]\!]$) are embedded;
- white vertices have valency at most q, with incident half-edges of different colours;
- all non-root vertices (white or coloured) have valency at least 3; and
- each half-edge carries a colour $i \in [\![1,q]\!]$, and is incident either to a white vertex or to a vertex of colour i.

We call *kernel diagrams* the (connected) graphs satisfying these properties. The excess of such a diagram K is as usual defined as $E - V + 1$, with E its number of edges and V its number of vertices. An important property is that a constellation and its kernel have equal excess. We let $\mathbb{K}^q$ be the family of kernel diagrams and $\mathbb{K}^q_\delta$ those of excess δ. Since every non-root vertex in a kernel diagram has valency at least 3, it is an easy exercise to show that K has at most $3\delta - 1$ edges (this calculation will however be detailed in Section G.2.2), so that $\mathbb{K}^q_\delta$ is a finite set.

An edge of K is called unicoloured (resp. bicoloured) if its two half-edges have the same colour (resp. have different colours). For $K \in \mathbb{K}^q$ we let $\mathcal{V}_\circ(K)$ and $\mathcal{V}_\bullet(K)$ be the sets of white vertices and of coloured vertices in K, and denote $V_\circ = \mathrm{Card}(\mathcal{V}_\circ)$ and $V_\bullet = \mathrm{Card}(\mathcal{V}_\bullet)$; we let $E(K), E^{=}(K), E^{\neq}(K)$ be the numbers of edges, of unicoloured edges, and of bicoloured edges in K; we also use refined notations $E_{\bullet\bullet}(K)$, $E^{=}_{\bullet\bullet}(K)$, $E^{\neq}_{\bullet\bullet}(K)$, $E_{\bullet\circ}(K)$, $E^{=}_{\bullet\circ}(K)$, $E^{\neq}_{\bullet\circ}(K)$, $E_{\circ\circ}(K)$, $E^{=}_{\circ\circ}(K)$, $E^{\neq}_{\circ\circ}(K)$ to denote the numbers of any/unicoloured/bicoloured edges whose extremities are coloured/coloured (resp. coloured/white, resp. white/white).

For $K \in \mathbb{K}^q_\delta$ we let $\mathbb{S}^q_{\delta,K}$ be the set of q constellations (all of excess δ) that have K as kernel. A constellation in $\mathbb{S}^q_{\delta,K}$ is generically obtained from K where each edge e is replaced by a core chain of the right type, i.e. a sequence of valency-two vertices of arbitrary length, alternatively coloured and white, which respects the boundary conditions, in the sense that extremal edges match the colours of the two half-edges that compose e, and an extremal vertex of the chain is white iff the incident extremity is white. coloured leaves are then added to white vertices so that they have one incident edge of each colour $i \in [\![1,q]\!]$. An arbitrary tree rooted at a colour-i corner is then inserted at every colour-i corner (Fig. G.5).

To obtain the generating function $\mathcal{G}_{\delta,K}(z)$ of the family $\mathbb{S}^q_{\delta,K}$, with z dual to the number of white vertices, one must therefore take the product of the generating functions of the core-chains whose types correspond to the colouring of vertices and of half-edges in K, together with a certain number of tree generating functions. The generating function $\mathcal{G}_T(z)$ of q-coloured stacked trees rooted on a colour-i corner (for any fixed $i \in [\![1,D]\!]$) and counted according to their number of white vertices is given by

$$\mathcal{G}_T(z) = 1 + z\mathcal{G}_T(z)^q. \tag{G.8}$$

Its coefficients are the Fuss–Catalan numbers: $[z^n]\mathcal{G}_T(z) = \frac{1}{qk+1}\binom{qk+1}{k}$.

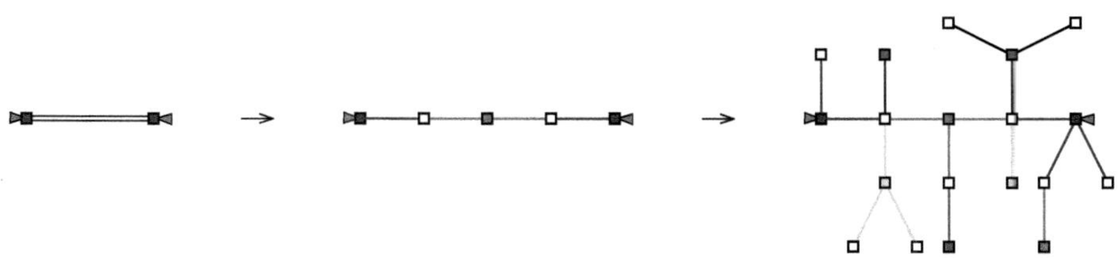

Figure G.5 *The constellations in $\mathbb{S}^q_{\delta,K}$ are obtained from K by replacing the edges by sequences of valency-two vertices and then attaching trees in the corners*

Proposition G.2.2 *For $q \geq 3$, the generating function of rooted bipartite $(q+1)$-edge-coloured graphs of fixed order δ is expressed as $\mathcal{G}_\delta(z) = \sum_{K \in \mathbb{K}^q_\delta} \mathcal{G}_{\delta,K}(z)$, where for each $K \in \mathbb{K}^q_\delta$,*

$$\mathcal{G}_{\delta,K}(z) = \left[\mathcal{B}_K(y)\left[\frac{y}{(1+y)\left(1-(q-1)y\right)}\right]^{E(K)} y^{V_\circ(K)}\right]_{y=z\mathcal{G}_T(z)^q}, \tag{G.9}$$

with the notation

$$\mathcal{B}_K(y) = [(q-1)y]^{E^{=}_{\bullet\bullet}(K)}[1/y - q + 2]^{E^{=}_{\bullet\circ}(K)+E^{=}_{\circ\circ}(K)}. \tag{G.10}$$

Proof Following the approach of Lionni (2017), we first compute the generating functions of core-chains of various types, in two variables $z_\circ, z_\bullet$ where $z_\circ$ (resp. $z_\bullet$) is dual to the number of non-extremal white (resp. coloured) vertices in the chain. We let $y = z_\circ z_\bullet$. For $i,j \in [\![1,q]\!]$ we let $\mathcal{G}^{ij}_{\bullet\bullet}(z_\circ, z_\bullet)$ be the generating function of core-chains whose extremal vertices are coloured and extremal edges have colours i,j respectively. By symmetry of the role played by the colours, for every $i \neq j$ the generating functions $\mathcal{G}^{ij}_{\bullet\bullet}(z_\circ, z_\bullet)$ are all equal to a common generating function denoted $\mathcal{G}^{\neq}_{\bullet\bullet}(z_\circ, z_\bullet)$, and for every $i \in [\![1,q]\!]$ the generating functions $\mathcal{G}^{ii}_{\bullet\bullet}(z_\circ, z_\bullet)$ are all equal to a common generating function denoted $\mathcal{G}^{=}_{\bullet\bullet}(z_\circ, z_\bullet)$. A decomposition by removal of the first white vertex along the chain yields the system

$$\mathcal{G}^{\neq}_{\bullet\bullet}(z_\circ, z_\bullet) = z_\circ + (q-2)y\mathcal{G}^{\neq}_{\bullet\bullet}(z_\bullet, z_\circ) + y\mathcal{G}^{=}_{\bullet\bullet}(z_\bullet, z_\circ), \quad \mathcal{G}^{=}_{\bullet\bullet}(z_\bullet, z_\circ) = (q-1)y\mathcal{G}^{\neq}_{\bullet\bullet}(z_\circ, z_\bullet),$$

whose solution is

$$\mathcal{G}^{\neq}_{\bullet\bullet}(z_\circ, z_\bullet) = \frac{z_\circ}{(1+y)(1-(q-1)y)}, \quad \mathcal{G}^{=}_{\bullet\bullet}(z_\bullet, z_\circ) = (q-1)y\mathcal{G}^{\neq}_{\bullet\bullet}(z_\bullet, z_\circ). \tag{G.11}$$

Similarly, we use the notations $\mathcal{G}^{ij}_{\bullet\circ}(z_\circ, z_\bullet), \mathcal{G}^{=}_{\bullet\circ}(z_\circ, z_\bullet), \mathcal{G}^{\neq}_{\bullet\circ}(z_\circ, z_\bullet)$ for the generating functions of core-chains whose extremal vertices are coloured/white. By deletion of the extremal white vertex we find

$$\mathcal{G}^{\neq}_{\bullet\circ}(z_\circ, z_\bullet) = z_\bullet\mathcal{G}^{\neq}_{\bullet\bullet}(z_\circ, z_\bullet), \quad \mathcal{G}^{=}_{\bullet\circ}(z_\circ, z_\bullet) = 1 + z_\bullet\mathcal{G}^{=}_{\bullet\bullet}(z_\circ, z_\bullet). \tag{G.12}$$

Finally, we use the notations $\mathcal{G}^{ij}_{\circ\circ}(z_\bullet, z_\circ), \mathcal{G}^{=}_{\circ\circ}(z_\bullet, z_\circ), \mathcal{G}^{\neq}_{\circ\circ}(z_\bullet, z_\circ)$ for the generating functions of core chains whose extremal vertices are white/white. By deletion of the extremal white vertices we find

$$\mathcal{G}^{\neq}_{\circ\circ}(z_\circ, z_\bullet) = z_\bullet^2\mathcal{G}^{\neq}_{\bullet\bullet}(z_\circ, z_\bullet), \quad \mathcal{G}^{=}_{\circ\circ}(z_\circ, z_\bullet) = z_\bullet(1 + z_\bullet\mathcal{G}^{=}_{\bullet\bullet}(z_\circ, z_\bullet)). \tag{G.13}$$

Given a kernel diagram $K \in \mathbb{K}^q$, let

$$\mathcal{A}_K(z_\circ, z_\bullet) := (\mathcal{G}^{=}_{\bullet\bullet})^{E^{=}_{\bullet\bullet}(K)}(\mathcal{G}^{\neq}_{\bullet\bullet})^{E^{\neq}_{\bullet\bullet}(K)}(\mathcal{G}^{=}_{\bullet\circ})^{E^{=}_{\bullet\circ}(K)}(\mathcal{G}^{\neq}_{\bullet\circ})^{E^{\neq}_{\bullet\circ}(K)}(\mathcal{G}^{=}_{\circ\circ})^{E^{=}_{\circ\circ}(K)}(\mathcal{G}^{\neq}_{\circ\circ})^{E^{\neq}_{\circ\circ}(K)}.$$

Using the expressions (G.11)–(G.13), we find

$$\begin{aligned}\mathcal{A}_K(z_\circ, z_\bullet) &= z_\bullet^{2E_{\circ\circ}(K)+E_{\circ\bullet}(K)}\mathcal{B}_K(y)\Big[\frac{z_\circ}{(1+y)\big(1-(q-1)y\big)}\Big]^{E(K)} \\ &= z_\bullet^{\sum_{v_\circ\in\mathcal{V}_\circ(K)}\mathrm{val}(v_\circ)}\mathcal{B}_K(y)\Big[\frac{z_\circ}{(1+y)\big(1-(q-1)y\big)}\Big]^{E(K)},\end{aligned}$$

where

$$\mathcal{B}_K(y) = [(q-1)y]^{E^{=}_{\bullet\bullet}(K)}[1/y - q + 2]^{E^{=}_{\bullet\circ}(K)+E^{=}_{\circ\circ}(K)}.$$

We then obtain

$$\mathcal{G}_{\delta,K}(z) = \mathcal{A}_K\big(z\mathcal{G}_T(z)^{q-2}, \mathcal{G}_T(z)^2\big) \prod_{v_\circ\in\mathcal{V}_\circ(K)} z\mathcal{G}_T(z)^{q-\mathrm{val}(v_\circ)} \prod_{v_\bullet\in\mathcal{V}_\bullet(K)} \mathcal{G}_T(z)^{\mathrm{val}(v_\bullet)},$$

where $z_\circ$ (resp. $z_\bullet$) has been replaced by $z\mathcal{G}_T(z)^{q-2}$ (resp. $\mathcal{G}_T(z)^2$), to account for tree attachments. This rearranges into

$$\mathcal{G}_{\delta,K}(z) = \Big[\mathcal{B}_K(y)\Big[\frac{y}{(1+y)\big(1-(q-1)y\big)}\Big]^{E(K)} y^{V_\circ(K)}\Big]_{y=z\mathcal{G}_T(z)^q}.$$

G.2.2 Singularity analysis

We can now obtain the singular expansion of $\mathcal{G}_{\delta,K}(z)$ for every given $K \in \mathbb{K}_\delta^q$, which yields the singular expansion of $\mathcal{G}_\delta(z)$ and the asymptotic estimate of $g_{n,\delta}$ stated in Theorem 15.3.2.

We start with the singularity expansion of the tree generating function $\mathcal{G}_T(z)$. From the equation $\mathcal{G}_T(z) = 1 + z\mathcal{G}_t(z)^q$ it is easy to find (see Bonzom et al. (2011)) that the dominant singularity of $\mathcal{G}_T(z)$ is

$$z_c = \frac{(q-1)^{q-1}}{q^q}, \qquad \text{with } \mathcal{G}_T(z_c) = \frac{q}{q-1}, \tag{G.14}$$

and we have the singular expansion

$$\mathcal{G}_T(z) = \frac{q}{q-1} - \sqrt{\frac{2q}{(q-1)^3}\big(1-\frac{z}{z_c}\big)} + o\big(\sqrt{z_c - z}\big). \tag{G.15}$$

Using (G.14), we have $z_c\mathcal{G}_T(z_c)^q = \mathcal{G}_T(z_c) - 1 = 1/(q-1)$, so that for any $K \in \mathbb{K}_\delta$ we have

$$\mathcal{G}_{\delta,K}(z) \sim \Big[\frac{1}{q\big(1-(q-1)z\mathcal{G}_T(z)^q\big)}\Big]^{E(K)} \frac{1}{(q-1)^{V_\circ(K)}}. \tag{G.16}$$

From (G.14) and (G.15), we have

$$(q-1)z\mathcal{G}_T(z)^q = 1 - \sqrt{\frac{2q}{q-1}\big(1-\frac{z}{z_c}\big)} + o\big(\sqrt{z_c - z}\big), \tag{G.17}$$

and therefore, using the expression of Prop. G.2.2 we find

$$\mathcal{G}_{\delta,K}(z) \sim \left[\frac{q-1}{2q^3(1-\frac{z}{z_c})}\right]^{\frac{E(K)}{2}} \frac{1}{(q-1)^{V_\circ(K)}}. \tag{G.18}$$

The singularity exponent is thus maximal for kernel diagrams that have maximal number of edges (at fixed excess δ). As a kernel diagram $K \in \mathbb{K}_\delta^q$ has no vertices of valency one or two, apart maybe from the root,

$$2E(K) = \sum_{v \in K} \mathrm{val}(v) \geq 3(V(K)-1)+1 = 3V(K)-2,$$

with equality if and only if the root vertex has valency one, and all the other vertices have valency three. This implies that

$$\delta = E(K) - V(K) + 1 \geq \frac{1}{3}(E(K)+1).$$

The maximal number of edges of a kernel diagram with fixed excess δ is therefore $3\delta - 1$, and we denote by $\overline{\mathbb{K}}_\delta^q$ the subset of those diagrams in $\mathbb{K}_\delta^q$, i.e. kernel diagrams with a root-leaf and all the other vertices of valency 3. We hence obtain

$$\mathcal{G}_\delta(z) = \left(\frac{q-1}{2q^3(1-\frac{z}{z_c})}\right)^{\frac{3\delta-1}{2}} \sum_{K \in \overline{\mathbb{K}}_\delta^q} \frac{1}{(q-1)^{V_\circ(K)}} + o\left(\frac{1}{z_c - z}\right)^{\frac{3\delta-1}{2}}. \tag{G.19}$$

We orient cyclically the edges at each white vertex by the natural order of the colours they carry. For each non-root white vertex, there are $3\binom{q}{3}$ ways of choosing the colours of the incident half-edges, so that they are ordered correctly. Moreover, there are q ways of choosing the colour $i \in [\![1, q]\!]$ of the half-edge incident to the root vertex, as well as the colour of each coloured trivalent vertex (this fixes the colour of the incident half-edges). Let $\mathbb{M}_\delta$ be the set of maps with one leaf (called the root) and $2\delta - 1$ other vertices all of valency 3; note that these maps have $3\delta - 1$ edges hence have excess δ.

From the preceding discussion we obtain

$$\sum_{K \in \overline{\mathbb{K}}_\delta^q} \frac{1}{(q-1)^{V_\circ(K)}} = \frac{q}{q-1} \sum_{M \in \mathbb{M}_\delta} \left(q + \frac{1}{q-1} 3\binom{q}{3}\right)^{2\delta-1} = \frac{q}{q-1} m_\delta \left(\frac{q^2}{2}\right)^{2\delta-1}, \tag{G.20}$$

where the factor raised to power $2\delta - 1$ corresponds to the choice for each vertex of valency 3 whether it is coloured or white.

Finally, we obtain the following expression:

$$\begin{aligned}
\mathcal{G}_\delta(z) &= \frac{q}{(q-1)} \left(\frac{q-1}{2q^3(1-\frac{z}{z_c})}\right)^{\frac{3\delta-1}{2}} \left(\frac{q^2}{2}\right)^{2\delta-1} m_\delta + o\left(\frac{1}{z_c - z}\right)^{\frac{3\delta-1}{2}} \\
&= \frac{2}{q(q-1)} \left(\frac{q-1}{2q^3(1-\frac{z}{z_c})}\right)^{\frac{3\delta-1}{2}} \left(\frac{q^4}{4}\right)^{\delta} m_\delta + o\left(\frac{1}{z_c - z}\right)^{\frac{3\delta-1}{2}}.
\end{aligned}$$

Using transfer theorems of singularity analysis Flajolet and Sedgewick (2009), we conclude that, for $\delta \geq 1$,

$$[z^n]\mathcal{G}_\delta(z) = \frac{1}{\Gamma\left(\frac{3\delta-1}{2}\right)} \frac{2}{q(q-1)} \left(\frac{q-1}{2q^3}\right)^{\frac{3\delta-1}{2}} \left(\frac{q^4}{4}\right)^{\delta} m_\delta \cdot n^{3(\delta-1)/2} \cdot z_c^{-n}. \tag{G.21}$$

This gives the asymptotic estimate of $g_{n,\delta} = [z^n]\mathcal{G}_\delta(z)$ in Theorem 15.3.2.

G.3 The connectivity condition and SYK graphs

In Subsections G.3.1, G.3.2, and G.3.3 we give sufficient conditions for a bipartite $(q+1)$-edge-coloured graph to be an SYK graph; we then deduce (using again singularity analysis) that the non-SYK graphs have asymptotically negligible contributions, which ensures that in Theorem 15.3.2, the coefficients $c_{n,\delta}$ have the same asymptotic estimate as the coefficients $g_{n,\delta}$ (the latter estimate having been established in the last section). In Subsection G.3.4, we adapt these arguments to non-necessarily bipartite graphs.

G.3.1 Preliminary conditions

Proposition G.3.1 *A coloured graph $G \in \mathbb{G}^q$ is an SYK graph if and only if for every edge of colour 0, the two incident vertices are linked by a path containing no edge of colour 0.*

Proof By assumption, a coloured graph $G \in \mathbb{G}^q$ is connected. Consider a colour-0 edge e in G, and its two extremities. If these vertices are linked in G by a path containing no colour-0 edge, deleting e does not change the connectivity. As it is the case for every colour-0 edge, the proposition follows.

Definition G.3.2 *We say that a white vertex of a constellation $S \in \mathbb{S}^q$ is admissible, if the corresponding two vertices in the coloured graph $\Psi^{-1}(S)$ are linked in the graph by a path containing no edge of colour 0.*

Proposition G.3.1 can be reformulated in terms of constellations using this last definition:

Corollary G.3.3 *A constellation is the image of an SYK graph if and only if all its white vertices are admissible.*

In both of the following subsections, we will need the following lemmas.

Lemma G.3.4 *A white vertex with at least one tree attached to it is admissible.*

Proof Consider such a white vertex $v_\circ$ in a map $S \in \mathbb{S}^q$, and a tree attached to it via an edge of colour $i \in [\![1,q]\!]$. We prove the lemma recursively on the size of the tree contribution.

If the tree is just a colour-i leaf, it represents in the coloured graph $G = \Psi^{-1}(S)$ an edge of colour-i between the corresponding two vertices in G, so that $v_\circ$ is admissible. If the tree has at least

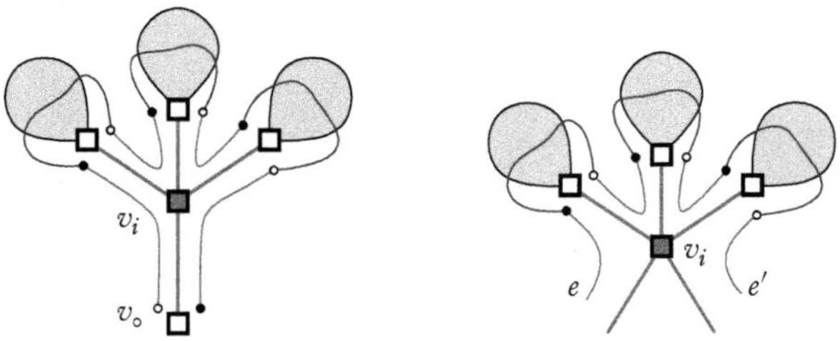

Figure G.6 *Concatenation of coloured paths in Lemmas G.3.4 and G.3.5*

one white vertex, then the colour-i neighbour v_i of $v_\circ$ has valency greater than one. All the other white vertices attached to v_i have a smaller tree attached, and from the recursion hypothesis, they are admissible. To each corner of v_i corresponds a colour-i edge in G, so that we can concatenate the coloured paths in G linking the pairs of vertices for all these white vertices, as illustrated on the left of Fig. G.6. This concludes the proof.

Consider two colour-i edges e and e' in a coloured graph G corresponding to two corners incident to the same colour-i vertex v_i in $S = \Psi(G)$. These corners split the edges incident to v_i into two sets V_a and V_b.

Lemma G.3.5 *With these notations, if all the edges in either V_a or V_b all lead to pending trees, then there exists a path in G containing both e and e', without any colour-0 edge.*

Proof Suppose that the condition of the lemma holds for V_a. All the white extremities of edges in V_a have a tree attached, so that from Lemma G.3.4, they are admissible. As previously stated in the proof of Lemma G.3.4, we can concatenate the coloured paths for all these white vertices, using the colour-i edges incident to the corners between the edges in V_a, as shown on the right of Fig. G.6.

Consider a q-constellation $S \in \mathbb{S}^q$, its core diagram C, and its kernel diagram $K \in \mathbb{K}^q$. Consider a white vertex $v \in S$. We will say that it also belongs to C if it is not internal to a tree contribution, and we will say that it also belong to K if, in addition, it is not a chain vertex of C.

Lemma G.3.6 *With these notations, if v is of valency smaller than q in K, then it is admissible in S.*

Proof If v is of valency $d < q$ in K, it means that $q - d > 0$ tree contributions have been removed in the procedure leading from a constellation S to its kernel diagram K. We conclude applying Lemma G.3.4.

Lemma G.3.7 *Let $G \in \mathbb{G}^q$ and let K be the kernel of the constellation associated to G. If K has no white vertex of valency q, then G is an SYK graph.*

Proof Let $S \in \mathbb{S}_{\delta,K}$. From Lemma G.3.6, the vertices of S which also belong to K are admissible, as they are of valency smaller than q. The other white vertices of S necessarily have a tree attached: either they are internal to a tree contribution, or they are chain-vertices of the corresponding core diagram. We conclude applying Lemma G.3.4 to every white vertex, and then Corollary G.3.3.

G.3.2 The case $q > 3$

For $\delta \geq 1$ and $q \geq 3$, a $(q+1)$-edge-coloured graph is called *dominant* if the kernel of its associated constellation belongs to $\overline{\mathbb{K}}_\delta$.

Corollary G.3.8 *For $q > 3$ and $\delta \geq 1$, a dominant $(q+1)$-edge-coloured graph is an SYK graph. Hence $g_{n,\delta} \sim c_{n,\delta}$ as $n \to \infty$.*

Proof The associated kernel K has a root vertex of degree 1 and all the other vertices of degree 3. Hence all the vertices of K have valency smaller than q. Hence, from Lemma G.3.7, G is an SYK graph.

We have seen in Section G.2.2 that the non-dominant $(q+1)$-edge-coloured graphs have asymptotically a negligible contribution. Hence, $g_{n,\delta} \sim c_{n,\delta}$.

G.3.3 The case $q = 3$

It now remains to show that for $q = 3$ we have $c_{n,\delta} \sim g_{n,\delta}$. Note that we cannot just apply Lemma G.3.7 as in the case $q > 3$, since all non-root vertices of a kernel diagram $K \in \mathbb{K}^3$ have valency at least $3 = q$.

Lemma G.3.9 *Let $S \in \mathbb{S}^q$ be a q-constellation, with C its core diagram and K its kernel diagram. Let $v_\circ$ be a white vertex which belongs to S, C, and K.*

If there is at least one core chain in C incident to $v_\circ$ and containing at least one internal white vertex, then $v_\circ \in S$ is admissible.

Proof Consider a core-chain in C incident to $v_\circ$ and denote $v'_\circ$ the closest white chain-vertex in the chain (see Fig. G.7). There necessarily is a colour-i chain vertex for some $i \in [\![1,q]\!]$ between $v_\circ$ and $v'_\circ$, which we denote v_i (v_i is in the chain, at distance one from both $v_\circ$ and $v'_\circ$).

The vertex $v'_\circ$ has $q-2>0$ trees attached in S, and using Lemma G.3.4, it is therefore admissible. We denote p' the corresponding path in $G = \Psi^{-1}(S)$.

The vertex v_i has two incident corners in C, both of which might have some tree contributions attached in S. These tree contributions are naturally organized in two groups V_a and V_b (which correspond to the two corners of $v'_\circ$ in C). Applying Lemma G.3.5 to both groups, we obtain two

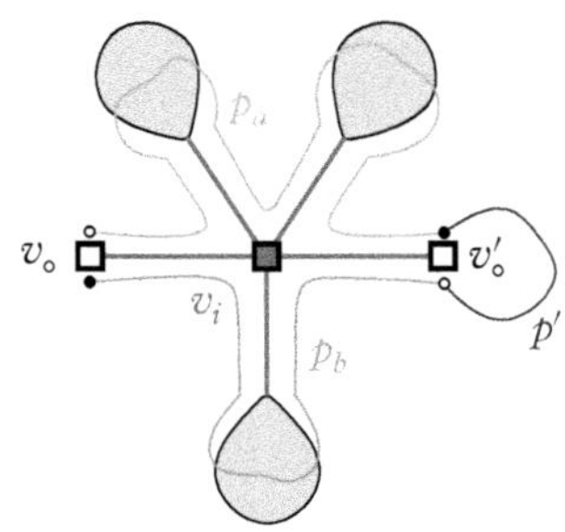

Figure G.7 *Concatenation of paths in the proof of Lemma G.3.9*

paths p_a and p_b in G. The concatenation of p', p_a and p_b, gives a coloured path between the two vertices corresponding to $v_\circ$ in G, so that $v_\circ$ is admissible.

Lemma G.3.10 *Consider a $(q+1)$-edge-coloured graph G, and the core-diagram C of the q-constellation $S = \Psi(G)$. If every white vertex of C either is of valency $d < q$ or has an incident core chain containing at least one white chain vertex, then G is an SYK graph.*

Proof This is a simple consequence of Lemma G.3.6, Lemma G.3.9, and Corollary G.3.3.

Lemma G.3.11 *For $q \geq 3$ and $\delta \geq 1$, let G be a random edge-coloured graph in $\mathbb{G}^q_{n,\delta}$, and let C be the core of the associated constellation S. Then, a.a.s. all the core chains of S contain at least one (internal) white vertex.*

Proof Let $r_{n,\delta}$ be the number of edge-coloured graphs from $\mathbb{G}^q_{n,\delta}$ with n vertices, such that one of the core-chains is distinguished (i.e. the kernel has a distinguished edge) with the condition that this distinguished core chain has no internal white vertex. Lemma G.3.10 ensures that $g_{n,\delta} - c_{n,\delta} \leq r_{n,\delta}$ hence we just have to show that $r_{n,\delta} = o(g_{n,\delta})$. We let $\mathcal{R}_\delta(z) = \sum_{n\geq 1} r_{n,\delta} z^n$ be the associated generating function. For every $K \in \mathbb{K}_\delta$, the contribution to $\mathcal{R}_\delta(z)$ in the case where the distinguished edge of K has two white extremities, and two half-edges of the same colour is (with the notations in the proof of Prop. G.2.2) equal to

$$\left[\frac{E^{=}_{\circ\circ}(K)\cdot(1+z_\bullet)}{\mathcal{G}_{\circ\circ}(z_\circ,z_\bullet)}A_K(z_\circ,z_\bullet)\right]\prod_{v_\circ\in\mathcal{V}_\circ(K)} z\mathcal{G}_T(z)^{q-\mathrm{val}(v_\circ)}\prod_{v_\bullet\in\mathcal{V}_\bullet(K)}\mathcal{G}_T(z)^{\mathrm{val}(v_\bullet)},$$

where $z_\circ = \mathcal{G}_T(z)^{q-2}$ and $z_\bullet = \mathcal{G}_T(z)^2$. One can then check that, due to the $\mathcal{G}_{\circ\circ}(z_\circ, z_\bullet)$ appearing in the denominator, the leading term in the singular expansion is $O((z-z_c)^{-(E(K)-1)/2})$. This also holds for all the other possible types of the distinguished kernel edge, so that we conclude that $r_{n,\delta} = O(z_c^{-n} n^{(3\delta-4)/2}) = o(g_{n,\delta})$.

Theorem G.3.12 *For $q \geq 3$ and $\delta \geq 1$, we have $c_{n,\delta} \sim g_{n,\delta}$ as $n \to \infty$.*

Proof From Lemma G.3.10 and Lemma G.3.11 it directly follows that (for $q \geq 3$ and $\delta \geq 1$) the random edge-coloured graph $G \in \mathbb{G}^q_{\delta,n}$ is a.a.s. an SYK graph. Hence, for $q \geq 3$ we have $c_{n,\delta} \sim g_{n,\delta}$.

G.3.4 The non-bipartite case

Let us go through the arguments of the last section, to adapt them in the case of generic coloured graphs.

Firstly, choosing an orientation for every colour-0 edge and colour-$0i$ cycle does not change the number of 0-residues, so that we can work with signed coloured graphs and signed constellations. Proposition G.3.1, Corollary G.3.3 are obviously true for signed coloured graphs and signed constellations. Lemma G.3.4, Lemma G.3.6, and Lemma G.3.7 also hold for signed constellations and signed coloured graphs, as tree contributions represent bipartite (melonic) subgraphs. Therefore, Corollary G.3.8 is also valid for non-necessarily bipartite $(q+1)$-coloured graphs, with $q > 3$ and $\delta > 0$.

Similarly, chains (and their attached trees) represent bipartite subgraphs of the coloured graphs, so that for $q = 3$, Lemmas G.3.5 and G.3.9 and G.3.10 can still be used without modification. It remains to adapt Lemma G.3.11 for signed coloured graphs, i.e. to prove that a.a.s, all core chains of a signed constellation with n vertices and excess δ contain at least one internal white vertex. This is true, as choosing a sign $\pm$ for every one of the $\delta + n - 1$ edges does not modify this property.

Thus, Theorem G.3.12 generalizes for non-necessarily bipartite graphs in $\tilde{G}^q$: For $q \geq 3$ and $\delta \geq 1$, we have $\tilde{c}_{n,\delta} \sim \tilde{g}_{n,\delta}$ as $n \to \infty$. Theorem 15.3.2 follows from this, as well as (G.7) and (G.21).

Appendix H
Proof of Theorem 16.1.1

We will prove that non-bipartite graphs are at best NNLO, i.e. they have reduced degree $\delta(G) \geq q-2$. The proof proceeds by induction on the number of colours q. Since our proof uses classical tools of random tensor/coloured graph theory, we will use the more common variable

$$D = q - 1, \tag{H.1}$$

so that the graphs are edges carrying the colours $1, \ldots, D+1$, and we want to prove that $\delta(G) \geq D-1$ for non-bipartite graphs.

We start with $D = 3$, i.e. $q = 4$. We recall (Gurau 2017; Witten 2019) that the degree of a 4-coloured graph G is defined as

$$\omega(G) = \sum_{\mathcal{J}} g(G_{\mathcal{J}}), \tag{H.2}$$

where there are 3 ribbon graphs $G_{\mathcal{J}}$ called *jackets* and $g(G_{\mathcal{J}})$ is the genus of $G_{\mathcal{J}}$. Each $G_{\mathcal{J}}$ has all the vertices and edges of G, but only a subset of faces determined by a permutation up to cyclic ordering and orientation reversal.[1] Equation (16.5) implies that $\omega(G)$ is an integer. This implies that jackets with half-integral genera come in pairs.

Let us now show that if G is non-bipartite, $\omega(G) \geq 2$. Assume that G has a planar jacket, i.e. $g(G_{\mathcal{J}}) = 0$ for some $\mathcal{J}$. Moreover, all faces of $G_{\mathcal{J}}$ have even length since they are bicoloured. As a result, $G_{\mathcal{J}}$ is a planar ribbon graph with faces of even length. A classical result in graph theory then ensures that $G_{\mathcal{J}}$, hence G is bipartite. Therefore, a non-bipartite graph has no planar jackets. To minimize the degree, it must have two jackets with genera 1/2 and one of genus 1, i.e. $\omega(G) \geq 2$. Since $\delta(G) = 2\omega(G)/(D-2)!$, we conclude that $\delta(G) \geq D-1$ at $D = 3$.

As the previous argument does not work for larger values of D, we now perform an induction. To do so, we will distinguish a colour $i \in \{1, \ldots, D+1\}$ and control the number of faces of colours $(ij)_{j \neq i}$ using the results we derived for the SYK model, where the role of the colour 0 is played by the colour i. The obstacle is however that in the SYK model, removing the colour 0 does not disconnect the graph, while removing the colour i in the present model typically does disconnect the graph. We will show how to overcome that obstacle so that indeed adding the colour i to a graph with D colours has the same diagrammatic effect as averaging over disorder in the SYK.

Therefore, we introduce an operation on the graph, known as the $1-$dipole contraction, see Gurau (2012). If G is connected, deleting every edge of colour $i \in \{1, \cdots, D+1\}$, one obtains a graph with D colours only, which we denote $G_{\hat{i}}$ and whose connected components are called

[1] The three permutations are here $\mathcal{J} = (1234), (1243), (1423)$. The faces of $G_{\mathcal{J}}$ are then the bicoloured cycles with colours $(\mathcal{J}^i(1)\mathcal{J}^{i+1}(1))$.

bubbles. If an edge of colour i separates two distinct bubbles, a 1-dipole contraction can be performed as follows. One first contracts the edge as described in (15.35). Deleting the resulting vertex one is left with a set of un-contracted half-edges, two for each colour, which are then contracted while respecting the colours, as follows

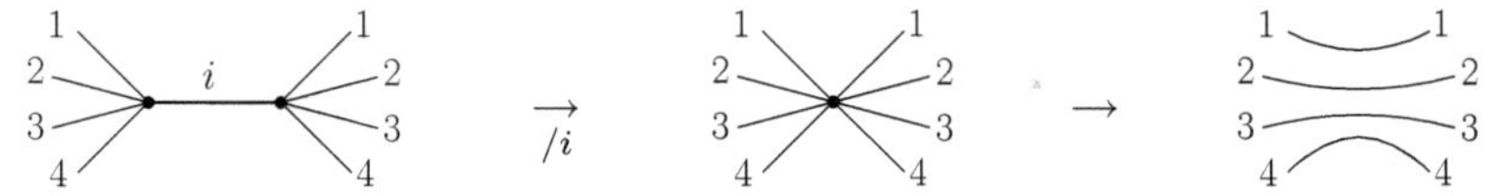

(H.3)

Crucially, an 1-dipole contraction does not change the degree (nor the topology of the underlying D-dimensional cell complex), as shown in Gurau (2012) (Lemma 2), and also preserves (non-) bipartiteness.[2]

A series of 1-dipole contractions on edges of a fixed colour $i \in \{1,\ldots,D+1\}$ can turn a connected graph G into a new graph T with a single bubble $T_{\hat{i}}$, i.e. such that the graph $T_{\hat{i}}$, having all edges of colour i removed, is connected. Indeed, we perform a 1-dipole contraction for some edge of colour i between two connected components of $G_{\hat{i}}$. The result has one connected component less and we can perform a 1-dipole contraction on another edge between two of the remaining connected components and so on, until only one connected component $T_{\hat{i}}$ remains.

The reduced degree of T is that of G, $\delta(G) = \delta(T)$ and moreover,

$$\delta(G) = \delta(T) = \delta_i(T) + \delta(T_{\hat{i}}), \tag{H.4}$$

where $\delta(T_{\hat{i}})$ is the degree of the bubble as a coloured graph with D colours,

$$\delta(T_{\hat{i}}) = D - 1 + \frac{(D-1)(D-2)}{4} V(T) - \sum_{\substack{j,k\neq i \\ j<k}} F_{jk}(T), \tag{H.5}$$

and $\delta_i(T)$ is

$$\delta_i(T) = D - F_i(T) + (D-1)(V(T)/2 - 1). \tag{H.6}$$

Note that $\delta_i(T)$ can be interpreted as a reduced degree in the sense of the SYK model, that is it only takes into account the number $F_i(T)$ of faces of colours $(ij)_{j\neq i}$ for fixed i (in the SYK model, $i = 0$). Indeed, since T has a unique bubble after removing the colour i, it can be interpreted as a graph from the SYK model with i replacing 0. Moreover, we recall that in the SYK model, the

[2] Clearly, a 1-dipole contraction of a bipartite graph is still bipartite. The other way around, consider a (connected) bipartite bubble and a D-cut of that bubble, i.e. a set of D edges $e_1,\ldots,e_D$ with the colours $1,\ldots,D$ which can be cut to create several connected components. Denote v_k the white vertex of e_k and v'_k its black vertex for $k = 1,\ldots,D$. All vertices (v_k) are in the same connected component after the cut is performed. Indeed, since each colour is incident exactly once on each vertex, except the colour k on v_k and v'_k for all k, there is a path alternating the colours 2 and 1 starting from v_1 and which must end on v_2 or v'_2. Since the path is alternating, it actually ends on v_2, and similarly with the others v_k. Therefore all (v_k) are in one connected component and all (v'_k) are in another after the cut. It is then clear that one can perform a 1-dipole insertion, by inserting a new edge with colour $D+1$, while preserving bipartiteness.

reduced degree can be defined as $D - \chi_0(T) = D + (D-1)V(T)/2 - F_0$ (see (15.33) with $q = D$ in this case). Here $\delta_i(T)$ only differs from the latter by a constant term $-(D-1)$.

We can therefore use the induction hypothesis on $\delta(T_{\hat{i}})$ and the results of Section 15.3.2 for $\delta_i(T)$. Suppose G is a non-bipartite graph with $D+1 > 4$ colours, and suppose we have proven that non-bipartite graphs with D colours have a reduced degree $\delta \geq D-2$. We know from section (15.3.2) that if $\delta_i(T) = 0$, then $T_{\hat{i}}$ is melonic and the edges of colour i must connect its canonical pairs of vertices, therefore T is also melonic and so is G. In particular it is bipartite, which is excluded, so that $\delta_i(T) \geq 1$.

If $\delta(T_{\hat{i}}) \geq D-2$, then there is nothing to prove. From the induction hypothesis (telling us that graphs with $\delta(T_{\hat{i}}) < D-2$ are bipartite and thus classified in Gurau and Schaeffer (2016)), we know that we have to look at the cases where $\delta(T_{\hat{i}}) = 0, D-3$.

When $\delta(T_{\hat{i}}) = D-3$, only the case $\delta_i(T) = 1$ is non-trivial. For $\delta_i(T) = 1$, we showed in the previous section that T is an NLO SYK graph as in (15.48) and (15.49), with colour i connecting the canonical pairs, as was the colour 0 then. As furthermore $\delta(T_{\bar{i}}) = D-3$, we know from the induction hypothesis that $T_{\hat{i}}$ is bipartite, which leaves only the cases on the left of (15.48) and (15.49). But then, T itself is bipartite, which is excluded as the inverse of (H.3) preserves bipartiteness.

Only the case $\delta(T_{\hat{i}}) = 0$, i.e. $T_{\hat{i}}$ is melonic, remains to be investigated. We will do so by proving the more general property (P): for a graph G with $D+1$ colours having a melonic bubble, either there exists a colour k and a bubble $B_{\hat{k}}$ such that $\delta(B_{\hat{k}}) > 0$, or $\delta(G) = 0$, or $\delta(G) \geq D-1$. This way, if $T_{\hat{i}}$ is melonic and G is not melonic, then either $\delta(G) \geq D-1$ as desired, or we can choose another colour k so that $\delta(T_{\hat{k}}) > 0$. From the induction hypothesis, $\delta(T_{\hat{k}}) \geq D-3$, which is the case we already dealt with.

We prove the property (P) inductively on the number of vertices. It is obvious if G only has two vertices as it is necessarily melonic. Consider a larger graph G, and the melonic bubble $B_{\hat{i}}$ with D colours. Recall that D-coloured melonic graphs are defined as the recursive insertion of pairs of vertices connected by $D-1$ edges as on the left of (H.7), and in particular always contain such a pair of vertices. In G, the colour i can be as any one of the cases below in (H.7): *i*) between the vertices of the pair, *ii*) as two edges reaching another connected component so that when cutting those two edges, G separates into two connected components, *iii*) or as two edges which do not separate G.

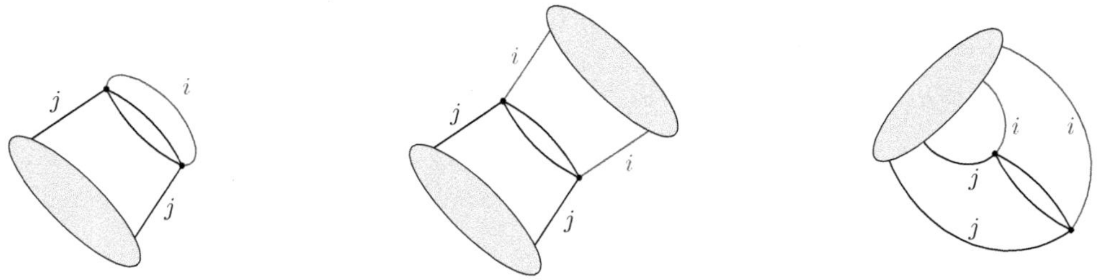

(H.7)

In the case *i*), by deleting the two vertices of the pair and reconnecting the two pending edges of colour j, one recovers a smaller graph to which the property (P) applies. As inserting a pair of vertices connected by D edges does not change neither $\delta(G)$ nor $\delta(B_{\hat{k}})$, (P) applies to G too. In the case *ii*), one may cut the two edges of colour i and reconnect the pending half edges so that an edge of colour i connects the two vertices of the pair. This creates two connected components G_1, G_2, to which (P) applies. The inverse operation does not change $\delta(G)$ (one looses D faces and one connected component), and if there is a non-melonic bubble $B_{\hat{k}}$ in G_1 or G_2, it gives rise to a non-melonic bubble in G. The property (P) thus applies to G. In the last case *iii*), one may cut the two edges of colour i and reconnect them as done for case *ii*). The number of faces then increases by $D-2$ if the vertices of the pair belong to two different faces of colour ij, and by $D-1$ or D otherwise, in which cases one finds $\delta(G) \geq D-1$ directly. On the other hand, if the vertices of the pair belong to two different faces of colour ij, two such faces of colour ij would still be in a bubble $B_{\hat{k}}$ with $k \neq i, j$, so that it cannot be melonic. This proves the property (P) and concludes the proof.

Appendix I
Summary of results on the diagrammatics of the coloured SYK model and of the Gurău–Witten model

In this appendix we give a list of the results obtained on the diagrammatics of the coloured SYK model (see subsection 15.3.2) and of the Gurău–Witten model (see section 16.1).

Coloured SYK model

Connected vacuum graphs

contribution	O(N)	O(1)	O(1/N)	...	$O(1/N^{k-1})$
real SYK	closed melonic	(15.46)	(15.52),(15.53),(15.54),(15.55) and twisted	...	$\ell_m(G_{/0}) = k$
complex SYK	"	left of "	(15.53),(15.54),(15.55)	...	", bipartite

Two-point function

contribution	O(1)	O(1/N)	$O(1/N^2)$
real SYK	melonic Fig.15.4	(15.48) and (15.49)	open an edge in (15.53),(15.54),(15.55) and twisted
complex SYK	"	left of "	open an edge in (15.53),(15.54),(15.55)

Four-point function

contribution	O(1/N)	$O(1/N^2)$	$O(1/N^3)$
real SYK	(15.45)	(15.60), (15.61), (15.62), (15.63), (15.64)	open two edges in (15.53),(15.54),(15.55) 2] and twisted
complex SYK	"	non-twisted in "	open two edges in (15.53),(15.54),(15.55) 2]

Cut edges e and e' such that $\eta(G_{e,e'}) = 2$.

Gurau–Witten model

Connected vacuum graphs

contribution	$O(N^{q-1})$	$O(N^2)$	$O(N)$	$O(1)$
real Gurau–Witten	closed melonic	left of (15.46), without colour 0, and non-separable "U" chain	right of (15.46), without colour 0, and non-separable "U" chain	(15.46) with broken "B" chains instead of "U"
complex Gurau–Witten	"	"	$\emptyset$	left of "

For $D > 4$, the previous contributions are the LO, NLO, NNLO, and NNNLO. We do not prove that those NNLO and NNNLO are the only ones. For $D = 3, 4$, there are other obvious contributions as well to the NNLO (e.g. non-twisted (24), (25), (26) without the colour 0 edges and such that non-separable chains are of the unbroken type) and to the NNNLO (e.g. (24), (25), (26) without the colour 0 edges and such that non-separable chains are of the unbroken type, one of them containing a twist).

Two-point function

contribution	$O(1)$	$O(1/N^{q-3})$	$O(1/N^{q-2})$	$O(1/N^{q-1})$
real Gurau–Witten	melonic Fig.16.1	(16.8)	(16.10)	(16.8) and (16.10), with broken "B" chains instead of "U"
complex Gurau–Witten	"	"	$\emptyset$	(16.8), with broken "B" chains instead of "U"

Four-point function $q \geq 6$—two colours

contribution	$O(1/N^{q-2})$	$O(1/N^{q-1})$	$O(1/N^{2q-5})$
real Gurau–Witten	unbroken chain left of (16.15)	broken chain right of (16.15)	(16.19) and (16.20)
complex Gurau–Witten	"	"	"

Four-point function $q = 4$—two colours

contribution	$O(1/N^2)$	$O(1/N^3)$
real Gurau–Witten	unbroken chain left of (16.15)	broken chain, right of (16.15) (16.19) and (16.20)
complex Gurau–Witten	”	”

Four-point function $q \geq 6$—one colour

contribution	$O(1/N^{q-2})$	$O(1/N^{q-1})$	$O(1/N^{2q-6})$
real Gurau–Witten	unbroken chain left of (16.15)	broken chain right of (16.15)	(16.21)
complex Gurau–Witten	”	”	”

Four-point function $q = 4$—one colour

contribution	$O(1/N^2)$	$O(1/N^3)$
real Gurau–Witten	unbroken chain left of (16.15)	broken chain, right of (16.15) (16.19), (16.20), (16.22)
complex Gurau–Witten	”	” except (16.22)

Four-point function $q = 4$—four colours

contribution	$O(1/N^{3/2})$	$O(1/N^{5/2})$
real Gurau–Witten	A single vertex with four external legs dressed with propagators	(16.23)
complex Gurau–Witten	”	”

Bibliography

Abdelmalek Abdesselam. (2003). The Jacobian conjecture as a problem of perturbative quantum field theory. *Annales Henri Poincaré*, 4:199–215.

Abdelmalek Abdesselam. (2003). Feynman diagrams in algebraic combinatorics. *Séminaire Lotharingien de Combinatoire*, 49:45.

Abdelmalek Abdesselam. (2004). The Grassmann–Berezin calculus and theorems of the matrix-tree type. *Adv. App. Math.*, 33:51–70.

Sergei Alexandrov, Marc Geiller, and Karim Noui. (2012). Spin foams and canonical quantization. *SIGMA*, 8:055.

Paolo Aluffi and Matilde Marcolli. (2009). Feynman motives of banana graphs. *Commun. Num. Theory Phy.*, 3:1–57. arXiv:0807.1690.

Paolo Aluffi and Matilde Marcolli. (2010). Parametric Feynman integrals and determinant hypersurfaces. *Adv. Theor. Math. Phys.*, 14:911–963. arXiv:0901.2107.

Paolo Aluffi and Matilde Marcolli. (2011). Feynman motives of banana graphs. *Intern. J. Geom. Methods Mod. Phys.*, 8:203–237.

Paolo Aluffi and Matilde Marcolli. (2011). Graph hypersurfaces and a dichotomy in the Grothendieck ring. *Lett. Math. Phy.*, 95:223–232. arXiv:1005.4470.

Walter Appel. (2002). *Mathématiques pour la physique et les physiciens!* H & K Éditions.

Remi Cocou Avohou, Vincent Rivasseau, and Adrian Tanasa. (2015). Renormalization and Hopf algebraic structure of the five-dimensional quartic tensor field theory. *J. Phys.*, A48(48):485204.

C. W. Bass, E. H. Connell, and D. Wright. (1982). The Jacobian conjecture: Reduction of degree and formal expansion of the inverse. *Bull. Amer. Math. Soc.*, 7(2):287–330.

M. C. Bergere and F. David. (1979). Integral representation for the dimensionally regularized massive Feynman amplitude. *J. Math. Phys.*, 20:1244.

M. C. Bergere and F. David. (1981). Integral representation for the dimensionally renormalized Feynman amplitude. *Commun. Math. Phys.*, 81:1.

Spencer Bloch, Helene Esnault, and Dirk Kreimer. (2009). On motives associated to graph polynomials. *Commun. Math. Phys.*, 267:181–225. arXiv:math/0510011.

F. A. Berezin. (1966). *The Method of Second Quantization*. New York: Academic Press.

Olivier Bernardi. (2008). A characterization of the Tutte polynomial via combinatorial embeddings. *Annals of Combinatorics*, 12:139–153.

T. A. Brody, J. Flores, J. B. French, P. A. Mello, A. Pandey, and S. S. M. Wong. (1981). Random matrix physics: Spectrum and strength fluctuations. *Rev. Mod. Phys.*, 53:385–479.

Daniel N. Blaschke, Francois Gieres, Erwin Kronberger, Thomas Reis, Manfred Schweda, et al. (2008). Quantum corrections for translation-invariant renormalizable, non-commumicative holidy. *JHEP*, 0811:074. arXiv:0807.3270.

Joseph Ben Geloun, Thomas Krajewski, Jacques Magnen, and Vincent Rivasseau. (2010). Linearized group field theory and power counting theorems. *Class. Quant. Grav.*, 27:155012.

Joseph Ben Geloun, Jacques Magnen, and Vincent Rivasseau. (2010). Bosonic colored group field theory. *Eur. Phys. J.*, C70:1119–1130. arXiv:0911.1719.

Joseph Ben Geloun and Vincent Rivasseau. (2011). A renormalizable 4-dimensional tensor field theory. *Commun. Math. Phys.*, in press.

Valentin Bonzom, Razvan Gurau, and Vincent Rivasseau. (2012). Random tensor models in the large N limit: Uncoloring the colored tensor models. *Phys. Rev.*, D85:084037.

Valentin Bonzom, Razvan Gurau, Aldo Riello, and Vincent Rivasseau. (2011). Critical behavior of colored tensor models in the large N limit. *Nucl. Phys.*, B853:174–195.

Valentin Bonzom, Razvan Gurau, James P. Ryan, and Adrian Tanasa. (2014). The double scaling limit of random tensor models. *JHEP*, 09:051.

Joseph Ben Geloun and Adrian Tanasa. (2008). One-loop β functions of a translation-invariant renormalizable noncommutative scalar model. *Lett. Math. Phy.*, 86:19–32.

David J. Broadhurst and Dirk Kreimer. (2001). Exact solutions of Dyson–Schwinger equations for iterated one loop integrals and propagator coupling duality. *Nucl. Phys.*, B600:403–422.

Spencer Bloch and Dirk Kreimer. (2008). Mixed Hodge structures and renormalization in physics. *Commun. Num. Theor. Phys.*, 2:637–718. arXiv:0804.4399[hep-th].

Daniel N. Blaschke, Erwin Kronberger, Arnold Rofner, Manfred Schweda, Rene I.P. Sedmik, et al. (2010). On the problem of renormalizability in non- commutative gauge field models: A critical review. *Fortsch. Phys.*, 58:364–372. arXiv:0908.0467.

Valentin Bonzom, Luca Lionni, and Adrian Tanasa. (2018). Diagrammatics of a colored SYK model and of an SYK-like tensor model, leading and next-to-leading orders. *J. Math. Phys.*, 58(5):052301.

Valentin Bonzom, Victor Nador, and Adrian Tanasa. (2018). Diagrammatic proof of the large N melonic dominance in the SYK model.

Valentin Bonzom, Victor Nador, and Adrian Tanasa. (2018). Diagrammatic proof of the large N melonic dominance in the SYK model. *Lett. Math. Phys.*, 109(12):2611–2624.

Aristide Baratin and Daniele Oriti. (2010). Group field theory with non-commutative metric variables. *Phys. Rev. Lett.*, 105:221302. arXiv:1002.4723.

Valentin Bonzom. (2013). Revisiting random tensor models at large N via the Schwinger–Dyson equations. *JHEP*, 03:160.

D. V. Boulatov. (1999). A Model of three-dimensional lattice gravity. *Mod.Phys.Lett.*, A 7:1629–1646. arXiv:hep-th/9202074.

N. N. Bogoliubov and O. S. Parasiuk. (1957). On the multiplication of the causal function in the quantum theory of fields. *Acta Math.*, 97:227–266.

B. Bollobás and O. Riordan. (1999). A Tutte polynomial for coloured graphs. *Combin. Probab. Comput.*, 8:45–93.

Bela Bollobás and Olivier Riordan. (2001). A polynomial invariant of graphs on orientable surfaces. *Proc. London Math. Soc.*, 83:513–531.

Bela Bollobás and Olivier Riordan. (2002). A polynomial of graphs on surfaces. *Math. Ann.*, 323:81–96.

Francis Brown. (2009). The massless higher-loop two-point function. *Commun. Math. Phys.*, 287:81–96.

Daniel N. Blaschke, Arnold Rofner, Manfred Schweda, and Rene I. P. Sedmik. (2009). One-loop calculations for a translation invariant non-commutative gauge model. *Eur. Phys. J.*, C62:433–443. arXiv:0901.1681.

Valentin Bonzom and Matteo Smerlak. (2010). Bubble divergences from cellular cohomology. *Lett. Math. Phys.*, 93:295–305.

Sylvain Carrozza. (2003). *Tensorial methods and renormalization in Group Field Theories*. PhD thesis. Paris: University of Paris XI and Albert Einstein Institute.

Alain Connes, Michael R. Douglas, and Albert S. Schwarz. (1998). Noncommutative geometry and matrix theory: Compactification on tori. *JHEP*, 9802:003.

Guillaume Chapuy. (2008). *Combinatoire bijective des cartes de genre supérieur*. PhD thesis. École Polytechnique.

S. Chmutov. (2009). Generalized duality for graphs on surfaces and the signed Bollobás–Riordan polynomial. *J. Combinat. Theory*, B 99:617–638. arXiv:0711.3490.

Alain Connes and Dirk Kreimer. (1998). Hopf algebras, renormalization and noncommutative geometry. *Commun. Math. Phys.*, 199:203.

Alain Connes and Dirk Kreimer. (2000). Renormalization in quantum field theory and the Riemann–Hilbert problem. 1. The Hopf algebra structure of graphs and the main theorem. *Commun. Math. Phys.*, 210:249–273. hep-th/9912092.

Abhijit Champanerkar, Ilya Kofman, and Neal Stoltzfus. (2011). The multivariate signed Bollobás–Riordan polynomial. *Bull. Lond. Math. Soc.*, 43:972–984.

Sylvain Carrozza, Thomas Krajewski, and Adrian Tanasa. (2016). Using Grassmann calculus in combinatorics: Lindström–Gessel–Viennot Lemma and Schur functions. In *Proceedings of Gascom 2018*, 2016, forthcoming in *Electronic Notes in Discrete Mathematics*, arXiv:1604.06276 [math.CO].

Alain Connes and Matilde Marcolli. (2008). *Noncommutative Geometry, Quantum Fields and Motives*. Paris: World Scientific.

S. Chmutov and I. Pak. (2007). The Kauffman bracket of virtual links and the Bollobás–Riordan polynomial. *Moscow Math. J.*, 7:409–418.

Sylvain Carrozza and Victor Pozsgay. (2008). SYK-like tensor quantum mechanics with Sp(N) symmetry. Nucl. Phys. B 941 (2019) 28–52.

Henry Crapo and William R. Schmitt. (2005). A free subalgebra of the algebra of matroids. *Eur. J. Combinatorics*, 26:1066–1085.

Sergio Caracciolo, Alan D. Sokal, and Andrea Sportiello. (2014). Noncommutative determinants, Cauchy–Binet formulae, and Capelli-type identities II. Grassmann and quantum oscillator algebra representation. *Annales Inst. H. Poincaré* **D**, *Comb., Physics and their Interactions*, 1:1–46.

Sylvain Carrozza and Adrian Tanasa. (2016). $O(N)$ random tensor models. *Lett. Math. Phys.*, 106(11):1531–1559.

Sylvain Carrozza and Adrian Tanasa. (2018). Pfaffians and nonintersecting paths in graphs with cycles: Grassmann algebra methods. *Adv. Appl. Math.*, 93:108–120.

F. David. (1985). Planar diagrams, two-dimensional lattice gravity and surface models. *Nucl.Phys.*, B257:45.

G. Duchamp, P. Blasiak, A. Horzela, K.A. Penson, and A.I. Solomon. (2008). Hopf algebras in general and in combinatorial physics: A practical introduction. arXiv:0802.0249[quant-ph].

C. de Calan and A.P.C. Malbouisson. (1980). Complete Mellin representation and asymptotic behaviors of Feynman amplitudes. *Annales Poincare Phys. Theor.*, 32:91–107.

P. Di Francesco. (2004). 2D quantum gravity, matrix models and graph combinatorics. In *Application of random matrices in physics. Proceedings, NATO Advanced Study Institute, Les Houches, France, June 6–25*, pp. 33–88.

P. Di Francesco, Paul H. Ginsparg, and Jean Zinn-Justin. (1995). 2-D gravity and random matrices. *Phys. Rept.*, 254:1–133.

S. Doplicher, K. Fredenhagen, and J.E. Roberts. (1994). Space-time quantization induced by classical gravity. *Phys. Lett.*, B331:39–44.

Axel de Goursac. (2009). *Noncommutative geometry, gauge theory and renormalization*. PhD thesis, University Orsay and University Munster. arXiv:0910.5158.

M. Disertori, R. Gurau, J. Magnen, and V. Rivasseau. (2007). Vanishing of beta function of non commutative phi**4(4) theory to all orders. *Phys. Lett.*, B 649:95–102. hep-th/0612251.

Stéphane Dartois, Razvan Gurau, and Vincent Rivasseau. (2013). Double Scaling in Tensor Models with a Quartic Interaction. *JHEP*, 09:088.

Thibault Delepouve, Razvan Gurau, and Vincent Rivasseau. (2016). Universality and Borel summability of arbitrary quartic tensor models. *Ann. Inst. H. Poincare Probab. Statist.*, 52(2):821–848.

Axel de Goursac, Andrea Sportiello, and Adrian Tanasa. (2916). The Jacobian conjecture, a reduction of the degree to the quadratic case. *Annales Henri Poincare*, 17(11):3237–3254.

Axel de Goursac, Adrian Tanasa, and Jean-Christophe Wallet. (2008). Vacuum configurations for renormalizable non-commutative scalar models. *Euro. J. Phys.*, 53:459–466.

Axel de Goursac, Jean-Christophe Wallet, and Raimar Wulkenhaar. (2007). Noncommutative induced gauge theory. *Eur. Phys. J.*, C51:977–987.

Gérard H. E. Duchamp, Florian Hivert, Jean-Christophe Novelli, and Jean-Yves Thibon. (2011). Noncommutative symmetric functions vii: Free quasi-symmetric functions revisited. *Annals of Combinatorics*, 15:655–673.

Gérard H. E. Duchamp, Florian Hivert, and Jean-Yves Thibon. (2002). Noncommutative symmetric functions vi: Free quasi-symmetric functions and related algebras. *Int. J. Algebra Comput.*, 12:671–717.

Margherita Disertori. (2011). Modèles de matrices aléatoires et supersymétrie. Habilitation à diriger des recherches. Paris: University Paris VII (in French).

Philippe Dennery and André Krzywicki. (1967). *Mathematics for Physicists*. Harper International Editions.

Margherita Disertori and Vincent Rivasseau. (2008). *Random matrices and the Anderson model*, pages 139–177. In Margherita Disertori, Werner Kirsch, Abel klein, Frédéric Klopp and Vincent Rivasseau (ed.): *Random Schrodinger Operators, Panorama et Synthèses*, 25.

P. Doubilet, G.-C. Rota, and J. Stein. (1976). On the foundation of combinatorial theory ix: Combinatorial methods in invariant theory. *Studies in Appliced Math.*, 53:185–216.

Stephane Dartois, Vincent Rivasseau, and Adrian Tanasa. (2014). The $1/N$ expansion of multi-orientable random tensor models. *Annales Henri Poincare*, 15:965–984.

G. Duchamp and C. Tollu. (2007). Sweedler's dual and Schutzenberger's calculus. *Contemp. Math.*, 539:67–78. Conference on Combinatorics and Physics, Bonn.

F. J. Dyson. (1952). Divergence of perturbation theory in quantum electrodynamics. *Phys. Rev.*, 85:631–632.

Kurusch Ebrahimi-Fard and Frederic Patras. (2018). Exponential renormalization. *Annales Henri Poincare*, 11:943–971. arXiv:1003.1679[math-ph].

Kurusch Ebrahimi-Fard and Frederic Patras. (2011). Exponential Renormalization II: Bogoliubov's R-operation and momentum subtraction schemes. arXiv:1104.3415[math-ph].

Johanna Ellis-Monaghan and Criel Merino. (2010). *Graph Polynomial and their Applications. II. Interrelations and Interpretations*, pp. 1–42. Birkhauser. arXiv:0806.4699, invited chapter for Structural Analysis of Complex Networks.

J. Ellis-Monaghan and Sarmiento. (2011). A recipe theorem for the topological tutte polynomial of Bollobás and Riordan. *Euro. J. Combinatorics*, 32:782–794.

J. Ellis-Monaghan and L. Traldi. (2006). Parametrized Tutte polynomials of graphs and matroids. *Combin. Probab. Comput.*, 15:835–854.

Jonathan Engle, Roberto Pereira, and Carlo Rovelli. (2007). The Loop-quantum-gravity vertex-amplitude. *Phys. Rev. Lett.*, 99:161301. arXiv:0705.2388.

Bertrand Eynard. (2016). *Counting Surfaces*. Springer.

H. Figueroa and J. M. Gracia-Bondia. (2001). On the antipode of Kreimer's Hopf algebra. *Mod. Physic. Lett.*, A 16:1427.

Laurent Freidel, Razvan Gurau, and Daniele Oriti. (2009). Group field theory renormalization - the 3d case: Power counting of divergences. *Phys. Rev.*, D80:044007.

J. Feldman, Knörer H., and E. Trubowitz. (2002). *Fermionic Functional Integrals and the Renormalization Group*, volume 16. CRM Monograph Series.

Laurent Freidel and Kirill Krasnov. (2008). A new spin foam model for 4d gravity. *Class. Quantum Grav.*, 25:125018. arXiv:0708.1595.

Laurent Freidel and David Louapre. (2003). Nonperturbative summation over 3-D discrete topologies. *Phys. Rev.*, D68:104004.

Laurent Freidel and Etera R. Livine. (2006). Effective 3-D quantum gravity and non-commutative quantum field theory. *Phys. Rev. Lett.*, 96:221301. hep-th/0512113.

Éric Fusy, Luca Lionni, and Adrian Tanasa. (2108). Combinatorial study of graphs arising from the Sachdev–Ye–Kitaev model. accepted for publication in EJC.

Philippe Di Francesco, Pierre Mathieu, and David Sénéchal. (1997). *Conformal Field Theory*. Springer.

L. Foissy. (2010). Classification of systems of Dyson–Schwinger equations in the Hopf algebra of decorated rooted trees. *Adv. Math.*, 224:2094–2150.

L. Foissy. (2012). Lie algebra associated to systems of Dyson–Schwinger equations. Adv. Math. (in press).

Laurent Freidel. (2005). Group field theory: An overview. *Int. J. Theor. Phys.*, 44:1769–1783.

Philippe Flajolet and Robert Sedgewick. (2008). *Analytic Combinatorics*. Cambridge: Cambridge University Press.

P. Flajolet and R. Sedgewick. (2009). *Analytic Combinatorics*. Cambridge: Cambridge University Press.

Eric Fusy and Adrian Tanasa. Oxford, Ogdens, Asymptotic expansion of the multi-orientable random tensor model. *Electron. J. Comb.*

Eric Fusy. (2011). Private communication.

J. B. French and S. S. M. Wong. (1970). Validity of random matrix theories for many-particle systems. *Phys. Lett.*, 33B:449–452.

A. Guionnet, G. Anderson, and O. Zeitouni. (2009). *An Introduction to Random Matrices*. Cambridge: Cambridge University Press.

Jose M. Gracia-Bondia and Joseph C. Varilly. (1988). Algebras of distributions suitable for phase space quantum mechanics. 1. *J. Math. Phys.*, 29:869–879.

Razvan Gurau, Adolfo P. C. Malbouisson, Vincent Rivasseau, and Adrian Tanasa. (2007). Non-commutative complete Mellin representation for Feynman amplitudes. *Lett. Math. Phys.*, 81:161–175. arXiv:0705.3437.

Razvan Gurau, Jacques Magnen, Vincent Rivasseau, and Adrian Tanasa. (2009). A translation-invariant renormalizable non-commutative scalar model. *Comm. Math. Phys.*, 287:275–290.

Razvan Gurau, Jacques Magnen, Vincent Rivasseau, and Fabien Vignes-Tourneret. (2006). Renormalization of non-commutative phi(4)**4 field theory in x space. *Commun. Math. Phys.*, 267:515–542. hep-th/0512271.

G. Gallavotti and F. Nicolò. (1985). Renormalization theory in four-dimensional scalar foelds. *Comm. Math. Phys.*, 100:545.

Razvan Gurau and Vincent Rivasseau. (2011). The 1/N expansion of colored tensor models in arbitrary dimension. *Europhys. Lett.*, 95:50004.

Razvan Gurau and James P. Ryan. (2012). Colored tenser models: A review. *SIGMA*, 8:020. arXiv:1109.4812.

Razvan Gurau and James P. Ryan. (2014). Melons are branched polymers. *Annales Henri Poincare*, 15(11):2085–2131.

David J. Gross and Vladimir Rosenhaus. (2017). A generalization of Sachdev–Ye–Kitaev. *JHEP*, 1702:093.

Razvan Gurau, Vincent Rivasseau, and Alessandro Sfondrini. (2004). *Renormalization: An Advanced Overview.*

Razvan Gurau and Gilles Schaeffer. (2016). Regular colored graphs of positive degree. *Ann. Inst. Henri Poincaré D Comb., Phys. Interactions*, 3:257–320.

Razvan Gurau and Adrian Tanasa. (2008). Dimensional regularization and renormalization for non-commutative quantum field theory. *Annales Henri Poincaré*, 9:655–683.

Razvan Gurau, Adrian Tanasa, and Donald R. Youmans. (2015). The double scaling limit of the multi-orientable tensor model. *EPL*, 111(2):21002.

Razvan Gurau. (2010). Topological graph polynomials in colored group field theory. *Annales Henri Poincare*, 11:565–584.

Razvan Gurau. (2011). Colored group field theory. *Commun. Math. Phys.*, 304:69–93.

Razvan Gurau. (2011). The 1/N expansion of colored tensor models. *Annales Henri Poincare*, 12:829–847. arXiv:1011.2726.

Razvan Gurau. (2012). The complete 1/N expansion of colored tensor models in arbitrary dimension. *Annales Henri Poincare*, 13:399–423.

Razvan Gurau. (2014). Universality for random tensors. *Ann. Inst. H. Poincare Probab. Statist.*, 50(4):1474–1525.

Razvan Gurau. (2017). Quenched equals annealed at leading order in the colored SYK model. *EPL*, 119(3):30003.

Razvan Gurau. (2017). *Random Tensors*. Oxford: Oxford University Press.

Razvan Gurau. (2917). The complete $1/N$ expansion of a SYK–like tensor model. *Nucl. Phys.*, B916:386–401.

I. Gessel and X Viennot. (1985). Binomial determinants, paths, and hook length formulae. *Adv. Math.*, 58:300–321.

Harald Grosse and Raimar Wulkenhaar. (2005). Renormalization of phi**4 theory on noncommutative R**4 in the matrix base. *Commun. Math. Phys.*, 256:305–374. hep-th/0401128.

Klaus Hepp. (1866). Proof of the Bogolyubov–Parasiuk theorem on renormalization. *Commun. Math. Phys.*, 2:301–326.

Claude Itzykson and Jean-Bernard Zuber. (1995). *Quantum Field Theory*. Dover Publications.

Christian Kassel. (1995). *Quantum Groups*. Springer-Verlag.

O. H. Keller. (1939). Ganze cremona transformations. *Monats. Math. Phys.*, 47:299–306.

T. Krajewski, M. Laudonio, R. Pascalie, and A. Tanasa. (2019). Non-Gaussian disorder average in the Sachdev-Ye-Kitaev model. *Phys. Rev.*, D99(12):126014.

Thomas Krajewski and P. Martinetti. (2011). Wilsonian renormalization, differential equations and Hopf algebras. *Contemp. Math.*, 539:187. arXiv:0806.4309[hep-th].

Thomas Krajewski, Jacques Magnen, Vincent Rivasseau, Adrian Tanasa, and Patrizia Vitale. (2010). Quantum corrections in the group field theory formulation of the engle-pereira-rovelli-livine/freidel-krasnov models. *Physical Review*, D 82(124069). arXiv:1007.3150.

M. Kontsevich. (1992). Intersection theory on the moduli space of curves and the matrix Airy function. *Commun. Math. Phys.*, 147:1–23.

V. Krushkal and D. Renardy. (2010). A polynomial invariant and duality for triangulations. arXiv:1012.131.

Thomas Krajewski. (2012). Bijective proof of the Postnikov formula. unwritten.

Dirk Kreimer. (2006). Anatomy of a gauge theory. *Annals Phys.*, 321:2757–2781. arXiv:hep-th/0509135.

Dirk Kreimer. (2007). Dyson–Schwinger equations: From Hopf algebras to number theory. *Fields Institute Comm.*, 50:225–248. arXiv:hep-th/0609004.

Dirk Kreimer. (2013). The core Hopf algebra. *Clay Math. Proc.*, 11:313–322. arXiv:0902.1223[hep-th].

Thomas Krajewski, Vincent Rivasseau, and Adrian Tanasa. (2013). Combinatorial Hopf algebraic description of the multiscale renormalization in quantum field theory. Séminaire Lotharingien de Combinatoire (in press), p. 22.

T. Krajewski, V. Rivasseau, A. Tanasa, and Zhituo Wang. (2010). Topological graph polynomials and quantum field theory, Part I: heat kernel theories. *J. Noncommut. Geom.*, 4:29–82. arXiv:0811.0186 [math.CO].

Thomas Krajewski, Vincent Rivasseau, Adrian Tanasa, and Zhituo Wang. (2010). Topological graph polynomials and quantum field theories, part i: Heat kernel theories. *J. Noncommut. Geom.*, 4:29–82.

V. Krushkal. (2011). Graphs, links and duality on surfaces. *Combinatorics, Probability and Computing*, 20:267–287.

Thomas Krajewski, Vincent Rivasseau, and Fabien Vignes-Tourneret. (2011). Topological graph polynomials and quantum field theory. Part II. Mehler kernel theories. *Annales Henri Poincare*, 12:483–545.

Hagen Kleinert and Verena Schulte-Frohlinde. (2001). *Critical Properties of* Φ^4 *Theories*. World Scientific.

Bernd A. Kniehl and Oleg V. Tarasov. (2011). Finding new relationships between hypergeometric functions by evaluating Feynman integrals. arXiv:1108.6019.

Igor R. Klebanov and Grigory Tarnopolsky. (2011). Uncolored random tensors, melon diagrams, and the Sachdev–Ye–Kitaev models. *Phys. Rev.*, D95(4):046004.

Dirk Kreimer and Walter D. van Suijlekom. (2009). Recursive relations in the core Hopf algebra. *Nucl. Phys.*, B820:682–693. arXiv:0903.2849[hep-th].

Pierre Lalonde. (1996). A non-commutative version of Jacobi's equality on the cofactors of a matrix. *Discrete Math.*, 158:161–172.

B. Lindström. (1973). On the vector representations of induced matroids. *Bull. London Math. Soc.*, 5:85–90.

L. Lionni. (2017). *Colored discrete spaces: Higher dimensional combinatorial maps and quantum gravity*. PhD thesis. Paris: Univer. Paris-Saclay: Springer. arXiv:1710.03663.

Matteo Laudonio, Romain Pascalie, and Adrian Tanasa. (2020). *Combinatorial Aspects of the Sachdev-Ye-Kitaev Model*. Proceedings of the Bucharest Conference on Geometry and Physics, IMAR September 2019.

Etera R. Livine and Simone Speziale. (2007). A New spinfoam vertex for quantum gravity. *Physical Review*, D 76:084028. arXiv:0705.0674.

E. Langmann, R.J. Szabo, and K. Zarembo. (2004). Exact solution of quantum field theory on noncommutative phase spaces. *Journal of High Energy Physics*, 0401:017. arXiv:hep-th/0308043.

Juan Martin Maldacena. (1999). The large N limit of superconformal field theories and supergravity. *Int. J. Theor. Phys.*, 38:1113–1133. [Adv. Theor. Math. Phys. 2, 231(1998)].

Matilde Marcolli. (2010). *Feynman Motives*. World Scientific.

M. L. Mehta. (2004). *Random Matrices*. Elsevier.

K. K. Mon and J. B. French. (1975). Statistical properties of many-particle sspectra. *Annals Phys.*, 95:90–111.

J. Schwarz, M. Green, and E. Witten. (2012). *Superstring Theory*. Cambridge: Cambridge University Press.

Jacques Magnen, Karim Noui, Vincent Rivasseau, and Matteo Smerlak. (2009). Scaling behaviour of three-dimensional group field theory. *Class. Quant. Grav.*, 26:185012.

Bogdan Morariu and Alexios P. Polychronakos. (2001). Finite noncommutative Chern-Simons with a Wilson line and the quantum Hall effect. *JHEP*, 0107:006. hep-th/0106072.

Jacques Magnen, Vincent Rivasseau, and Adrian Tanasa. (2009). Commutative limit of a renormalizable noncommutative model. *Europhysics Lett.*, 86(11001).

Juan Maldacena and Douglas Stanford. (2016). Comments on the Sachdev–Ye–Kitaev model. *Physical Review D*, 94.

Shiraz Minwalla, Mark Van Raamsdonk, and Nathan Seiberg. (2000). Noncommutative perturbative dynamics. *JHEP*, 0002:020. hep-th/9912072.

Karim Noui. (2008). A model for the motion of a particle in a quantum background. *Phys. Rev.*, D78:105008. arXiv:0807.0969.

Jean-Christophe Novelli and Jean-Yves Thibon. (2010). Free quasi-symmetric functions and descent algebras for wreath products, and noncommutative multi-symmetric functions. *Discrete Mathematics*, 310:3584–3606.

Dhritiman Nandan, Anastasia Volovich, and Congkao Wen. (2012). On Feynman rules for mellin amplitudes in AdS/CFT. *JHEP*, 1205:129. arXiv:1112.0305.

S. Oda. (1980). The Jacobian problem and the simply-connectedness of $\mathbf{a}^n$ over a field k of characteristic zero. Osaka University preprint.

Daniele Oriti. (2006). The group field theory approach to quantum gravity. In Approaches to Quantum Gravity - toward a new understanding of space, time, and matter, D. Oriti ed., forthcoming. Cambridge: Cambridge University Press, gr-qc/0607032.

Daniele Oriti. (2011). The microscopic dynamics of quantum space as a group field theory. In *Proceedings, Foundations of Space and Time: Reflections on Quantum Gravity: Cape Town, South Africa*, pp. 257–320.

Dine Ousmane Samary, Carlos I. Pérez-Sánchez, Fabien Vignes-Tourneret, and Raimar Wulkenhaar. (2015). Correlation functions of a just renormalizable tensorial group field theory: The melonic approximation. *Class. Quant. Grav.*, 32(17):175012.

James Oxley. (1992). *Matroid Theory*. Oxford: Oxford University Press.

A. M. Perelomov. (2006). *Generalized Coherent States and Their Applications*. Springer.

Alejandro Perez. (2013). The spin foam approach to quantum gravity. *Living Rev. Rel.*, 16:3.

O. Parcollet and A. Georges. (1999). Non-fermi-liquid regime of a dopped mott insulator. *Phys. Rev. B*, 59:5341–5360.

J. Polchinski. (1998). *String Theory*. Cambridge: Cambridge University Press.

Alexios P. Polychronakos. (2001). Quantum Hall states as matrix Chern–Simons theory. *JHEP*, 0104:011. hep-th/0103013.

Joseph Polchinski and Vladimir Rosenhaus. (2016). The spectrum in the Sachdev-Ye-Kitaev model. *J. High Energy Phys.* arxiv:1601.06768.

Michael Edward Peskin and Daniel V. Schroeder. (1995). *An Introduction to Quantum Field Theory*. Addison-Wesley Pub. Co.

W. Pauli and F. Villars. (1949). On the invariant regularization in relativistic quantum theory. *Rev. Mod. Phys.*, 21:434–444.

J. Riordan. (1975). The distribution of crossings of chords joining pairs of 2n points on a circle. *Math. Comput.*, 29:215–222.

Vincent Rivasseau. (1992). *From Perturbative to Constructive Renormalization*. Princeton: Princeton University Press.

Vincent Rivasseau. (2002). An introduction to renormalization, pp. 139–177. In *Duplantier, B. (ed.) *Poincaré* (Bourbaphy) Seminar.

Vincent Rivasseau. (2007). *Non-commutative renormalization*, pages 15–95. Xth Poincaré (BourbaPhy) Seminar. arXiv:0705.0705.

Vincent Rivasseau. (2012). *Advanced Quantum Field Theory*. Theoretical Physics Masters course, École Normale Supérieure Paris.

Vladimir Rosenhaus. (2003). An introduction to the SYK model.

Carlo Rovelli. (2004). *Quantum Gravity*. Cambridge University Press.

G.-C. Rota and J. Stein. (1976). Applications of Cayley algebras. *Accademics NAzionale dei Lincei atti dei Convegni Lincei 17, Colloquio Internazionale sulle Teorie Combinatorie, Roma*, Tomo 2.

Vincent Rivasseau and Adrian Tanasa. (2008). Parametric representation of covariant noncommutative QFT models. *Communications in Mathematical Physics*, 279:355–379. arXiv:math-ph/0701034.

Matti Raasakka and Adrian Tanasa. (2013). Combinatorial Hopf algebra for the Ben Geloun-Rivasseau tensor field theory. [Sem. Lothar. Combin.70,B70d(2014)].

Vincent Rivasseau and Adrian Tanasa. (2015). Generalized constructive tree weights. *J. Math. Phys.*, 55:043509.

Matti Raasakka and Adrian Tanasa. (2015). Next-to-leading order in the large N expansion of the multi-orientable random tensor model. *Annales Henri Poincare*, 16(5):1267–1281.

Vincent Rivasseau, Fabien Vignes-Tourneret, and Raimar Wulkenhaar. (2006). Renormalization of noncommutative phi**4-theory by multi-scale analysis. *Commun. Math. Phys.*, 262:565–594.

Vincent Rivasseau and Zhituo Wang. (2010). How are Feynman graphs resumed by the loop vertex expansion? arXiv:1006.4617.

Vincent Rivasseau and Zhituo Wang. (2010). Loop vertex expansion for Phi**2K theory in zero dimension. *J. Math. Phys.*, 51:092304.

Vincent Rivasseau and Zhituo Wang. (2014). How to resum feynman graphs. *Annales Henri Poincare*, 15(11):2069–2083.

James P. Ryan. (2012). Tensor models and embedded Riemann surfaces. *Phys. Rev.*, D85:024010.

Manfred Salmhofer. (2006). *Renormalization: An Introduction*. Springer.

Gabor Sarosi. (2018). AdS_2 holography and the SYK model. *PoS*, Modave2017:001.

William R. Schmitt. (1994). Incidence hopf algebras. *J. Pure and Applied Algebra*, 96:299–330.

Gilles Schaeffer. (2009). Habilitation à diriger les recherches. Paris: École Polytechnique.

Alan D. Sokal. (2009). *The multivariate Tutte polynomial (alias Potts model) for graphs and matroids*, pp. 173–226. Cambridge University Press, 2005. contribution to Surveys in Combinatorics.

E. R. Speer. (1969). *Generalized Feynman Amplitudes*. Princeton: Princeton University Press.

J. Stembridge. (1990). Nonintersecting Paths, pfaffians, and plane partitions. *Adv. Math.*, 83:96–131.

Nathan Seiberg and Edward Witten. (1999). String theory and noncommutative geometry. *JHEP*, 9909:032. hep-th/9908142.

S. Sachdev and J. Ye. (1993). Gapless spin fluid ground state in a random, quantum Heisenberg magnet. *Phys. Rev. Lett.*, 70:3339.

Kelli Talaska. (2012). Determinants of weighted path matrices. arXiv:1202.3128 [math.CO].

Adrian Tanasa. (2008). Feynman amplitudes in non-commutative quantum field theory. Invited contribution to *Modern Enclycopoedia Math. Phys.*

Adrian Tanasa. (2008). Overview of the parametric representation of renormalizable non-commutative field theory. *J. Phys.: Conference Series*, 103(012012).

Adrian Tanasa. (2008). Scalar and gauge translation-invariant noncommutative models. *Romanian J. Phys.*, 53:1207–1212.

Adrian Tanasa. (2009). Parametric representation of a translation-invariant renormalizable noncommutative model. *J. Physics A: Math. Theoret.*, 42(365208).

Adrian Tanasa. (2010). Algebraic structures in quantum gravity. *Classical and Quantum Gravity*, 27(095008), 2010.

Adrian Tanasa. (2010). Combinatorial Hopf Algebras in (Noncommutative) Quantum Field Theory. *Romanian J. Physics, Numéro Spécial à l'honeur de l'academicien Oliviu Gherman*, 55(9):1142–1155, 2010. invited paper.

Adrian Tanasa. (2011). Generalization of the Bollobás–Riordan polynomial for tensor graphs. *J. Math. Phys.*, 52:073514.

Adrian Tanasa. (2012). Multi-orientable group field theory. *Journal of Physics A*, 45:165401. arXiv:1109.0694.

Adrian Tanasa. Random tensor models, a quantum field theoretical particularization. *Proceedings of the Romanian Academy A* (in press), 2012.

Adrian Tanasa. (2012). Some combinatorial aspects of quantum field theory. arxiv:1102.4231.

Adrian Tanasa. (2016). The multi-orientable random tensor model, a review. *SIGMA*, 12:056.

Adrian Tanasa. (2019). Combinatorial Quantum Field Theory and the Jacobian conjecture. In *Proc. Conf. Transient Transcend. Transylvania, Brasov.*

G. 't Hooft. (1973). Dimensional regularization and the renormalization group. *Nuclear Physics*, B 61:455.

G. 't Hooft. (1974). A planar diagram theory for strong interactions. *Nucl. Phys.*, B72:461 [337(1973)].

T. Thiemann. (2017). *Modern Canonical Quantum General Relativity.* Cambridge: Cambridge Monographs in Mathematical Physics.

G. 't Hooft and M. Veltman. (1972). Regularization and renormalization of gauge fields. *Nuclear Physics*, B 44:189.

Adrian Tanasa and Dirk Kreimer. (2012). Combinatorial Dyson-Schwinger equations in noncommutative field theory. *J. Noncommut. Geometry*, (in press). arxiv:0907.2182.

J. Touchard. (1959). Contributions à l'étude du problème des timbres postes. *Canad. J. Math.*, 2:385–398.

Adrian Tanasa and Patrizia Vitale. (2010). Curing the ultraviolet/infrared mixing for field theories with translation-invariant $\star$ products. *Phys. Rev.*, D81(065008).

Adrian Tanasa and Fabien Vignes-Tourneret. (2008). Hopf algebra of non-commutative field theory. *J. Noncommut. Geom.*, 2:125–139.

Andrew Vince. (1993). The classification of closed surfaces using colored graphs. *Graphs and Combinatorics*, 9:75–84.

Walter D. van Suijlekom. (2007). Renormalization of gauge fields: A Hopf algebra approach. *Commun. Math. Phys.*, 276:773–798.

F. Vignes-Tourneret. (2009). The multivariate signed Bollobás–Riordan polynomial. *Discrete Mathematics*, 20:5968–5961. arXiv:0811.1584.

S. Wang. (1990). A Jacobian criterion for separability. *J. Algebra*, 65:453–494.

E. Wigner. (1995). Characteristic vectors of bordered matrices with infinite dimensions. *Annals Math.*, 62:548–564.

Edward Witten. (1990). Gauge theories, vertex models and quantum groups. *Nucl. Phys.*, B330:285–346.

Edward Witten. (2019). An SYK-like model without disorder. *J. Phys.*, A52(47):474002.

D. Wright. (1989). The tree formulas for reversion of power series. *J. Pure Applied Algebra*, 57(2):191–211.

Karen Yeats. (2006). *A Combinatorial Perspective on Quantum Field Theory.* Springer Briefs in Mathematical Physics.

Karen Yeats. (2006). *Growths Estimates for Dyson–Schwinger Equations*. PhD thesis, Univ. Boston. arXiv:0810.2249[math-ph].

T. Zaslavsky. (1992). Strong Tutte functions of matroids and graphs. *Trans. Amer. Math. Soc.*, 334:317–347.

W. Zimmermann. (1969). Convergence of Bogolyubov's method of renormalization in momentum space. *Commun. Math. Phys.*, 15:208–234.

Jean Zinn-Justin. (2003). *Quantum Field Theory and Critical Phenomena*. Oxford: Oxford University Press (4th edition).

P. Zinn-Justin. (2009). *Integrability and Ccombinatorics: Selected Topics*. Lecture notes for Les Houches Summer School.

A. Zvonkin. (1997). Matrix integrals and map enumeration: An accessible introduction. *Computers Math. Applications: Math. Comp. Model.*, 26:281–304.

B. Zwiebach. (2009). *A First Course in String Theory*. Cambridge: Cambridge University Press.

Index